CELL STRUCTURE AND FUNCTION

Cell Structure and Function

SECOND EDITION

ARIEL G. LOEWY Haverford College

PHILIP SIEKEVITZ Rockefeller University

HOLT, RINEHART AND WINSTON, INC.
New York   Chicago   San Francisco   Atlanta   Dallas
Montreal   Toronto   London   Sydney

Illustrations by George V. Kelvin • Science Graphics

Cover: Photograph of a liver cell from Batrachoseps attenuatus.
*Intracellular crystals, courtesy Don W. Fawcett, Harvard School
of Medicine.*

The harmonious co-operation of all beings arose,

not from the orders of a superior authority external to themselves,

but from the fact that they were all parts in a hierarchy of wholes forming a cosmic pattern,

and what they obeyed were the internal dictates of their own natures.

CHUNG TZU

(Third Century B. C.*)*

Preface
to the Second Edition

The past two decades have witnessed an explosive release of insights into the molecular machinery of the cell. We are beginning to discern a molecular pattern that includes such phenomena as the self-duplication of the cellular hereditary material and the control it exerts in the formation of the catalysts of the cell. Biological specificity, a property so characteristic of the world of life, is in the process of being related to the structure and interaction of macromolecules. The electron microscope has suddenly discovered a new world, rich and intricate in detail, which is being actively interpreted in molecular terms. The regulation of this very complicated cellular machinery is now being examined in a manner that, at the very least, has operational rigor.

These developments have strengthened our conviction that there is an inextricable relationship between cellular structure and cellular function. The more classical disciplines of cytology, cell physiology, biochemistry, and biophysics are becoming fused into one discipline that is often referred to as the "molecular biology of the cell."

The purpose of this book is to document these exciting developments and to make them accessible to the introductory student. It is our conviction that since the cell is the "common denominator" of living systems, it is extremely important that the beginning biology student should become acquainted with the major facts and theories of cell biology. This introduction will serve a dual purpose: it will provide the student with a firm basis from which to examine the other manifold phenomena of the world of life, and it will also serve to convince him, at an early stage of his development, of the crucial importance of the physical sciences to the study of living systems.

This book has grown considerably in its second edition because of a number of spectacular developments in cell biology in the last five years. We have added new material to topics such as cellular fine structure, protein and nucleic acid chemistry, properties of enzymes, nucleic acid and protein biosynthesis, muscle contraction, active transport, and cellular regulation. Furthermore, we have added two major topics: a chapter on the molecular basis of structure formation and a discussion of photosynthesis. We have sought to increase clarity by explaining in greater detail certain topics that had been treated in an overly condensed form in the first edition. Lastly, we have increased our emphasis on the visual aspects of the presentation by increasing the number and clarity of the micrographs, line drawings, graphs, and charts.

Thus, although the second edition is larger, we do not think that it represents a more advanced treatment of the subject. On the contrary, we hope that we have increased its usefulness to the beginning student who has been exposed to a year of high school chemistry and the equivalent of a "BSCS course" in biology.

We have continued in this edition to emphasize the experimental basis of the ideas we discuss. Part Four especially can make use of this approach in teaching biology because the first half of the book prepares the student by supplying him with some basic information on cell structure and function (Part Two) and on the chemical properties of the molecules of the cell (Part Three). One can think of the organization of this textbook as being helical in nature, each turn of the helix discussing structure and function at higher resolution and in more intimate detail. Some of the redundancy in the presentation is a necessary consequence of this approach.

We have referred frequently to the names of the contributors to biological theory not because we wish to develop a cult around their persons, but because we want to emphasize that science is not merely an abstract structure of thought but a definite social *activity* occurring in a finite and often crucial historical context. Science has become one of the least parochial, most universal forms of human interaction ever developed by man, capable of bringing together into a common framework of assumptions and accepted rules of procedure human beings from the most diverse social systems and cultures.

It is not the intention of the authors to make this an "elementary" book in the sense of limiting it necessarily to the simpler and more accessible aspects of cell biology. Yet we hope that it is a book for beginners because we believe it is precisely the beginner who deserves an initial statement that is representative of the contemporary quality and mode of the field.

We wish to thank the following for reading given chapters of the manuscript and making valuable suggestions: George E. Palade, Robert M. Gavin, Jr., Harmon C. Dunathan, Howard K. Schachman, Daniel E. Koshland, Jr., Neville R. Kallenbach, Edouard Kellenberger, Vivianne D. Nachmias, George L. Gerstein, Frederick C. Neidhardt, André T. Jagendorf, and Robert K. Crane.

We also wish to thank the following for their patience in reading and commenting on the entire manuscript: Hans Ris, James F. Bonner, Sarah Elgin, Elizabeth U. Green, and Grace M. Stoddard.

Haverford, Pennsylvania   A.G.L.
New York City   P.S.
May 1969

Contents

CELL STRUCTURE AND FUNCTION

CELL BIOLOGY — Part ONE

In this part living matter is viewed as a special segment
of the physical world, having an ancient evolutionary
history, "moving parts" of molecular dimension, and an
extensive capacity to adapt and evolve. As the common
denominator of all living phenomena, the machinery
of the cell is fundamental to all biological processes, and
although diversity is characteristic of the world of life,
it is only endless variation built on a universal scheme
whereby energy is utilized for the maintenance and
extension of intricate structure and complex behavior.

The Common Denominator of Living Matter

The study of the world of life is as old as man himself, for the classification and comparison of the prodigious diversity of living forms have been intimately connected with man's survival. Yet the idea that there is an underlying unity encompassing all phenomena of life is a relatively recent development in human thought. Our ancestors were aware of the differences between a bat and a bird long before they had an inkling that there were similarities between the tissues of a mushroom and a man.

Three major systems of thought, each barely a century old, have stressed different facets of the unity of living matter.

The Darwin-Wallace *theory of evolution* by natural selection (1858) suggested a universally applicable mechanism whereby the diversity of living forms could have "descended by modification" from a common ancestry.

Mendel's *laws of heredity* (1866) provided a general mechanism whereby organisms within a species combine, resegregate, and transmit the hereditary information which represents the accumulation of billions of years of evolutionary development.

The *cell theory*, first enunciated by Schleiden and Schwann in 1838, emphasized yet another aspect of the unity of the world of life. This theory stated that all living systems are composed of cells and of cell products. Today this statement may seem to be self-evident, but we must not forget that we have lived with this concept for some time. To Schleiden's and Schwann's predecessors the idea that the organism is the fundamental unit of biological activity was an easily observable, well-established, common-sense principle. That organisms are in turn composed of smaller subunits, each of which is endowed with an individual identity, must have been a revelation of first magnitude. As great a biologist as T. H. Huxley wrote in 1853 that cells ". . . are not the instruments but indications, . . . they are no more the producers of vital phenomena than the shells scattered in orderly lines along the seabeach are the instruments by which the

3

gravitational force of the moon acts on the oceans. Like these, the cells mark only where the vital tides have been and how they have acted."

The extent to which the cell has an individuality of its own and a capability of independent existence was by no means clear even to Schleiden and Schwann. It is only in recent years that we have begun to recognize the degree to which all cells share a large portion of the principles we call the "machinery of life." We have learned that cells, whether they are protozoans or human liver cells, duplicate their genetic material in the same way, utilize their hereditary information to synthesize proteins in the same way, handle the transfer of energy in the same way, regulate the exchange of materials in the same way, convert chemical energy into work in the same way, and so on. In fact, it has been disconcerting to those interested in differentiation or in the problem of cancer that so few fundamental biochemical differences can be detected between cells of various types.

The purpose of this book is to emphasize the *common* features of the machinery of the cell because we believe that the cell is the basic "module," the common denominator of all the immense variety of living forms. One must not forget, however, that all cells also play specialized roles over the entire range of diversity in biological form and function; the generalized cell we are about to describe is an abstraction that does not exist as such in nature. Therefore, before embarking on a description of this mythical entity, the "generalized cell," we must survey briefly some of the major differences in cellular organization found in the world of life.

The recent development of electron microscopy has made it clear that two basic plans of cellular organization occur, the simpler *procaryotic* and the more complex *eucaryotic*; these plans have many major differences. The transition from procaryotic to eucaryotic organisms is so discontinuous that one must conclude it represents one of the major evolutionary steps in the long history of the world of life.

The procaryotes (Fig. 1-1), which include the bacteria and the blue-green algae, are small cells (0.5–3μ), lacking (1) a membrane around their nucleus and (2) clearly defined, membrane-limited organelles such as *mitochondria, chloroplasts, Golgi body,* and *lysosomes*. The genetic information in procaryotes, at least in the organisms studied so far, is located on a single chromosome that consists of a circular double strand of DNA and that lacks the basic proteins called *histones* found in the chromosomes of eucaryotic cells. The nucleus also lacks a *mitotic apparatus* and *nucleoli*. Procaryotic cells are generally surrounded by *cell walls* constructed mostly of carbohydrates and amino acids. Their *plasma membrane*, which lies directly below the cell wall, often forms intrusions into the cytoplasm called *mesosomes* (Fig. 1–1A), which at times are relatively complex in structure. They do not exhibit cytoplasmic streaming or amoeboid motion, locomotion being frequently achieved by simple *flagella* which lack the multi-stranded complexity of the flagella of the eucaryotes.

The procaryotic world of life is characterized by its ubiquity in distribution, its rapid growth and short generation time, its tremendous biochemical versatility and genetic flexibility, and its consequent usefulness to experimental biologists, who in recent years have exploited these properties to great advantage.

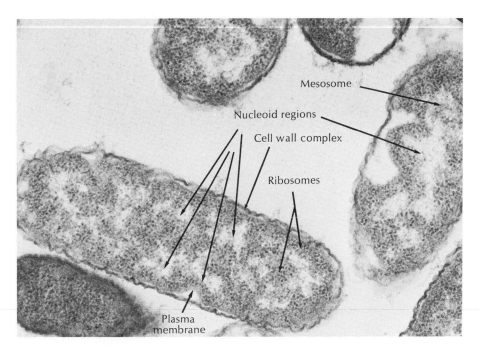

Mesosome

Nucleoid regions

Cell wall complex

Ribosomes

Plasma
membrane

Fig. 1-1. Electron micrograph of two procaryotic cells. A. A bacterium, *Escherichia coli* cells. (Courtesy of J. J. Jamieson, Rockefeller University.)

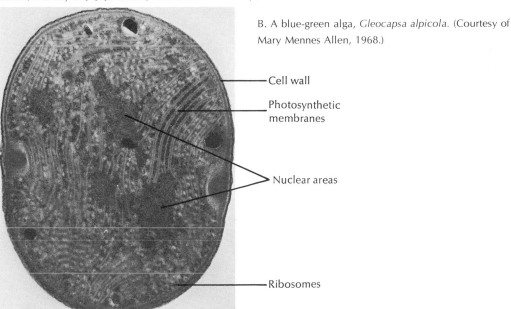

B. A blue-green alga, *Gleocapsa alpicola*. (Courtesy of Mary Mennes Allen, 1968.)

Cell wall

Photosynthetic
membranes

Nuclear areas

Ribosomes

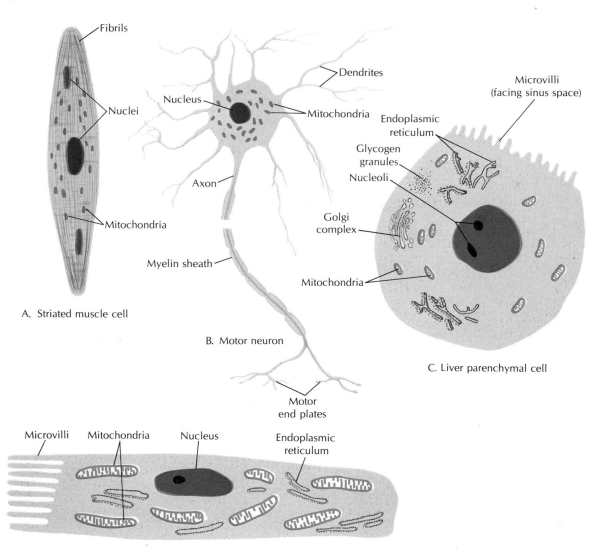

Fibrils

Nuclei

Mitochondria

A. Striated muscle cell

Dendrites

Nucleus

Mitochondria

Axon

Myelin sheath

B. Motor neuron

Motor
end plates

Microvilli
(facing sinus space)

Endoplasmic
reticulum

Glycogen
granules

Nucleoli

Golgi
complex

Mitochondria

C. Liver parenchymal cell

Microvilli   Mitochondria   Nucleus   Endoplasmic
reticulum

D. Kidney cell (proximal convoluted tubule)

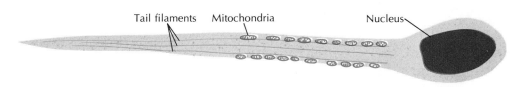

Tail filaments   Mitochondria   Nucleus

E. Sperm cell

The eucaryotes are the cells of the remaining groups of *protists* such as *algae, fungi,* and *protozoa* and of all plants and animals. They contain extensive internal membrane systems such as the *endoplasmic reticulum* and the Golgi body, as well as organelles surrounded by membranes such as the nucleus, the mitochondria, the chloroplasts, and the lysosomes. Their cytoplasm is capable of internal movement, and they contain, in addition, a number of more complex energy transducing systems such as the mitotic spindle, multistranded flagella, and the sliding filaments of *striated muscle* cells. The nuclei of eucaryotic cells contain nucleoli and more than one chromosome consisting of DNA and histones. Hence, there is the need for a mitotic apparatus which controls the complicated choreography of *mitosis* as it occurs in the eucaryotic cell.

Eucaryotic cells carry out most, if not all, of the functions of procaryotic cells; but, in addition, most of them have evolved the potential for their cooperative existence as subunits of multicellular differentiated organisms.

Animal cells (Fig. 1-2) lack a rigid cell wall surrounding their plasma membrane. Their mitotic apparatus includes centrioles, and division occurs by constriction of the cell. They lack chloroplasts and depend ultimately on plants as their source of food and therefore of energy. Animal cells are noted for their mobility and their ability to ingest food particles that are subsequently digested inside the cell.

Plant cells (Fig. 1-3) are surrounded by polysaccharide walls against which lies a thin layer of cytoplasm enclosing one or more large *vacuoles.* Structural rigidity is achieved by the "turgor pressure" of the cytoplasm against the cell wall (see *The Living Plant* by Peter Ray in the Holt, Rinehart and Winston Modern Biology Series). The mitotic apparatus of plant cells has a spindle, but in higher plants there are no centrioles, and cell division occurs by the growth of a new partition separating the two daughter cells. Many plant cells contain chloroplasts, which endow the cell with the important capability of converting light energy into chemical energy.

The plant and animal cell types described are of course differentiated into numerous forms, each specialized for particular functions which they perform in the organism. Within a single organism like the mammalian body, one finds many different cell types (Fig. 1-2) such as the elongated and fibrillar striated muscle cell, the slender

◀ **Fig. 1-2.** Various animal cell types. The diagrams show the great differences among shapes of animal cells, shapes conditioned by their functions. (Cells are not drawn to the same scale.) The striated muscle cell (A) is multinucleate, full of mitochondria to furnish energy necessary for the contraction of the fibrils which fill the cytoplasm (Chapter 16.) The motor neuron (B) has long extensions of its cytoplasm, extensions sometimes meters long, which are really membrane-mediated conductors of the electrical potential (Chapter 17). The liver parenchymal cell (C) is more like what we would consider a typical cell, with no morphological features that make it extraordinary. The kidney cell (D) is typified by its narrowness, in which long mitochondria lie adjacent to the cell membrane, ready to furnish the energy necessary to transport ions across that membrane (Chapter 17). The sperm (E) is a cell made up of a locomotory fibrillar part, surrounded by mitochondria, whose task it is to move an information-loaded nucleus (Chapters 14 and 16).

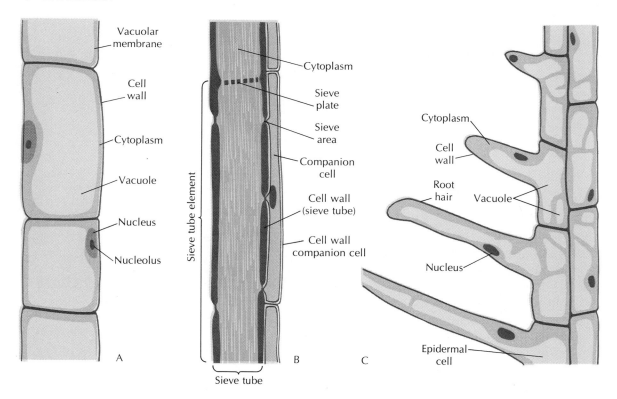

Fig. 1-3. A variety of plant cell types illustrating some of the diversity in structure which can be observed in the cells of higher plants. A. Parenchyma cells. B. Phloem sieve-tube and companion cells. C. Root hair cell.

and branched nerve cell, the metabolically active liver cell, the osmotically active kidney cell, and the free-swimming, protozoanlike sperm cell. The cells of the higher plants also vary a great deal in their structure and function (Fig. 1-3). There are the undifferentiated *parenchyma* cells of the growing regions of the plant body, the *phloem sieve tubes* with their special properties for food transport, and the *root hairs* adapted for water and mineral absorption.

Finally, mention should be made of a distinct evolutionary line of animals, the protozoa. Instead of developing into a multicellular organism composed of a number of different cell types, protozoa have managed to evolve a number of complicated intracellular organelles that in a simplified manner parallel some of the functions of higher organisms. Paramecium, for instance (Fig. 1-4), is an exquisitely complex little unicellular animal which contains, in addition to its other eucaryotic organelles, (1) a network of cytoplasmic fibrils that coordinates the beat of its cilia and that is even capable of reversing the direction of their beat (avoidance reaction); (2) a specialized organ for ingestion of food (oral groove and gullet); (3) organelles for digestion (food vacuoles);

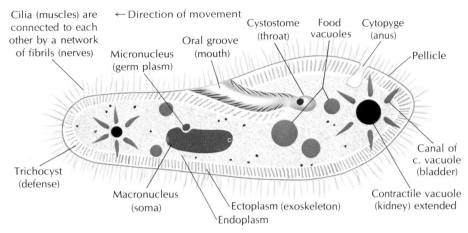

Cilia (muscles) are connected to each other by a network of fibrils (nerves)

← Direction of movement

Micronucleus (germ plasm)

Oral groove (mouth)

Cystostome (throat)

Food vacuoles

Cytopyge (anus)

Pellicle

Trichocyst (defense)

Macronucleus (soma)

Ectoplasm (exoskeleton)

Endoplasm

Canal of c. vacuole (bladder)

Contractile vacuole (kidney) extended

Fig. 1-4. The complex structure of the animalcule or unicellular animal *Paramecium*. This little animal has a variety of organelles with functions that are analogous to certain multicellular organs (indicated in brackets) of the metazoa.

(4) a device for excretion of solids (anal pore); (5) two "kidneys" (contractile vacuoles) for the excretion of liquid; (6) somatic DNA (macronucleus) presumably carrying information for the synthesis of the animal's proteins; (7) germ plasm DNA (one or more micronuclei) presumably responsible for the hereditary continuity of the animal; and (8) a "defensive" apparatus (trichocysts) composed of little spears which this unicellular animal shoots at its adversaries.

These distinct cell types represent specializations by which cells adapt to their specific roles. It is not within the scope of this book to do more than refer on occasion to these specialized functions. It is important, however, to recognize that many specialized cell functions find their origin in general cellular phenomena occurring in attenuated form throughout the cellular world. Nerve conduction is based on action potentials that occur in all cells. Muscular contraction is based on a mechanism of conversion of chemical energy into work, which appears to occur universally. Osmotic work found, for instance, in a highly active form in kidney cells also appears to be a universal property of living matter.

This brief excursion into some of the specialized cells of the animal and plant world should not be regarded as an attempt to catalog the cellular diversity found in nature. Rather, it should draw attention to the range of experimental materials available for the study of those cell phenomena that occur universally. A valuable practice in experimental cell biology is to select that system for examination which, since it is biologically efficacious in carrying out a certain function, proves to be singularly suited as experimental material for the analysis of that function. Even if no generalized or archetypal cell exists, it is convenient to create one to serve as a conceptual model within which most cell functions can be incorporated and discussed.

Our discussion of the archetypal cell will, like most experimental approaches to biology, rest on the presupposition that biological phenomena can be analyzed in terms of physical and chemical principles. This admittedly is an act of faith that cannot be justified on a priori grounds; yet, it is an eminently useful act of faith and represents the philosophical basis of an approach that is yielding communicable and verifiable results of ever-widening significance with ever-increasing rapidity. This approach, which in recent years has become known as "molecular biology," has been very successful not only because it has utilized a variety of complementary approaches, genetic, biochemical, and optical, but also because it has addressed itself to the central and most basic phenomena of cell biology. It is an exhilarating fact borne out by the research of the last two decades that there is a relationship between the importance of a cellular phenomenon and its susceptibility to analysis and clarification. This development is not as surprising as it might first appear since we know that the functioning of the cell is a highly predictable phenomenon. One would therefore expect that the central and most crucial aspects of this system must have properties of a high order of specificity that are especially susceptible to investigation. For instance, the duplication of the hereditary material, its transcription and translation into proteins, and the resulting specificity of proteins responsible for the regulation of cellular processes, are phenomena centrally important in the biology of the cell. At the same time those very properties of specificity which contribute so importantly to their function in the cell make them particularly susceptible to analysis.

Historically, the utilization of two areas of biological specificity constituted the first steps in this powerful approach. They were the employment of gene mutation in the study of heredity and the employment of enzyme specificity in the study of metabolism. The combination of genetic and biochemical approaches, and structural localization of biological events by high resolution optical techniques, have provided impetus for the more recent and dramatic successes of cell biology, but the success and ingenuity of modern cell biology depend ultimately on the immensely more successful integration of systems in the cell itself! It is then the purpose of this book to document the recent explosive growth of a molecular biology of the cell.

SUGGESTED READING LIST

Offprints

    Brachet, J., 1961. "The Living Cell." *Scientific American* offprints. San Francisco: W. H. Freeman & Co.

Articles, Chapters, and Reviews

    Stanier, R. Y., Doudoroff, M., and Adelberg, E., 1963. "The Internal Organization of Cells," ch. 4 in *The Microbial World*. Englewood Cliffs, N.J.: Prentice-Hall, Inc.

Books

    Bloom, W., and Fawcett, D. W., 1962. *Textbook of Histology*. Philadelphia: W. B. Saunders Co.

Bourne, G. H., 1962. *Division of Labor in Cells*. New York: Academic Press, Inc.

Novikoff, A. and Holtzman, E. *The Cell and Its Organelles*. New York: Holt, Rinehart and Winston, Inc. (In preparation.)

Picken, L. E. R., 1960. *The Organization of Cells and Other Organisms*. Oxford: Clarendon Press.

Chapter 2　Life and the Second Law of Thermodynamics

It has been estimated that between two and four billion years ago life started on our planet. Whether life started spontaneously from nonliving matter or whether it came here in the form of spores seeded from distant solar systems, we don't know. In either case we might suppose that life first began by the transformation of nonliving materials into living matter. There have been many speculations in recent years on the mechanism involved in such a remarkable transformation, a subject which was given great impetus by the imaginative writings of Oparin and the suggestive experiments by Urey and Miller (see the Suggested Reading List at the end of this chapter).

ON DEFINING LIFE

We have said that the living has come from the nonliving. This presupposes that there is a difference between these two states of matter, and it is therefore legitimate to ask what the difference might be. A "complete" definition of living matter is not yet possible, for this involves a complete knowledge and understanding of it. Nevertheless, since we are usually able to distinguish between "animal, vegetable, and mineral," it is logical to ask ourselves upon what criteria such judgments are based.

Are there any unique criteria of living matter? We usually answer this question by listing a number of characteristics that we associate with living matter such as movement, reproduction, metabolism, sensitivity, growth, and so forth. It is generally admitted that each of these properties can be found in the nonliving world, but it is argued that in living matter they are present simultaneously and find their most complex degree of expression. Thus, although it is true that locomotives work, computers think, and some automated machines can reproduce, we do not have a computer that can dig ditches, reproduce, enjoy poetry, and make moral judgments. If we are committed to the notion that living matter is subject to the laws of chemistry and physics, however, then we must accept the possibility that such robots can, in principle, be built.

Are we to conclude that there need be no essential difference between the man and the monster? We think yes, but with one crucial proviso: *the historical origins of man and robot are different.* Man built the robot and not vice versa. We emerge, then, with a criterion of living matter that is historical in character. Life is that process which started at an uncomplicated level and which has spontaneously gained in complexity through a series of events we call evolution. This series has taken a long time and finds its present most complex expression in the form of *Homo sapiens.* The above-mentioned ever-expanding organization is at present being extended by man to his physical surroundings so that different forms of organization, as concrete as buildings and machines and as abstract as knowledge and wisdom, are being accumulated at an increasing rate. Whether this new form of extraorganismic organization we call civilization is evolutionarily stable, leading to undreamed-of developments in the social evolution of man, or whether it is permeated with biologically rooted and historically transmitted contradictions incapable of solution at a social level, only the future will determine.

Yet, it is conceivable that once human social relations have developed to a certain level of social consensus, man will be able to apply successfully his understanding of biological principles to the regulation of human evolution itself. Should this happen, our evolutionary criterion for the definition of living systems may still be applicable. After all, the emergence of intelligence has been a gradual evolutionary process, and it should not be surprising that at some point this phenomenon itself might "feed back" and in time influence the evolutionary process from which it has emerged. Indeed, the evolution of the human mind is thus far the most significant and dramatic step in biological evolution, a development which will have a far-reaching influence not only on human evolution, but also on all other forms of biological evolution occurring on this planet.

In addition to the important historical or evolutionary criterion that we employ to characterize living matter, we might add two other properties which may be characteristic not only of life here on earth, but also of all other life as it may exist throughout the universe. We refer to the extraordinary properties of "compactness" and "energetic economy" of living matter.

The almost infinitesimally minute scale into which biological organization is capable of being compressed is one of the most dazzling wonders of nature. One need only compare the monstrous electronic computers powered by watts of energy with the infinitely more sophisticated human brain powered by mere microwatts of energy to recognize the remarkable spatial and energetic economy with which biological systems operate.

The tremendous compactness of living systems is of course achieved by building the components of the living machine on a molecular scale. As we shall see throughout this book, the cell is capable of replicating, transcribing, and translating information stored in the spatial arrangements of single molecules, which, although large if compared with the molecules generally encountered in the nonliving world, are still extraordinarily tiny when compared with the components of the machines we build. Furthermore, the enzymes that are the cell's regulatory devices are also single molecules, many orders of magnitude smaller than the components of our own machines. One of the astonishing consequences of the compactness of living matter is the tremendous amplification effects exhibited by living systems. The 5×10^{-15} grams of DNA present in the nucleus of a fertilized egg of a whale determines most of the hereditary qualities of an animal which will eventually weigh 5×10^7 grams! This is an amplification effect of 22 orders of magnitude, a feat which electronic engineers might well wish to emulate.

In addition to the criteria of (1) evolutionary continuity, (2) spatial compactness, and (3) energetic economy, there are many other characteristics of living systems one could enumerate that may or may not be universal for all life throughout the universe. One can speculate, for instance, about the universality of carbon as the backbone of biological molecules, of water as the solvent for the machinery of life, of high-energy phosphate compounds as mediators of energy transductions, of nucleic acids as the stable storers of hereditary information, of proteins as the catalysts and regulators, of lipid as the ubiquitous component of membranes, and so on. One of the fascinations of the developing field of *exobiology* is that we may soon be in a position to obtain answers to some of these questions. If and when we do, we will have raised the generality of biological phenomena to a higher level of universality.

A SHORT COURSE IN THERMODYNAMICS

We have said so far that the fundamental characteristic of the world of life is the evolution, maintenance, and extension of a tremendous degree of organization, capable of being compressed into tiny volumes. Is this process of organization, maintenance, and extension a unique property of living matter? The Second Law of Thermodynamics, a fundamental law of the physical universe, states that systems in isolation spontaneously tend toward states of greater disorganization. At first glance it would seem that the Second Law is not obeyed by living matter, and indeed this is what G. N. Lewis, one of the creators of the thermodynamic theory, suspected. However, in order to examine this problem more carefully, we must make a more precise statement of the Laws of Thermodynamics.

The First Law of Thermodynamics states that all forms of energy, chemical, mechanical, electrical, or radiant, are interconvertible. It also states that during the process of conversion, energy is neither created nor destroyed. Thus, the total energy content of an isolated system remains constant. It has been demonstrated repeatedly with very careful calorimetric measurements that living systems obey this law.

The First Law provides us with an overall balance sheet for energy conversions occurring in a process, but it does not permit us to predict whether a given process will indeed occur. According to the First Law the atoms of a gas in a given vessel could divide themselves into two regions, one hot and the other cold *as long as the total energy of the system remained unchanged.* It is the Second Law of Thermodynamics that predicts such a thing could not occur spontaneously. Thus, the Second Law of Thermodynamics is concerned with the "direction" of change and is capable of predicting whether or not a certain reaction can occur spontaneously, that is, without input of energy from the outside.

When a process occurs, it is possible to obtain some work from it. Another way of describing the Second Law of Thermodynamics, therefore, is to say it is able to distinguish between energy in a given system that is capable of doing work and energy that is not capable of doing work. To do this the Second Law makes use of four quantities, only one of which, the temperature T, can be measured directly on an absolute scale. The other three quantities (G, H, and S) are measured only as "changes" between two states of the process, and we shall therefore use the symbol Δ to signify "the change in a given quantity."

1. The *free energy change* or ΔG (often referred to as ΔF in biochemistry textbooks) is the energy which "drives" the reaction. It is often described as "energy capable of performing work." For a system in equilibrium at constant temperature and pressure $\Delta G = 0$. A spontaneous reaction will have a negative free energy change; the higher the negative value of ΔG, the greater the driving force of the reaction and the greater the amount of work that can be obtained from such a system.

The free energy change of a given reaction can be measured in a number of ways. In living systems we can usually assume no changes in pressure and temperature, and we usually define the "standard free energy change" ($\Delta G°$) as the gain or loss of calories when 1 mole of reactant is converted to 1 mole of product under standard conditions. It can be shown that if a chemical reaction is carried out at low concentration, the equilibrium constant K of the reaction is equal to the product of the concentrations of the reaction products divided by the product of the concentrations of the reactants. That is, in the reaction

$$A + B \quad \rightleftarrows \quad C + D$$

$$\frac{[C]\,[D]}{[A]\,[B]} = K$$

It can also be shown that

$$\Delta G° = -RT \ln K$$

where R (the gas constant) $= 1.987$ cal mole^{-1} deg^{-1} and T (absolute temp) measured in degrees Kelvin $= 273 +$ degrees centigrade.

This equation is a most useful tool for the evaluation of $\Delta G°$ in biochemical reactions. The following example shows how it is used.

glucose-1-phosphate   glucose-6-phosphate

The equilibrium of the above reaction has been studied at 25° and pH 7, and K was found to be 17. Thus, substituting into the preceding equation, we have

$$\Delta G° = -(1.987)(298.15)(2.303)(\ln 17)$$
$$= -1700 \text{ cal mole}^{-1}$$

We can therefore conclude that under standard conditions this transphosphorylation reaction will occur spontaneously, a fact which we will have the opportunity to observe in our studies of carbohydrate metabolism in Chapter 12.

2. The *change in enthalpy* or ΔH (sometimes referred to as the change in heat content) is a measure of the heat absorbed or given off in a reaction or process. We use the convention of characterizing heat given off as a negative enthalpy change. The enthalpy change of the reaction can be measured in a calorimeter, which is simply a container so well insulated from the surroundings that the heat produced or absorbed can be directly computed from the temperature change and the heat capacity of the system being studied.

3. As we have already said, a process can occur if free energy decreases. The amount of heat given off, however, can vary, and in some processes heat may not be given off but absorbed. Thus, for instance, when one dissolves urea in water, one finds that the temperature of the solution will drop. This apparent lack of direct relationship between free energy and enthalpy puzzled chemists and physicists until Clausius (1850) introduced a new concept that provided the missing connection between these quantities. Clausius coined the word *entropy* for a certain state of matter with its dimensions expressed in calories per degree. When the change in entropy is multiplied by the absolute temperature, a term $T\Delta S$ is obtained which helps relate enthalpy to free energy.

We can thus write the following equation:

$$\Delta G = \Delta H - T\Delta S$$

where ΔG (or ΔF) is the change in free energy measured in calories; ΔH is the change in enthalpy (or heat content) measured in calories; T is the absolute temperature measured in degrees Kelvin; and ΔS is the change in entropy expressed in calories per degree.

This generalization, which we call the Second Law of Thermodynamics, is one of the important general laws of nature. Historically, it began as an empirical generalization based on the behavior of heat engines, chemical processes, and electromotive cells. In recent years it has been possible to derive the Second Law by applying statistical mechanics to assumptions based on the particulate nature of matter; this is an analysis that has led to more general statements of the Second Law which the student will encounter when he studies thermodynamics at a more advanced level.

The Second Law of Thermodynamics concerns itself only with the initial and final states of a process, not with the particular path taken in traveling between these states. It can predict whether a certain reaction will occur spontaneously, that is, in isolation. It cannot, however, predict that a reaction involving a decrease in free energy will indeed occur at a measurable rate since all its reaction pathways may be "blocked" by energy barriers. Another aspect of the Second Law that must be recognized is that it is a "limiting" statement. Thus, the work that can be obtained from a process involving a given negative ΔG is at a maximum only if the process is "reversible," but in our real world of friction and "irreversible" phenomena the amount of work that can be obtained is always less than the amount predicted by the Second Law; this is why perpetual motion machines are not possible in our real world.

Numerous attempts have been made to give the concept of entropy a physical meaning. Although some purists object to this practice, entropy can be visualized as the degree of disorganization of a system. As entropy increases, so does the state of disorganization. Thus, when a process occurs, part of the energy may appear as an increase of $T\Delta S$, which can no longer be used to perform work. The Second Law predicts that energy must be invested to decrease entropy (increase organization), a fact which is familiar to every housewife, librarian, or laboratory worker. Thus, the increase in entropy is the directional arrow which seems to dominate the historical development of our universe as we know it at the present time.

If this is so, how can we explain the maintenance of order in living systems, much less the increase of order during the development of the fertilized egg to the mature individual? Furthermore, how can we reconcile the Second Law with the evolution of life from the simplest organic molecule to man?

WHY LIVING ORGANISMS DO NOT DEFY THE LAWS OF THERMODYNAMICS

The Second Law of Thermodynamics can predict that if a given process involves a free-energy increase, it cannot occur spontaneously. If we couple this process with another one which involves a free-energy decrease, then it is possible for one process to drive the other one. Thus, by putting the two processes together we create one single process, and this is thermodynamically acceptable because the only things that count are the initial and final states!

Let us, for instance, consider the following situation. The equilibrium between malic and fumaric acid is toward the side of malic acid. Nevertheless, it can be shown that if the enzyme aspartase is placed into a system consisting of malic acid and the enzyme fumarase, then the reaction will proceed in the direction of aspartic acid. Thus, we have

$$\text{malic acid} \xrightarrow[\substack{\Delta G = +700 \\ \text{cal/mole}}]{\text{fumarase}} \text{fumaric acid} \xrightarrow[\Delta G = -3720]{\text{aspartase}} \text{aspartic acid}$$

For the whole reaction of malic acid to aspartic acid, therefore, there is an overall decrease of G of 3020 cal/mole, and we can expect the reaction to proceed in the direction of producing aspartic acid. This effect can be visualized simply by imagining that the equilibrium between fumaric and aspartic acid is so far in the direction of aspartic acid that it lowers the fumaric concentration sufficiently to cause malic to react and form fumaric acid.

In the preceding example we have simply driven a reaction by coupling it to another. We have not demonstrated that by so doing we can lower the entropy, that is, cause an increase in organization. According to the Second Law, a reaction can still be spontaneous even though entropy decreases, provided the $-T\Delta S$ is "paid for" by a sufficiently large $-\Delta H$. This is indeed what must occur when large, highly organized macromolecules are synthesized from small monomers. The way in which the cell manages to perform this trick is to build into the monomer a great deal of free energy by synthesizing a high-energy phosphate configuration. These molecules, containing as much as 5000 cal/mole of free energy for each high-energy phosphate group, can be knitted together into a polymer even though this may entail an overall decrease in entropy since the decrease in enthalpy is large enough to "pay for" it.

(See Chapters 12 and 13.)

Now it is possible to ask "Where did the phosphate energy come from which was put into the monomer?" It can be shown that it came from another process that released energy such as the oxidation of glucose to CO_2 and H_2O. If one pursues the origin of this energy long enough, one eventually discovers a source of energy which is external to the cell. In the case of animals it is the variety of foods, which invariably leads one to the world of green plants, and thence to the light emanating from the sun, caused by the atomic fusion reactions occurring in it. Here we have the clue to the apparent disregard of the Second Law, which G. N. Lewis and others thought to be characteristic of living matter.

Because of the characteristic autonomous behavior of living matter, G. N. Lewis and others were led to think of living matter as being autonomous also from the point of view of energy when, in fact, nothing could be further from the truth. Living matter is capable of preserving its highly improbable-appearing individuality only at the expense of large amounts of free energy extracted from the environment. As soon as this supply of free energy is cut off, living systems proceed spontaneously to a greater state of disorganization (death). This type of labile system that is maintained at a certain level of organization by a continuous supply of free energy is often described as a *steady state*. It does not represent an *equilibrium* in which the system has achieved the lowest possible free energy and the highest possible disorganization. In fact, a steady state is a system *away* from equilibrium, which can only be maintained in this apparent constancy by the continuous supply of free energy. A nonliving example of a steady state is, for instance, a burning candle. The energy input is provided by the free energy present in the wax, and it is released to give a flame that, although characteristic and unchanging in appearance, experiences a rapid flux of energy and materials moving through it.

The Second Law, which we have described, is not directly applicable to non-equilibrium systems like the candle or the cell. An entirely new field of *nonequilibrium thermodynamics* applicable to steady-state phenomena has been developed in recent years, and this trend has provided further confirmation for the general applicability of thermodynamics to living systems.

In summary we can conclude that the Second Law tells us that if the entropy of an isolated system decreases, then the $T\Delta S$ term must be smaller than the ΔH term; and that if this is not so (as is often the case with biological systems), then we are dealing not with an isolated system but only with a portion of a larger system. Thus, when the energy source is included in the system, we can expect the Law of Thermodynamics to be obeyed.

A constant supply of free energy is only one requirement for the maintenance of a steady state. There must also be a minimal organization capable of absorbing and channeling the energy in a usable manner. Biological evolution is the phenomenon whereby a staggering variety of types of organization have slowly arisen that are capable of extracting free energy from our extremely variegated environment. As evolution has proceeded, the organization of the biosphere has become increasingly complex, and the amount of energy utilized has continually increased. This also should not be viewed as a violation of the Second Law. The machinery of natural selection, utilizing the phenomena of reproduction, sexuality, inherited change, and differential survival (see Levine's *Genetics* and Savage's *Evolution* in the Holt, Rinehart and Winston Modern Biology Series), provides a mechanism whereby the plentiful supply of free energy from the sun is being utilized to engender organizations of increasing complexity on this planet.

With the advent of human intelligence the evolution of various types of organization—be it a library, a scientific theory, or a computer—can no longer be considered quite as random as the genetic mutations are believed to be, but here too the general principle holds; namely, that free energy is necessary to permit the elaboration of these products of the human imagination.

Five billion years ago, before life had begun on our planet, all the free energy poured on it by the sun was rapidly dissipated as useless heat and radiated into outer space. Then tiny systems evolved, capable of trapping some of the free energy, which was used to maintain and extend the organization of these systems. As evolution proceeded, this trapping process became more efficient and extensive. Today, the trapped rays of the sun are supplying the free energy to build cities or to turn living matter on itself in the investigation of the principles by which it is governed. The steady state of life has acquired some of the free energy of the sun and is holding it in the form of an ever-expanding "biosphere."

There is a Third Law of Thermodynamics, which states that at absolute zero temperature the entropy of every substance is a minimum. We might enunciate a "Fourth Law," which would state that given plenty of time, the necessary atomic building blocks, the right temperature, and a steady supply of free energy possibly fluctuating in a diurnal cycle, a "bios" of increasing complexity will of necessity develop, which has the property of decreasing its own entropy at the expense of large amounts of free energy supplied by the sun. The current views held by astronomers that the suns of the universe are likely to be accompanied by planets of which a finite and very large number are similar to the planet earth lead us to the conviction that we are not alone in this universe. We no longer think of evolution as the "great coincidence," but as a full-fledged law of nature.

We conclude that living matter is not outside the physical world but an integral part of it. It is a fascinating special case of physical matter, distinguished by the long and characteristic history of its development and by the compact and economic functioning of its molecular organization.

SUGGESTED READING LIST

Offprints

Abelson, P. H., 1956. "Paleobiochemistry." *Scientific American* offprints. San Francisco: W. H. Freeman & Co.

Brown, H., 1957. "The Age of the Solar System." *Scientific American* offprints. San Francisco: W. H. Freeman & Co.

Miller, S. L. and Urey, H. C., 1959. "Organic Compound Synthesis on the Primitive Earth." *Science*, pp. 245–251. Bobbs-Merrill Reprint Series. Indianapolis: Howard W. Sams & Co.

Penrose, L. S., 1959. "Self-reproducing Machines." *Scientific American* offprints. San Francisco: W. H. Freeman & Co.

Urey, H., 1952. "The Origin of the Earth." *Scientific American* offprints. San Francisco: W. H. Freeman & Co.

Wald, G., 1954. "The Origin of Life." *Scientific American* offprints. San Francisco: W. H. Freeman & Co.

Wald, G., 1958. "Innovation in Biology." *Scientific American* offprints. San Francisco: W. H. Freeman & Co.

Articles, Chapters, and Reviews

Edsall, J. T. and Wyman, J., 1958. "Biochemistry and Geochemistry," ch. 2 in *Biophysical Chemistry*. New York: Academic Press, Inc.

Books

Baker, J. J. and Allen, G. E., 1965. *Matter, Energy and Life*. Palo Alto: Addison-Wesley Pub. Co., Inc.

Blum, H. F., 1955. *Time's Arrow and Evolution*. Princeton, N.J.: Princeton University Press.

Crick, F. H. C., 1967. *Of Molecules and Men*. Washington Paperbacks. Seattle and London: University of Washington Press.

Klotz, I., 1967. *Energy Changes in Biochemical Reactions*. New York: Academic Press, Inc.

Lehninger, A. L., 1965. *Bioenergetics*. New York: W. A. Benjamin, Inc.

Oparin, A. I., 1961. *Life, Its Nature, Origin and Development*. New York: Academic Press, Inc.

THE NATURAL HISTORY OF THE CELL

Part TWO

In this part we acquaint the student with broadly
observable facts of cell structure and cell behavior.
Although structure and function are inextricably connected, being but different facets of the same phenomenon
of life, it will suit our purpose to discuss the one in
terms of the other and vice versa. Thus, we hope that
the student will emerge with a picture of cell structure and
function into which he will be able to fit the numerous
details he will encounter in the second half of this book.

Cell Function, The Behavioral Basis for Biological Organization

The cell is a fundamental unit of biological activity. Although it is divisible into subcellular organelles that retain some of the properties of the intact functioning unit, the cell nevertheless represents the smallest module of *relatively independent* activity. It is the smallest portion of the organism that exhibits the *range of properties* we have come to associate with living matter — properties like reproduction, mutation, metabolism, and sensitivity. This is not surprising since from an evolutionary standpoint the cell is the autonomous unit from which multicellular organisms developed. In this chapter we shall deal with the range of properties that the cell possesses to give the reader a feeling for the "personality" of the cell.

In the previous chapter we presented a formal or overall view of the cell. We said that the cell is a piece of machinery that utilizes energy to maintain its individuality in the face of an ever-changing environment; but, although the use of energy is a thermodynamic necessity, it is far more interesting to begin to consider what the cell does with the energy.

Let us therefore begin to discuss the activities in which the cell engages and how these are related to each other to bring about the integrated phenomenon of cell function. We shall describe the functioning of a cell from two points of view, which are by no means mutually exclusive but which represent two ways of looking at the same process. One point of view concerns itself with the "flow of energy," the other with the "flow of information."

THE FLOW OF ENERGY

Since the cell maintains itself by utilizing energy, it must have one or more mechanisms for the input of energy, and it must have a number of machines or energy *transducers* capable of changing energy into *work*.

Figure 3-1 is a flow diagram, summarizing in very general terms the energy transductions of the cell. From the point of view of energy input it should be noted that plants and animals resemble each other closely, except with respect to the ultimate source of energy upon which they must rely. The plant is an *autotroph* that uses light energy to manufacture carbohydrates, fats, and proteins; whereas, the animal is a *heterotroph* that ingests these substances as foods and then transforms them into the particular carbohydrates, fats, and proteins it requires. Some plants such as fungi and a majority of the procaryotic organisms are also heterotrophs, but they do not ingest whole food particles. Instead, they extrude enzymes into their surroundings. These digestive enzymes break down the food into small, soluble molecules, which then enter the microbial cell by diffusion.

The carbohydrates, fats, and proteins or their breakdown products are not the immediate fuels that run the cell's machines. The ubiquitous ATP (adenosine triphosphate) performs this function. The *respiratory metabolism* of the cells consists of a multitude of very precisely regulated chemical steps whereby the food stuffs or "crude"

(See Chapters 7 and 12.)

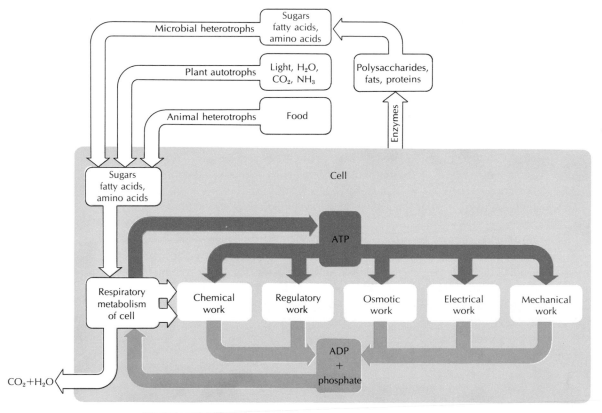

Fig. 3-1. The flow of energy.

fuels are broken down and the energy thus released is utilized for the synthesis of ATP, the "high-grade" fuel of the cell. In our industrialized society the high-grade fuel driving most of our sophisticated machines is, of course, electric power, which is generated by a variety of transducers that use water power, coal, fuel oil, and atomic energy. The latter are analogous to the carbohydrates, fats, and proteins, the crude fuels of the cell.

The respiratory metabolism of the cell is not only a source of energy, but also a source of material for the cell. The cell uses many breakdown products of the carbohydrates, fats, and proteins to build the large variety of compounds of which it is composed, including other carbohydrates, fats, and proteins. It is this close interweaving between the utilization of materials for generating energy and materials for the construction of other materials that makes the regulation of the cell's metabolism such a special and fascinating process.

The energy of the high-grade fuel ATP is used to "drive" most energy-requiring processes of the cell. These processes usually represent a *transduction* of one kind of energy into another, or in more general terms, a change of one thermodynamic parameter into another. We often refer to such processes as a form of work carried out by specialized machines. The cell then uses a variety of machines as energy transducers to carry out different forms of work. Let us now examine what these energy transducers are.

Chemical Work

(See Chapters 12 and 13.)

Figure 3-1 shows a machine that is capable of using the energy from ATP to perform chemical work. In reality the cell contains many such machines, only a few of which we understand in any detail. The cell performs chemical work whenever it utilizes energy to synthesize compounds that contain more free energy (G), more enthalpy (H), or less entropy (S) than the building blocks of which these compounds are composed. The cell has biosynthetic mechanisms for nucleic acids, proteins, fats, polysaccharides, and so on, and each of these includes the chemical machinery for transducing some of the energy of ATP into chemical work.

(See Chapter 14.)
(See Chapter 15.)
(See Chapter 12.)

Unlike the machines we build, the cell has to carry out the combustion of carbohydrates, fats, and proteins in a structural matrix composed of essentially the same kinds of materials it is breaking down. Thus, the cell must constantly replace components that have become altered during their function. Furthermore, one of the major facts of life is that living matter always extends itself. The cell not only repairs itself, it also duplicates. We do not know whether this is a physiological necessity or an evolutionary consequence of natural selection, but whichever the origin, the fact remains that a great deal of the energy funneled into a cell is utilized for growth and cell duplication. The concept of the *hereditary material* is necessary to the understanding of how the cell performs these vital functions and will be discussed in the next section.

(See Chapter 14.)

(See Chapter 17.)   **Osmotic Work**

We have already said that living material differs greatly from its environment in the relative abundance and the absolute concentration of its component materials. To bring about and maintain this situation, the cell must perform osmotic work. It must perform work both to accumulate materials found in low concentrations in its environment as well as to extrude materials found in high concentration in its surrounding medium. Not only do accumulation and elimination require work, but also the maintenance of this thermodynamically improbable situation requires work. The amount of energy required to maintain this improbable distribution of materials is reduced by the presence of the plasma membrane, which represents an extremely tight barrier to diffusion, thus restricting severely the movement of molecules and ions into and out of the cell. If it were not for this property of the plasma membrane, the retention or exclusion of molecules and ions by the cells would be as wasteful of energy as trying to store water in a sieve.

(See Chapter 17.)   **Electrical Work**

The cell's plasma membrane is never equally permeable to specific cations and anions, nor are they ever accumulated to the same extent. For these reasons all cells will exhibit a slight separation of charge across their membrane, leaving the outside positive with respect to the inside or vice versa. This electrical difference is utilized in nerve cells for communication by the transmission of an electric impulse which takes the form of a breakdown in charge separation. The separation of charges across the membrane, like any other form of work, requires energy. It is especially pronounced in organisms like the electric eel, which utilizes a great deal of energy to generate 1/8 amp at 10,000 v as a special protective adaptation.

(See Chapter 16.)   **Mechanical Work**

All of life is connected with motion of some sort, whether it is the contraction of muscle cells, the beat of cilia, the flow of the amoeba, the cyclosis of the protoplasm in a plant cell, or the movement of chromosomes. Motion requires energy funneled into a machine that can convert chemical energy into work.

(See Chapter 18.)   **Regulatory Work**

The cell must invest energy in the biosynthesis of complex, highly organized macromolecules, but it must also regulate their interactions once they are synthesized. We are only at the threshold of our understanding of the mechanism of cellular regulation and know very little about the particular steps in these processes where energy must be applied. Nevertheless, we can predict that since cellular regulation is an order-producing (entropy-decreasing) phenomenon, special points in the metabolism of the cell will be found at which energy is funneled into the regulatory process.

In addition to these five general categories of energy-requiring activities, cells may engage in a number of other such functions that are not quite so universal and which no doubt also require an energy input. During the growth and development of an organism, cells interact with each other extensively (see *Interacting Systems in Development* in the Holt, Rinehart and Winston Modern Biology Series). The physiological regulation of multicellular systems is also an area of intense interaction (see *Animal Structure and Function* in the Holt, Rinehart and Winston Modern Biology Series). Some cells even produce light.

From an evolutionary standpoint, perhaps the most significant recent development in the thermodynamics of life is that some of the organization of the cell is transferred outside its cellular limits, that is, living systems have learned to utilize energy to organize their surroundings. This social phase of biological evolution occurs most extensively at the organismic and population levels and is most evident in human activities although it must be said that we are at a stage of social evolution at which it is not clear whether organizing activities will, in the near future, outweigh the disorganizing activities which appear to be escalating around us.

THE FLOW OF INFORMATION

We have pointed out on several occasions that the growth of a cell or the development of an organism represents an increase in organization which is "paid for" by free energy supplied to the cell from its surroundings. It is important to recognize, however, that not all of the organization is generated *de novo* in the cell, but that some organization is maintained and transmitted from cell to cell. The material that carries this organization, which we shall call the *hereditary material*, represents the evolutionary "wisdom" that has slowly accumulated over billions of years of evolution. We shall define the hereditary material as those molecules or structures of the cell which are, at least in part, involved in their own reproduction. The hereditary material represents the starting point of the flow of information in the cell. There is by now little question that most of the hereditary information is stored by cells in their nuclei or nucleoid regions in the form of deoxyribose nucleic acid (DNA). As we shall see in later chapters, however, a number of cellular structures such as chloroplasts and mitochondria also contain DNA, and it is likely that DNA will be found in association with other cellular structures such as centrioles and basal bodies. Furthermore, certain plant and animal viruses store their hereditary information in ribose nucleic acid (RNA).

The question arises whether DNA and, in some cases, RNA represent the only structures that store hereditary information, or whether the cell possesses other levels of information storage that pass biological information on from generation to generation. This is an important question which we will discuss in Chapter 11. For the moment we shall content ourselves with the above general definition of the hereditary material.

To date, the only system of information flow we are beginning to understand is the one that starts with DNA. The elucidation of this system of information flow represents a triumph of modern molecular biology. Its simplicity and beauty is a tribute to the cell, which over billions of years of evolution has become the magnificent thing that it is.

We shall not attempt to trace the history of this exciting development. A perceptive history of this accomplishment of the human intellect has yet to be written. Let us simply trace in nonchemical language the flow of information through the cell and from cell to cell.

The basic and most important facts about DNA are:

1. it is a stable (chemically inert) material ideally suited for the storage of information;
2. it is a linear structure so that the information included in it is arranged in a linear array;
3. it is capable, in the presence of the appropriate enzymes, of self-duplication (replication); and
4. it is at the initiation point of a process which culminates in the synthesis of proteins (catalytic or structural), the chemical properties of which are based on three-dimensional organization.

Thus, we have here a flow of information in which there is a transformation from linear to three-dimensional organization. Finally, it must be emphasized that DNA seems to be also capable of "responding" to information reaching it from other portions of the cell. We are learning that DNA is not the absolute beginning of a biochemical flow of information but that it can be regulated by messages reaching it from other parts of the cell. This would not have been surprising to the Taoist philosopher, Chuang Tzu (see the Preface), who saw the world as a harmonious interplay of its components rather than as a rigid, hierarchical array of subordinate sets.

The process of information flow of DNA to protein occurs over several stages, and each stage is likely to be regulated by "feedback loops" or other regulatory messages from the cell. As Fig. 3-2 shows, DNA transfers its message to messenger RNA by a straightforward mechanism of linear "transcription," which is capable of being regulated by the cell. Then follows a process of "translation," which involves changing the linear code of nucleic acid based on four letters into the linear code of proteins based on 20 letters. The 20-lettered linear message is now capable of arranging itself in a three-dimensional array: a unique arrangement for each different protein. It is this latter process which generates a new property not encountered in the linear topology of the DNA and RNA. This property is the highly specific chemical reactivity of the proteins, which endows the cell with its uncanny ability to promote and regulate its chemical processes and endows the millions of biological species with their distinctness and the billions of biological organisms with their individuality. The three-dimensional arrangement that brings about the chemical specificity of the proteins is the immediate and most important "goal" of the information flow from the DNA molecule. Once the protein has been synthesized and its three-dimensional structure generated, it can become an enzyme catalyzing a chemical reaction, a structural protein to be incorporated in a particular cell structure, a "repressor" that combines with the DNA to

(See Chapter 15.)

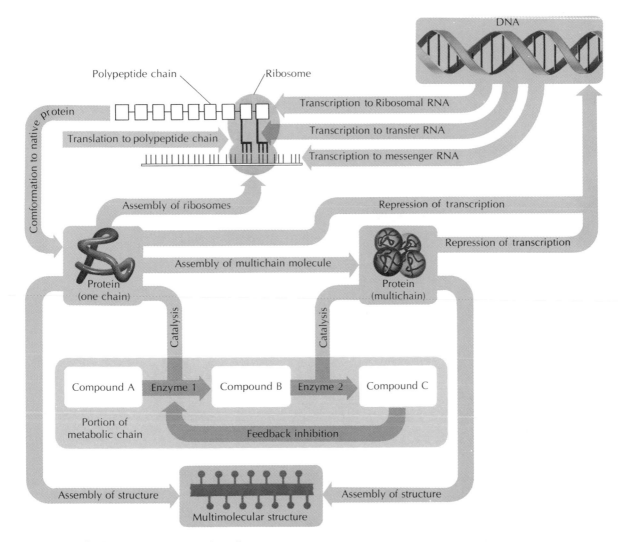

Polypeptide chain

Ribosome

DNA

Comformation to native protein

Translation to polypeptide chain

Transcription to Ribosomal RNA

Transcription to transfer RNA

Transcription to messenger RNA

Assembly of ribosomes

Repression of transcription

Protein (one chain)

Assembly of multichain molecule

Protein (multichain)

Repression of transcription

Catalysis

Catalysis

Compound A · Enzyme 1 · Compound B · Enzyme 2 · Compound C

Portion of metabolic chain

Feedback inhibition

Assembly of structure

Assembly of structure

Multimolecular structure

Fig. 3-2. The flow of information in the cell.

prevent it from synthesizing a messenger, or some other kind of regulatory protein, the existence of which we may as yet not even suspect. With these many different kinds of proteins — 500–1000 in a bacterium and probably more than 10,000 in a human cell — an integrated piece of machinery is being maintained. The purpose of modern molecular biology is to study the properties of these component pieces and to understand the mechanism of their interactions.

SUGGESTED READING LIST

Books

Allen, J. M., ed., 1962. *The Molecular Control of Cellular Activity*. New York: McGraw-Hill Book Co., Inc.

Davson, H., 1959. *A Textbook of General Physiology*. Boston: Little, Brown & Co., Inc.

Giese, A. C., 1962. *Cell Physiology*, 2d ed. Philadelphia: W. B. Saunders Co.

Kennedy, D., 1965. *The Living Cell*. Readings from *Scientific American*. San Francisco: W. H. Freeman & Co.

Thompson, D'Arcy W., 1966. *On Growth and Form*, abridged ed. Cambridge: Cambridge University Press.

Cell Structure, The Organizational Basis for Biological Function

The previous chapter emphasized that the cell uses energy to drive the processes that maintain its highly individual structure in the midst of an ever-changing environment. It is also possible, however, to look at this phenomenon from the other direction and say that without an existing structure energy could not be properly transduced so as to maintain, extend, and duplicate the living system. Thus, energy is necessary to *maintain* structure, but structure is in turn necessary to *utilize* the energy.

A dramatic demonstration of the importance of biological structure was provided by the experiments of Skoultchi and Morowitz, who cooled the eggs of the brine shrimp *Artemia* to temperatures below 2°K (−271°C) and showed that upon rewarming their hatch rate was the same as that of control eggs held at room temperature. Since at 2°K we have structure but presumably not process, it is reasonable to conclude that structure is not only a necessary condition, but even a sufficient condition for initiating biological function. It would thus appear that living processes could be generated by putting together the proper structures, the synthesis of life becoming "merely" a very complicated exercise in organic chemistry. The purpose of this chapter, then, is to describe the fine structure of living cells in order to provide a conceptual framework within which the molecules and the molecular processes of living systems can be located and visualized.

CELL SIZE AND SHAPE

In mathematics an equation remains unchanged if both sides are enlarged equally. This is not true in the physical world, where relationships between objects are very much influenced by their absolute dimensions, their physical structures being dependent for organization on physical forces which are after all finite.

It is obvious that the expanding and shrinking world of Lewis Carroll's Alice is the product of the imagination of a mathematician, not that of a physicist or biologist. One of the early investigators of size and shape in biological systems was the British biologist, D'Arcy Thompson, who pointed out, for example, that an elephant of twice the weight would have to be supported by leg bones of four times the cross-sectional area. In spite of D'Arcy Thompson's widely read, interesting book, biologists on the whole have been slow in recognizing the significance of absolute dimensions. Yet, the world of life falls into a size range that from a cosmic viewpoint is restricted to a narrow band of magnitude. Figure 4-1 is a summary of the size relationships observed on a cosmic, biological, and atomic level. It illustrates the fact that biological systems fall into definite, yet overlapping size ranges. For example, the largest cell is larger than the smallest mammal, the largest virus larger than the smallest cell, and the largest macromolecule larger than the smallest virus.

If cells fall into a certain size range, what are the laws governing this fact? We shall, for the sake of convenience, define a *cell* as a *unit of biological activity delimited by a semipermeable membrane and capable of self-reproduction in a medium free of other living systems. A virus* we shall define as a *self-reproductive biological unit that does not have a finite semipermeable membrane and is capable of self-reproduction only within a living cell.*

It is not surprising, in view of the continuity of the evolution of living systems, that forms exist that appear intermediate between viruses and cells. These systems, called *rickettsias*, appear to have membranes but are incapable of growing outside living cells. It is likely, however, that they have been derived from more independent bacteria, which have been converted to parasitism and have lost some of their biochemical potentialities.

The smallest living cells are found among the bacteria. The bacterium *Dialister pneumosintes*, for instance, has the dimensions of 0.4–$0.6 \times 0.5 \times 1.0\mu$ [1000μ (microns) $= 1$ mm (millimeter); 1000 mμ (millimicron) $= 1\mu$; 10A (angstrom units) $= 1$ mμ.] Assuming a 75 percent water content, this amounts to 2.8×10^{-14} grams of dry weight per cell. Table 4-1 shows the composition of this bacterium in terms of the major classes of compounds generally found in living systems.

Table 4-1   COMPOSITION OF THE BACTERIUM
DIALISTER PNEUMOSINTES (from Morowitz)

	MASS PER CELL		
	($\times 10^{-14}$ GRAMS)	($\times 10^{-8}$ DALTONS)	PERCENT
Dry weight	2.80	160	
DNA	0.13	7.8	4.66
RNA	0.30	18	10.33
Protein	1.20	72	43.00
Carbohydrate	0.47	28	16.45
Lipid	0.61	37	21.70

A weight of 0.13×10^{-14} grams of DNA is equivalent to 6.5×10^8 daltons (one dalton is a molecular weight of 1, or 1 gram is approximately 6×10^{23} daltons).

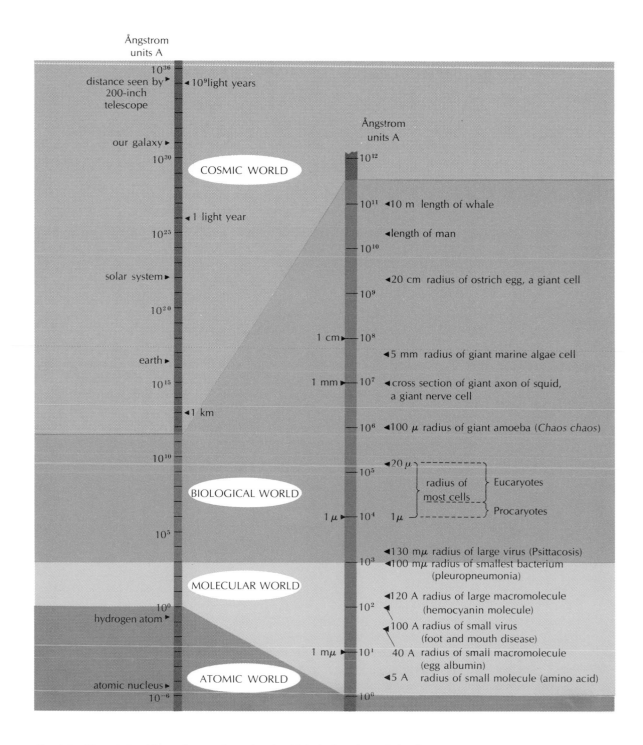

Fig. 4-1. Size relationships at the atomic, molecular, biological, and cosmic levels.

As will be seen in Chapters 14 and 15, double-stranded DNA must have approximately 20 times the weight of the protein for which it provides information. Thus, 6.5×10^8 daltons of DNA will carry the information for 3.3×10^7 daltons of protein. If one assumes an average molecular weight for the proteins of 40,000, one obtains an estimate of $\frac{3.3 \times 10^7}{4 \times 10^4} = 800$ different kinds of proteins in this cell. Thus, if we ignore for a moment the presence of structural proteins which do not carry any enzymatic functions, there are likely to be in the cells of the bacterium *Dialister pneumosintes* some 800 separate enzymes that must catalyze an equal number of biochemical reactions. This is not a large number when we consider that we already know of some 500 biochemical reactions that occur in the larger and more complicated *Escherichia coli.* Anything very much smaller than the bacterium we have been considering is likely to have insufficient DNA to make the variety of proteins required to sustain the 500–1000 biochemical reactions that probably constitute a minimum number of different reactions necessary for the maintenance of life. It is possible, of course, that smaller free-living organisms will be discovered in time. There are, for instance, a group of organisms called *pleuropneumonialike organisms* (PPLO) or mycoplasma, some of which have a DNA content of only 5×10^8 daltons and a dry weight of only 4.8×10^{-15} grams. It is possible that even smaller independent organisms will be discovered, but it is very likely that they will not be very much smaller because a minimum would soon be reached below which there would be an insufficient diversity of enzymes to catalyze the number of reactions necessary for the support of an independent cellular existence. In the viruses, on the other hand, which rely for much of their self-duplication on the biochemical versatility of their living hosts, the magnitude of the DNA genome tends to be much smaller. Thus, one of the largest viral DNA genomes is found in Vaccinia, being of the order of 1.6×10^8 daltons; the DNA of the bacteriaphage lambda is as small as 3.2×10^7 daltons.

The problems affecting the upper limit in cell size are very different. First of all, it must be recognized that although the vast majority of cells lies in the range of 0.5 to 20μ in diameter, there are some truly giant cells, the existence of which is related to some very special biological circumstance. The eggs of reptiles and birds, for example, are gigantic cells which store large amounts of food materials and water for the developing embryo. As soon as development begins, however, they subdivide into smaller cells. The conditions that affect the upper limit in cell size seem to be relational ones. First, there is the relation between the nucleus and the rest of the cell. As we shall see later, there is evidence that the nucleus releases certain "messengers" that determine the synthesis of the cell's proteins. Clearly this will determine the amount of cytoplasm a given nucleus can "control." Some cells manage to transcend this limitation by being multinucleate. Examples are giant cells of the amoeba *Chaos chaos* or the green alga *Nitella.* Second, there is the relationship between the various parts of the cell. The larger the cell the greater the problem of diffusion. In large multinucleate cells like *Nitella* or in the *slime molds* this problem is solved by a very active form of cytoplasmic streaming. Third, there is the relationship between the cell and its environment. The plasma membrane is an extremely impermeable layer. The larger the cell the smaller the surface-volume ratio becomes. Thus, as the cell enlarges, it has a tendency to become increasingly isolated from its environment, a handicap that the cell sometimes overcomes by a number of anatomical modifications. The enlargement of the surface area by extensive invagination is one device. Another anatomical feature, found in nerve cells, which in larger animals can be several feet long, is the development of a threadlike geometry. In a long, thin thread the increase in surface area remains proportional to volume, and, thus, the degree of contact between the cell and its environment is not reduced.

If one sets aside for the moment the special cells that attain unusually large dimensions through a series of special anatomical modifications, one finds that the average eucaryotic cell is several hundred times larger than the average procaryotic cell. This startling fact dramatizes the idea that the evolution of the eucaryotic cell was an evolutionary step of major importance. The increase in size and complexity obviously had to be accompanied by an increase in DNA content, requiring an entirely new technique for reapportioning the hereditary material (the mitotic apparatus). Problems of diffusion had to be dealt with by separating biochemical functions in special organelles. Since the eucaryotic cell represents an evolutionary culmination of intracellular complexity, we shall concern ourselves primarily with it in the succeeding pages even though we shall refer to viruses and procaryotes whenever appropriate.

The rest of this chapter will be a trip through the animal and plant cell (Figs. 4-2 and 4-3). Starting outside we shall explore the extensive membrane system of the cell after which we shall pass through the membrane into the cytoplasmic matrix of the cell. Having explored a number of structures of the cytoplasmic matrix, we shall examine the various cellular organelles which are held in the cytoplasmic matrix but are separated from it by one or two membranes. Throughout our trip we shall discuss the relationship between the structures we observe and their cellular function.

THE NEW MICROSCOPY

The Membrane System

The cell surface is delimited by a definite, very sharply defined "skin." As we shall see in Chapter 17, this is the osmotically active *plasma membrane*, which lowers the rate of movement of molecules into and out of the cell in a differential or selective manner. Thus, the plasma membrane helps to determine what molecules are allowed into the cell as well as what molecules are to be excluded from entering; it also determines which molecules are to be kept in the cell and which molecules are to be allowed to escape. This function of acting as a permeability barrier made cell physiologists fully aware of the plasma membrane three decades before it was visually revealed by the electron microscope. We now know that the plasma membrane is not only a passive diffusion barrier, but also carries *catalytically active* regions and that the machinery which utilizes energy for carrying out *osmotic work* is also located in or

(See Chapter 17.)

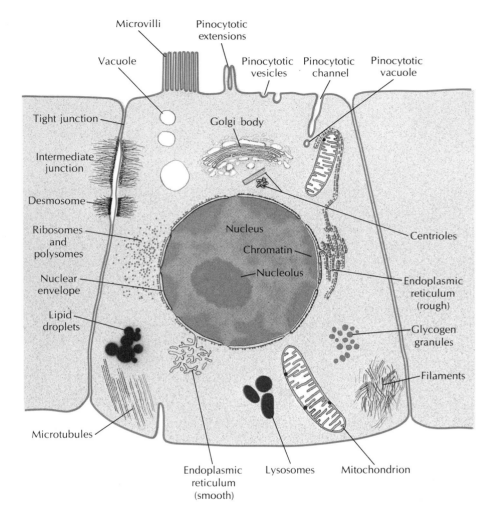

Fig. 4-2A. Drawing of generalized animal cell.

near the membrane. Under the electron microscope the plasma membrane appears to be a deceptively simple structure. It is about 80–100 A thick and is composed of three layers, which differ in their electron density and in their behavior toward stains, the central region being less electron dense and staining less strongly than the two external layers (Fig. 4-4).

Although we usually think of the plasma membrane as the outermost portion of the cell, most cells extrude some material that forms a layer outside the plasma membrane. This layer can be a relatively simple coating of *mucopolysaccharide* found outside many animal cells or between them in the intercellular space. In most procaryotic cells and in most plant cells this extracellular layer is a complex, often multilayered structure, called the cell wall (Fig. 4-5). In bacteria it appears that the cell wall actually lies between two cell membranes. The bacterial and plant cell wall is a tough elastic structure built by the cell with the aid of special enzymes that link the subunits into a continuous saclike envelope. In higher plants the constituents of the cell wall are polysaccharides, whereas in bacteria a complex structure is made from simple sugars, amino sugars, amino acids, and lipids. We usually think of these layers lying outside the plasma membrane as being secreted products of cellular activity, which can at times be removed experimentally leaving behind a cell called a spheroplast which still retains a number of cellular functions.

(See Chapter 10.)

Fig. 4-2B. Animal Cell. A liver cell from a new-born rat (× 11,000). (Courtesy of G. E. Palade, Rockefeller University.)

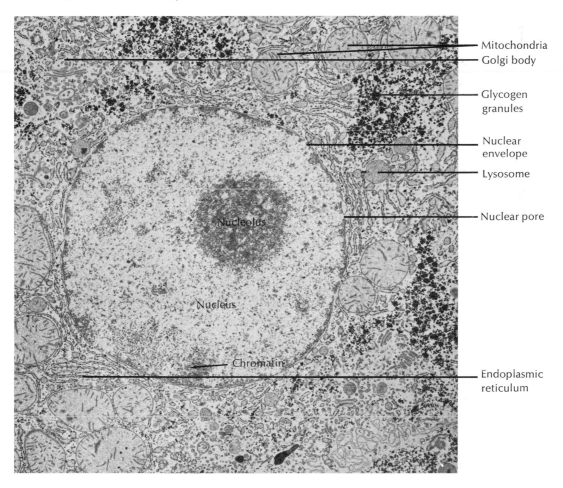

- Mitochondria
- Golgi body
- Glycogen granules
- Nuclear envelope
- Lysosome
- Nuclear pore
- Endoplasmic reticulum

Nucleolus

Nucleus

Chromatin

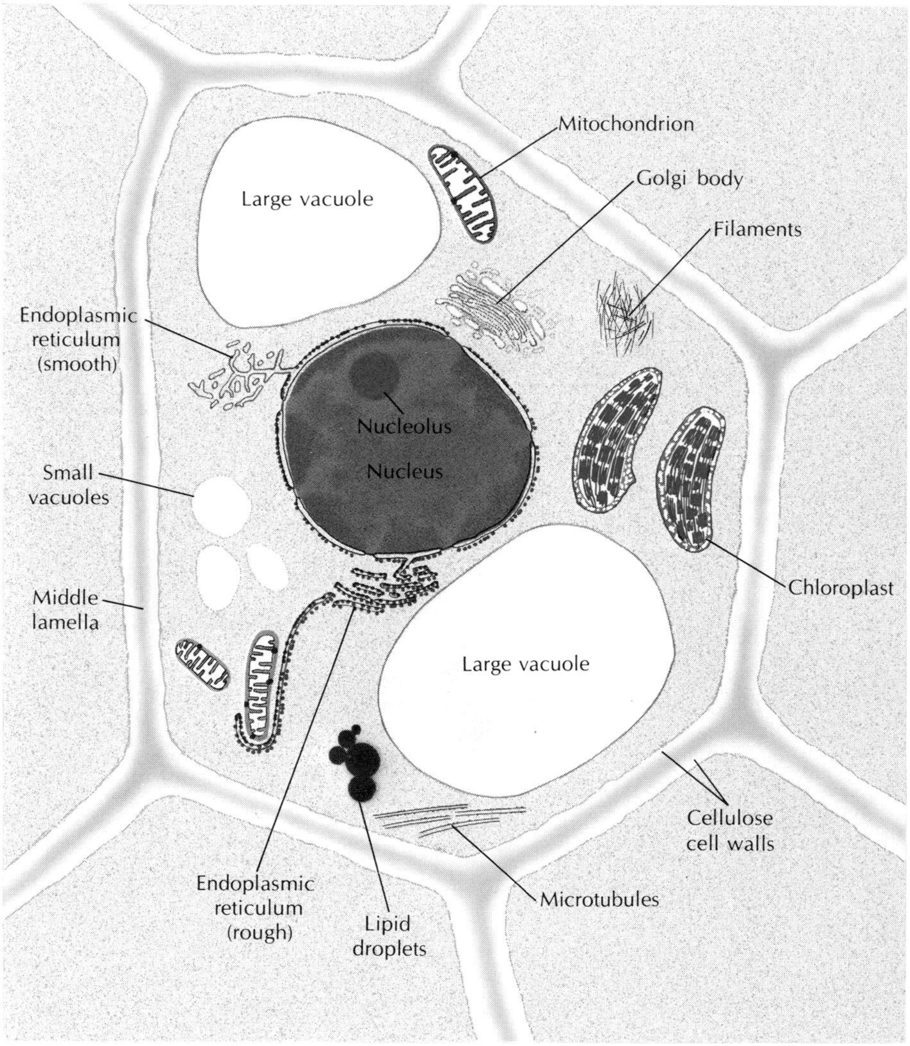

Fig. 4-3A. Drawing of generalized plant cell.

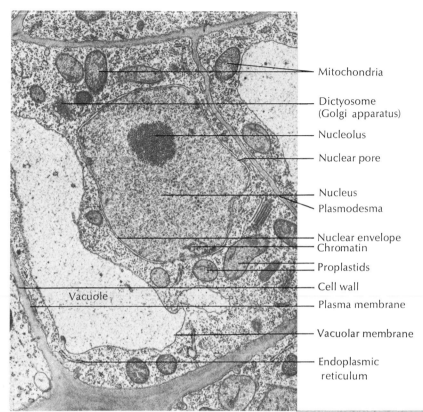

Mitochondria

Dictyosome
(Golgi apparatus)

Nucleolus

Nuclear pore

Nucleus
Plasmodesma

Nuclear envelope
Chromatin

Proplastids

Cell wall

Plasma membrane

Vacuolar membrane

Endoplasmic
reticulum

Vacuole

Fig. 4-3B. Plant cell. A root tip cell from *Arabidopsis thaliana* (× 21,000). (Courtesy of M. C. Ledbetter, Brookhaven National Laboratory.)

Fig. 4-4. Plasma membrane. Boundary between two glial cells of the annelid, *Aphrodite* (× 260,000). (From Fawcett, *The Cell: An Atlas of Fine Structure*. W. B. Saunders Co., 1966.)

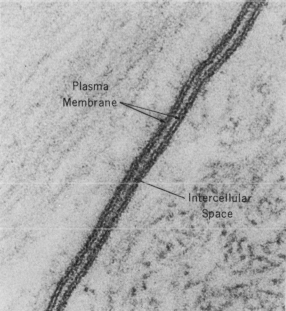

Plasma
Membrane

Intercellular
Space

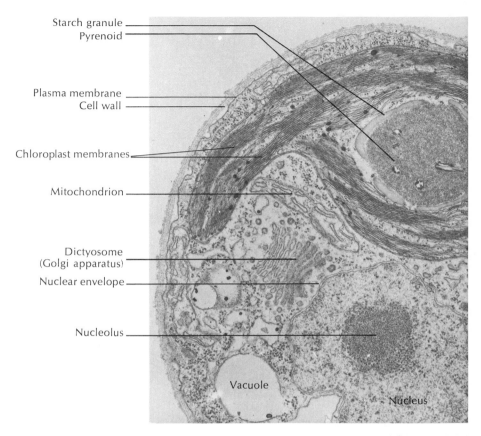

Starch granule
Pyrenoid
Plasma membrane
Cell wall
Chloroplast membranes
Mitochondrion
Dictyosome
(Golgi apparatus)
Nuclear envelope
Nucleolus
Vacuole
Nucleus

Fig. 4-5. Cell wall. An alga, *Chlamydomonas reinhardii* (× 30,000). (Courtesy of G. E. Palade, Rockefeller University.)

Before the advent of the electron microscope we used to think of the plasma membrane as a thin coat, stretched tightly over the cell as it indeed appears to be in the case of a few cells such as the *erythrocyte*. Now we know that this is often not the case. The plasma membrane, for instance, will frequently amplify its surface area by forming elongated slender cell processes called *microvilli* (Fig. 4-6). These are found in cells that are active in the transport of material; they are absorptive cells as in the intestinal epithelium or secretory cells. Besides such stable enlargement of the cell surface as represented by the microvilli the plasma membrane also engages in transitory evaginations and invaginations of the cell surface. Thus, the cell extends its surface to form folds or blunt protrusions, cross sections of which can be readily seen under the electron microscope (Fig. 4-7) and in some cells are large enough to be seen even under the light microscope. These protrusions can be seen to coalesce, thus entrapping some liquid that moves into the cell in the form of a *vacuole* or vesicle. This type of transport, which involves "bulk drinking" of the external solution and is called *pinocytosis*, has

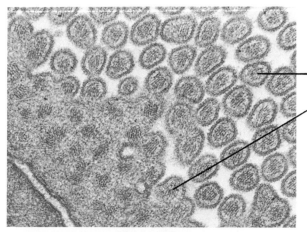

Cross sections

Tangential section

Fig. 4-6. Microvilli. Brush border of intestinal epithelial cell. Microvilli are shown in tangential and cross sections (× 84,000). (Courtesy of G. E. Palade, Rockefeller University.)

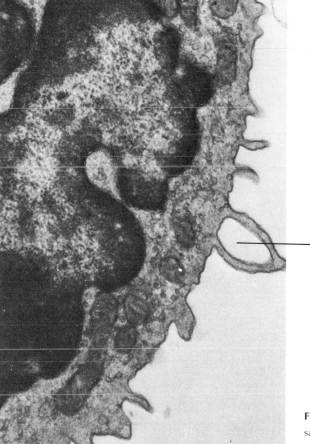

Pinocytotic vesicle

Fig. 4-7. Extensions of cell membrane. Leucocyte in salamander (× 26,000). (From Fawcett, *The Cell: An Atlas of Fine Structure.* W. B. Saunders Co., 1966.)

(See Chapter 17.)

attracted widespread attention in recent years, though it was first observed by Warren Lewis in 1935. Through pinocytosis materials became transported into the cell in a package that is surrounded by the cell membrane. Once the vesicle is inside the cell the entry of materials into the cell sap is controlled by the vesicular membrane. The pinocytotic mechanism begins to operate by the invagination of the plasma membrane. In some instances the plasma membrane develops a fuzzy or filamentous appearance at the site of invagination, and we suspect that this represents an adaptive modification to promote surface binding of specific molecuies such as proteins, which then enter the cell by pinocytosis. Should these materials be of a macromolecular nature, they must first be digested to smaller molecules. Pinocytosis can therefore be regarded as a device by which a heterotrophic cell acquires metabolites in concentrated form (as macromolecules). In other instances, as in the case of the blood capillaries, pinocytotic vesicles can be seen forming in the surface membranes and also as free vesicles in the cytoplasm (Fig. 4-8); thus, it would appear that bulk transport of solution across this endothelial cell (which is the capillary wall) occurs inside these vesicles.

Finally, the plasma membrane is intimately connected with special structures that are responsible for cell-to-cell attachment in multicellular animals. Although

Fig. 4-8. Pinocytotic vesicles. Blood capillary cell, showing the entire width of the cell (× 315,000). (Courtesy of G. E. Palade, Rockefeller University.)

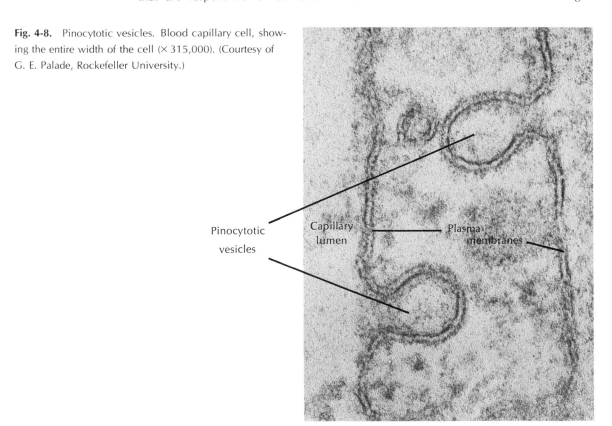

Pinocytotic vesicles

Capillary lumen

Plasma membranes

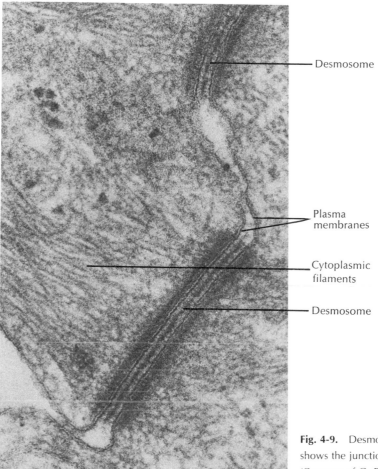

Desmosome

Plasma membranes

Cytoplasmic filaments

Desmosome

Fig. 4-9. Desmosome. Amphibian skin. The picture shows the junction between two cells ($\times$ 210,000). (Courtesy of G. E. Palade, Rockefeller University.)

cell-to-cell attachment probably occurs all along the surface membrane, these special structures are likely to provide additional "strong points," thus anchoring cells to each other on a more permanent basis. Of these various structures the *desmosome* is the most complex in appearance (Fig. 4-9). It includes a thickening of the cell membranes running strictly parallel to each other, separated by an enlarged intercellular space (250A) as well as bundles of fine cytoplasmic filaments extending into the cytoplasm. We should also mention that in many cases there occurs fusion of cell membranes, probably causing an increase in permeability and thus allowing small molecules to move more freely from cell to cell.

As we shall see, the plasma membrane is continuous with the eucaryotic cell's internal membrane systems, which it resembles closely in fine structure. In fact, one has the impression that the membrane systems of all cells have a universal molecular plan. (See Chapter 17.)

Thus, we find, for instance, that in osmotically active cells the plasma membrane is increased in amount rather than modified in its fundamental composition or construction in order to allow a more rapid exchange of materials. The term *unit membrane* was coined by Robertson for this notion of a universal molecular architecture of all cell membranes. In the very near future, as we shall become more familiar with the molecular structure of the surface membrane, we shall no doubt find that a basic plan does occur even though this basic plan may well be modified to allow some degree of variation in function.

If one explores the cell surface in greater detail, one finds that in a few cases the invaginations are not bounded by a vesicle but may instead lead right into the depth of the cell. These channels connect with a complex set of vesicles that interlace the interior of the cell, but we think this communication is intermittent in character, giving the cell some regulatory control over the movement of materials in and out of the endoplasmic reticulum. The vesicles were first discovered with the electron microscope in thinly spread tissue culture cells in 1945 and were named *endoplasmic reticulum* by Porter in 1953. By the use of ultrathin sections and improved fixation techniques developed by Palade and Porter (1954), it was finally recognized that the endoplasmic reticulum represented cavities of a great variety of shapes and dimensions surrounded

Fig. 4-10. Rough (granular) endoplasmic reticulum. Acinar cell from the pancreas. The picture shows membranes lining flattened vesicles or cisternae; ribosomes are attached to the membranes (× 90,000). (Courtesy of D. W. Fawcett, Harvard Medical School.)

Vesicles

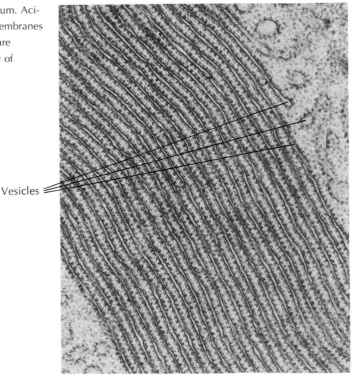

by membranes. Subsequent work showed that the endoplasmic reticulum is found almost universally in eucaryotic cells and that it can exist in "rough" or granular form (Fig. 4-10) and in "smooth" or agranular form (Fig. 4-11). The rough form of the endoplasmic reticulum carries   on its cytoplasmic side numerous closely spaced granules (diameter 150 A) called ribosomes which, as we shall see in Chapter 15, are the sites of protein synthesis. The systems of vesicles that we call the endoplasmic reticulum vary greatly in shape from loose irregular networks to flat saclike *cisternae* lying in parallel or concentric array stacked against each other. (Compare Fig. 4-10.) At times the endoplasmic reticulum becomes greatly distended so as to take up a large proportion of the volume of the cell (Fig. 4-12). At other times it becomes packed with granules, droplets of lipid, and even with protein crystals. In the muscle cells the endoplasmic reticulum is a highly developed structure called *sarcoplasmic reticulum,* which surrounds the myofibrils (Fig. 4-13) and is very intimately connected with the

(See Fig. 15-2.)

Fig. 4-11.   Smooth (agranular) endoplasmic reticulum. Interstitial cell from testes (× 51,000). (From Fawcett, *The Cell: An Atlas of Fine Structure.* W. B. Saunders Co., 1966.)

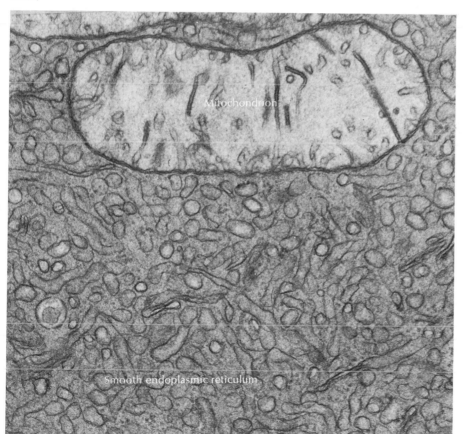

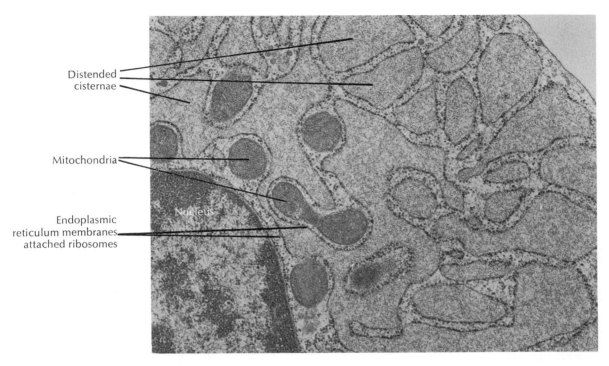

Distended cisternae

Mitochondria

Nucleus

Endoplasmic reticulum membranes attached ribosomes

Fig. 4-12. Endoplasmic reticulum. Plasma cell from bone marrow (× 26,000). (From Fawcett, *The Cell: An Atlas of Fine Structure.* W. B. Saunders Co., 1966.)

Fig. 4-13. Sarcoplasmic reticulum. Muscle from swim bladder of *Opsanus tau* (× 30,000). (From Fawcett, D. W., and J. P. Revel, *Journal Cell Biology,* Supplement. 10:89–109. 1961.)

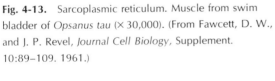

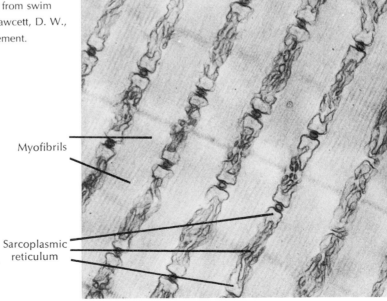

Myofibrils

Sarcoplasmic reticulum

process of contraction. These numerous variations in structure and content of the endoplasmic reticulum occur not only from one cell type to another, but also in different developmental and physiological stages of the same cell type, thus giving the impression that it is a labile system capable of rapid modifications in structure and function. In spite of its physical interconnection with the plasma membrane, the endoplasmic reticulum appears to be slightly thinner (50–60 A) than the plasma membrane.

The function of the endoplasmic reticulum is still not perfectly understood. As we have seen, it can often be shown to contain a variety of materials. Furthermore, by the use of time studies it has been possible to infer, in some cases, that directional transport from one part of the cell to another and even out of the cell appears to occur in the endoplasmic reticulum. Thus, one might conclude that the endoplasmic reticulum is a major adaptive modification of the eucaryotic cell, supplementing free diffusion with a more directed or channeled form of transport and providing for segregation and concentration of various molecules such as proteins in pancreas, triglycerides in intestinal epithelium, and even ions as (Ca^{2+}) in sarcoplasmic reticulum. Furthermore, the importance of membrane systems as such in the metabolism of cells has become increasingly apparent in recent years. Thus, one might also view the endoplasmic reticulum as permitting a wider deployment of those metabolic properties requiring the participation of membranes.

We have shown that the endoplasmic reticulum is continuous with the plasma membrane; it can also be shown that it is continuous with the *Golgi complex* and with the membranes around the nucleus called the nuclear envelope.

The Golgi complex is composed of a system of membrane-bound vesicles that are sufficiently characteristic in organization, location, and function to warrant classification as a separate cell organelle. This structure was discovered in 1898 by Golgi, who used a silver-impregnation technique to stain it. In the ensuing years there was considerable disagreement about the existence of this organelle, many cytologists believing that it was an artifact of fixation or staining procedures. Studies with the *phase contrast microscope* in the early 1940s showed that there was a region of different refractive index in living cells usually close to the nucleus, precisely the area in which the Golgi complex had been discovered by the heavy-metal staining method. The use of the electron microscope for the study of thin sections of cells developed in the 1950s finally proved, without a doubt, that a special organelle existed in the precise location assigned to it by Golgi.

Figure 4-14 shows an electron micrograph of a cell that had been previously stained with osmium. It shows a series of tightly packed smooth vesicles in parallel or often semicircular array. The outer vesicles alone are heavily impregnated with electron dense osmium stain, a differential pattern of staining that seems to occur in general. Thus, it appears that Golgi had discovered only the outer part of the Golgi complex. It should be noted in Fig. 4-14 that the outer vesicles of the Golgi complex are flat whereas the inner ones have a tendency to be more extended, which appears to be a property of this organelle when it is metabolically active.

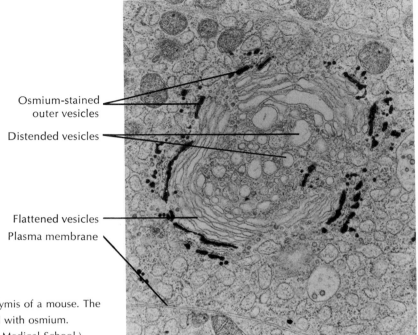

Osmium-stained
outer vesicles

Distended vesicles

Flattened vesicles

Plasma membrane

Fig. 4-14. Golgi complex. Epididymis of a mouse. The preparation was previously stained with osmium. (Courtesy of D. S. Friend, Harvard Medical School.)

The function of the Golgi complex is still not perfectly understood. In secretory cells the Golgi complex is the site of segregation and accumulation of the product secreted by the particular cell (Fig. 4-15). In cells specialized for absorption such as the cells of intestinal epithelia, which absorb from the intestinal tract, the vesicles of the Golgi complex become filled with lipid; these are triglycerides that have been synthesized from monoglycerides and fatty acids in the endoplasmic reticulum membranes. These latter compounds have been previously "absorbed" by the cell and reach the endoplasmic reticulum. Yet, secretion and absorption may not be the most general function of the Golgi complex since it is found almost universally in eucaryotic cells. It is probable that the function of the Golgi complex is to bring into the interior and close to the nucleus many biochemical reactions for which membranes are required. It is becoming increasingly clear that many enzyme systems are localized on membranes, and it is possible that their biochemical function depends, in part, on their localization at an interface. Thus, we can conclude tentatively that the Golgi complex plays a role both in transport and in metabolism. It is possible the *mesosome*, the folded membrane structure that is usually an infolding of the plasma membrane near the nuclear area in some procaryotic cells (Fig. 4-16), has a function that might be related to cell wall formation and to oxidative phosphorylation.

The last membrane system is the so-called *nuclear envelope*, consisting of two membranes surrounding the nucleus of the eucaryotic cell. During nuclear division

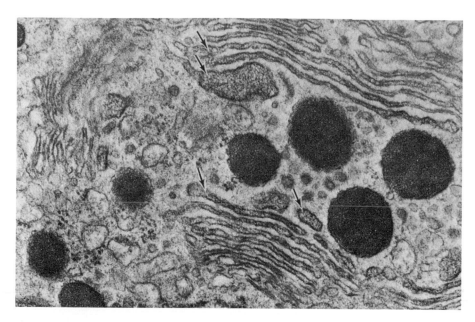

Fig. 4-15. Golgi apparatus. Epithelial cell of Brunner's gland of a mouse. Arrows point to secretory products within vesicles of Golgi apparatus (× 54,000). (Dr. D. Friend, from Fawcett, *The Cell: An Atlas of Fine Structure.* W. B. Saunders Co., 1966.)

Fig. 4-16. Mesosome, *Diplococcus pneumoniae* (× 200,000). (Courtesy of A. Tomasz, J. Jamieson, and E. Ottolenghi, Rockefeller University.)

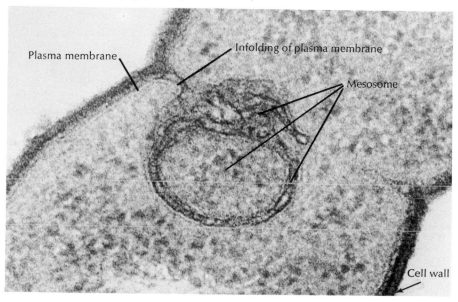

when the chromosomes go through the complicated dance that results in the equi-distribution of the hereditary material into the two daughter cells, the nucleus is not surrounded by a membrane system. During telophase, the last stage of nuclear division, a number of flat vesicles or sacs of the endoplasmic reticulum arrange themselves around the chromosomes and fuse extensively, leaving only small circular regions called "pores" (Figs. 4-17 and 4-18). At times these pores look optically empty, and at other times a septum or even a plug can be clearly seen within them (Fig. 4-19).

Fig. 4-17. Nuclear pores. Acinar cell from pancreas. Arrows point to pores in cross section at nucleus (× 22,000). (From Fawcett, *The Cell: An Atlas of Fine Structure*. W. B. Saunders Co., 1966.)

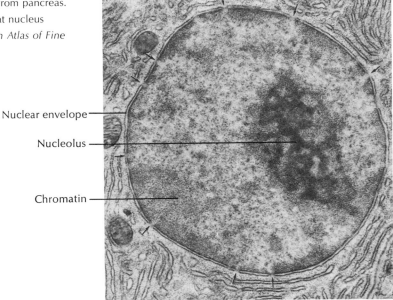

Nuclear envelope

Nucleolus

Chromatin

The permeability properties of these pores are as yet not understood. Studies utilizing radioactive amino acids seem to indicate that there is some restriction of free diffusion of amino acids into the nucleus. Furthermore, isolated nuclei have been shown to swell or shrink osmotically, which also indicates that the pores are not freely permeable to diffusion of small molecules. It is possible that the nuclear pores are merely "spacers" or structural devices for maintaining the equidistant, parallel spacing (400–700 A) of the two membranes of the nuclear envelope; the function of the septum would then be to seal off the nuclear pore. It is also possible that the nuclear pores are the sites of transfer of macromolecules between nucleus and cytoplasm. The inner and outer membranes of the nuclear envelope are not always alike in appearance. The outer membrane is often lined with ribosomal particles (Fig. 4-19) and is often in continuity with similar membranes of the endoplasmic reticulum, whereas the inner membrane is sometimes lined with a dense, filamentous layer or *fibrous lamina*.

In summary it should be emphasized that the eucaryotic cell is extensively inter-laced by a membrane system that interconnects the nucleus with the Golgi complex

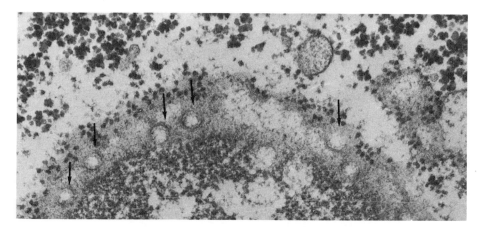

Fig. 4-18A. Nuclear pores. Liver cell. Arrows point to pores, which are seen looking down on nuclear envelope; the picture indicates that the pores are not just openings in a nuclear envelope but are distinct structures ($\times$ 59,000). (Courtesy of G. E. Palade, Rockefeller University.)

Nuclear pores

Nuclear membrane

Fig. 4-18B. Nuclear pores from the onion root tip as revealed by the freeze-etch technique ($\times$ 75,000). (See Fig. 17-27 for the explanation of this method.) (Courtesy of D. Branton, University of California, Berkeley.)

and in turn connects these with the cell exterior. As we shall see in Chapter 16, the membrane system is rich in lipid and relatively labile in structural organization, having the properties of a macromolecular liquid phase that is not miscible with water but

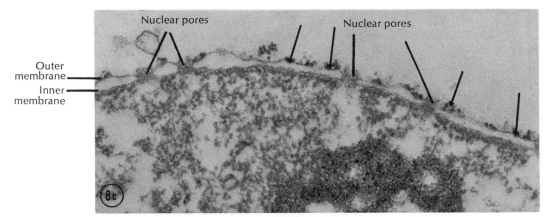

Fig. 4-19. Nuclear pores. Liver cell. Arrows point to ribosomal particles on outer membrane (× 75,000). (Courtesy of R. Maggio, P. Siekevitz, and G. E. Palade, Rockefeller University.)

instead will arrange itself in thin layers or sheets, forming a variety of interfaces in the cell. Thus, the cell is a system composed of two phases not miscible with each other: (1) an aqueous phase, with a consistency varying from that of a liquid to that of a solid, which we shall call the *cytoplasmic matrix* or *ground substance,* and (2) the membrane system, a relatively fluid phase of lipoprotein covering and interpenetrating the cytoplasmic matrix.

An important feature of the membrane system is that it topologically subdivides the cell into compartments that enclose aqueous phases. Thus, the cell is really composed of a variety of aqueous phases each of which is separated from the others by one or even two membranes. In the case of the latter, such as in mitochondria (compare Fig. 13-1) or in chloroplasts, we have in reality two aqueous phases separated from each other by a membrane.

The Cytoplasmic Matrix

So far it has been possible for us to wander through the cell without once crossing through a membrane. Were we a small molecule, we could have entered the endoplasmic reticulum through one of the surface pores where it joins the membrane, or we could have moved through "smooth" or "rough" vesicles into the Golgi complex and thence into the nuclear envelope. Eventually, we might have moved out of the cell at the opposite end from our point of entry. We might thus have traveled across the cell without once having passed through a membrane, having remained entirely in the "external compartment" of the cell. Let us now move through a membrane and enter the "internal compartment" of the cell, which we have called the *cytoplasmic matrix*. This aqueous phase, composed mainly of protein macromolecules in various states of aggregation, is the supporting medium of the cell organelles such as the nucleus, mitochondria, chloroplasts, ribosomes, and lysosomes. The ground substance has the unusual property of being capable both of viscous flow like a liquid and of elastic deformation like a solid. Furthermore, depending on the particular physiological or developmental state of the cell, it can vary in the degree to which it is one or the other. Generally, the cytoplasmic matrix near the outer membrane, often referred to as the *ectoplasm,* tends to be more like a solid whereas the interior of the cell, or *endoplasm*, is generally in a more fluid state. The viscous properties of the ground substance were studied extensively in the 1930s by (1) measuring the rate of Brownian movement of particles in various portions of the cell, (2) measuring the rate of movement of cytoplasmic particles in the gravitational field developed by centrifugation, and (3) other ingenious techniques such as causing the movement of small particles of iron through a magnetic field. As a result of these measurements, lively discussions ensued among the cell physiologists of that period regarding the submicroscopic or "colloidal" properties of the ground substance. Most workers agreed that although the ground substance appeared optically homogeneous in the light microscope, it must nevertheless contain a submicroscopic skeleton responsible for its elastic properties. These early workers conceived of a *cytoskeleton* of highly elongated particles interacting with each other to form a "brush heap" or gel. As recently as 1950 Francis Crick wrote about protoplasm being like "mother's sewing basket" filled with spools of thread, buttons, and knitting needles in untidy array. Until recently it looked as if no "knitting needles" or cytoskeleton of elongated particles was likely to be found in the ground substance except for the very obvious fibrillar organization of special cells such as striated muscle cells or special cellular structures such as the spindle, the centrioles, or the cilia. In the last few years, however, electron microscopists have begun to discover *filaments* and *microtubules* in a large variety of cell types, and it now seems clear that these elongated rods are a universal component of the ground substance.

Cytoplasmic filaments are rods of indefinite length and 40–50 A in thickness. Their presence is, of course, most obvious when they are numerous and well organized as in the striated muscle (Fig. 4-20), near and seemingly attaching to the desmosome (Fig. 4-21), or in the axoplasm of nerve cells. Once cytoplasmic filaments were recognized in their abundant and more organized manifestations, however, they were observed even in the more generalized cells such as the amoeba or undifferentiated plant cells.

It is still premature to assign well-defined functions to all filaments encountered in cells. In the case of muscle we know that the two types of filaments present are involved in the contractile mechanism of the muscle fiber. We suspect that in the case of the amoeba the filaments may also be part of the mechanochemical machinery converting chemical energy into work and thus bringing about the rapid cytoplasmic movement or "streaming" exhibited by these cells. In other instances it is entirely possible that filaments have a more passive function of providing a certain amount of rigidity or tensile strength for particular cells or parts of cells.

At times one can observe that filaments arrange themselves in such a manner as to form tubular structures called microtubules. In recent years more and more examples

I filaments

A filaments

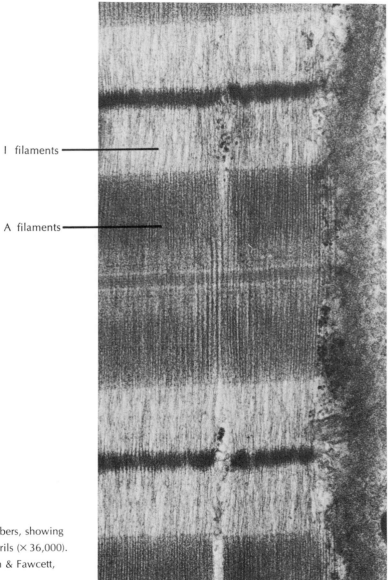

Fig. 4-20. Filaments. Cardiac muscle fibers, showing highly organized filaments in muscle fibrils ($\times$ 36,000). (Courtesy of D. W. Fawcett, from Bloom & Fawcett, *Textbook of Histology*, 9th ed.)

of microtubules have been discovered in a wide variety of both plant and animal cells so that they can now be considered a universal component of the eucaryotic cell. Microtubules are 200–270 A in external diameter, having an electron dense wall some 50–70 A thick (Fig. 4-22). In certain cases it has been possible to resolve the wall into a collection of some 10 to 14 filamentous subunits. In view of this and the presence of

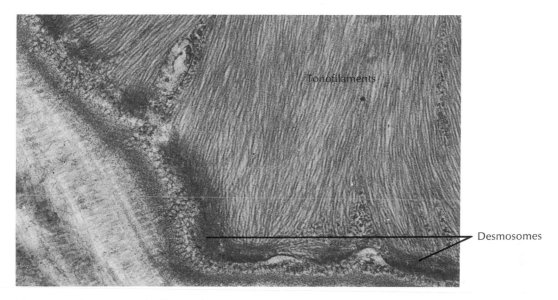

Fig. 4-21. Filaments. Basal cell in epidermis (× 80,000). (From Fawcett, *The Cell: An Atlas of Fine Structure.* W. B. Saunders Co., 1966.)

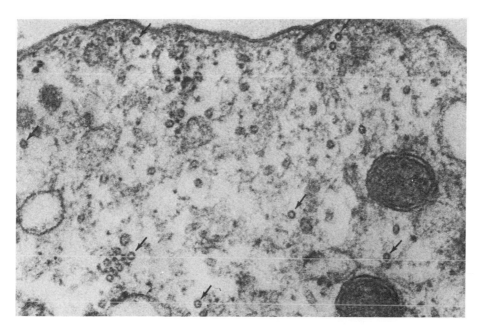

Fig. 4-22. Microtubules. Podocyte in a kidney. Arrows point to cross section of tubules (× 120,000). (Dr. A. Schechter, from Fawcett, *The Cell: An Atlas of Fine Structure.* W. B. Saunders Co., 1966.)

occasional microtubules among collections of cytoplasmic filaments, it seems reasonable to assume that the microtubule is an aggregate of microfilaments. On the other hand, it is also possible that the subunits responsible for the formation of the filaments can aggregate in such a manner so as to form microtubules without first forming microfilaments. It is also, of course, entirely possible that the two structures are not related to each other at all.

Microtubules are often found in portions of cells requiring some stiffness as at the outer edge of nucleated erythrocytes of fish where they probably help maintain the flattened shape characteristic of these cells. Thus, like the filaments, they are probably part of the cytoskeleton of the cytoplasmic matrix. Also like the filaments, microtubules seem to be involved with the machinery of motion, a conclusion that will become plausible when we consider the ubiquitous presence of microtubules in the mitotic spindle, the kinetosomes, and the cilia.

The *spindle* (Fig. 4-23) is a structure that is formed during nuclear division or mitosis and is part of the machinery responsible for the equidistribution of the hereditary material into the two daughter cells. It is a structure that is sufficiently cohesive to be separated from the rest of the cytoplasmic matrix by differential centrifugation. Its structure has been studied intensively by Inoue, who utilized a polarizing microscope to reveal the presence of the spindle's organized submicroscopic elements. Recent electron microscopic studies have established that these elements are microtubules lying in parallel array (Fig. 4-24). As we shall see later, movement of the chromosomes appears to result from the interaction of special regions on the chromosomes (centromeres) with the microtubules.

Fig. 4-23. Mitotic spindles as seen by a polarizing microscope. The mitotic spindle fibers that pull the chromosomes apart are seen as bright streaks. A. Oöcyte of marine worm *Chaetopterus pergamentaceous.* B. Endosperm cell from the fruit of African blood lily *Haemanthus katherinae.* Chromosome pairs are just about to be separated (× 800). (Photographs courtesy of S. Inoue, University of Pennsylvania, and of *Chromosoma.*)

 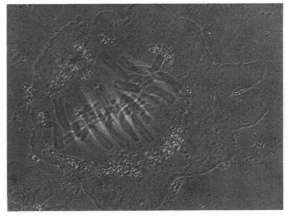

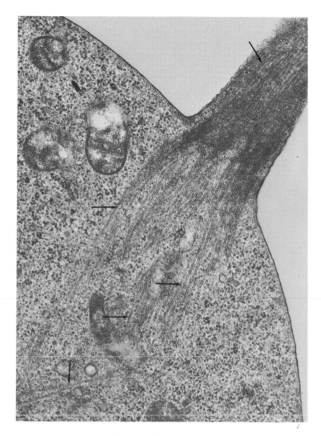

Fig. 4-24. Spindle of a dividing erythroblast in bone marrow. Arrows point to spindle microtubules (× 44,000). (From Fawcett, *The Cell: An Atlas of Fine Structure.* W. B. Saunders Co., 1966.)

Located near the nucleus or at each pole of the spindle during mitosis are two *centrioles*. These are cylindrical bodies (Fig. 4-25B) usually lying at right angles to each other about 150 mμ in diameter and 300–500 mμ in length. Centrioles are self-replicating structures, which seem to be involved with the organization of the spindle although the temporal and spatial association of these structures may be secondary in nature. The internal structure of the centriole is precise and universal in character, composed of nine groups of triplet microtubules evenly spaced around the circumference of a cylinder and oriented at a characteristic 30° angle of inclination (Fig. 4-25A). High resolution studies suggest that the innermost tubule has two arms, one extended toward the center and the other toward the tubule of the neighboring group. The centriole is closed at one end (Fig. 4-25B) and thus is not a symmetrical cylinder in the longitudinal direction; the centriole also does not have radial symmetry since the microtubules are inclined at an angle in a clockwise direction when one looks at it from the open end. Centrioles, especially in animal cells, are often surrounded by a number of structures consisting of spherical bodies, fibrous arms, or microtubules radiating out from them. These structures produce an effect that can be seen under the light microscope and are called asters.

Fig. 4-25. Centrioles. A. Embryonic chick pancreas; cross section (× 180,000; Dr. Jean Andre). B. Embryonic chick epithelium; longitudinal section (× 41,000; Dr. Sergei Sorokin). (From Fawcett, *The Cell: An Atlas of Fine Structure.* W. B. Saunders Co., 1966.)

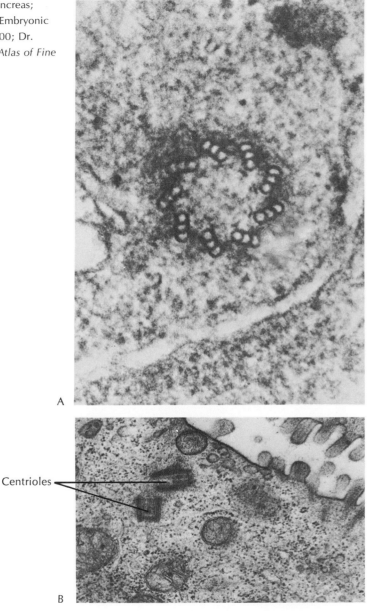

A

Centrioles

B

In certain cells the centrioles will replicate repeatedly in a manner which we do not understand at all; but unlike the replication manner of chromosomes, the centrioles move toward the surface membrane of the cell and form *basal bodies* or *kinetosomes,* which in turn give rise to the *cilia* or *flagella* (Fig. 4-26). The cilium is precisely the

same in all eucaryotic cells from paramecium to man, not only in its general organization, but even down to its last detail of fine structure and dimension. This constitutes
a dramatic piece of visual evidence for the unity of life at the eucaryotic cellular level.
It would seem that nature has been satisfied to use the cilium essentially without
modification since it was developed by some lowly unicellular green alga several
billion years ago.

A cross section of the cilium shows an organization that differs somewhat from the
arrangement of elements found in the basal body and in the centrioles. Since the cilium
juts out from the surface of the cell, it is surrounded by the plasma membrane. The
fibrillar units are arranged in eleven groups, two single tubular elements at the center
and nine double tubules around the circumference (Fig. 4-26). These tubules are fused

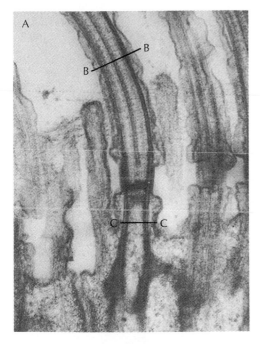

Fig. 4-26. A. Cilia on the gill of the fresh-water mussel,
Anodonta cataracta. The shaft of the cilia extends outward
from the basal bodies of the epithelial cell cytoplasm
(× 34,000). B. Cross section of a flagellum from the
protozoan, *Trichonympha;* in position B–B of part A
(× 127,500). C. Cross section of a basal body in
Trichonympha; in position C–C of part A (× 127,500).
(Courtesy of I. Gibbons, University of Hawaii.)

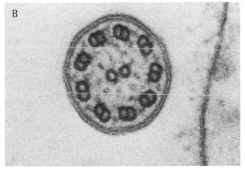

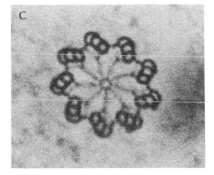

closely to each other and share part of their "walls." Ciliary microtubules and cytoplasmic microtubules are similar in appearance and dimension (Fig. 4-27); like cytoplasmic microtubules ciliary microtubules are composed of some 10–14 filaments. One of the tubules in each of the nine peripheral fibrillar units has a pair of arms (Figs. 4-26 and 27) pointing toward the tubule of the neighboring pair. Gibbons demonstrated these arms are composed of a protein, which he named *dynein,* that appears to be involved in the production of mechanical work. The architecture of the cilium is intimately related to the ciliary beat, the direction of which is thought to be perpendicular to a plane passing through the two central tubules. Except for the two central tubules the tubular elements in the cilium and those in the basal body are continuous with each other. At the base of the cilium one finds a transition zone in which the organization of the tubular elements of the basal body changes into that of the cilium.

We have discussed the cytoplasmic matrix as the site of two important functions, those of chemical energy transduction to mechanical work and of protein synthesis. In addition the cytoplasmic matrix, as its name implies, has the obvious function of supporting all the cell organelles such as the nucleus, the plastids, and the lysosomes.

THE CELL ORGANELLES

Having entered the cytoplasmic matrix by passing through a membrane, we can now visit a number of highly organized cellular bodies that can be entered only by pene-

Fig. 4-27. Body of flagellum of *Giardia muris.* Cross section through shaft of flagellum. Arrows point to arms coming out of one of the pairs of tubules (× 300,000). (Dr. D. Friend, from Fawcett, *The Cell: An Atlas of Fine Structure.* W. B. Saunders Co., 1966.)

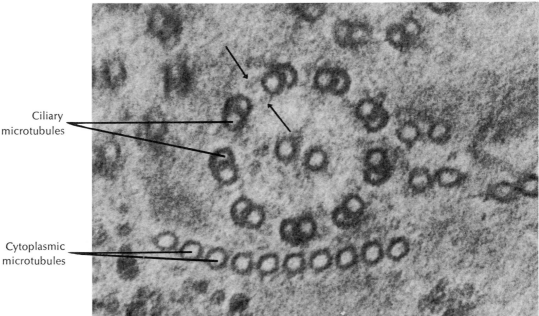

Ciliary microtubules

Cytoplasmic microtubules

trating through one or two membranes. These cellular bodies we shall call organelles.

The nucleus, as the name suggests, is a central or crucial cell organelle. It is the site of storage and replication of most of the cell's hereditary material. Its immediate importance in the metabolic events of the cell is demonstrated by the instantaneous disappearance of important activities such as cytoplasmic motion when the nucleus is artificially removed from the cell of an amoeba. Furthermore, enucleated cells will stop dividing and will eventually die.

(See Chapter 13.)

In the procaryotic cell's nucleus, often referred to as the nucleoid, the hereditary material is not surrounded by a membrane and consists of a double strand of DNA, usually forming a closed circle. Although the DNA of the procaryotic chromosome is generally considered to be "naked" as compared with the DNA of the eucaryotic chromosome, which is rich in proteins, there is no doubt that some structural materials must be associated with it to account for its tightly condensed structure in the cell. Furthermore, there must be a number of enzymes associated with it to account for its self-duplicating properties. In eucaryotic cells the hereditary material is found in chromosomes, which are composed of DNA in association with basic proteins called *histones*, with acidic proteins, and probably with some small amounts of RNA. When the eucaryotic nucleus is not dividing, the chromosomes are long, slender threads packed into the nucleus. Nuclear division (mitosis) is an ingenious piece of choreography whereby the eucaryotic cell manages to physically separate its hereditary material into two exactly equal parts. Nuclear division is generally correlated with the division of the entire cell, and it is therefore appropriate to consider for a moment the phenomenon of cell division as a whole.

(Fig. 8-10.)

(See Chapter 14.)

Growth in the world of life usually occurs through division of cells and their subsequent enlargement. This principle generally applies whether the cells are part of a colony of microorganisms or a multicellular system like an oak tree or a man. Since growth by division is an exponential phenomenon (1, 2, 4, 8, 16 . . .), the numbers will quickly become very large; therefore, growth cannot remain in an accelerating phase for long. The exponential phase is soon followed by a decelerating phase ending in an eventual leveling off (Fig. 4-28). During the exponential or "logarithmic phase" of growth, cells are in their most active metabolic condition; consequently, most experimental studies are performed with cells "harvested" in the exponential phase of their growth cycle. By making direct or indirect measurement of the growth characteristics of cell culture or organisms, it is possible to calculate a "generation time," which is the average time that has elapsed between successive cell doublings during the exponential phase of growth. The generation time under favorable growth conditions for the bacterium *Escherichia coli* can be as low as 20 min while for mammalian cells in tissue cultures it may be as many hours.

Although there are a number of variations on the main theme, the "mitotic cycle" of nuclear division correlates with the overall cycle of cell division in the following way. Figure 4-29 shows a number of phases through which the cell generally must pass during its growth cycle. Let us begin with the end of nuclear division. Two daughter nuclei have formed, and the parent cell has been divided into two daughter cells.

There follows a period of growth during which both the nucleus and cytoplasm enlarge. Biochemically, it is a period of active protein and RNA synthesis and often represents some 30–40 percent of the generation time of dividing cells (see the G_1 phase in Fig. 4-29). During this phase chromosomes are fully extended, slender structures lying intertwined in the nucleus. They usually do not absorb much stain, and the nucleus will appear under the light microscope to contain a fine mesh of slender threads, which used to be referred to as the chromatin network. A few characteristic regions of the chromosomes, however, do stain more intensely at this stage. These "heterochromatic regions" have shown a number of interesting correlations with the expression of the genetic material (see Levine's *Genetics* in the Holt, Rinehart and Winston Modern Biology Series) and have therefore attracted the attention of cytologists and geneticists for several decades. Much writing and theorizing have been done about their possible significance, but unfortunately we do not have as yet a sound experimental basis to support any theory.

Fig. 4-28. The bacterial growth curve. The lower part of the figure shows how the bacterial mass increases with time. The upper part of the figure shows how the specific growth rate changes; it is constant only during the exponential phase of growth. (Courtesy of Sistrom, *Microbial Life*.)

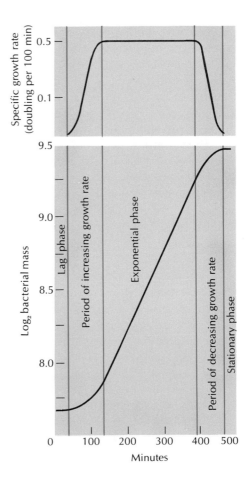

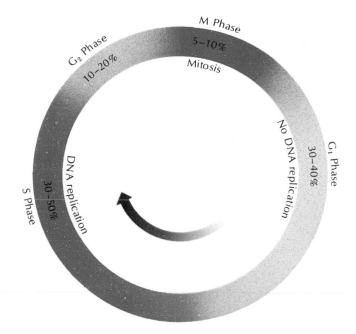

Fig. 4-29. The diagram shows the separation of phases during one cycle of cell division and that mitosis and DNA replication are generally separated in time by a G_1- and G_2-phase.

The G_1 phase is followed by the S phase during which DNA and histone synthesis can be shown to occur, resulting finally in the doubling of the slender, fully extended chromosomes. Thus, it is at this stage that the cell's hereditary material becomes duplicated. In their fully extended form the chromatids cannot possibly be separated from each other because all the chromosomes are extensively intertwined about each other. The S phase may last 30–50 percent of the generation time of the cell. It is followed by the G_2 phase; we do not as yet understand the metabolic significance of this phase, and thus we refer to it in our ignorance as "preparation for mitosis." The G_2 phase lasts some 10–20 percent of the generation time and is followed by the M phase (mitosis), which normally lasts some 5–10 percent.

Because chromosomes in a dividing nucleus take up stains and can therefore be seen very clearly in the light microscope, the structural events occurring during mitosis have been under intense scrutiny by light microscopists during the first half of this century. These studies, in which the sequence of mitosis has been pieced together from observations on dead material, have been dramatically corroborated by recent studies on living material with the phase contrast microscope. By utilizing differences in refractive index in the material to be observed, this microscope reveals beautifully the structure of the chromosomes of the living cell during mitosis (Figs. 4-23B and 4-30).

The details of the behavior of the chromosomes, the nuclear envelope, the spindle, the centrosomes, and the centriole during mitosis are numerous indeed, and many that have been observed in the last 40 years are purely descriptive in nature, having so

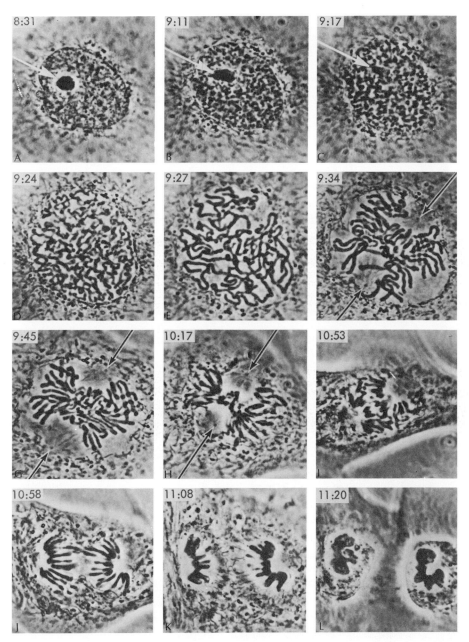

Fig. 4-30. Mitosis in an animal cell (× 800). Arrows point to separating centrioles. (Dr. W. Bloom, from Bloom and Fawcett, *Textbook of Histology,* 9th ed.)

far not been related to metabolic or molecular events. We shall therefore restrict ourselves to a broad outline of the most salient aspects of nuclear division.

The process of mitosis has traditionally been divided into four main phases, which are arbitrary divisions of a continuous process (Fig. 4-30).

We start with *prophase* (Fig. 4-30 A–E) in which the chromosomes gradually shorten and thicken, a phenomenon brought about by the extensive coiling of the two chromatids that were produced in the *S* phase and are still lying close and parallel to each other. The chromatids coil on themselves rather than around each other so that eventually they will be able to separate. This double character of the chromosomes begins now to be seen and becomes increasingly obvious during prophase. At the same time one also sees the gradual disappearance of the nuclear envelope and of the nucleolus. As the shortening and thickening of the chromosomes proceeds, one can observe that the chromatids are held together at one particular point called the *centromere*. Toward the end of prophase the chromosomes migrate to the *equatorial plane* of the dividing nucleus while the spindle begins to be organized. In animal cells the spindle starts to form between the centrioles while they still lie to one side of the nucleus. As the centrioles move to the opposite poles of the nucleus, the spindle enlarges and moves through the nuclear region. In plant cells the centrioles appear to be absent, and the spindle is immediately organized to lie between the poles of the dividing nucleus.

At *metaphase* (Fig. 4-30 F–H) the spindle is fully developed, the chromosomes are poised at the equatorial plane, and the nuclear envelope and nucleoli have fully disappeared. At this point the number of chromosomes can easily be counted, and their size and shape can be determined. Thus, it can be shown that every species has a characteristic number of chromosomes. If one is observing a mitotic division during the diploid stage of a given species' life history, it can be shown that every chromosome is found in duplicate. Each of these duplicates is called a chromatid as long as it remains in association with the other. A chromosome can most easily be recognized in metaphase by the relative length of its "arms" on either side of the centromere, which is located in a constant position in each particular chromosome.

Anaphase (Fig. 4-30 I–K) begins when the chromatids of each chromosome start moving apart, the centromeres of each chromosome having split and the daughter centromeres now leading the way in opposite directions. It can be shown by the use of *micromanipulation techniques* that the centromeres are attached to the spindle. We do not as yet understand the mechanism whereby the chromosomes move.

Once the chromosomes have reached the opposite poles of the spindle, *telophase* (Fig. 4-30 L) begins and consists of the reversal of all the changes occurring in prophase. The spindle disappears, chromosomes lengthen by uncoiling, the nucleoli slowly appear again, and the nuclear envelope is once again laid down.

As these nuclear events unfold, the whole cell divides into two daughter cells, in the animal cell by the process of active constriction of the cell and in the plant cell by the growth of a cellulose wall. Figure 4-31 depicts the chromosomal changes during mitosis in a plant cell.

This dramatic series of events, which has fascinated cytologists for many decades, is nature's first piece of choreography invented at the birth of the eucaryotic cell. Its function, clearly, is that of apportioning accurately the hereditary material, which in

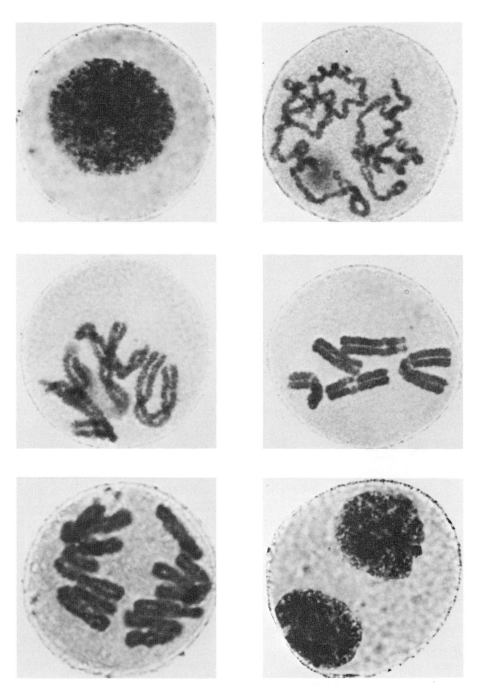

Fig. 4-31. Several stages of the mitotic cycle in microspores of *Trillium erectum* (× 800). (Courtesy of A. H. Sparrow and R. F. Smith, Brookhaven National Laboratory.)

the eucaryotic cells comes in several packages or "linkage groups." As every chore-ographer knows, the best way of separating dancers on a stage with a minimum of confusion is to string them out in a line, after which they can easily "fall apart" into two groups. The shortening of the chromosomes makes the disentanglement and migra-tion possible. On the other hand, the significance of the disappearance of the nuclear envelope and of the nucleolus during mitosis is not yet understood. Since the disappear-ance of the nuclear envelope, however, is not necessary for chromosomal replication, as for instance in protozoa, it is possible that this phenomenon is not of universal importance.

The use of the electron microscope has contributed disappointingly little to precisely those areas of cytology that were at their most developed stage during the heyday of the light microscope. The best electron micrographs of the metaphase chromosomes show only homogeneous granular masses.

The heterochromatic regions of the chromosomes can be seen in the light micro-scope as optically dense regions of "condensed chromatin." Such regions are often seen in the electron microscope to be disposed mostly around the periphery of the nucleus adjacent to the nuclear envelope or next to the nucleolus (Fig. 4-2).

We must conclude that the precise structure of the eucaryotic chromosome, the molecular organization of the DNA-histone complex, is as yet one of the unsolved, important problems of modern cell biology.

The nucleoli, which form during telophase in association with certain portions ("nucleolar organizer regions") of certain chromosomes, can be seen clearly in stained sections because they bind basic dyes. In the electron microscope they often appear to be composed of two regions. The central homogeneous region is surrounded by a more filamentous region (Compare Fig. 4-17) that carries dense granules (150 A diameter) resembling ribosomes, although they appear less dense and somewhat smaller than ribosomes. These particles may indeed be ribosomes in an early stage of assembly. The nucleolus, rich in ribonucleic acid (RNA), plays an important role in the synthesis of ribosomal RNA and possibly of transfer RNA.

(See Chapter 14.)

The chromosomes and nucleolus are supported in a "nuclear matrix" (*nucleo-plasm*), which seems to contain granules of various sizes and densities and is as yet of unknown composition and function. During mitosis the nucleoplasm is of course continuous with the cytoplasmic ground substance.

The *mitochondria* are the "powerhouses of the cell" in which the foodstuffs or "low-grade" fuels of the cell are oxidized and the energy thus released is utilized for the synthesis of ATP, the "high-grade" fuel of the cell. As indicated before, ATP is the major high-energy source that is utilized for driving the various energy-requiring machinery of the cell.

Mitochondria have been observed as slender, short threads in the light microscope since the middle of the last century, but it is only since the technique of differential ultracentrifugation has been utilized to separate them from other cell constituents that their biochemical function has been clearly identified. It is also only since electron microscopy has been applied to fixed, stained, and thin-sectioned material that the fine structure of mitochondria has become apparent.

Mitochondria are composed of two membrane systems, an outer membrane that is stretched over the organelle and an inner membrane that is thrown into folds and penetrates the interior of the organelle (Fig. 4-32). In lower forms such as protozoa (and strangely enough in the adrenal cortex of mammals) the inner membrane forms tubules (Fig. 4-33), whereas in the cells of more advanced animals and plants they form flattened vesicles or *cristae* (Fig. 4-32).

The space between the two membranes is generally optically homogeneous, whereas the space in the interior contains a dense matrix, a variety of electron-dense granules. Mitochondria contain RNA, and recently it has been shown that they also

Fig. 4-32. Mitochondria. Acinar cell of pancreas (× 70,000). (Courtesy of G. E. Palade, Rockefeller University.)

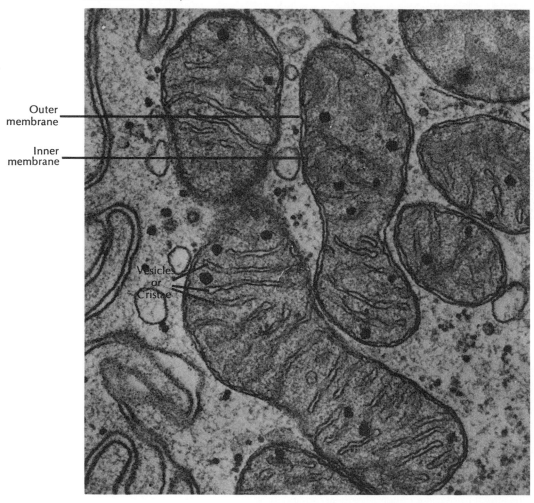

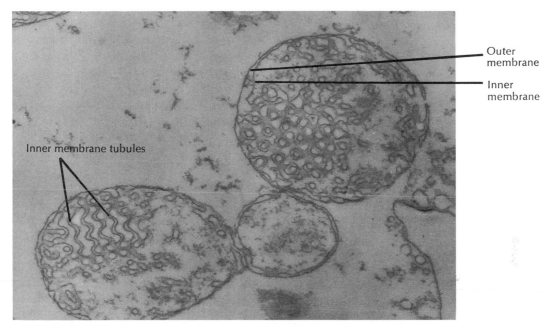

Outer
membrane

Inner
membrane

Inner membrane tubules

Fig. 4-33. Mitochondria. Amoeba, *Chaos chaos* (× 31,500). (Courtesy of G. D. Pappas, Albert Einstein College of Medicine.)

contain DNA and are self-duplicating organelles. It is now thought that the DNA in the mitochondria contains at least some of the hereditary information required for the process of mitochondrial replication and growth.

The surface of the inner membrane adjoining the interior of the organelle is lined with granules that are attached to the membrane with slender stalks. It is postulated that each granule, together with the stalk and the immediate portion of the membrane to which it is attached, forms a "coupling system" which is active in the closely integrated metabolic reactions involved in the energy metabolism of the cell.

(Fig. 13-1.)

(See Chapter 13.)

The disposition of the mitochondria in the cell constitutes a beautiful visual documentary of their function and of the general phenomenon involving relationship of structure and function in the cell. Mitochondria are often found in close juxtaposition to the cellular machinery they supply with ATP. In muscle, for instance, mitochondria are neatly sandwiched between the contractile elements of the muscle cell (Fig. 4-34); in the sperm tail they can be shown to surround the longitudinal fibrils responsible for the motion of the sperm flagellum; in epithelia engaged in active transport mitochondria can be seen in close proximity to the enlarged, highly folded surface membrane (Fig. 4-35); in cells that actively synthesize proteins "rough," ribosome-lined endoplasmic reticulum profiles can often be seen to be closely wrapped around the mitochondria (Fig. 4-36); while, finally, in other cells mitochondria can be seen in close juxtaposition to droplets of lipid.

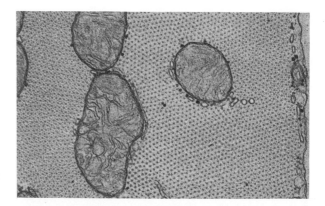

Fig. 4-34. Mitochondria (cross section) embedded between contractile elements in a cardiac muscle cell (× 65,000). (Courtesy of D. W. Fawcett, Harvard Medical School.)

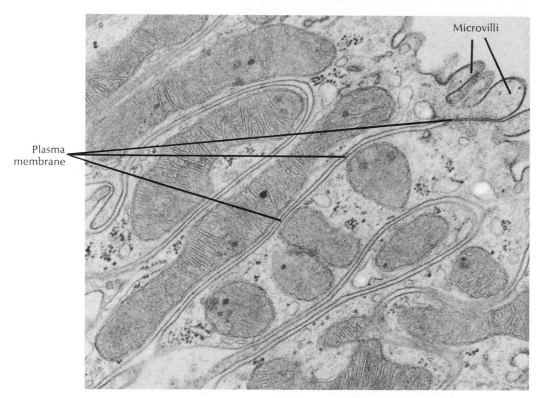

Fig. 4-35. Mitochondria in proximity to surface membrane of distal convoluted tubule of kidney (× 45,000). (Courtesy of M. Farquhar and G. E. Palade, Rockefeller University.)

The mitochondrion has often been viewed as a special example of a more general category of organelle called *plastids*. Plant cells, in addition to mitochondria, contain *leucoplasts*, which are plastids concerned with the transformation of glucose to starch;

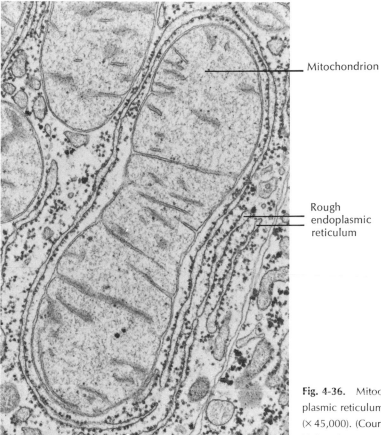

Mitochondrion

Rough
endoplasmic
reticulum

Fig. 4-36. Mitochondria in relation to a rough endoplasmic reticulum of a liver cell of a new-born rat
(× 45,000). (Courtesy of G. E. Palade, Rockefeller
University.)

chromoplasts, which are the site of synthesis and localization of many plant pigments;
and, most importantly, *chloroplasts*, which are the organelles in which photosynthesis
takes place.

Chloroplasts in the eucaryotic plant cell are surrounded by a double-membraned
envelope, with other membranes in the interior forming a highly laminated internal   (Compare Fig. 13-10.)
structure. In the higher plants most of the laminated structure is concentrated in
flattened sacs (discs or thylakoids), frequently fused in piles called grana, with many
grana per plastid (Fig. 4-37). In lower plants such as the green algae discs are as large
as the plastid and are usually fused into a single granum (Fig. 4-38). In procaryotic
photosynthetic cells such as the photosynthetic bacteria and blue-green algae chloroplasts do not exist as separate organelles, and the membrane system, either laminated
or vesicular, is distributed throughout the cell although it is concentrated near the
cell membrane. As shall be seen in Chapter 13, the chemical mechanism of certain
steps in the photosynthetic chain of reactions requires the presence of two immiscible

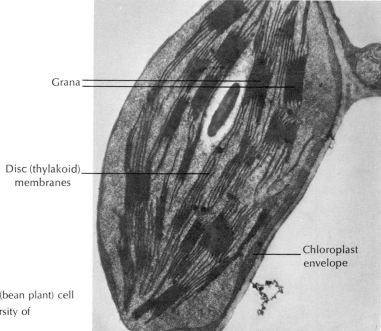

Grana

Disc (thylakoid) membranes

Chloroplast envelope

Fig. 4-37. Chloroplast of a higher plant (bean plant) cell (× 20,000). (Courtesy of D. Weier, University of California.)

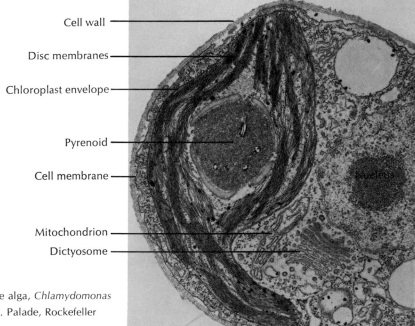

Cell wall

Disc membranes

Chloroplast envelope

Pyrenoid

Cell membrane

Mitochondrion

Dictyosome

Nucleus

Fig. 4-38. Chloroplast of the alga, *Chlamydomonas* (× 18,000). (Courtesy of G. E. Palade, Rockefeller University.)

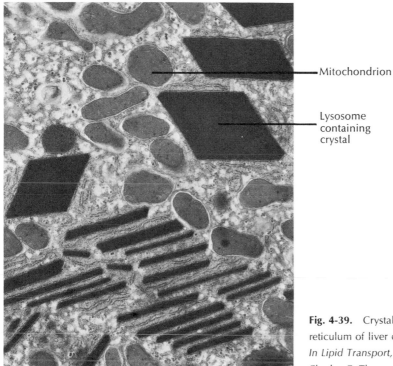

Mitochondrion

Lysosome
containing
crystal

Fig. 4-39. Crystalline arrays within endoplasmic reticulum of liver cell (× 19,000). (From Fawcett, *In Lipid Transport,* H. C. Meng, Ed., Springfield, Ill.: Charles C. Thomas, 1964.)

phases. The membrane system associated with the photosynthetic process is therefore a necessary and universal component. A second component has recently been discovered in chloroplasts with high resolution electron microscopy. It consists of granules with the shape of oblate spheres 100 A in diameter and 110 A thick. These granules seem to be buried in the lamellae of the membranes either evenly distributed as in the chloroplasts of the lower plants or in packets or grana in the chloroplasts of higher plants. These granules, or *quantosomes,* possibly contain the chlorophyll and the enzymes necessary for the initial steps of the photosynthetic process although this is not certain at present. (Compare Fig. 13-12.)

Chloroplasts, like mitochondria, are self-replicating organelles, and they also have recently been shown to contain DNA.

The nucleus and the plastids are surrounded by an "envelope" or double membrane although in the case of mitochondria the inner membrane invaginates to form cristae, and possibly in the case of the chloroplasts it invaginates to form lamellae. There are, however, some cellular organelles that are surrounded by a single membrane only. To date, two types of such structures have been described, the lysosomes and the microbodies.

Lysosomes are a variety of densely staining bodies surrounded by a membrane and containing hydrolytic enzymes. Surprisingly, they were not discovered initially with the electron microscope but by *differential centrifugation*. It was found that when homogenized cells were centrifuged, a certain cellular fraction that sedimented more slowly than mitochondria contained a number of hydrolytic enzymes, which become active only after severe mechanical disruption. This suggested that these enzymes were surrounded by a membrane, and careful electron microscopic and cytochemical studies on these fractions identified a number of membrane-bound bodies now known as lysosomes (compare Fig. 4-2). The material in the lysosomes takes on various appearances (Fig. 4-41), although mostly the visible contents are in the form of partly digested bodies or digestive residues. It seems clear that the function of the lysosome membrane is that of separating the hydrolytic enzymes from other parts of the cell, thus protecting it from self-digestion. Upon cell death it can be observed that enzymes are released from the lysosomes, whereupon they very rapidly digest the remainder of the cell (Fig. 4-40). Thus, the lysosome is analogous to the digestive vacuole in unicellular organisms. It appears that in addition to this scavenging function lysosomes also have a metabolic role during the life of the cell.

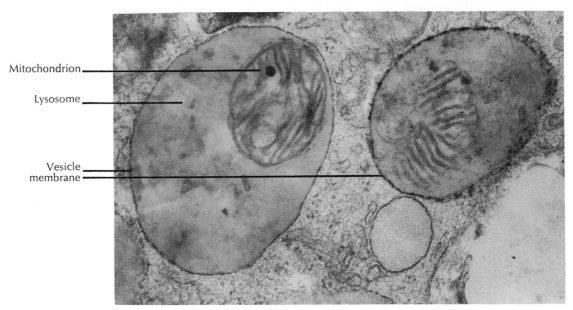

Mitochondrion

Lysosome

Vesicle membrane

Fig. 4-40. Lysosomes. Kidney cell. Heterophagic vacuole surrounded by a membrane showing a mitochondrion inside ($\times$ 57,000). (Courtesy of F. Miller and G. E. Palade, Rockefeller University.)

Microbodies of various kinds and peroxysomes (Fig. 4-42) are bodies surrounded by a membrane and containing dense granular or sometimes crystalline material. Recently, they have been found to contain certain enzymes; however, their overall function is poorly understood. They are not necessarily found in all cells although they are quite widely distributed. Lysosomes and microbodies appear to be the cell's

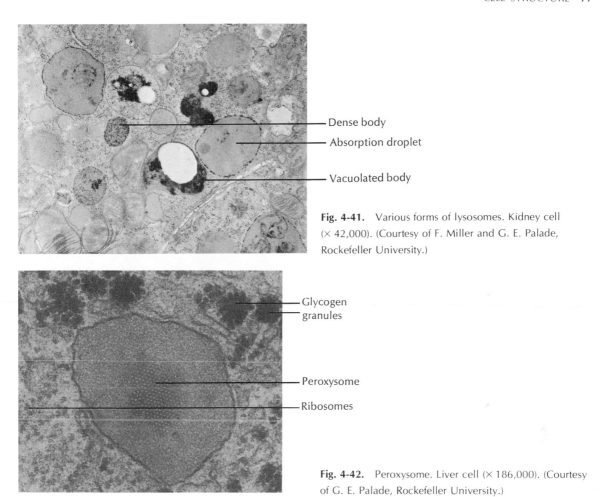

Fig. 4-41. Various forms of lysosomes. Kidney cell (× 42,000). (Courtesy of F. Miller and G. E. Palade, Rockefeller University.)

— Dense body

— Absorption droplet

— Vacuolated body

— Glycogen granules

— Peroxysome

— Ribosomes

Fig. 4-42. Peroxysome. Liver cell (× 186,000). (Courtesy of G. E. Palade, Rockefeller University.)

way of storing enzymes in large quantities while at the same time keeping the bulk of these enzymes out of circulation. It is likely that in the long run a greater diversity of such bodies will be discovered and will be classified according to the chemical role of the enzymes which they store. In addition, various types of cells contain inclusions specific to that cell such as the zymogen granules of the pancreas, which are storage depots for secretory enzymes (Fig. 4-43).

This concludes our electron-microscopic tour of the cell. We have entered the cell and have moved through its extensive outer compartment bounded by the endoplasmic reticulum, the Golgi body, and the nuclear envelope. We then passed through one of the membranes into the cytoplasmic matrix. Moving through the cytoplasmic matrix, we have inspected various types of fibrillar organizations such as the centrioles, the basal bodies, the cilia, and a variety of microtubules and filaments. Finally, we

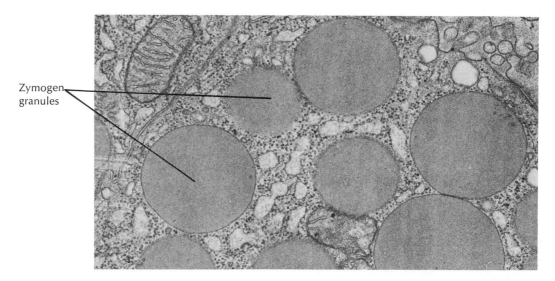

Zymogen granules

Fig. 4-43. Zymogen granules. Acinar cell of pancreas (× 42,000). (Courtesy of G. E. Palade, Rockefeller University.)

entered a variety of organelles bounded by double or single membranes. We have indicated in very general terms the functional roles each cellular structure appears to play. It should therefore be clear by now that there is an intimate and reciprocal relationship between structure and function, each being responsible for and serving the other. On the other hand, it should also now be obvious that a deeper understanding of cellular phenomena can only be gained by increasing our resolution down to the molecular level. Before we can do this, we must in Part III provide the student with an adequate background in the chemistry of the cell's molecules. We can do this, of course, only briefly, and the student is urged to consult the references suggested in the reading list for further amplification of his experience. Armed with these new insights into the chemistry of life, we shall then retrace our steps, and for the remaining portion of this book (Part IV) re-examine the relationship between structure and function in the "high resolution language" of modern molecular biology.

SUGGESTED READING LIST

Offprints

Baserga, R. and Kisieleski, W. E., 1963. "Autobiographies of Cells." *Scientific American* offprints. San Francisco: W. H. Freeman & Co.

Brachet, J., 1961. "The Living Cell." *Scientific American* offprints. San Francisco: W. H. Freeman & Co.

de Duve, C., 1963. "The Lysosome." *Scientific American* offprints. San Francisco: W. H. Freeman & Co.

Horne, R. W., 1963. "The Structure of Viruses." *Scientific American* offprints. San Francisco: W. H. Freeman & Co.

Inoue, S. and Bajer, A., 1961. "Birefringence in Endosperm Mitosis," *Chromosoma*, pp. 48–63. Bobbs-Merrill Reprint Series. Indianapolis: Howard W. Sams & Co.

Morowitz, H. J. and Tourtellotte, M. E., 1962. "The Smallest Living Cells." *Scientific American* offprints. San Francisco: W. H. Freeman & Co.

Articles, Chapters, and Reviews

Palade, George E., 1958. "A Small Particulate Component of the Cytoplasm," ch. 10 in *Frontiers in Cytology*. New Haven: Yale University Press.

Books

Bloom, W. and Fawcett, D. W., 1968. *A Textbook of Histology*, 9th ed. Philadelphia: W. B. Saunders Co.

Brachet, J., 1957. *Biochemical Cytology*. New York: Academic Press, Inc.

Brachet, J. and Mirsky, A. E., eds., 1959–1961. *The Cell*. New York: Academic Press, Inc.

De Robertis, E. D. P., Nowinski, W. W., and Saez, F. A., 1960. *General Cytology*, 3d ed. Philadelphia: W. B. Saunders Co.

Engstrom, A. and Finean, J. B., 1958. *Biological Ultrastructure*. New York: Academic Press, Inc.

Fawcett, D. W., 1966. *An Atlas of Fine Structure—the Cell, Its Organelles and Inclusions*. Philadelphia: W. B. Saunders Co.

Finean, J. B., 1962. *Chemical Ultrastructure in Living Tissues*. Springfield, Ill.: Charles C. Thomas.

Haggis, G. H., 1967. *The Electron Microscope in Molecular Biology*. New York: John Wiley & Sons, Inc.

Jensen, W. A., 1964. *The Plant Cell*. Belmont, California: Wadsworth Publishing Co.

Jensen, W. A. and Park, R. B., 1967. *Cell Ultrastructure*. Belmont, California: Wadsworth Publishing Co.

Porter, K. R. and Bonneville, M. A., 1963. *An Introduction to the Fine Structure of Cells and Tissues*. Philadelphia: Lea & Febiger.

Wilson, E. B., 1928 (reprinted in 1953). *The Cell in Development and Heredity*. New York: The Macmillan Co.

In this part we discuss the molecules that are the moving parts of the living machine, emphasizing those properties which are important in explaining significant aspects of cell structure and function. The emphasis, of course, will be on the properties of macromolecules, which more than any other cell components provide the molecular insights for the understanding of the machinery of life.

Life and the Periodic Table

A characteristic property of living matter is that it is selective in its relationship to the environment. An example of this selectivity is the considerable difference between the abundance of available elements on the earth and those found in living organisms. The earth's crust is made mostly of oxygen, silicon, aluminum, sodium, calcium, iron, magnesium, and potassium, the remaining elements constituting less than one percent of the total. Living organisms, on the other hand, are made mostly of hydrogen, oxygen, carbon, and nitrogen, with the remaining elements constituting less than one percent of the total (Table 5-1). Thus, with the exception of oxygen there is no overlap between the most abundant elements in the earth's crust and those in living matter. Oxygen, with an atomic number of eight, is the heaviest of the major elements of living matter and at the same time the lightest of the major elements of the earth's crust. We conclude that living matter selects preferentially for the light elements made available to it in the earth's crust and atmosphere.

It is remarkable that in this respect living matter is very much like the cosmos as a whole. The stars as well as the interstellar matter are composed mostly of light elements. Our earth is but a "mineral ash" of heavy elements that remained after the light elements distilled into space owing to the weak gravitational pull of our tiny earth. Figure 5-1 shows the parallelism that, with the exception of helium, is found between the first 31 elements in the cosmos and in the living organism. Whether this striking parallelism is coincidental or a form of biological conservation of the chemical composition of the surroundings present when life first originated is as yet only a matter of speculation.

Although living matter often contains traces of all the elements found in its surroundings, only some 20 elements have been demonstrated to be essential for life. They are classified in three major categories according to their relative concentration in the cell (Table 5-2): major constituents, trace elements, and ultratrace elements.

Table 5-1 RELATIVE ABUNDANCES OF 14 MAJOR ELEMENTS IN
UNIVERSE, EARTH'S CRUST, AND HUMAN BODY
(Recalculated from Edsall and Wyman)

ELEMENT	ATOMIC NUMBER	RELATIVE ABUNDANCE IN ATOMS PERCENT		
		UNIVERSE	EARTH'S CRUST	HUMAN BODY
Hydrogen	1	90.79		60.3
Carbon	6	9.08		10.5
Nitrogen	7	0.0415		2.42
Oxygen	8	0.0571	62.55	25.5
Sodium	11	0.00012	2.64	0.73
Magnesium	12	0.0023	1.84	0.01
Aluminum	13	0.00023	6.47	
Silicon	14	0.026	21.22	0.00091
Phosphorus	15	0.00034		0.134
Sulfur	16	0.0091		0.132
Chlorine	17	0.00044		0.032
Potassium	19	0.000018	1.42	0.036
Calcium	20	0.00017	1.94	0.226
Iron	26	0.0047	1.92	0.00059

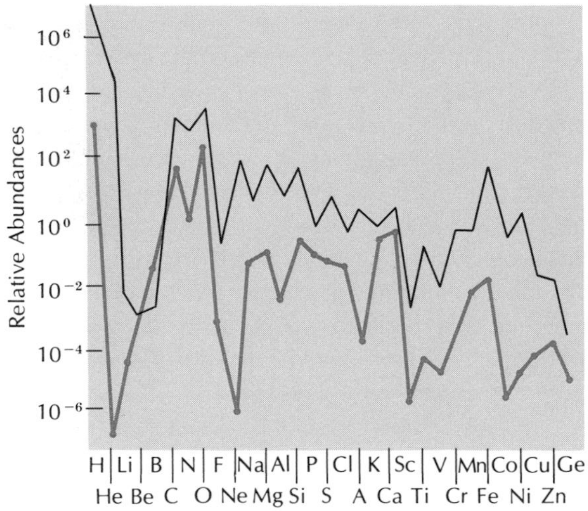

Fig. 5-1. Relative abundances of the first 31 elements in the cosmos (———, per 100 atoms Si) and in living organisms on the earth's surface (●———●, per 100 atoms C). (Courtesy of G. E. Hutchinson, 1943; uncorrected for recent data.)

Not all these elements have been shown to be required by all species. Some of them we know to be of universal importance (H, C, N, O, Na, Mg, P, S, Ca, K, and Cl), whereas others have been shown to be required in a large number of species and hence are probably also of general importance (Fe, Cu, Mn, and Zn). The universality of the remaining elements (B, Si, V, Co, and Mo) has so far not been established.

The study of the essential ultratrace elements is a continuing problem: one can never be certain that a particular element is *not* required, for it is a very laborious and technically complicated process to prove that it *is* required. The procedure involves the use of ultrapure water, chemicals, and even glassware. Thus, the nutritional necessity of some ultratrace elements was not suspected until their near absence in certain soils

Table 5-2 ELEMENTS USED IN LIVING SYSTEMS
AND SOME OF THEIR FUNCTIONS

CATEGORY	ELEMENT	SYMBOL	ATOMIC NUMBER	SOME KNOWN FUNCTIONS
Major constituents 2–60 atoms percent	Hydrogen	H	1	Universally required for organic compounds of cell
	Carbon	C	6	
	Nitrogen	N	7	
	Oxygen	O	8	
Trace elements 0.02--0.1 atoms percent	Sodium	Na	11	Important counterion involved in action potential, most of it being in the fluids outside the cell
	Magnesium	Mg	12	Cofactor of many enzymes
	Phosphorus	P	15	Universally involved in energy transfer reactions; necessary constituent of nucleic acids
	Sulfur	S	16	Found in proteins and other important substances
	Chlorine	Cl	17	One of the major anions
	Potassium	K	19	Universally found inside cells; important counterion involved in nerve conduction, muscle contraction, etc.
	Calcium	Ca	20	Cofactor in enzymes; important constituent of membranes and regulator of membrane and muscle activity
Ultratrace elements less than 0.001 atoms percent	Boron	B	5	Important in plants, probably as cofactor of enzymes
	Silicon	Si	14	Found abundantly in many lower forms
	Vanadium	V	23	Found in certain pigments of lower forms
	Manganese	Mn	25	Cofactor of many enzymes
	Iron	Fe	26	Cofactor of many oxidative enzymes and used in oxygen transport
	Cobalt	Co	27	Constituent of vitamin B_{12}
	Copper	Cu	29	Cofactor of many oxidative enzymes and used in oxygen transport of many marine organisms
	Zinc	Zn	30	Cofactor of many enzymes
	Molybdenum	Mo	42	Cofactor of a few enzymes

Elements in dark shade are known to be universally required; those in lighter shade are probably universally required.

caused the appearance of diseases and abnormalities in plants or animals. The absence of copper from certain regions in Australia caused a disease of sheep involving permanent effects on the nervous system, anemia, and deterioration of the wool with loss of kinkiness. A deficiency of boron in the soil has been observed to cause "heart rot" in beets, "cracked stems" in celery, "internal cork" in apples, and a host of other abnormalities in plants. The nutritional requirement of ultratrace elements such as boron can best be shown by curing the deficiency disease through the addition of these elements to the soil. One hundredth of a part per billion parts of soil will cause the disease, one tenth of a part per billion will cure it, and one part per billion is a sufficiently high concentration to poison the plant.

What is the role of the 20-odd elements in the living machine? The elements H, C, N, O, P, and S are the building blocks of the cell's organic compounds. Most carbohydrates and lipids contain H, C, and O whereas proteins also contain N and S, and nucleic acids contain N and P. Thus, these six elements are the main constituents of living matter. The ability of different permutations and combinations of these elements to produce the molecular diversity of the cell makes them unique in their fitness to support life. Cellular molecules range in nature from the gas carbon dioxide through the liquid water to the solid cellulose, and from the highly polar amino acid through the less polar glucose to the nonpolar fat. It has, at times, been suggested that on another planetary system silicon might replace carbon. This seems unlikely when one considers the full range of compounds and properties found in the world of carbon that could hardly be matched by a presumed world of silicon.

Many of the elements are usually found as ions; thus, we have Na^+, Mg^{2+}, PO_4^{2-}, SO_4^{2-}, Cl^-, K^+, and Ca^{2+}, and from the major elements we have CO_3^{2-} and NO_3^-. The importance of these ions to the well-being of the cell has been recognized for a long time. The physiological literature of the 1920s and 1930s is replete with observations on the effect of ions such as Na^+, Mg^{2+}, K^+, and Ca^{2+} on the physical consistency and the functioning of a large variety of cells. The absolute amounts and the relative balance of these ions are maintained in living systems within narrow limits, and any experimental variation of these quantities results in marked changes in biological properties such as cellular permeability, irritability, contractility, protoplasmic viscosity, and cell division. The importance of the balance among these cations can be understood by a consideration of the fact that pairs of these ions often have been observed to have antagonistic effects on each other. Thus, K^+ is known to decrease protoplasmic viscosity and to cause muscle relaxation, whereas Ca^{2+} has been observed to cause gelation of the cytoplasm as well as to initiate muscle contraction. An interesting aspect of the Na^+–K^+ balance is that in animals Na^+ is located mostly outside the cells in the tissue fluids while K^+ is inside the cells. The cell must of course utilize energy to bring about and maintain this difference in distribution.

The importance of the ionic composition and balance in living systems can also be illustrated by the remarkable conservation of this parameter throughout biological evolution. Table 5-3 gives the ionic composition of a number of organisms of different evolutionary types along with the ionic composition of seawater. A. B. Macallum was the first to conclude that the parallelism shown here meant that life originated in the sea and that subsequent evolution did little to change the ionic balance. After one billion years of evolution on land, though our body fluids are less concentrated than is seawater or even less concentrated than seawater was one billion years ago, we still carry the ionic balance of seawater in our cells and body fluids!

In spite of the overall conservatism of the ionic balance, the data in Table 5-3 also show a gradual change in the concentration of certain ions during the course of evolution. The most marked change is in the absolute concentrations, especially those of Na^+, Mg^{2+}, Cl^-, and SO_4^{2-}. One can also observe, however, a gradual decrease in

Table 5-3 IONIC COMPOSITION OF SEAWATER AND
THE BODY FLUIDS OF SEVERAL SPECIES

	Na$^+$	K$^+$	Ca^{2+}	Mg^{2+}	Cl$^-$	SO$_4^{2-}$
VERTEBRATES						
Man	145	5.1	2.5	1.2	103	2.5
(mammal)	100	3.5	1.7	.83	71	1.7
Rat	145	6.2	3.1	1.6	116	
(mammal)	100	4.2	2.1	1.1	80	
Frog	103	2.5	2.0	1.2	74	
(amphibian)	100	2.4	1.9	1.2	72	
Lophius	228	6.4	2.3	3.7	164	
(fish)	100	2.8	1.0	1.6	72	
INVERTEBRATES						
Hydrophilus	119	13	1.1	20	40	0.14
(insect)	100	11	.93	17	34	.13
Lobster	465	8.6	10.5	4.8	498	10
(arthropod)	100	1.9	2.3	1.0	110	2.2
Venus	438	7.4	9.5	25	514	26
(mollusk)	100	1.7	2.2	5.7	120	5.9
Sea cucumber	420	9.7	9.3	50	487	30
(echinoderm)	100	2.3	2.2	12	120	7.2
Seawater	417	9.1	9.4	50	483	30
	100	2.2	2.3	12	120	7.2

Black numbers are expressed in millimoles (mM) per liter; color numbers are relative, expressed in terms of 100 units of Na$^+$.

the relative concentrations of Mg^{2+} and SO$_4^{2-}$.

There is as yet very little detailed understanding on a molecular basis of the role of these ions. Both proteins and nucleic acids (which are the main macromolecular components of the cell) are negatively charged polyvalent ions and require cations as counterions. Furthermore, special relationships have been shown to exist between certain macromolecules and specific cations. Thus, the concentration of Mg^{2+} has been shown to affect the state of aggregation of the ribosomes, probably through its effect on the RNA constituents of these bodies. These little cytoplasmic organelles, which we shall discuss later, have been shown to break into two smaller components when the Mg^{2+} concentration is lowered. Calcium is almost certainly a counterion for the phospholipid constituents of the cell's membrane systems, and the effect of Ca^{2+} on lowering the threshold of nerve excitability as well as its release from the membrane system in muscle contraction show the role calcium plays in membrane phenomena.

Finally, potassium has been shown to interact selectively with myosin, the contractile protein of muscle. These are only a few examples from a field of cell biology that has yet to be fully explored on the molecular level.

But what about the ultratrace elements? What conceivable role could an element play if it is present at a concentration of 10^{-8} M? The answer seems simple and clear. In all cases that have been examined in detail, these ions turned out to be necessary "cofactors" for certain biological enzymes. Since enzymes, because of their catalytic nature, are generally required only in very low concentrations, it follows that their cofactors are also required in very low concentrations. A small percentage of the trace elements are also involved in enzyme activation. Table 5-4 gives a few examples of reactions utilizing enzymes that require certain specific metallic cofactors.

Table 5-4 SOME METAL-REQUIRING ENZYMATIC REACTIONS

glucose + ATP	$\xrightarrow[\text{Mg}^{2+}]{\text{hexokinase}}$	glucose-6-P + ADP
tryptophan + H_2O	$\xrightarrow[\text{K}^+ \text{ or } \text{Rb}^+]{\text{tryptophanase}}$	indole + pyruvate + NH_3
soluble fibrin	$\xrightarrow[\text{Ca}^{2+}]{\text{plasma transglutaminase}}$	insoluble fibrin
arginine + H_2O	$\xrightarrow[\text{Co}^{2+} \text{ or } \text{Mn}^{2+} \text{ or } \text{Ni}^{2+}]{\text{arginase}}$	urea + 2,5-diaminovaleric acid
histidine	$\xrightarrow[\text{Fe}^{3+} \text{ or } \text{Al}^{3+}]{\text{histidine decarboxylase}}$	histamine + CO_2
catechol	$\xrightarrow[\text{Cu}^{2+}]{\text{phenol oxidase}}$	o-benzoquinone
lactic acid + NAD^+	$\xrightarrow[\text{Zn}^{2+}]{\text{lactic dehydrogenase}}$	pyruvic acid + NADH + H^+
nitrate + NADPH + H^+	$\xrightarrow[\text{Mo}^{2+}]{\text{nitrate reductase}}$	nitrite + $NADP^+$ + H_2O

Some of the metallic ions can be prevented from functioning by the presence of certain inhibitors. The four iron atoms of the protein hemoglobin are the site of the oxygen-carrying function of that molecule. Cyanide or carbon monoxide can preferentially bind with the iron and hence interfere with the protein's ability to carry oxygen. Such compounds are poisons and even in very low concentrations are capable of causing the death of the cell.

We have seen that living matter selects from its environment some 20 specific elements with which to build its fabric. In the next four chapters we shall examine a number of molecules that are elaborated by the cell and that perform the most important functions of the machinery of life.

SUGGESTED READING LIST

Offprints
   Fowler, W. A., 1956. "The Origin of the Elements." *Scientific American* offprints. San Francisco: W. H. Freeman & Co.
Articles, Chapters, and Reviews
   Edsall, J. T. and Wyman, J., 1958. "Biochemistry and Geochemistry," ch. 1 in *Biophysical Chemistry*. New York: Academic Press, Inc.
Books
   Gamow, G., 1960. *Mr. Thompkins Explores the Atom*. Ithaca, N.Y.: Cornell University Press.
   Henderson, L. J., 1958. *The Fitness of the Environment*. Boston: The Beacon Press.
   Pauling, L., 1956. *General Chemistry*. San Francisco: W. H. Freeman & Co.

Chapter 6

Water and Life

Water is the dispersion medium of living matter. Even land organisms that at first glance appear to thrive in a gaseous medium are found to live in a watery medium when examined at the cellular level. Actively living cells consist of 60–95 percent water, and the significance of water can be appreciated by the observation that even dormant cells and tissues like spores and seeds have water contents of 10–20 percent.

The ubiquity of water should not distract us from examining its very special and unique properties. It is this uniqueness that makes water especially suited for the biological role it plays. The great physiologist, L. J. Henderson, once wrote a book entitled *The Fitness of the Environment* in which he demonstrated that the phenomenon of biological adaptation could be considered not only from the point of view of organisms adapting to their environment, but also from the reverse direction—namely, the suitability of the physical environment for the support of life.

A HYDRIDE OF UNUSUAL PROPERTIES

Water is a hydride of oxygen that possesses a uniquely strong degree of interaction among its molecules. Figure 6-1 shows the heat of vaporization of a number of hydrides. The heat of vaporization, which is a measure of the energy necessary to separate the molecules of the liquid from one another, is therefore used here as a measure of the strength of intermolecular forces. We see that in the case of the carbon series there is a rough proportionality between the atomic weight of the element forming the hydride and its heat of vaporization. This is not the case in the other series in which the first member is atypical, with water being the most atypical of all.

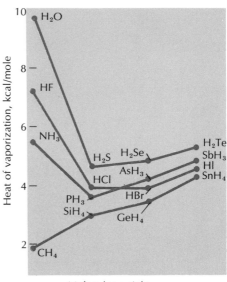

Fig. 6-1. Heats of vaporization of isoelectronic sequences of hydride molecules. Notice the atypical behavior of the hydrides of N, F, and O, water being the most atypical and also structurally the most unusual of the hydrides. (From *General Chemistry*, Second Edition, by Linus Pauling. W. H. Freeman and Company. Copyright © 1953.)

The reason for these unusually strong degrees of interaction is the fact that oxygen, fluorine, and nitrogen form hydrides in which only some of their electrons are involved in covalent bonds with the hydrogen; the remaining electrons stay close to the atom, rendering it strongly electronegative. Thus, these molecules have a high degree of electrical polarity, which in turn causes very strong intermolecular forces.

Let us consider the case of the water molecule. Of the eight outer electrons only four are involved in electron pair formation with the two hydrogen atoms. The oxygen nucleus exerts a strong attractive pull on the remaining four electrons, leaving the hydrogen atoms electropositive. The centers of negative charge, that is, the regions of highest concentration of electrons, are located in space away from the O-H bonds. The precise geometry of this arrangement is such that if the oxygen is placed in the center of a tetrahedron and the two hydrogens occupy two of the corners, then the centers of negative charge are concentrated in the direction of the two other corners of the tetrahedron (Fig. 6-2). Thus, water is not only an electrically polar molecule, but also a structure in which the two centers of positive charge and the two centers of negative charge fit a precise tetrahedral pattern of organization. In the liquid state water molecules exert orienting effects on each other, centers of negative charge attracting centers of positive charge and vice versa. Each water molecule attracts four other molecules that tend to arrange themselves in tetrahedral fashion around it. In liquid water this tetrahedral arrangement is in a dynamic state, clusters of oriented "crystalline" water being in equilibrium with more randomly distributed water molecules. When water freezes, however, the entire structure becomes ordered in the manner shown in Fig. 6-3. The tetrahedral structure of water in which each oxygen

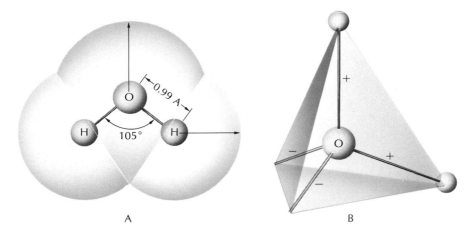

Fig. 6-2. The structure of water, illustrating the precise geometry of its polar properties. A. The bond distances and bond angle of the water molecule. B. Because of the strong attractive force exerted by the oxygen atom on the four unpaired outer electrons, there is a separation of positive and negative centers of charge that point exactly at the four corners of the tetrahedron. The consequence of this for the structure of ice can be seen in Fig. 6-3.

attracts two neighboring protons and each proton attracts the oxygen of a neighbor is therefore a direct consequence of the tetrahedral geometry of the centers of positive and negative charge of the water molecule.

THE HYDROGEN BOND

Using X-ray diffraction methods, it is possible to determine the exact dimensions of the ice crystal lattice. As shown in Fig. 6-3, the distance of the O-H bond is 0.99 A, and the oxygen-to-oxygen distance in ice is 2.76 A. It follows that the distance between protons and oxygen atoms in neighboring water molecules is 1.77 A. These weaker interactions or linkages are called *hydrogen bonds*. We shall encounter these bonds throughout this book, for they play a fundamental role in many important biological processes.

Hydrogen bonds also form when nitrogen and fluorine are substituted for oxygen, and this is why hydrogen fluoride and ammonia (see Fig. 6-1) are also endowed with very high heats of vaporization. In the case of ammonia and hydrogen fluoride, however, the geometry of interactions is such that only rings and chains are formed, thus making it impossible for them to form continuous, three-dimensional lattices. It follows that the properties of water are absolutely unique, and it would seem unlikely that ammonia or hydrogen fluoride would be suitable as a dispersion medium for living systems in other planets.

Hydrogen bonds form between hydrogen and strongly electronegative atoms such as F, O, or N, and these bonds are weak when they are compared with truly covalent bonds. In water, for instance, Pauling determined the strength of the hydrogen bonds

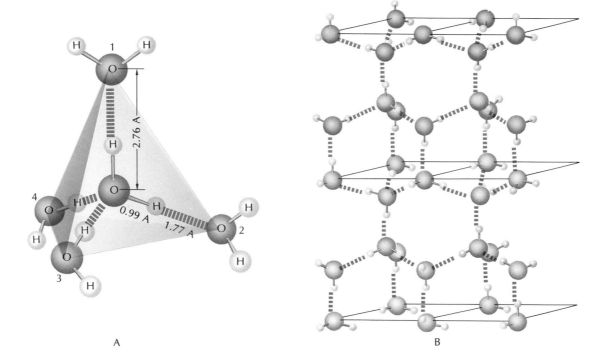

Fig. 6.3. The structure of ice. A. Tetrahedral coordination of water molecules in ice. Molecules 1 and 2 as well as the central H_2O molecule lie entirely in the plane of the paper. Molecule 3 lies above this plane, and 4 below, so that oxygens 1, 2, 3, and 4 lie at the corners of a regular tetrahedron. (From Edsall and Wyman, 1958.) B. The arrangement of water molecules in ice. This diagram shows how the water molecules are arranged in a loose network of adjacent tetrahedrons, forming a structure which is actually less tightly packed than water in the liquid state. (From *General Chemistry*, Second Edition, by Linus Pauling. W. H. Freeman and Company. Copyright © 1953.)

by measuring the heat of sublimation of ice and subtracting from this value the heat of sublimation of CH_4 (methane), which has similar general molecular properties (see Fig. 6-1) but does not form hydrogen bonds. He obtained a value of 4.5 kcal mole^{-1} which is less than one-twentieth of the value of the covalent O-H bond. Thus, hydrogen bonds are weaker than covalent interactions but stronger than the usual van der Waals forces. Their relative strength endows them with the ability to stabilize many complex macromolecular structures, and their relative weakness endows them with the mobility necessary to permit the occurrence of certain important changes in structure. These properties give them the preeminent position they occupy in the structure and function of the most important constituents of living matter.

The packing of water molecules in ice crystal lattice is a loose one. When ice melts, this regular and loose arrangement changes into a more random but more compact arrangement, causing an increase in density. Upon melting, the breakdown in structure is only partial, leaving a large portion of the liquid in a "paracrystalline" state, a unique and interesting property of water. As the temperature is slowly raised, more and more of the regular structure breaks down, and the increase in density continues, reaching a maximum at 4°C. Thereafter, because of the effect of temperature on the increase in average distance between water molecules, the density of water decreases. This special property of water is responsible for the freezing of bodies of water from the top down, rather than the other way around. The layer of ice formed insulates the water below from heat loss, thus slowing down the freezing of lakes and rivers and protecting the life in them.

OTHER UNIQUE PROPERTIES OF WATER

The structural features we have discussed so far are the cause of other unique properties of water such as unusual melting point, boiling point, latent heat of fusion, heat capacity, surface tension, and dielectric constant (Table 6-1).

Some of these properties have special consequences for living matter. The high latent heat of vaporization, for instance, provides the cooling capacity of evaporating water used by organisms to regulate their body temperatures. The high heat capacity, without doubt, acts as a "heat buffer" to protect the labile structures of the cell from thermal destruction by local, short-lived releases of thermal energy. The high dielectric constant of water acts like an "electrostatic charge buffer," screening out the intensity of interaction between charges and contributes, as we shall see, to the extensive solvent power of water. These are only a few examples of the many ways the unusual properties of water contribute to the functioning of the machinery of living things.

There are three other aspects of the behavior of water that are of the most fundamental importance. They are (1) the solvent power, (2) the dissociation, and (3) the dissociating effects of water.

The Solvent Power of Water

Water is an excellent solvent for compounds ranging from uncharged organic compounds to salts that are completely dissociated into ions even in the solid, crystalline state.

Hydrocarbons are notoriously insoluble in water, but as soon as one substitutes a hydroxyl group (—OH), one forms an alcohol, which is much more soluble in water. This is because the protons of hydroxyl groups form hydrogen bonds with the oxygen atoms of water.

hydroxyl group

alcohol

Table 6-1 PHYSICAL PROPERTIES OF WATER
COMPARED WITH OTHER LIQUIDS

| | LIQUIDS | | | |
	H_2O	ETHANOL	HEXANE	CHLOROFORM
Heat capacity (cal gram^{-1})	1	0.6	0.5	0.24
Heat of vaporization (cal gram^{-1})	596 at 0°C	262 at 64°C	79 at 68°C	59 at 61°C
Latent heat of fusion (cal gram^{-1})	79.7 at 0°C	24.9 at −114.4°C	—	
Surface tension (liquid-air, erg cm^{-2})	76	22	18	11
Dielectric constant	79	24	1.9	5

The following other organic groupings also enhance the solubility of organic compounds by forming hydrogen bonds with water:

carboxyl group

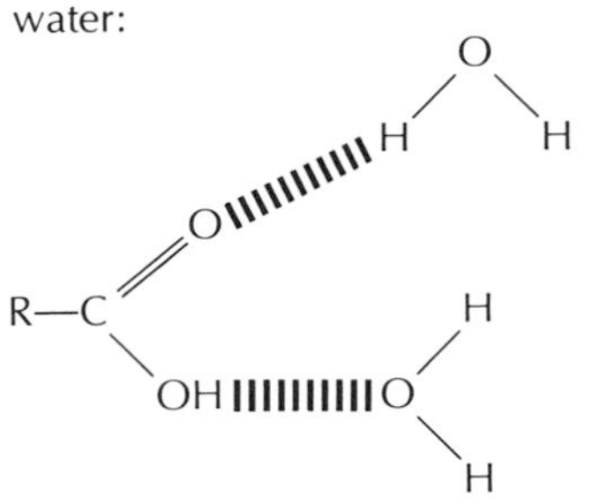

amino group

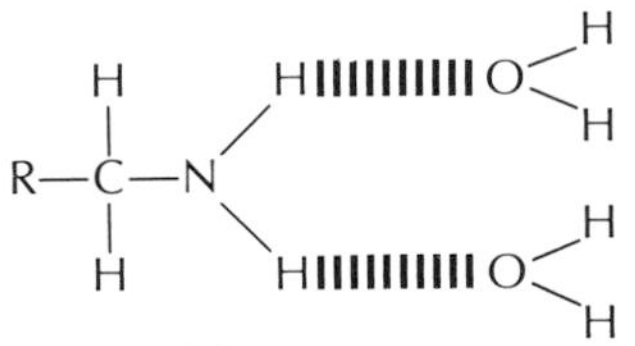

keto group

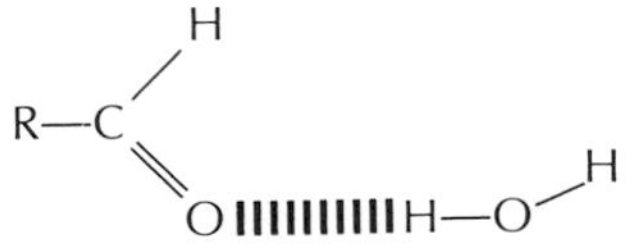

Because of water's strong dielectric properties, it is also a very good solvent for salts that are already fully dissociated into ions in the crystalline state. Thus, water molecules screen out the interaction between the oppositely charged ions, permitting the ions to be dispersed in aqueous solution. Water molecules tend to become oriented and packed around ions, forming an aqueous layer of special properties called *water of hydration.*

The Dissociation of Water

Water does not only act as a solvent for other charged substances, but is itself capable of dissociating to a slight degree in the following manner.

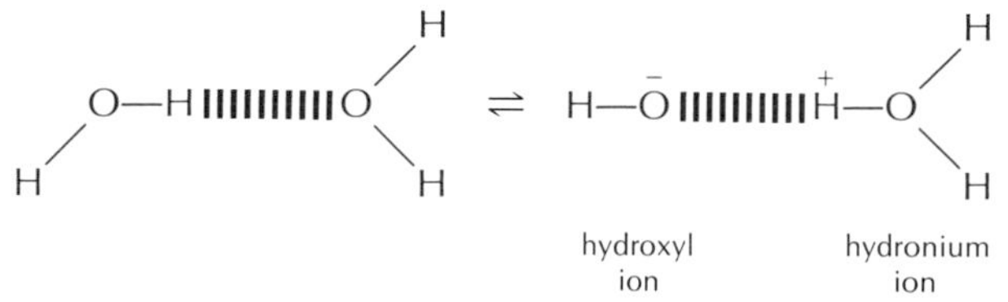

hydroxyl
ion

hydronium
ion

Since protons are relatively light and mobile particles, they manage at infrequent intervals to hop over to the oxygen of a water molecule, thus leaving a negatively charged hydroxyl ion (OH^-) and forming a positively charged hydronium ion (H_3O^+). At 25°C this event occurs in one out of 5.5×10^8 molecules of water, which is quite infrequent but much more frequent than the occurrence of free hydrogen ions or protons (H^+); therefore, when we normally write H^+, we really mean it to be a shorthand form for H_3O^+.

By dissociating into ions, water acts both as an acid and a base. This can best be understood if we consider the Brönsted theory of dissociation, which defines an acid as a substance that dissociates into a proton and a conjugate base.

$$A \rightleftharpoons H^+ + B$$

acid conjugate
base

An acid is therefore a proton donor; a conjugate base is a proton acceptor; and water, as seen in the following equation, acts as both an acid and a base, one molecule accepting a proton, the other donating a proton.

$$H_2O + H_2O \rightleftharpoons H_3O^+ + OH^-$$

The dissociation of water can be studied quantitatively by measuring the equilibrium constant of the reaction, which we now shall write in shorthand:

$$H_2O \rightleftharpoons H^+ + OH^-$$

According to thermodynamics, a true equilibrium constant that is independent of concentrations is determined by measuring *thermodynamic activities*. This, however, is not necessary in most biological systems because the concentrations of the molecules and ions in solution are low, and the temperature of the systems studied does not vary greatly. Under these conditions the concentrations of ions and molecules are numerically very close to their thermodynamic activities.

According to the laws of mass action, therefore, the equilibrium constant is

$$K_A = \frac{[H^+]\,[OH^-]}{[H_2O]}$$

where the symbol [ ] stands for concentration.

Since the concentration of water is so much larger than that of the ions H^+ and OH^-, it is for all practical purposes constant with respect to the concentrations of the ions. It is therefore customary to combine the equilibrium constant K_A with the concentration of water and obtain a new constant K_{WA}.

$$K_{WA} = [H^+][OH^-]$$

K_{WA} has been determined by a variety of methods and turns out to be 1×10^{-14} at 25°C. In pure water the $[H^+] = [OH^-] = 1 \times 10^{-7}$.

The acidity of a solution is dependent on $[H^+]$. Since the exponential character of the above quantities is inconvenient to deal with, Sörensen defined the pH scale, a more useful way of expressing acidity, as follows:

$$pH = -\log[H^+] = \log\frac{1}{[H^+]}$$

It is important that the student familiarize himself with this definition. First, it must be recognized that the pH scale is inversely related to the hydrogen ion concentration $[H^+]$, that is, the lower the pH the higher the hydrogen ion concentration. Secondly, the pH scale is exponential rather than arithmetical, which means that when a solution is taken from pH 7 to pH 6, the hydrogen ion concentration increases tenfold, and when it goes from pH 7 to pH 5, it increases a hundredfold.

One convenient property of the pH scale is that for solutions of strong acids such as HCl, the pH is approximately equal to the logarithm of the reciprocal of the acid concentration. Thus, the pH of a 0.001 N HCl solution is close to 3.

The most generally accepted standard for pH measurements is the hydrogen electrode, which is used to calibrate the glass electrode, the latter being much more convenient for routine pH determinations. Interestingly, these electrodes come much closer to measuring hydrogen ion activities than hydrogen ion concentrations, thus endowing our everyday pH measurements with undeserved thermodynamic rigor.

The Dissociating Effects of Water

Besides being capable itself of dissociating, water is able to enhance the dissociation of other substances, which we call *weak electrolytes* precisely because in pure form they occur largely in undissociated form and become more and more dissociated as they are increasingly diluted in water.

A weak acid such as acetic acid experiences the following equilibrium:

$$\underset{\substack{|\\H}}{\overset{\substack{H \quad O\\|\quad\parallel}}{H-C-C}}-OH + H_2O \;\rightleftharpoons\; \underset{\substack{|\\H}}{\overset{\substack{H \quad O\\|\quad\parallel}}{H-C-C}}-\bar{O} + H_3O^+$$

which shows how water competes with acetic acid for the proton of the hydroxyl group. This we write in shorthand fashion:

$$CH_3COOH \rightleftharpoons CH_3COO^- + H^+$$

The equilibrium constant for this reaction at 25°C is:

$$K_A = \frac{[CH_3COO^-][H^+]}{[CH_3COOH]}$$

$$= 10^{-4.75}$$

Here again the exponential form is inconvenient, and it is useful to express the dissociation of the weak acid in terms of the negative logarithm of the equilibrium constant.

$$pK_{A(acetic\ acid)} = -\log K_A = \log \frac{1}{K_A}$$

$$= 4.75$$

According to the Brönsted's theory, it is possible to express all weak electrolytes in the form of weak acids and their corresponding conjugated bases. Therefore, all weak electrolytes have one or more pK_A values, and the convention used to denote the most acid dissociation is pK_1, the next one pK_2, and so forth. Most authors use the term pK to mean pK_A, and we shall also do so from now on.

Table 6-2 lists a number of acids and their conjugate bases as well as their pK values. As we shall see presently, a knowledge of these values is useful for the understanding of the properties of buffer systems.

Table 6-2 ACIDS, CONJUGATE BASES, AND THEIR pK VALUES

ACID	CONJUGATE BASE	pK_A	
CH_3COOH	CH_3COO^-	4.75	
H_3PO_4	$H_2PO_4^-$	2	(pK_1)
$H_2PO_4^-$	HPO_4^{--}	7	(pK_2)
HPO_4^{--}	PO_4^{---}	12	(pK_3)
NH_4^+	NH_3	9.3	

If a weak acid is titrated with a strong base, we obtain a *buffer system*, which has the property, illustrated in Fig. 6-4, of "resisting" pH change. The midpoint of the titration curve, that is, the point where the pH change for a given amount of NaOH added is least, occurs at a pH value that is numerically equal to the pK value of the acetic acid. This can be demonstrated by applying the law of mass action to the dissociation of a weak acid in the following manner.

Let us consider the dissociation reaction of a weak acid.

$$HA \ \rightleftharpoons \ A^- + H^+$$

Then,

$$K_A = \frac{[A^-][H^+]}{[HA]}$$

or

$$K_A = \frac{[\text{Conjugate Base}][H^+]}{[\text{Acid}]}$$

$$\frac{1}{[H^+]} = \frac{1}{K_A} \times \frac{[\text{Conjugate Base}]}{[\text{Acid}]}$$

taking logarithms

$$pH = pK_A + \log \frac{[\text{Conjugate Base}]}{[\text{Acid}]}$$

When a strong base is added to a weak acid, a salt is formed with the conjugate base. Since such a salt is almost completely dissociated in solution, the concentration of salt is equal to the concentration of the conjugate base, so that we can write

$$pH = pK_A + \log \frac{[\text{Salt}]}{[\text{Acid}]}$$

This relationship, called the *Henderson-Hasselbach equation,* is of great value for predicting the properties of buffers. It shows what we have already demonstrated in Fig. 6-4, namely, that midway through the titration when $[\text{Salt}] = [\text{Acid}]$, pH is numerically equal to pK_A $\left(\text{when } \frac{[\text{Salt}]}{[\text{Acid}]} = 1, \text{ then } \log 1 = 0 \text{ and } pH = pK_A\right)$. Since

Fig. 6-4. Titration curve of acetic acid. This curve can easily be established by adding known quantities of NaOH to a known quantity of acetic acid and measuring the pH values obtained. The buffer capacity can be obtained by computing the derivative of this curve, that is, the rate of NaOH added per rate of pH change. This curve shows a maximum at the midpoint of the titration curve, that is, the point at which the addition of a given amount of base has the least effect on pH. The pH at which this occurs is numerically equal to the pK_A of the acid. It is also the pH at which the buffer has maximum buffering action.

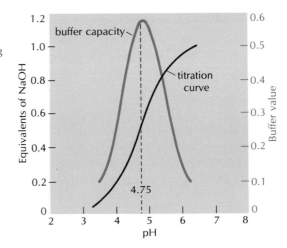

buffers are used both in nature and in the laboratory as devices for stabilizing pH, it follows that if one wants to stabilize pH of a solution at a given value, one chooses a buffer in which the weak acid component has a pK_A that is as close as possible to the pH value one wishes to maintain. Thus, if one wishes to maintain the pH of a solution at 4.7, a buffer containing acetic acid is well suited for the purpose. An inspection of Fig. 6-4 shows that the buffering efficiency of such a system is at its maximum over a range of one pH unit above and below the pK value of the weak acid.

As we shall see in succeeding chapters, the macromolecules, which are both the structural and functional entities of the cell, are weak polyelectrolytes that owe their state of dissociation, and hence many of their physical characteristics, to the presence of water or dilute salt solutions. Water, therefore, is not merely the dispersion medium of the cell, but also a major influence on the properties of the molecules it disperses. The properties of water that we have discussed have innumerable connections with the functioning of the living machine. It is not any *one* but the simultaneous presence of *all* these properties that makes water the unique solvent for the world of life.

SUGGESTED READING LIST

Offprints

Buswell, A. M. and Rodebush, W. H., 1956. "Water." *Scientific American* offprints. San Francisco: W. H. Freeman & Co.

Articles, Chapters, and Reviews

Edsall, J. T. and Wyman, J., 1958. "Water and Its Biological Significance," Ch. 2 in *Biophysical Chemistry*. New York: Academic Press, Inc.

Books

Henderson, L. J., 1958. *The Fitness of the Environment*. Boston: The Beacon Press.
Pauling, L., 1956. *General Chemistry*. San Francisco: W. H. Freeman & Co.
Pauling, L., 1960. *The Nature of the Chemical Bond*, 3d ed. Ithaca, N.Y.: Cornell University Press.

Chapter 7   The
Small Molecules
of the Living Machine

Organic chemistry, the chemistry of the living world, is the chemistry of carbon. An outstanding property of the carbon compounds utilized by the world of life is that they are remarkably inert, reacting only at infinitesimally low rates with each other, with water, and with atmospheric oxygen. This is so in spite of the fact that large amounts of energy are often released when these compounds do react. In Chapters 10 and 18 we shall discuss how the cell overcomes this sluggishness of organic compounds and utilizes it for purposes of molecular control. A second outstanding property of carbon is its versatility in forming a virtually unlimited number of compounds that vary widely in their properties. Both inertness and this versatility in the organic compounds of the cell, which "adapt" the compounds so well to the requirements of living processes, are due to the following properties of carbon.

CARBON, A UNIQUE ELEMENT IN THE PERIODIC TABLE

First, carbon is a light element with the atomic number 6. It shares, with other elements in the second horizontal row of the periodic table, the inability to expand its octet. Thus, compounds of the tetravalent carbon atom are not complexed or solvated to any appreciable extent. Since reactions of other elements usually proceed via coordinated intermediates, it is not surprising that most reactions of carbon compounds are slow even though some of these reactions may involve the release of a great deal of energy.

Second, carbon can form four covalent bonds with other carbon atoms. This property permits the formation of the *carbon skeleton,* the fundamental framework around which organic molecules are constructed. Furthermore, the strength of the C—C bond—which is comparable to that of the C—O, C—H, and C—N (Table 7-1) bonds—ensures the stability of the basic carbon skeleton and permits the existence of organic compounds containing long chains of carbon atoms. Since a carbon atom can combine with 1, 2, 3, or 4 other carbon atoms, great structural versatility is possible in the basic skeleton, which can include straight or branched chains, rings, networks, and combinations of these structures.

Third, carbon, being centrally located in the periodic table, is capable of reacting both with electronegative elements like oxygen, nitrogen, phosphorus, sulfur, and chlorine, and with the electropositive element hydrogen. Thus, compounds of carbon are known in which formal oxidation states of -4, -2, 0, $+2$, and $+4$ can be assigned to the carbon atom. This ability to react with a variety of electropositive and electronegative elements is further expanded by the ability of carbon to form single, double, or triple bonds with carbon and double bonds with oxygen and nitrogen and endows carbon compounds with the enormous versatility that we encounter in the world of life.

In summary, carbon shares with other elements in the same (horizontal) row the property of forming compounds which react very slowly; it also shares the property of versatility with the elements in the same (vertical) column of the periodic table. Carbon is therefore the only element which possesses both slow reactivity and almost infinite versatility. One is forced to conclude that carbon is a truly unique element and that it is unlikely a bios based on a different element such as silicon would ever evolve in the universe. In this respect a compound of carbon that deserves special mention is carbon dioxide, a compound of unique significance in the world of life. Unlike silicon dioxide, which is a solid, carbon dioxide is a gas that is sufficiently soluble in water to distribute itself equally between the gas phase and aqueous solution. Carbon dioxide can hydrate itself reversibly to form the weak carbonic acid. Carbon dioxide is of course the primary source of carbon from which all of the living material on this earth is built.

Table 7-1 AVERAGE VALUES FOR BOND ENERGIES FOUND IN
ORGANIC COMPOUNDS, KCAL

C—C	C—H	C—O	C—N
82	99	82	65
C=C	O—H	C=O	N—H
145	110	170	103

Bond energy is defined here as the energy (in kcal) needed to break the bond.

In this chapter we shall concern ourselves with a brief survey of the functional groups of the organic compounds, after which we shall introduce briefly a number of organic molecules that are of importance in living systems and that we shall encounter in succeeding chapters. For further detail on molecular properties the student is referred to the beautiful little book entitled *Molecules* by J. C. Speakman.

THE FUNCTIONAL GROUPS OF ORGANIC COMPOUNDS

The properties of organic molecules are a composite of a limited number of functional groups, which are linked in a variety of permutations to the carbon skeleton. It is customary to catalog these functional groups by beginning with a saturated hydrocarbon, that is, with carbon in its most reduced state, and then proceeding to states of higher oxidation. Figure 7-1 is such a scheme showing the relationships of the most important *aliphatic* functional groups encountered in molecules found in living systems. It should be emphasized that living matter synthesizes these compounds by routes that are entirely different from the ones shown in Fig. 7-1, a fact that will become clear when we study the biosynthesis of organic compounds in later chapters.

The functional groups listed in Fig. 7-1 are normally written in a more convenient shorthand form. Table 7-2 lists some of these, as well as some of the compounds that are frequently encountered in nature.

The compounds shown in Fig. 7-1 belong to a general category we call *aliphatic,* which comprises all compounds that are not *aromatic.* We can thus define both these categories by describing what we mean by the term aromatic in the following manner.

If we take benzene (Fig. 7-2) and measure its heat of combustion, we find that it is some 36 kcal less than the value one would calculate for a compound with the formula C_6H_6. This lowering of the observed heat of combustion is called the *stabilization energy* and is paralleled by the remarkable lack of reactivity of benzene. These properties have been shown to be due to the "resonance structure" of benzene, which, as Fig. 7-2 shows, has a conjugated double bond system. The student of modern organic chemistry will discover that such compounds have "π electrons which are delocalized," a property benzene shares with other aromatic ring systems containing conjugated double bonds, some of which are depicted in Fig. 7-2.

One of the main reasons for classifying organic molecules in terms of their functional groups is that these groups determine the chemical reactivity and physical behavior of the compounds. This is the subject matter of organic chemistry, and the student is referred to an appropriate textbook for a detailed study of the field. We shall limit outselves here to a short treatment of a few phenomena that are directly applicable to the remaining portions of this book.

THE COVALENT BOND

Carbon shares with other elements such as H, O, N, S, and P the ability to form covalent bonds. A covalent bond is formed when two atoms can achieve a stable electronic grouping by sharing a pair of electrons. A stable electronic grouping is

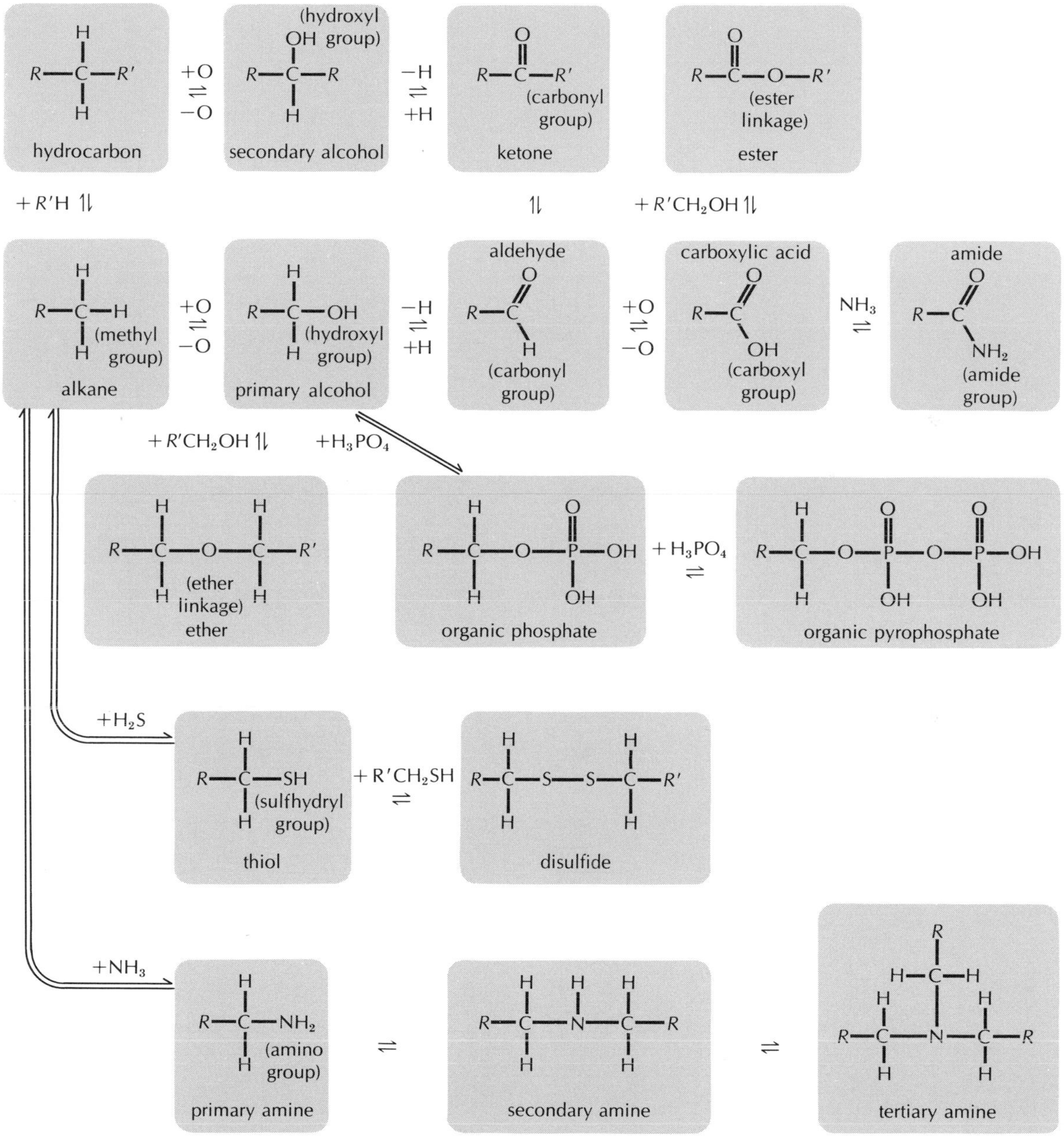

Fig. 7-1. Major aliphatic groups.

usually reached by forming an outer shell of two or eight electrons as found in the inert gases such as helium and neon. In methane (CH_4), for instance, hydrogen, which has

Table 7-2 A NUMBER OF FUNCTIONAL GROUPS AND
REPRESENTATIVE COMPOUNDS CONTAINING THEM

STRUCTURAL FORMULA	SHORTHAND FORM	GENERAL NAME	COMPOUND	COMMON NAME
H | R—C—H | H	RCH_3	Alkane	CH_4 C_2H_6 C_3H_8	Methane Ethane Propane
H | R—C—OH | H	RCH_2OH	Alcohol	CH_3OH C_2H_5OH C_3H_7OH	Methanol Ethanol Propanol
O ‖ R—C | H	RCHO	Aldehyde	HCHO CH_3CHO	Formaldehyde Acetaldehyde
O ‖ R—C—R′	RCOR′	Ketone	CH_3COCH_3	Acetone
O ‖ R—C | OH	RCOOH	Carboxylic acid	HCOOH CH_3COOH	Formic acid Acetic acid
O ‖ R—C—O—R′	RCOOR′	Ester	$CH_3COOC_2H_5$	Ethyl acetate
O ‖ R—C—NH$_2$	$RCONH_2$	Amide	CH_3CONH_2	Acetamide
H H | | R—C—O—C—R′ | | H H	$RCH_2OCH_2R′$	Ether	$C_2H_5OC_2H_5$	Diethyl ether
H | R—C—SH | H	RCH_2SH	Sulfhydryl	C_2H_5SH	Ethanethiol
H | R—C—NH$_2$ | H	RCH_2NH_2	Amine	$C_2H_5NH_2$	Ethylamine

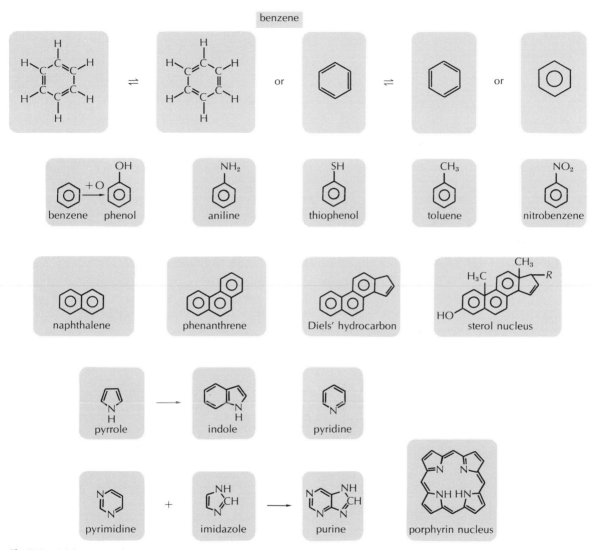

Fig. 7-2. Major aromatic groups.

one electron, and carbon, which has an outer shell of four electrons, can form a compound in which four pairs of electrons are shared. Thus, by this sharing process hydrogen becomes heliumlike and carbon becomes neonlike in their respective electronic groupings.

$$4H \cdot + \cdot \overset{\cdot}{\underset{\cdot}{C}} \cdot \quad \rightarrow \quad H \overset{H}{\underset{\cdot\cdot}{\overset{\cdot\cdot}{C}}} H$$

methane

In carbon dioxide, for instance, all three atoms achieve octets, becoming neonlike.

$$2 \; \ddot{\ddot{O}}\colon + \cdot\dot{\underset{\cdot}{C}}\cdot \; \rightarrow \; \colon\!\ddot{O}\colon\colon C\colon\colon\!\ddot{O}\colon$$

carbon dioxide

There are, of course, deeper reasons than the ones outlined above that explain the phenomenon of electron sharing. The field of *wave mechanics*, which is concerned with this problem, uses a highly mathematical language. For a simple nonmathematical approach to this as well as other problems of molecular structure the student is again referred to the Speakman book, *Molecules*.

Covalent compounds can be described in a number of ways that help illuminate the structure and function of classes of organic molecules.

Bond distances, the distances between the centers of two atoms held together by covalent bonds, have proved to be remarkably regular quantities. Thus, the C—H bond length in methane is 1.09 A; in benzene, 1.08; and in ethylene, 1.07. The C—C bond distance in ethane or cyclohexane is 1.54 but in ethylene is 1.34; however, this large difference turns out to be due to the fact that in ethylene the C—C bond is double! In addition to this regularity of covalent bonds we find that they are also additive. Thus, it is possible to obtain an estimate of bond distances in molecules by adding the covalent radii of the respective atoms. Table 7-3 lists some of these covalent radii.

Table 7-3 COVALENT RADII OF ATOMS FOUND IN ORGANIC MOLECULES (A)

BOND	H	C	N	O	P	S
Single	0.37	0.77	0.70	0.66	1.10	1.04
Double		0.67	0.60	0.55		
Triple		0.60	0.55			

Bond energies, which we have already listed in Table 7-1, also prove to have additive characteristics. Since it is possible to estimate the heat of formation of an organic compound by summing the respective bond energies, for instance, using the data in Table 7-1, one can calculate the energy of formation of ethyl alcohol (C_2H_5OH) to be 769 kcal/mole; the experimental value is 765 kcal/mole.

Bond angles are another parameter whereby the geometry of organic molecules can be described. As we have already seen, the three atoms of water do not lie in a straight line; the two hydrogen atoms form an angle of 105° with the oxygen atom. This nonlinear arrangement is also true for H_2S (93°) and NO_2 (134°). However, the most important geometric fact is the tetrahedral arrangement of the valences of the carbon atom. Thus, although we conventionally write methane (CH_4) in the manner shown in Fig. 7-3A, the actual orientation of the atoms in space is such that carbon fits in the center of a tetrahedron, and the four atoms of hydrogen fit in the four corners of the tetrahedron (Fig. 7-3B). We shall have ample occasion to refer to bond angles

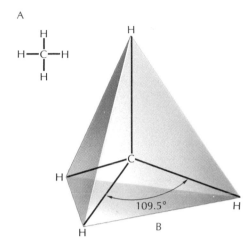

Fig. 7-3. The tetrahedral structure of carbon. A. Conventional formula. B. Tetrahedral arrangement in space.

when we discuss macromolecules and will, therefore, not list any more examples here. However, one consequence of the tetrahedral structure of carbon is the dissymmetry of many organic molecules. When a carbon atom in an organic molecule is bonded to four different atoms, we find that two different spacial arrangements can be built, one being to the other as an object is to its mirror image (Fig. 7-4). A solution of one member of a mirror image pair will rotate the plane of polarized light in one direction; whereas, its isomer will rotate the plane of light equally but in the opposite direction. This is why such isomers are referred to as *optical isomers,* and the carbon atom responsible for this dissymmetry is referred to as an *asymmetric*

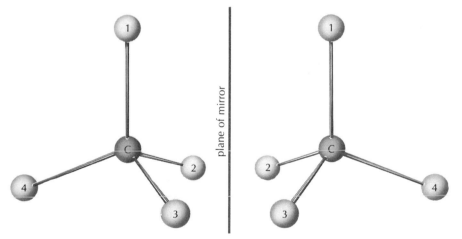

Fig. 7-4. The dissymetry of two molecules built from one-carbon atom on which four different atoms or groups are substituted.

(See Chapters 8 and 9.)

(See, for instance, Fig. 9-12.)

carbon atom. By using X-ray diffraction methods, it has been recently possible to discover the absolute spatial configuration of tartaric acid (COOHCHOHCHOHCOOH), and from this it has been possible to infer the spatial configurations of all other dissymmetric molecules such as amino acids. The convention that has been adopted to describe the absolute spatial configuration of an optically active compound is to draw the tetrahedron in such a way that two of its sides face the reader while the other two sides lie away from him.

Normally, carbon atoms can rotate freely about the C—C single bond. However, there is another form of isomerism that is due to restriction of rotation of two neighboring atoms with respect to each other. This usually occurs when the two carbon atoms are linked by a double bond, as in the cis- and trans-isomers maleic and fumaric acid.

maleic acid
cis-isomer form

fumaric acid
trans-isomer form

Here, rotation is prevented so that in one molecule the hydrogen atoms are always on the same side of the C═C bond (cis-); whereas, in the other isomer the hydrogens are always on opposite sides (trans-).

In the above case, the isomerism is due to a restriction caused by a double bond, which "freezes" one arrangement or the other; however, in molecules that consist of a chain of three or more singly bonded carbon atoms, where rotation is possible around each bond, one can easily see that an infinite number of arrangements or conformations is possible. In spite of the many possibilities, as we shall examine later, most molecules assume preferred conformations that are energetically more favorable than others because of weak — but important — forces, which will be discussed shortly.

(See Chapter 9.)

Because of these weak forces, a straight-chain hydrocarbon of the formula CH_3—$(CH_2)_x$—CH_3 prefers a *planar zigzag* conformation: that is, the carbon atoms are in a plane, but the C—C bond angles are 114° rather than 180° (see Fig. 7-5A). Because of this zigzag, if one puts six singly bonded carbon atoms into a ring, the ring will not lie flat but will twist, giving rise to still another kind of isomerism. As shown in Fig. 7-5B, the twisted ring of cyclohexane can exist in two conformations, called the *chair* and *boat* conformations. Normally, the chair form is more stable because it allows the hydrogen atoms to avoid each other as much as possible, but there are occasions when the boat form is forced on the ring by the remainder of the molecule of which it may be a part.

A

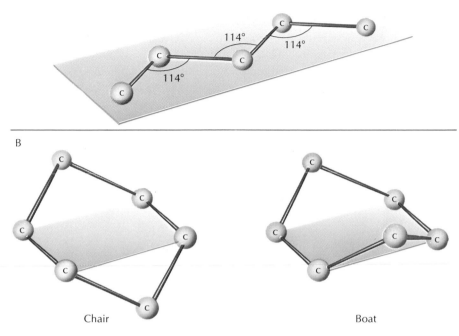

B

Chair   Boat

Fig. 7-5. Conformations of singly bonded carbon atoms. A. In an unrestricted chain. B. In a six-membered ring. The chair form is more stable.

THE WEAK INTERACTIONS IN ORGANIC COMPOUNDS

As we have seen, a covalent interaction is a strong one, involving energies of about 100 kcal mole^{-1}. There are also a number of weak interactions that are called intermolecular simply because organic molecules are usually defined as being held together by covalent bonds. For macromolecules, however, these weak interactions may be intramolecular and bring about the preferred conformations to which we referred before, which are responsible for the biological specificity of the macromolecule. These interactions are collectively classified as van der Waals interactions because they can be evaluated from the van der Waals corrections of the ideal gas law:

$$PV = nRT$$

where P is pressure, V is the volume, T is the absolute temperature, n is the number of moles, and R (the gas constant) is 1.987 cal deg^{-1} mole^{-1}. The van der Waals equation is

$$\left(P + \frac{a}{V^2}\right)(V - b) = nRT$$

in which a/V^2 is a correction accounting for the attraction between molecules and b is a correction accounting for their volume.

The intermolecular forces responsible for the deviation from the ideal gas law differ from each other in intensity and the rate at which they decrease with increasing distance. We shall list them here in order of strength, starting with the strongest (Fig. 7-6), but before we do, we must introduce the concept of *polarity* of molecules.

As we have seen in our discussion of water, a molecule, although electrically neutral, may be polar in that its "center of gravity" of positive charge does not coincide with the "center of gravity" of negative charge. Such a molecule has a *dipole moment*, which can be determined by measuring the tendency of the molecule to orient itself in an electric field. Therefore, the greater the charge separation, that is, polarity, of the molecule, the greater is the dipole moment.

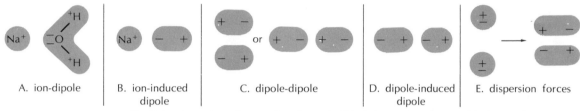

A. ion-dipole   B. ion-induced dipole   C. dipole-dipole   D. dipole-induced dipole   E. dispersion forces

Fig. 7-6. A number of categories of van der Waals interactions responsible for weak attractions between molecules. The interactions are listed in order of decreasing strength.

The strongest of the weak forces is the *ion-dipole* interaction, which is due to the interaction of an ion and a polar molecule such as Na^+ and H_2O (Fig. 7-6A). This force is frequently responsible for the hydration of ions and plays an important role in many biological phenomena.

The *ion-induced dipole* interaction is due to the effect of an ion on a nonpolar molecule (Fig. 7-6B). Here, the charge of the ion induces a dipole in a nonpolar molecule, and the result is a somewhat weaker interaction.

An attraction force may also occur between two dipoles (Fig. 7-6C). This *dipole-dipole* interaction brings about the orientation of molecules with respect to each other and may be of great importance in the assembly processes discussed later.

(See Chapter 11.)

A dipole may also induce a dipole in a nonpolar molecule. *Dipole-induced dipole* forces are among the weakest interactions (Fig. 7-6D).

Finally, two nonpolar molecules are capable of mutual attraction because of mutually induced dipoles. Although the electrons of a neutral molecule may be evenly distributed about the protons when viewed for a long period of time, at any single moment it is quite likely that the center of negative charge does not coincide with the center of positive charge. Consequently, there exists an instantaneous dipole that, temporary though it is, can induce a similar dipole in a nearby molecule. The dipoles thus formed generate a weak attraction called *dispersion forces,* which explain the fact that nonpolar molecules of gases such as H_2, N_2, or He can form liquids under proper conditions of temperature and pressure.

In summary, the important feature to recognize about van der Waals interactions between organic molecules is that they represent a family of weak, short-range interactions—significant only when the molecules are close together, but of utmost importance to biological structure, specificity, and reactions. The overall effect is, of course, a summation of the various individual types of interaction.

Since molecules that are not covalently bonded cannot come too close to each other without exerting strong repulsive forces on each other and since the van der Waals interactions fall off rapidly with distance, it is possible to establish van der Waals radii for atoms, functional groups, and molecules, which define the minimum distance between noncovalently bonded atoms. This is, of course, the distance over which the van der Waals attractive forces will be greatest (Table 7-4).

Table 7-4 VAN DER WAALS RADII (A)

H	N	O	F	P	S	—CH$_3$
1.2	1.5	1.4	1.35	1.9	1.85	2.15

THE HYDROGEN BOND

We have already discussed the anomalous behavior of H_2O, HF, and NH_3, which can be explained by the assumption that a bond forms between the hydrogen atom of one molecule and an electronegative atom (F, O, N) of a neighboring molecule. By studying crystals of a variety of organic molecules with X-ray and neutron diffraction techniques, it has been possible to demonstrate that the distance between the hydrogen and the electronegative atom is less than the van der Waals radius of the electronegative atom. In fact, when one first used X-ray diffraction techniques, which were unable to locate hydrogen atoms, one observed distances between oxygens ranging between 2.75 and 2.45 A. Such a distance is shorter than the 2.8 A van der Waals distance between two oxygens and much greater than the 1.4 A distance between two covalently bonded oxygens. The term *hydrogen bond* arose when this anomalous distance could be explained by placing a hydrogen between the oxygens. (See Fig. 7-7.)

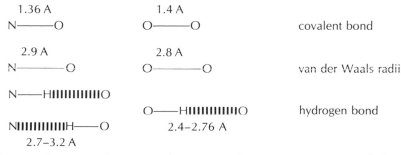

Fig. 7-7. Three types of interactions between two electronegative atoms. Henceforth, we shall use the above notation for the noncovalent part of the hydrogen bond.

Hydrogen bonds are weak when compared with covalent bonds but fairly strong when compared with van der Waals interactions. The energy needed to break a covalent bond lies between 50 and 100 kcal mole^{-1}; whereas, the energy to break hydrogen bonds ranges from 0.5 to 12 kcal mole^{-1}. As we shall see, hydrogen bonds are of enormous importance in biology, especially in relation to structure of macromolecules and the interactions between them. Although these bonds may not supply the bulk of the energy stabilizing the conformation of macromolecules, they are primarily responsible for the specificity of the interactions. (Most of the energy stabilizing the conformation of macromolecules is supplied by yet another interaction called the *hydrophobic bond.* Since this bond occurs primarily in macromolecules

(See Chapters 8 and 9.) we shall describe it later.)

We shall not attempt to explain the hydrogen bond in wave mechanical terms. It suffices to say that the bond can, in part, be explained in electrostatic terms, that is, in terms of an attraction between a proton with residual positive charge and an atom like oxygen with residual negative charge.

PARAMETERS OF MACROMOLECULES

Knowledge of the interactions between organic compounds gives us an insight into three parameters of macromolecules that are important to understanding living matter.

(See Chapters 8 and 9.) The first is the internal organization (conformation) of macromolecules; the second
(See Chapters 10 and is the interaction between macromolecules; and the third parameter is the interaction
11.) of the molecules with their aqueous solvent, a property which can be divided into solubility and dissociation.

Solubility of Organic Compounds

As we have already indicated in our discussion of water, most organic compounds are rendered soluble in water by the substitution of various organic groupings on the carbon backbone. These groupings are (1) nonpolar groups in which dipoles can be induced by the highly polar water molecules; (2) polar groups; and (3) polar groups that become dissociated by the action of the aqueous solvent. The following groupings are listed in order of increasing polarity and, hence, solubility in water:

| $(-CH_2)_n$ | $-CH_3$ | ether | ketone | aldehyde | ester | amide |
| alkyl | methyl | | | | | |

| alcohol | carboxyl | amino | organophosphoric | Zwitterion (amino acid) |

The more such groups are substituted on the carbon backbone, the more soluble the compound. Glucose, for instance, which carries five hydroxyl groups and one keto group, is extremely soluble in water. In high concentrations glucose forms very viscous solutions because the molecules form large complexes by hydrogen bonding with water molecules.

Dissociation of Organic Compounds

Water brings about or enhances the dissociation of a number of organic groupings. As we have already shown, the pK_A of such a dissociation is a measure of how strong an acid (proton donor) such a grouping is. It should be remembered that the conjugate base of a very weak acid is a strong base. The following is a list of the most important dissociating groups occurring in organic molecules of biological importance.

One of the strongest organic acids is the phosphate group, the dissociation of the first hydroxyl having a pK_A of 2. It is interesting to note that this strongly acidic group is present in nucleic acids.

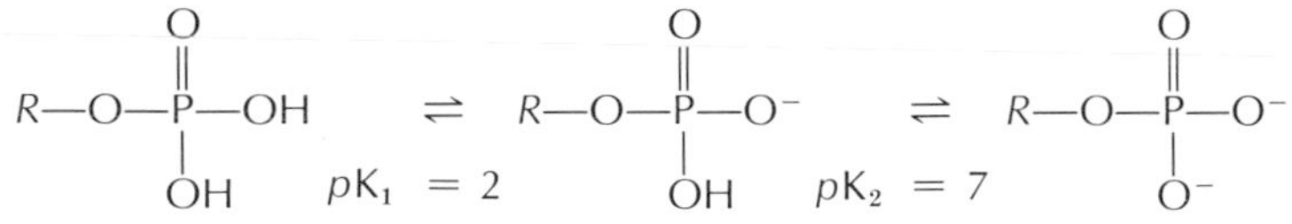

$$R—O—\overset{\overset{\textstyle O}{\|}}{\underset{\underset{\textstyle OH}{|}}{P}}—OH \quad \rightleftharpoons \quad R—O—\overset{\overset{\textstyle O}{\|}}{\underset{\underset{\textstyle OH}{|}}{P}}—O^- \quad \rightleftharpoons \quad R—O—\overset{\overset{\textstyle O}{\|}}{\underset{\underset{\textstyle O^-}{|}}{P}}—O^-$$

$$pK_1 = 2 \qquad pK_2 = 7$$

organic phosphate

The carboxyl group is probably the most ubiquitous acidic group of biological organic molecules, especially of proteins.

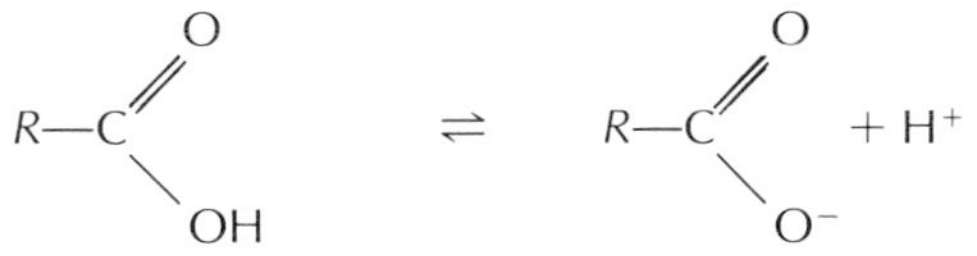

$$R—C\overset{\displaystyle O}{\underset{\displaystyle OH}{\big<}} \quad \rightleftharpoons \quad R—C\overset{\displaystyle O}{\underset{\displaystyle O^-}{\big<}} \; + H^+$$

carboxyl group $\quad pK_{COOH} = 3\text{–}4$

The imidazole group, also found in proteins, has a pK_A at 6–7; thus, like water, it is neither a strong acid nor a strong base.

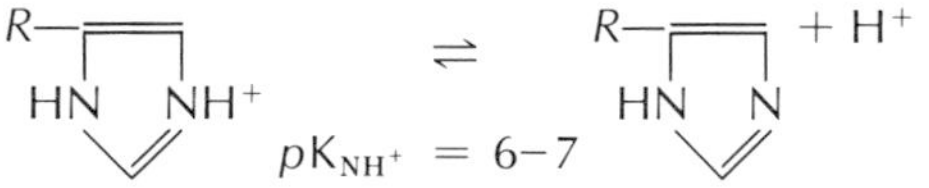

$$\underset{\underset{\text{HN} \quad \text{NH}^+}{}}{R—\boxed{}} \quad \rightleftharpoons \quad \underset{\underset{\text{HN} \quad \text{N}}{}}{R—\boxed{}} \; + H^+$$

$$pK_{NH^+} = 6\text{–}7$$

imidazole group

The amino group is the most frequently occurring basic group, especially in proteins.

$$R—NH_3^+ \quad \rightleftharpoons \quad R—NH_2 + H^+$$

$$pK_{NH_3^+} = 9\text{–}10$$

amino group

The guanidinium group is one of the most basic groups. It is also found on protein molecules.

$$
\underset{\substack{R—N—C—NH_2 \\ pK_{NH_2^+} = 12-13}}{\overset{\substack{H \quad NH_2^+ \\ | \quad \| }}{}} \rightleftharpoons \underset{\substack{R—N—C—NH_2 + H^+ \\ \text{guanidinium group}}}{\overset{\substack{H \quad NH \\ | \quad \|}}{}}
$$

The importance of dissociation in biological systems cannot be overemphasized since the structure and function of macromolecules are strongly dependent on pH. The presence of positive and negative charges on the backbone of macromolecules plays an important role in the spatial organization (conformation) of the molecules, a subject which we will discuss in considerable detail later.

(See Chapters 9 and 10.)

ULTRAVIOLET LIGHT ABSORPTION

A very practical use of multiple bonds (which may be considered a functional group), and particularly of two or more adjacent multiple bonds, is found in measuring the concentration of biologically important compounds. A number of organic compounds are colorless in visible light but intensely absorb wavelengths in the near ultraviolet. This absorption is due to excitation of electrons in the molecule to higher energy levels by photons of the ultraviolet light. Such "excited state" molecules are usually so short-lived that they cannot react chemically before losing their energy to surrounding molecules, although in some life processes such as photosynthesis energy from excited states is used to carry out important biochemical reactions.

The ease with which an electron in a particular molecule can be excited depends on the structure of that molecule. Compounds without double or triple bonds usually require considerable energy for electron excitation, and these absorb only in the far ultraviolet region where the wavelength is shorter and hence the energy higher; compounds containing multiple bonds absorb in the near ultraviolet or visible range. Biologically important molecules whose concentrations can be measured by ultraviolet absorption include the nitrogen bases of the nucleic acids and the amino acids tyrosine, tryptophan, and phenylalanine. Fortunately for the biologist, the absorption maximum is at 260 mμ for the nucleic acids and at 280 mμ for these three amino acids; this means that we can distinguish between proteins and nucleic acids by spectrophotometric means.

(See Chapter 8.)
(See Chapter 9.)

So far we have discussed the general properties of organic compounds. We shall now turn to a brief description of some of the more important small organic molecules found in living systems.

CARBOHYDRATES

Carbohydrates are a group of universally occurring compounds having the general formula $(CH_2O)_n$. Carbohydrates make up the bulk of living matter in plants since they perform both structural and food storage functions.

D-glucose, a hexose, is the living world's most widespread six-carbon sugar. It contains five hydroxyl groups and one aldehyde group. In solution it is found mostly in the form of a six-membered ring, a form that is in equilibrium with a small amount of the open-chain form. The ring (pyranose) form causes the aldehyde group to be relatively unreactive.

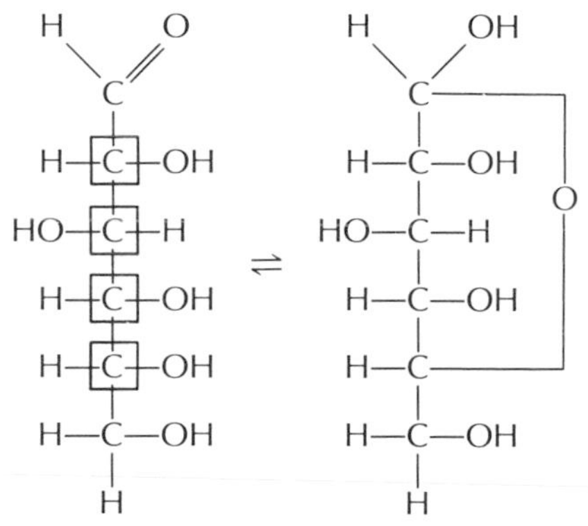

open-chain form   6-membered ring form of α-D-glucose
                              (pyranose)

D-glucose

The six-membered ring of glucose is mostly found in the chair conformation below:

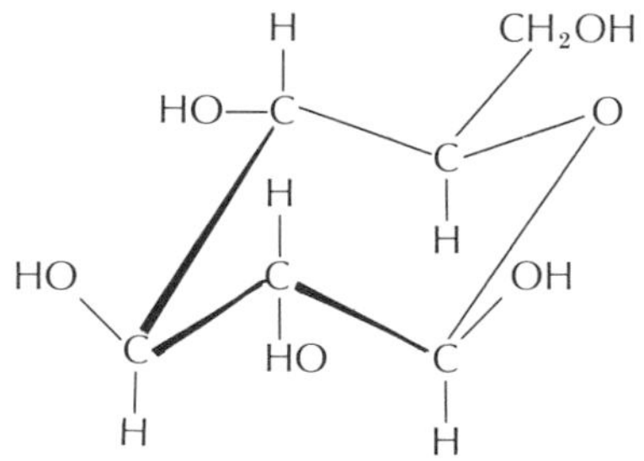

β-D-glucose

Since glucose has four asymmetric carbon atoms (see boxed carbon atoms above), 16 different isomers of glucose can in principle be formed, but only three of them are found in nature. The short representation of D-glucose shown next provides a convention whereby a particular isomer can be specified. The heavy part of the ring is nearest to the reader, and the orientation of the hydroxyl groups is either up or down. The hydrogen atoms are not generally shown but are assumed to be at the empty ends of the vertical lines.

short representation of D-glucose
(The convention for numbering carbon
atoms is shown.)

Although five-membered rings are on the whole less stable than six-membered ones, the five-membered ring of the five-carbon sugar (pentose) D-ribose is an important and universal constituent of living matter. As we shall see, D-ribose and its derivative deoxyribose are involved both in energy and in information-transfer reactions.

D-ribose   deoxyribose

Of the possible isomers, two other pentoses (D-xylose and L-arabinose) exist in nature, but they do not play as prominent a role in nature as does D-ribose.

Nature is very generous in its utilization of the pentose and hexose structures. Numerous derivatives of these basic *monosaccharide* structures are known, involving substitutions of methyl groups (L-fucose), carboxyl groups (glucuronic acid), hydroxyl groups (D-sorbitol), amino groups (D-glucosamine), and phosphate groups (glucose-6-phosphate).

L-fucose   D-glucuronic   D-sorbitol   D-glucosamine   glucose-6-phosphate
                acid

Monosaccharides can link together to form a chain—called an *oligosaccharide* —composed of 2–9 residues. The most important ones are all *disaccharides*:

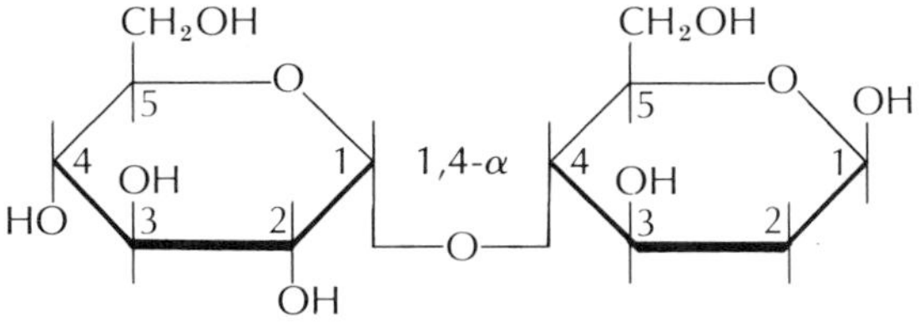

CH$_2$OH ... CH$_2$OH

5 ... O ... 5 ... O ... OH

4 ... OH ... 1 ... 1,4-α ... 4 ... OH ... 1

HO ... 3 ... 2 ... 3 ... 2

... O ...

OH

Residues: D-glucose   D-glucose

maltose
(obtained when starch is degraded)

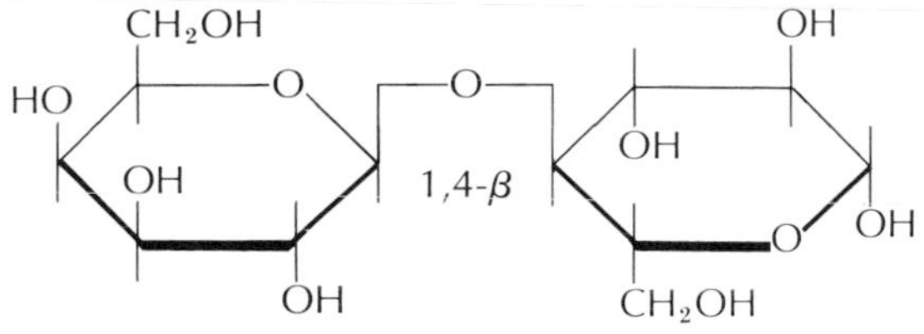

CH$_2$OH ... OH

HO ... O ... O

... OH

OH ... 1,4-β ... O ... OH

... CH$_2$OH

OH

Residues: D-glucose   D-galactose

lactose
(found in milk of mammals)

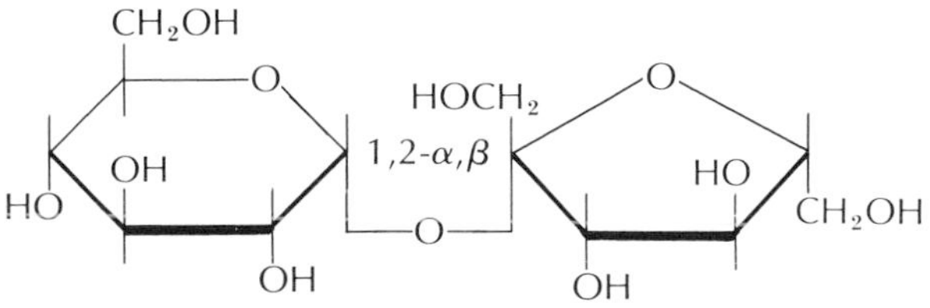

CH$_2$OH ... HOCH$_2$ ... O

... O

OH ... 1,2-α,β ... HO ... CH$_2$OH

HO ... O

OH ... OH

Residues: D-glucose   D-fructose

sucrose

Sucrose is found in all plants carrying on photosynthesis and is the main low-molecular weight food source in animals.

It should be noted that the *glycosidic bond* joining the residues in oligo- and polysaccharides can form in two ways, but nature usually utilizes only one of these isomers. In the above structures we designate the types of linkages occurring in maltose, lactose, and sucrose. For a detailed understanding of the conventions used the student should consult an advanced biochemistry textbook.

Polysaccharides are chains consisting of more than nine monosaccharide residues. Nature does not usually stop with a few residues, but synthesizes gigantic molecules that serve either food storage or structural roles in the organism. We shall discuss these large molecules at this point even though by doing so we are departing from this chapter's subject of small molecules.

Glycogen and *starch* are the animal and plant kingdoms' major food storage products, respectively. A large molecule is a useful food storage product because for a given amount of material it is present in low molar concentrations and therefore does not contribute significantly to the osmotic pressure of the cell.

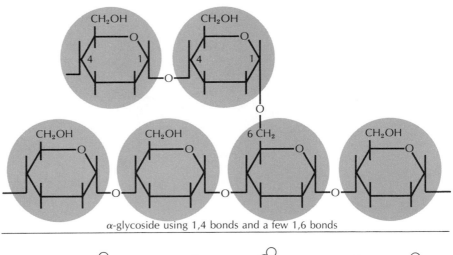

α-glycoside using 1,4 bonds and a few 1,6 bonds

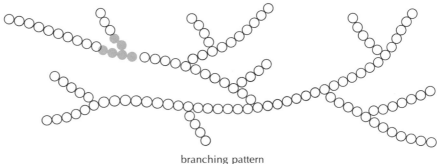

branching pattern

Glycogen

Liver glycogen is a polymer of about 30,000 glucose residues (α-glycosidic linkage) forming long branched chains. The chain length between points is usually 10–14 glucose residues, and the whole structure is a "bushy" construction forming huge water-soluble molecules (molecular weight = 5×10^6) having the general shape of a flattened ellipsoid.

Starch, the plant storage product, is deposited as huge microscopically visible granules in the plant cell. The starch granule is made up of two types of molecules: (1) an unbranched chain of a 1,4-α-glycoside of 250–300 glucose residues, usually arranged in space in the form of a helix; and (2) a molecule much like glycogen but smaller (1000 residues) and less frequently branched (on the average of every 25 residues). The precise arrangement of these two types of molecules in the starch grain is not yet understood. However, the optical birefringence (turning the plane of polarized light) of starch grains suggests that precise structural principles are involved.

Cellulose is the major structural component of plants, being present either by itself or with other compounds such as lignin in the cell wall of all plants from green algae to oak trees. Wood is essentially a collection of empty cell walls consisting of cellulose and lignin. The cellulose molecule is a long, unbranched chain of a 1,4-β-glycoside of some 14,000 glucose residues, with a molecular weight of 2.3×10^6. This gigantically long thread is folded in some manner to give long fibrils that are visible under the electron microscope. The folded molecule is insoluble because the hydroxyl groups react with each other to cross-link the molecule, rather than hydrogen bonding with the water and dissolving.

portion of cellulose molecule

The secondary plant cell wall is made of cellulose fibrils arranged in alternate layers lying at different angles to each other (Fig. 7-8), the whole structure being interlaced by lignin acting apparently as a cement. The cell wall, a saclike envelope of the plant cell, has tremendous tensile strength exceeding even that of the highest quality steel.

Chitin is a structural substance of the cell wall of fungi and the exoskeleton of arthropods (crabs and insects). It is also a linear, unbranched molecule held together by 1,4-β-linkages, but in this case the sugar residue is a derivative of glucose (N-acetyl-glucosamine).

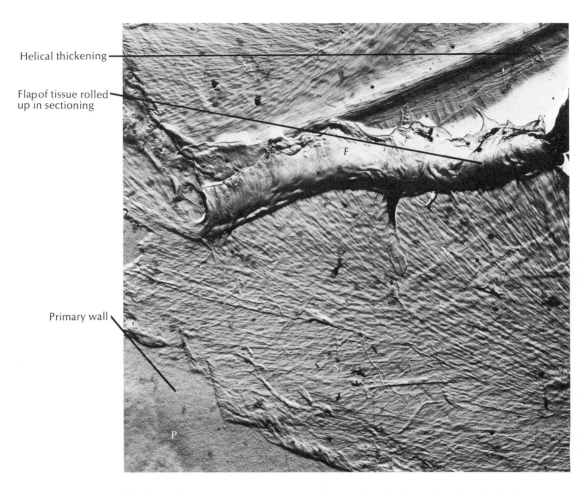

Helical thickening

Flap of tissue rolled up in sectioning

Primary wall

Fig. 7-8. Electron micrograph of direct carbon replica of plant cell wall (tracheid of Douglas fir). P, primary cell wall; S_1, S_2, and S_3, are successive layers of secondary cell wall with cellulose fibers at different angles; F is flap of tissue rolled up; $\times$ 3550. (Courtesy of Cote, SUNY College of Forestry, Syracuse, New York.)

The structural materials that form the cell walls of bacteria are among the most complex macromolecular edifices found in nature. The bacterial cell wall is constructed of chains that are interconnected with covalent cross-links by shorter side branches. It can therefore be argued, as indeed Weidel has, that the bacterial cell wall is a single, gigantic, bag-shaped macromolecule.

The fundamental building block of the long polysaccharide chain in the bacterial cell wall is a disaccharide consisting of glucose derivatives (N-acetylglucosamine and N-acetylmuramic acid) held together by a 1,4 glycosidic linkage. The disaccharide units are linked to each other by 1,4 glycosidic bonds. These bonds are susceptible to hydrolysis by the enzyme lysozyme, a phenomenon that both bacteriophages and higher animals use in their predatory or defensive interactions with bacteria.

(See Chapter 10.)

The polysaccharide chains in the bacterial cell wall are cross-linked by short runs of *polypeptide* chains, which are polymers of amino acids. Some of the specific details of the cross-linking have yet to be worked out. This structure is one of several that we know to occur. There is a great deal of variation, especially among different types of bacteria, and even within a given bacterium the cell wall appears to be composed of multiple layers varying in their structural details.

(See Chapter 9.)

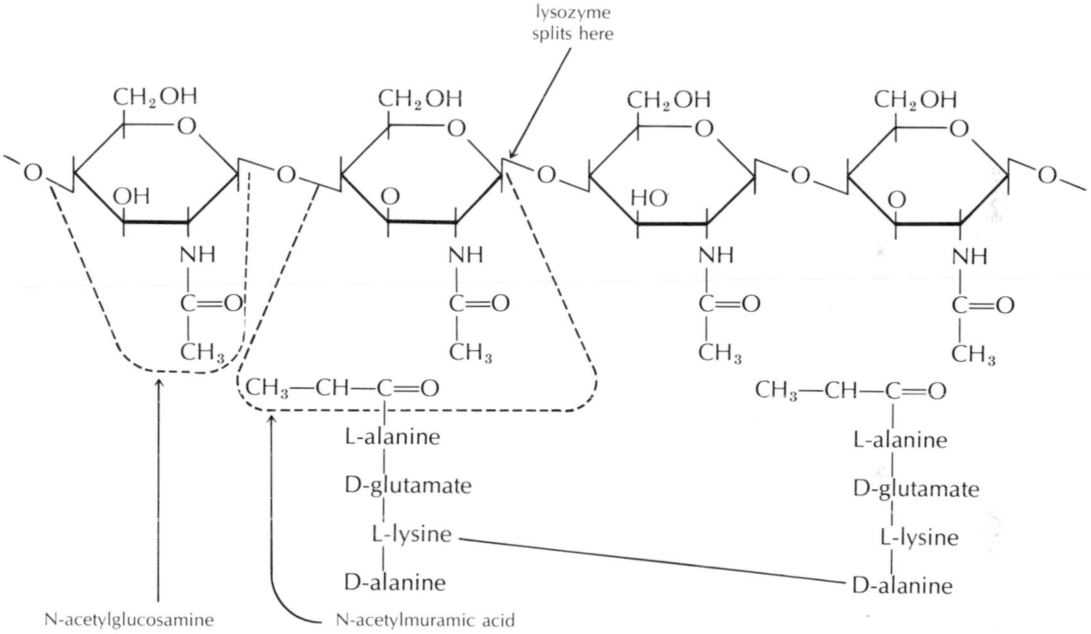

the polysaccharide structure of the bacterial cell wall

Since the bacterial cell wall appears to be held together by a network of covalent bonds, the growth of the cell must involve a specific dissolution of part of its structure followed by a highly controlled rebuilding of it. How this process is controlled by the bacterium to result in a synchronized process of high-structural specificity is a subject of a great deal of contemporary work.

LIPIDS

Lipids are not necessarily related in chemical structure, for they are defined by the common physical property of solubility in nonpolar, organic solvents. A number of classes of compounds belong here, the most important of which are fatty acids, fats, glycerophosphatides, sphingolipids, and steroids.

Fatty acids are compounds consisting of a long hydrocarbon chain with a carboxyl group at one end. The general formula is

$$CH_3(CH_2)_nC\underset{O^-}{\overset{O}{\diagup}}$$

Since fatty acids are synthesized from 2-carbon units (acetyl radicals), all naturally occurring fatty acids have an even number of carbon atoms, the most frequent numbers being 16 or 18.

Fatty acids can be either "saturated," when their hydrocarbon chain contains only single bonds, or "unsaturated," when one or more double bonds are present.

Typical saturated fatty acids are

$CH_3(CH_2)_{14}COO^-$ palmitic acid

$CH_3(CH_2)_{16}COO^-$ stearic acid

Typical unsaturated fatty acids are

$CH_3(CH_2)_7CH{=}CH(CH_2)_7COO^-$ oleic acid

$CH_3(CH_2)_4CH{=}CH(CH_2)CH{=}CH(CH_2)_7COO^-$ linoleic acid

Fatty acids have a split personality in that they consist of both an extremely polar, water-soluble end and an extremely nonpolar, water-insoluble portion. Their schizophrenia is beautifully demonstrated by their interaction with water, that is, they both do and do not dissolve in water. This they achieve by forming a monomolecular layer at the surface of water, with the carboxyl group sticking into the water and the hydrocarbon chains waving above the surface.

The monomolecular layer of fatty acids at the surface lowers the surface tension

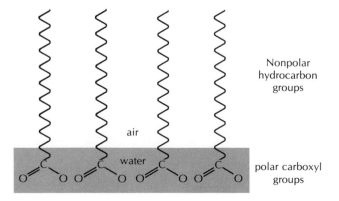

of water, an effect that increases the "wetting" power of water. This helps explain the cleaning action of soaps, which are sodium or potassium salts of fatty acids.

Fats are esters of glycerol and fatty acids.

$$
\begin{array}{l}
CH_2O\,H \\
CHO\,H \\
CH_2O\,H
\end{array}
\;+\;
\begin{array}{l}
HO\,OC(CH_2)_n\,CH_3 \\
HO\,OC(CH_2)_{n'}CH_3 \\
HO\,OC(CH_2)_{n''}CH_3
\end{array}
\;\rightleftharpoons\;
\begin{array}{l}
H-\overset{\displaystyle H}{\underset{}{C}}-O-\overset{\displaystyle O}{\overset{\|}{C}}-(CH_2)_n\,CH_3 \\[4pt]
H-\overset{}{C}-O-\overset{\displaystyle O}{\overset{\|}{C}}-(CH_2)_{n'}\,CH_3 \;+\;3H_2O \\[4pt]
H-\overset{}{\underset{\displaystyle H}{C}}-O-\overset{\displaystyle O}{\overset{\|}{C}}-(CH_2)_{n''}CH_3
\end{array}
$$

glycerol   fatty acids   fat

As the formula above indicates, the fatty acids need not be the same ones — a fat may contain one, two, or three different fatty acids. A fat is hydrolyzed (saponified) to fatty acids and glycerol by heating it in strong alkali. This is how our ancestors converted animal fat to soap for thousands of years.

The melting point of fats depends on the degree of saturation of its fatty acids. The higher the degree of unsaturation, the lower the melting point. When a fat has a sufficiently high degree of unsaturation to be a liquid at room temperature, it is called an oil. Oils can be converted to "hard" fats by saturating the fatty acids. This is how margarine is manufactured from vegetable oils.

Fats play an important role in the cell as storers of chemical energy. Because they have a long hydrocarbon chain, their overall state of oxidation is very low; thus, they have much further to go than carbohydrates and proteins before they are "burned" to CO_2 and H_2O.

Glycerophosphatides and *sphingolipids* comprise a rich variety of compounds that play an important structural and presumably functional role in membranes. The chemical structures of a large number of these compounds have been elucidated, but their exact organization in membrane systems is not yet understood. Many of these compounds such as the lecithins have the ability to form bimolecular leaflets when added to water. Consequently, they will form "myelin" figures, which, under the light microscope look like a highly convoluted membrane system and under the electron microscope show a fine structure of zones of differing electron density similar in appearance to cellular membranes. In recent studies proteins were added to a variety of phospholipids, thus generating artificial membranes the appearance of which, in the electron microscope, is even more similar to natural membranes than myelin figures composed of phospholipids alone.

Glycerophosphatides are all phosphate esters of glycerol, with the following general formula:

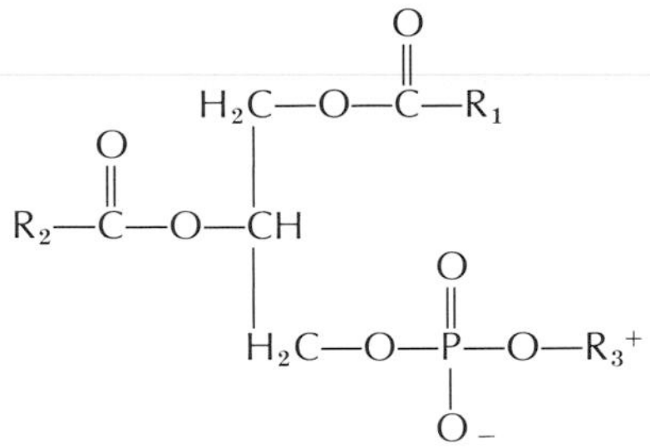

$$H_2C-O-\overset{\overset{O}{\|}}{C}-R_1$$

$$R_2-\overset{\overset{O}{\|}}{C}-O-CH$$

$$H_2C-O-\overset{\overset{O}{\|}}{\underset{\underset{O_-}{|}}{P}}-O-R_3{}^+$$

The important common property of all the glycerophosphatides is that one end of the molecule (R_1 and R_2) is extremely hydrophobic while the other end is hydrophilic because of the negative charge of the strong phosphoric acid residue and the positive charge contributed by a variety of R_3 residues.

The material we have brought together in Table 7-5 shows the complexity and variability of these important compounds. We introduce this material here because we feel certain that in the very near future the role of these compounds in membrane function will be studied with increasing success.

Sphingolipids are another category of lipids found in membranes. Here again, we have a great variety of complex compounds. The structure that is common to all of the compounds is a *sphingosine* residue on which two other residues are substituted as follows:

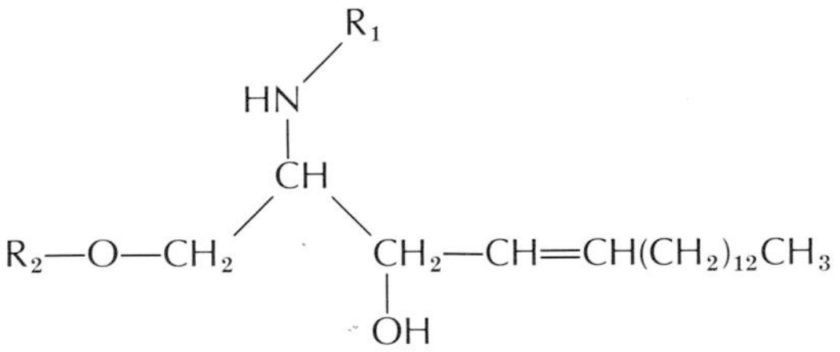

$$
\begin{array}{c}
R_1 \\
\diagup \\
HN \\
| \\
CH \\
\diagup \qquad \diagdown \\
R_2-O-CH_2 \qquad CH_2-CH{=}CH(CH_2)_{12}CH_3 \\
| \\
OH
\end{array}
$$

R_1 may be a variety of fatty acids while R_2 can vary to a great extent. Rather than listing these, let us illustrate a typical example:

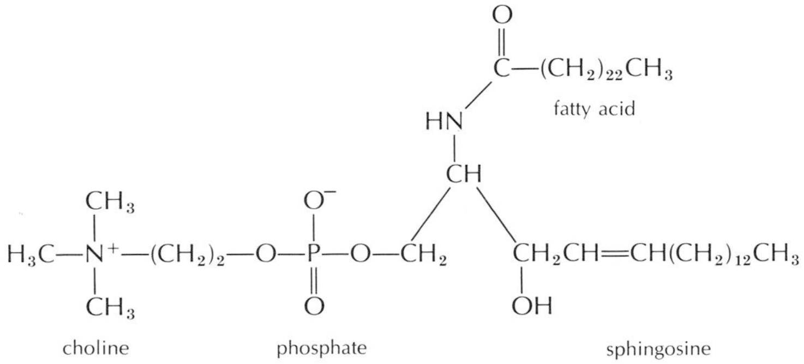

$$
\begin{array}{c}
O \\
\| \\
C-(CH_2)_{22}CH_3 \\
\text{fatty acid}
\end{array}
$$

$$
\begin{array}{c}
CH_3 \qquad\qquad O^- \qquad\qquad HN \\
| \qquad\qquad\qquad | \qquad\qquad\quad | \\
H_3C-N^+-(CH_2)_2-O-P-O-CH_2 \qquad CH \\
| \qquad\qquad\qquad \| \qquad\qquad\qquad \diagup\quad\diagdown \\
CH_3 \qquad\qquad O \qquad\quad CH_2CH{=}CH(CH_2)_{12}CH_3 \\
\qquad\qquad\qquad\qquad\qquad\qquad\qquad OH
\end{array}
$$

choline   phosphate   sphingosine

Sphingomyelin

Table 7-5 A PARTIAL LIST OF GLYCEROPHOSPHATIDES

NAME	RESIDUE 1	RESIDUE 2	RESIDUE 3
Phosphatidic acids	Fatty acid	Fatty acid	—H
Lecithins	Fatty acid	Fatty acid	$-(CH_2)_2-\overset{+}{N}\overset{CH_3}{\underset{CH_3}{\diagup}}-CH_3$ choline
Cephalins	Fatty acid	Fatty acid	$-(CH_2)_2-NH_3^+$ ethanolamine or $-CH_2-CH-NH_3^+$ $\mid$ COO^- serine
Inosides	Fatty acid	Fatty acid	inositol
Plasmalogens	Enol ether of fatty acid	Fatty acid	ethanolamine or choline

Notice that here again we have a structure consisting of two hydrophobic tails tied to a hydrophilic portion containing *both* a positive and a negative charge.

Finally, we should mention an entirely different class of lipids that appear to be built from a basic *isoprene* skeleton.

$$\underset{\text{isoprene skeleton}}{C-\overset{\overset{\displaystyle C}{\mid}}{C}=C-C}$$

The *steroids* are universally occurring compounds, many of which have been known for a number of years to play regulatory (hormonal) roles in multicellular animals. Their cellular function is being studied actively at present, and probably we shall find that steroids play a universal role in most cellular systems since steroids are found to occur in many cell membranes. Cholesterol is a well-known, widely distributed (nonhormonal) steroid.

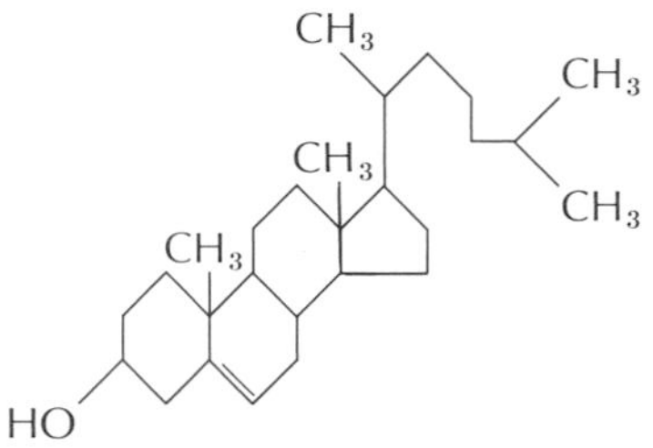

cholesterol

Carotenoids are another group of isoprene derivatives. They are widely distributed in the living world, but only procaryotes and plants can synthesize them. Animals require them for many physiological functions such as vision, but animals ultimately rely on the plant world for their manufacture. Carotene, one of the best-known compounds in this class, is shown below.

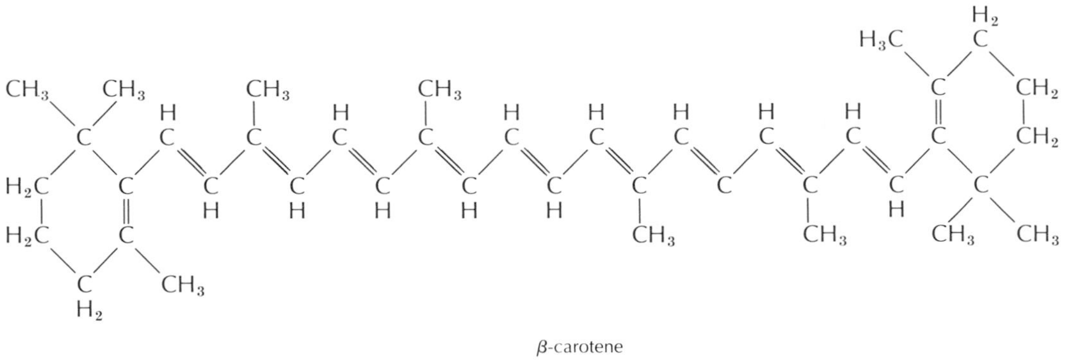

β-carotene

Carotenoid visual pigments feature in one of the most fascinating stories in the molecular physiology of vision. Because this topic (which has been worked out by Wald and his associates) represents a specialized piece of machinery used only by animals, we cannot, unfortunately, consider it here.

HIGH-ENERGY PHOSPHATE COMPOUNDS

In conclusion, we shall discuss a number of phosphoorganic compounds that play an important role in the energy transfer reactions of the cell.

The name high-energy phosphate was given to those phosphate esters whose hydrolysis leads to a uniquely high release of energy in the form of heat. Among these substances is ATP, the most universally important of the high-energy compounds, as well as the triphosphates of uridine, cytidine, and guanosine (UTP, CTP,

GTP), phosphocreatine, phosphoenolpyruvic acid, amino acid adenylates, and uridine diphosphate glucose, which we shall have occasion to discuss in later chapters.

The hydrolysis of ATP leads to ΔH values of about 9000 cal/mole; whereas, other compounds such as glucose-6-phosphate have ΔH values of about one-half of this. Actually, the high-energy content of the compound is more correctly given by the free energy of hydrolysis, the ΔG. The ΔG values are obtained from measurements of concentrations of the high-energy compounds in reactions in which they participate; they are a measure of differences between the free energies of reactants and of products. If the latter is higher, energy must be supplied for the reaction to proceed. (See Chapter 2.)

The reasons for the "high-energy" nature of these compounds is not precisely known, but it is thought to be due to several factors that make these compounds unique: there are striking differences between the compounds and the products of their hydrolysis such as in their resonance stabilities, their ionizations, and the intramolecular electrostatic repulsions. It was noted by Lipmann and Kalckar in 1941 that a prominent feature of the high-energy compounds is that they are anhydrides of phosphoric acid with a second acid such as a substituted phosphoric acid to form the nucleoside di- and triphosphates; or with a carboxylic acid compound (acetic acid) to form acetyl phosphate; or even with an enol to form phosphoenolpyruvic acid. In another case, however, phosphoric acid combines with a basic nitrogen compound such as in phosphocreatine. In all cases the formation of the anhydride bond reduces the number of resonance groups in the molecule, and since the thermodynamic stability of a substance is increased by its ability to assume many resonating forms, the high-energy compounds are less stable and their hydrolysis will yield a large amount of heat. Figure 7-9 shows some of these properties.

1. Inorganic phosphate resonating forms:

2. Carboxylate-phosphate anhydride, as in acetyl phosphate:

3. Phosphate-phosphate anhydride, as in the nucleoside di- and triphosphates; ADP and ATP:

4. Basic nitrogen-phosphate link, as in phosphocreatine:

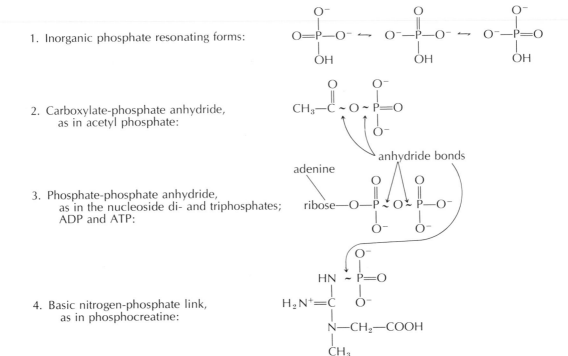

Fig. 7-9. Properties of "high-energy" compounds.

The above discussion includes but a limited sample of the immense variety of small organic compounds found in nature. We already know of about five hundred compounds involved in the metabolism of the bacterium *E. coli*. In this simple cell there are probably some 500 or 1000 additional compounds that we must still unravel from the complex interactions involved in its metabolism. In the cells of higher organisms there are probably many thousands, possibly tens of thousands, of reactions to be identified. In later chapters of this book we shall, of course, restrict ourselves to just a few examples of the roles the small organic molecules play in the "metabolic flux" of the cell.

As we have seen in Chapter 4, the cell is not a droplet of liquid containing a mixture of small molecules in solution. It is, rather, matter in a colloidal state of aggregation, possessing both the elastic properties of a solid and the viscous properties of a liquid. Furthermore, the cell is not made of one "protoplasmic" material but of a number of distinct organelles suspended in a ground substance that is interlaced by a hydrophobic network of membranes. Clearly, the cell has an intricate fine structure within which the metabolic flux of the small molecules is delicately regulated. The building blocks giving rise to this fine structure are the macromolecules of the cell, the nucleic acids, and the proteins. Therefore, in order to understand the structure and function of the cellular organelles, it is important to study in some detail the structural chemistry of the cellular macromolecules and then to relate the structure of the macromolecules to some of the functions observed in the cell.

(See Chapters 8 and 9.)
See Chapters 10 and 11.)

SUGGESTED READING LIST

Offprints

Cori, C. F., 1952. "The Enzymatic Synthesis and Molecular Configuration of Glycogen," *Carbohydrate Metabolism,* V. A. Najjar (ed.). The Johns Hopkins Press, pp. 3–13. Bobbs-Merrill Reprint Series. Indianapolis: Howard W. Sams & Co.

Derjaguin, B. V., 1960. "The Force Between Molecules." *Scientific American* offprints. San Francisco: W. H. Freeman & Co.

Roberts, J. D., 1957. "Organic Chemical Reactions." *Scientific American* offprints. San Francisco: W. H. Freeman & Co.

Strominger, J. L., 1962. "Biosynthesis of Bacterial Cell Walls." *Fed. Proc.,* pp. 134–143. Bobbs-Merrill Reprint Series. Indianapolis: Howard W. Sams & Co.

Books

Cram, D. J. and Hammond, G., 1964. *Organic Chemistry.* New York: McGraw-Hill Book Co., Inc.

Herz, W., 1966. *The Shape of Carbon Compounds.* New York: W. A. Benjamin, Inc.

Karlson, P., 1968. *Introduction to Modern Biochemistry.* New York: Academic Press, Inc.

King, E. L., 1964. *How Chemical Reactions Occur.* New York: W. A. Benjamin, Inc.

Pauling, L., 1960. *The Nature of the Chemical Bond,* 3d ed. Ithaca, N.Y.: Cornell University Press.

Speakman, J. C., 1966. *Molecules.* New York: McGraw-Hill Book Co., Inc.

Chapter 8

Nucleic Acids, Carriers of Biological Information

In 1868 Friedrich Miescher, a 22-year-old Swiss physician, extracted from the nuclei of pus cells a macromolecular substance that he knew was not protein and that he called *nuclein*. At that time the significance of the nucleus was barely understood, for it was only in 1876 that Oskar Hertwig was able to demonstrate that fertilization of the sea urchin egg involved the fusion of two nuclei, one from the egg and one from the sperm. Nevertheless, the influence of Miescher's work on nuclein was such that by 1884 Hertwig was able to write that "nuclein is the substance that is responsible not only for fertilization but also for the transmission of hereditary characteristics. . . ." Toward the end of his life Miescher appears to have doubted that this was true, and the rest of the biological world did also, for the concept that nuclein is the hereditary substance sank into oblivion toward the end of the century, only to be resurrected in the early 1940s.

Even if Miescher was not aware that the substance he isolated was the hereditary material, he approached his problem in a manner that was half a century ahead of his contemporaries. To begin with, he chose a biological material well suited for the isolation of nuclei: the pus cells obtained from the discarded bandages of wounded soldiers, made abundantly available by the Franco-Prussian War. Second, he used an amazingly modern technique for the disruption of cells under conditions that prevent most enzymatic activity—digestion with the enzyme pepsin in the presence of HCl followed by ether extraction, which layered his nuclei at the ether-water interface. Thus, without using the modern technique of differential centrifugation, Miescher was able to isolate a cellular organelle for the first time. Third, Miescher not only proceeded to study the nuclei under the microscope and identify them by a number of cytological techniques, but he also embarked on an extended chemical study of the materials found in them. This simultaneous cytochemical and biochemical approach to the study of cell organelles was truly visionary, for it was not to be used again by cell biologists before the 1940s.

Miescher dissolved most of the material obtained from his nuclei in a salt solution. Upon acidification he obtained a precipitate that had a property hitherto observed only in some lipid fractions—it contained large amounts of phosphorus. The material that Miescher called *nuclein* was later renamed *deoxyribonucleic acid* (DNA) and is now recognized as being the chemical structure that stores the cell's hereditary information.

Miescher followed his discovery with many years of careful experimentation. By fractionating the sperm heads of the Rhine salmon at low temperature, he was able to show that the nuclein was a material of high-molecular weight and that it was associated with an unusually basic protein which he called *protamine*. His measurement of the phosphorus content of "nuclein" (9.95 percent) agrees remarkably well with our modern values for DNA (9.22–9.24 percent) and is a tribute to the excellence of his preparations. When Miescher died, he left to his student Altmann the firm foundations of an entirely new field in biology that he, a single individual, had initiated. Thus, our understanding of the molecular basis of heredity was born just at the time when Mendel was discovering some of its biological manifestations.

THE ORGANIC CHEMISTRY OF THE NUCLEIC ACIDS

The forty years following Miescher's death saw the elucidation of the organic chemistry of the nucleic acids. It became apparent that there were two classes of these macro-molecular acids. Deoxyribonucleic acid (DNA) was found to be composed of (1) the purine nitrogen bases *adenine* and *guanine,* (2) the pyrimidine nitrogen bases *cytosine* and *thymine,* (3) the pentose sugar *deoxyribose,* and (4) *phosphoric acid* (Fig. 8-1). *Ribonucleic acid* (RNA) was found to be composed of the same building blocks, except that the pyrimidine *uracil* is substituted for *thymine* and the pentose *ribose* for *deoxy-ribose* (Fig. 8-1).

In recent years some additional nucleotide components have been found in both DNA and RNA in smaller amounts. These include a variety of methylated bases, sulfur-containing bases, the compound pseudouridine in which an abnormal base–sugar linkage occurs, and other modified bases. Although the role of these special bases is not yet fully understood, it is likely that they are not a caprice of nature but rather that they play an important part in the transfer of information. In the last 25 years, with the help of special enzymes called *nucleases,* some of which have the ability to split either DNA or RNA at various specific points in their macromolecular structure, it has been possible to find out how these building blocks of DNA and RNA are arranged inside the molecule. In both DNA and RNA the purine and pyrimidine nitrogen bases are linked to the deoxyribose or ribose at the 1'-carbon atom to form *nucleosides,* whereas phosphate is attached to the sugars at the 5'-carbon atom to form *nucleotides* (Fig. 8-2).

How then are the nucleotides tied together in the polymer? Again, from enzymatic degradation studies with specific nucleases it was found that the phosphates were

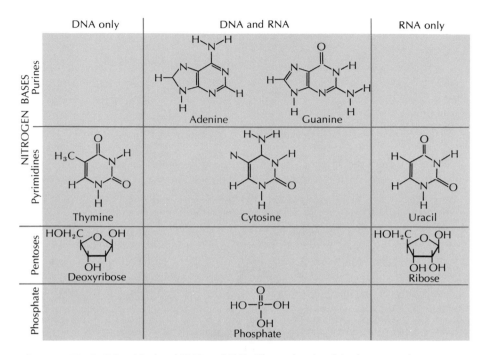

Fig. 8-1. The building blocks of DNA and RNA. The student is advised to memorize the structure of these important units from which the language of heredity is constructed.

(Fig. 11-5)

linked to both the 3' and 5' carbons of the ribose. In the case of DNA where there is no possibility of linkage with the 2'-carbon (since the oxygen is missing on the 2'-carbon of the deoxyribose) one could predict on purely structural grounds that DNA must be a linear, unbranched polymer. This was found to be indeed the case when it became possible to observe DNA directly under the electron microscope (Fig. 8-19). The unbranched, linear character of RNA was much harder to determine, and it is only in the last 15 years that convincing evidence has accumulated to that effect. The biggest stumbling block to these investigations proved to be the high reactivity of the hydroxyl group on the 2'-carbon atom so that under a variety of conditions of hydrolysis, phosphate would turn up on the 2'-carbon atom of ribose in hydrolysates of RNA. Very ingenious enzymatic and chemical studies showed, however, that this was an artifact, and with additional evidence from electron microscopy the linear, unbranched nature of RNA was finally established. We are therefore now in a position to make one of the most important biochemical generalizations about the chemistry of living systems; namely, that *DNA and RNA are linear polymers in which successive nucleoside residues are linked to one another with 3',5'-phosphodiester bridges* (Fig. 8-3). This generalization appears to be true for all biological systems from virus to man, and one wonders whether it will not also turn out to be true for all of living matter throughout the universe.

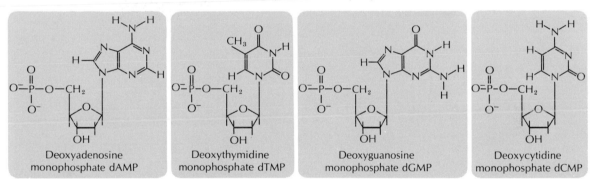

Fig. 8-2. (*Above*) The nucleoside and the nucleotide—relationship between the building blocks of the nucleic acids. (*Below*) The four nucleotides of DNA. Note that the bases are attached on the 1'-carbon atom of the ribose. The phosphate is attached to the 5'-carbon atom of the ribose. The nucleotides of RNA have the same structure except that ribose rather than deoxyribose and uracil rather than thymine are involved.

Figure 8-3 shows that the nucleic acids are acids because one phosphoric hydroxyl group is still available for dissociation. In some viruses and in procaryotic cells the negative charges on the DNA molecule are counterbalanced by divalent cations like Mg^{2+} and by basic polyamines. In other viruses and in the chromosomes of eucaryotic cells the *amino groups* of proteins act as the counterions, bringing about the formation of even larger and more complex macromolecules called *nucleoproteins*. The chromosomes of eucaryotic cells, for instance, contain histones, which, because they are especially rich in basic amino acids (such as lysine and arginine), are well suited to

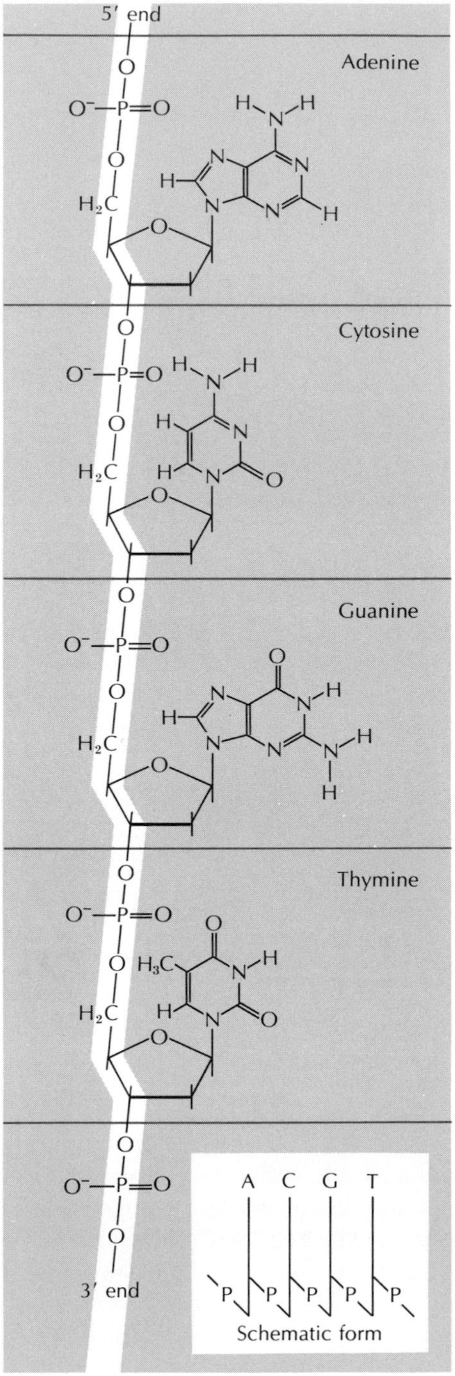

Fig. 8-3. A segment of a DNA polynucleotide, showing how the phosphates form 3′, 5′-phosphodiester bridges between the deoxyribose nucleosides. The RNA polynucleotide has the same structure except that ribose and uracil are present instead of deoxyribose and thymine.

act as counterions to the DNA of the chromosomes. It is this nucleoprotein complex that Miescher dissociated with his salt solution, thus permitting the DNA to be precipitated by acidification.

INDICATIONS OF BIOLOGICAL FUNCTION OF THE NUCLEIC ACIDS

While these painstaking investigations on the organic chemistry of the nucleic acids were in progress, several important developments led to a clarification of their biological function. These developments were based on a number of technical achievements contributed by workers in the 1930s and early 1940s such as methods for separating DNA from RNA and for purifying nucleic acids. We shall not dwell on these techniques but simply point out that improvement in purification techniques has continued up to the present and represents the basis on which most of the important theoretical developments rest. Of equal importance were some of the analytical discoveries such as (1) that DNA after hydrolysis reacts with Schiff's reagent to give a brilliant purple color (the Feulgen reaction), (2) that both DNA and RNA react with basic dyes, and (3) that they absorb ultraviolet light with great intensity in the 240–280 mμ region, the absorption maximum being at about 260 mμ (Fig. 8-4). By utilizing these discoveries a large number of ingenious methods were developed to localize the DNA and RNA in specific regions of the cell and to follow their relative abundance during various phases of cell growth and development. Thus, cytochemists such as Caspersson and Brachet observed in the 1940s numerous instances of correlation between the abundance of RNA in cells and their protein-synthesizing activity. These

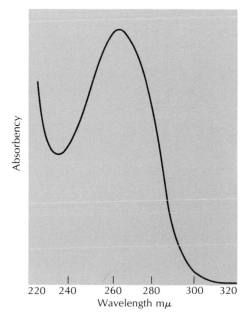

Fig. 8-4. Absorption spectrum of nucleic acids showing maximum in 260 mμ region. Proteins have a peak at 280 mμ so that when proteins contaminate a nucleic acid preparation, a small "hump" or "shoulder" will appear in this region.

classical studies have recently been supplemented by autoradiographic methods. We now know that (1) DNA is localized primarily in the chromosomes although recent work has shown that DNA is also found in mitochondria, chloroplasts, and possibly in other probably self-reproducing cell organelles, and (2) the bulk of the RNA is found in the cytoplasm although some is found in the nucleus—primarily in the nucleolus with smaller amounts on the chromosomes and even in the nuclear sap.

These generalizations, based on studies with eucaryotic cells, may in themselves not have been so wildly enlightening had not a number of microbial geneticists and microbiologists also entered the fray, and by the early 1950s several dramatic experiments with microorganisms had firmly established two important generalizations—(1) DNA (rarely RNA) was the hereditary material and (2) RNA was in some way involved in protein synthesis.

Evidence for the missing connection between these two important generalizations —that is, *that DNA was involved in RNA synthesis*—became available only a full decade later.

Since the experiments to which we refer above are treated in detail in R. P. Levine's *Genetics* (Second Edition) and W. R. Sistrom's *Microbial Life* (Second Edition) in the Holt, Rinehart and Winston Modern Biology Series as well as in articles listed in the bibliography, we shall simply refer to them here. They are:

1. the isolation and purification of many viruses and the demonstration that these self-replicating systems are composed of deoxyribonucleoprotein or ribonucleoprotein;
2. the work of Avery, MacLeod, and McCarty on "bacterial transformations" in which they showed that it was the DNA from the *R* or "rough" (nonencapsulated) strain of pneumococcus that could completely and permanently transform cells of the *S* or "smooth" (encapsulated) type into the *R* type;
3. the "Waring Blendor Experiment" by Hershey and Chase in which they utilized the radioactive isotope ^{32}P (to label the DNA) and ^{35}S (to label the protein) of the bacterial virus T2 and showed that during infection of the bacterial cell by the virus the *DNA entered the cell while most of the protein remained outside*;
4. the work on tobacco mosaic virus (TMV) carried out independently by Fraenkel-Conrat and by Schramm, showing that the RNA of the virus (the protein having been stripped away from it) can infect tobacco plants and produce complete viruses composed of both RNA and protein. To this one must add the work of Fraenkel-Conrat in which he showed that in the case of two different strains of TMV, which could be chemically dissociated into RNA and protein and then "reconstituted" so as to switch their respective proteins and nucleic acids, *it was the RNA of the "hybrid" virus rather than the protein that upon infection determined the strain of the progeny.*

These and other experiments convinced biologists by the middle 1950s that DNA is the hereditary material of the cell (with only a few exceptions such as certain viruses in which RNA acts as the hereditary material). This generalization today seems self-evident since it has, in these few intervening years, become a part of our general scientific culture. However, the student need only read the literature of the late 1940s and even the early 1950s to discover how widespread the belief had been that genes were in reality proteins. It would appear that the 60 intervening years had indeed buried Miescher's brilliant intuitions about the nature of the hereditary material!

Probably the single most important impetus for the concept that DNA was the hereditary material was provided by Watson and Crick, who in 1953 proposed a theory of the structure of DNA. Although the determination of the structure as such did not "prove" anything about the function of DNA, the structure did provide a useful conceptual framework for understanding DNA replication and protein synthesis. Ever since 1953 it has become increasingly apparent that the understanding of molecular structure is necessary to the understanding of biological function, especially at a molecular level. This is why we shall be studying in some detail the structure and molecular properties of the nucleic acids for the remainder of this chapter.

THE STRUCTURE OF DNA

We have learned so far that DNA and RNA are polynucleotides that form long, unbranched chains. This information does not tell us, however, how these molecules are organized three-dimensionally. As we shall see again and again, the precise three-dimensional configuration of biological macromolecules is of prime importance to the understanding of their biological function, and it is in this area of cell biology that some of the most exciting developments have taken place during the last decade. A major impetus was provided in 1953 by the formulation of the Watson-Crick hypothesis on the structure of DNA, an event in the history of science that has played a pivotal role in the emergence of cell biology as a molecular science. The precise history and the scientific logic behind the discovery of the structure of DNA have yet to be assembled with dispassion and precision. However, we do now have one personal account—that of J. D. Watson, *The Double Helix*—which for the moment represents our point of entry into a momentous period in the history of biology.

The account of the discovery of the structure of DNA that we shall give attempts to describe the facts and the reasoning that must have or might have preceded the publishing of the theory. The identification of the parts of the underlying process with particular persons is included for general interest but is perhaps not crucial to the argument. It is unlikely that we shall ever know "just like it was," but such uncertainty has yet to deter any historian!

In 1950 it was pointed out by Chargaff that in DNA the amount of adenine equals the amount of thymine and the amount of guanine equals the amount of cytosine. This important observation has become a universal biological law which holds true for every replicative form of DNA found in nature (Table 8-1).

Notice that within the limits of experimental error the ratios of $\frac{A}{T}$ and $\frac{G}{C}$ are equal to 1, except for the isolated coliphage ϕX 174. The reason for this exception

Table 8-1   BASE COMPOSITION OF DNA FROM VARIOUS SOURCES*

SPECIES	SOURCE	A	G	C	T	$\dfrac{A+T}{G+C}$	$\dfrac{A}{T}$	$\dfrac{G}{C}$
Calf	Thymus	29.0	21.2	21.2	28.5	1.35	1.01	1.00
Hen	Erythrocyte	28.8	20.5	21.5	29.2	1.39	0.99	0.95
Salmon	Sperm	29.7	20.8	20.4	29.1	1.42	1.02	1.02
Marine crab	All tissues	47.3	2.7	2.7	47.3	8.75	1.00	1.00
Euglena gracilis	Cell DNA	22.6	27.7	25.8	24.4	0.88	0.93	1.07
Euglena gracilis	Chloroplast	38.2	12.3	11.3	38.1	3.23	1.00	1.09
Escherichia coli W		24.7	26.0	25.7	23.6	0.93	1.04	1.01
Coliphage T2		32.5	18.2	16.8	32.5	1.88	1.00	1.12
Coliphage ϕX 174	Replicative form	26.3	22.3	22.3	26.4	1.18	1.00	1.00
Coliphage ϕX 174	Isolated virus	24.6	24.1	18.5	32.7	1.34	0.75	1.30

* From *Biological Chemistry* by Henry R. Mahler and Eugene H. Cordes. Copyright © 1966 by Henry R. Mahler and Eugene H. Cordes. Reprinted by permission of Harper & Row, Publishers.

will become apparent later. The ratio $\dfrac{A+T}{G+C}$ is not equal to 1 and has been used to characterize the DNA from a particular source.

(See Chapter 9.)

    While the work on the base composition of DNA was proceeding in Chargaff's laboratory, Franklin and Wilkins in London were using X-ray diffraction to obtain some precise measurements of DNA pulled into fibers from gels of high concentration. The technique of X-ray crystallography, which we shall describe at greater length later, is a powerful method for the analysis of macromolecular fine structure and is playing an increasingly important role in the understanding of the relationship between structure and function in cell biology. Earlier work by Astbury on DNA fibers had shown a 3.4 A "repeat," that is, a regularity in the molecule that has a period of 3.4 A. To this Franklin and Wilkins added a 34 A repeat, which was intriguing because nothing in the chemical structure of the polynucleotide appeared to correspond with this dimension—the distance from one phosphate to the next, for instance, is only about 7 A. The X-ray diagrams of Franklin and Wilkins also showed that the molecule is a very long, thin rod about 20 A in diameter.

    An important question concerned the number of chains in the molecule; there seemed to be a general inclination both in Pauling's group at the California Institute of Technology and in the London group to favor the number *three*. Given three chains, it was then important to decide whether the bases pointed toward the outside or toward each other in the center of the molecule. Pauling suggested that they stick out, but Franklin felt she had evidence that the phosphates are outside and that the bases consequently are in the center.

This was the situation in 1951 when James Watson, a 22-year-old American postdoctoral fellow, arrived in Cambridge and met Francis Crick, a physicist working for his Ph.D. in biophysics. Although ostensibly they were meant to work on other problems, they decided to collaborate on the study of DNA. Crick and Watson had two things in common: a deep interest in DNA, which they were convinced is the genetic material and therefore the most important of substances, and a great respect for Linus Pauling and his approach to the study of macromolecules. Pauling's approach combined a detailed knowledge of the stereochemistry of the relevant small molecules with the building of precise atomic models of these small molecules. This enabled Pauling and his associates to build atomic models of macromolecules that were consistent with the laws of stereochemistry. The building of models of macromolecules had turned out in Pauling's hands to be more than mere speculation, for by following certain rules of the stereochemical game, the building of a model became a *stereochemical experiment in space,* often yielding as much information as a direct chemical experiment might. To be sure, finding a stereochemically feasible structure does not constitute ultimate proof of its reality since it may be possible for other structures to satisfy the stereochemical requirement, but this limitation holds for other forms of experimentation also. The important thing is that both model-building and other forms of experimentation permit one to predict the consequences of a certain result or hypothesis.

It turned out that Watson and Crick had more to offer than simply Pauling's approach and Franklin's data. Crick had become enamored with helices and together with Cochran had worked out a theory regarding the scattering of X-rays by helical structures. According to this theory, helical structures produce a "crossways pattern" (Fig. 8-5)—which to the great excitement of Watson and Crick actually appeared in Franklin's most recent X-ray diagrams (which, we are told, though closely guarded nevertheless came into the possession of the Cambridge group). Another contribution, this time by Watson, was that of *twoness.* Watson as a biologist knew that things related to sex come in twos. Whether he had Chargaff's data regarding base-pairing in mind or whether Chargaff's important information was fitted into place when it was all over, we may never know. In any case, it seems clear that Watson preferred to build a model with two strands twisting around each other and with the bases pointing toward the center. But even though Watson now had a model that was almost complete, nothing much happened until Jerry Donohue pointed out to Watson that everyone was writing the wrong tautomeric forms of the bases. Although the evidence up to this point had been scanty, Donohue felt confident that the bases exist in the keto rather than the enol form. At this point Watson cut out models of the keto forms from cardboard and pushed them around each other on a flat surface; the solution to the structure of the gene "fell out." Watson had discovered that it was possible to form hydrogen bonds between A and T and between G and C in such a manner that the two base pairs had very similar dimensions (Fig. 8-6). Thus all at once, Watson laid bare the principle

Fig. 8-5. X-ray diffraction pattern from a crystalline fiber of the lithium salt of DNA. This "modern" picture shows far more detail than the original patterns obtained by Franklin and Wilkins. The crossway pattern of reflections going from 11 o'clock to 5 o'clock and from 1 o'clock to 7 o'clock and which provided Watson and Crick with the evidence for the helical arrangement of DNA can be seen very clearly. The elongated regions at the very top and bottom of the pattern provide the evidence for the regular stacking of purine and pyrimidine bases at 3.4 A and perpendicularly to the axis of the DNA molecule. (Courtesy of M. H. F. Wilkins, Medical Research Council Biophysics Research Unit, University of London King's College, England.)

which explains the stereochemical stabilization of the double helix as well as gene duplication. It remained for Crick to point out that the structure required that the strands run in opposite directions (Fig. 8-7) and to calculate that many features of the structure were consistent with Franklin's X-ray photographs. Thus, the model confirmed Franklin's suspicion that the 3.4 A repeat represents the stacking of bases along the helical axis, and the 34 A repeat represents the distance of a single twist of the double helix (Fig. 8-8).

The model of DNA that Watson and Crick built was soon examined with skepticism by X-ray crystallographers and structural organic chemists. Franklin and Wilkins checked the X-ray scattering properties of the model with their own X-ray photographs of DNA, and it was apparent to all concerned that the "essential features" of the structure of DNA had been discovered.

And so in 1953 the April 25 issue of the British magazine "Nature" carried three articles under the heading "Molecular Structure of Nucleic Acids." First came the Watson-Crick article that began: "We wish to suggest a structure for the salt of deoxyribose nucleic acid (D.N.A.). This structure has novel features which are of considerable biological interest." Then followed an article by Wilkins, Stokes, and Wilson, and finally one by Franklin and Gosling.

We have indulged in this long historical description not only because of the enormous intrinsic importance of the event, but also because it dramatically illustrates two major features of the process of scientific discovery. One has to do with the interdependence of the scientific community. It shows how much a theory is dependent on the work and thought of a large community, both past and present. X-ray crystallography from Bragg to Crick, organic chemistry of DNA from Miescher to Todd, the identification of the biological role of DNA from Griffin to Watson—all these thoughts and activities formed the background of the Watson-Crick theory.

In addition, however, there was also the other major feature of the process, and here we must give unique credit to Watson and Crick, who approached the problem with the conviction that the structure of DNA must be related to its biological function. In this relationship lies the key to the theory; and in the influence of this theory—that biological function is related to the structure of the cell's macromolecules—the Watson-Crick theory had its greatest impact and is rightly considered a keystone of the New Biology.

Fig. 8-6. Hydrogen bonding between adenine-thymine and guanine-cytosine bases in DNA. Note that the A-T pair has two hydrogen bonds while the G-C pair has three. The pairing between the respective purines and pyrimidines is so accurate that the bond distance does not deviate from the 2.83–2.90 average, still allowing the two sugar-phosphate backbones to have equivalent helical conformations. This property is the single most surprising fact regarding the DNA double helix.

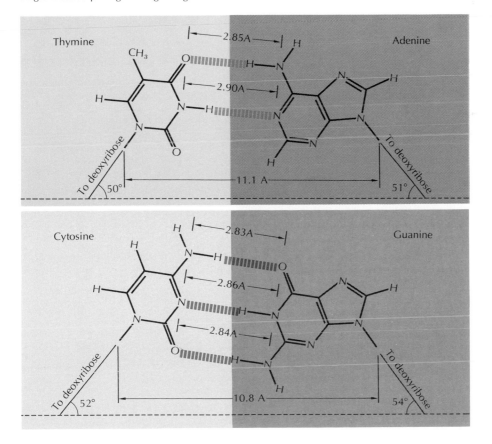

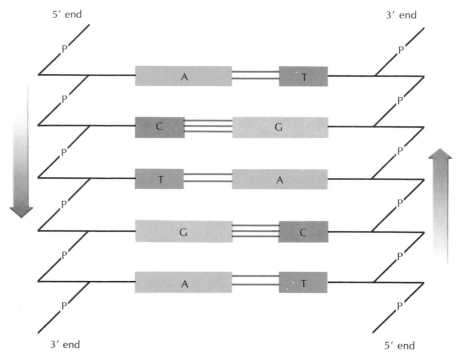

Fig. 8-7. Diagram of a flattened portion of the DNA double helix showing the reverse polarity of the sugar phosphate backbones. The diagram also shows how the similarity in the dimensions of the A═T and G≡C base pairs insures that the sequence of the bases does not place any restrictions on the structure.

Figure 8-8 shows the double helix of DNA in detail; the model is indeed 20 A in diameter with two polynucleotide chains twisted around each other, phosphate backbone on the outside, bases on the inside. The model also shows that the chains twist around each other in such a way as to create a major groove and a minor groove. Note also that the strands run in opposite directions. Figure 8-7 shows that the strands of DNA, irrespective of the order of the bases, are not the same if read from one end as they are if read from the other end. The convention has been established of writing the structural formula of a single nucleic acid chain by starting with the 5′ phosphate on the left. If this is done, then the sequence of some of the atoms along the chain is

$$P - 5'C - 3'C - P - 5'C - 3'C - P$$

or as is now generally written

This simple structural formula illustrates clearly the directional character of the nucleic acid strand.

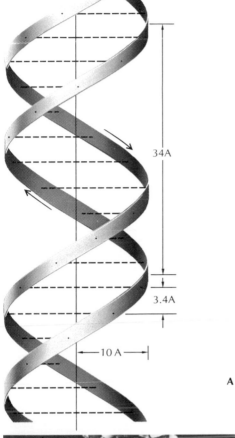

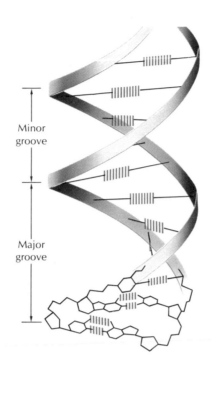

34A

3.4A

10 A

Minor groove

Major groove

A

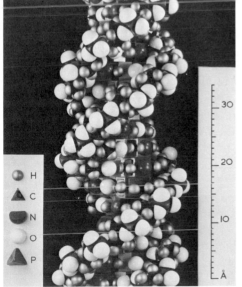

H
C
N
O
P

30

20

10

Å

B

Fig. 8-8. A. The double helix. Open models of DNA double helix showing hydrogen bonding of A=T and G≡C bases at the center of the molecule. B. Scale atomic model of DNA molecule. Note (1) how interior of molecule is filled by the bases and (2) the presence of a narrow (minor) and a wide (major) helical groove. (Courtesy of M. H. F. Wilkins, Medical Research Council Biophysics Research Unit, University of London King's College, England.)

One of the most important aspects of the model is that it allows tight stacking of the bases, which, when hydrogen-bonded in the proper way (Fig. 8-7), have exactly the right dimensions to permit the strands to form a regular helix. Thus, the space-filling model (Fig. 8-8B) demonstrates clearly the internal compactness of the double helix, the precision with which the proper base pairs fit inside the double helix. An important consequence of this compactness is that the structure places *no restrictions on the sequence* in which the base pairs follow each other. Thus, it is possible to construct different DNA molecules that are identical in their gross architecture and are distinguished only by the specific sequence of their base pairs.

Biological Significance of the Structure of DNA

It has been suggested long ago that the genetic or hereditary material of the cell must have two separate functions. It must be capable of self-duplication and also of initiating actions that ultimately find expression in a given cell structure or function. As a result of the work in biochemical genetics started by Ephrussi and by Beadle and Tatum (see *Genetics* in the Holt, Rinehart and Winston Modern Biology Series), it is now clear that the expression of gene action is the formation of a protein, be it an enzyme or a structural protein. DNA must therefore be capable both of duplicating itself and of providing the necessary information for protein synthesis. The structure of DNA provides a convenient device whereby a particular molecule with a particular sequence of base pairs could be duplicated. Thus, each strand of the molecule could determine the laying down of a "complementary strand," resulting in the formation of two identical molecules as follows. The significant property of the hydrogen bond formation between the complementary bases A$=$T and G$\equiv$C is that these interactions are highly specific. A sequence of ATGC in one strand will direct the enzymatic "laying down" of a sequence TACG in the complementary strand; thus nature has solved one of the major problems of biological self-duplication. Before this important insight it was difficult to imagine how a large molecule would go about being a biological template for the synthesis of another molecule just like itself. The answer appears to be that it *just doesn't* but that, instead, half of the molecule acts as a template for the synthesis of its complementary half which in turn acts as a template for a molecule like the original half and so on. This is, after all, how the sculptor reproduces three-dimensional structure—from original to mold to cast. An important difference is that in DNA the cast and the mold coexist in the same molecule. A precise geometrical model whereby the duplication of a DNA molecule could occur in a continuous manner without requiring the prior separation of strands was suggested by Levinthal and Crane (Fig. 8-9). They suggested that duplication could begin at one end, thus opening up the strand and providing the energy for the rotation of the two lengthening daughter strands as well as the shortening parent strand. They concluded that enough energy was available to overcome the viscous drag opposing these rotations and that there was enough mechanical strength in the helix to withstand the necessary torque without seriously stretching the bonds.

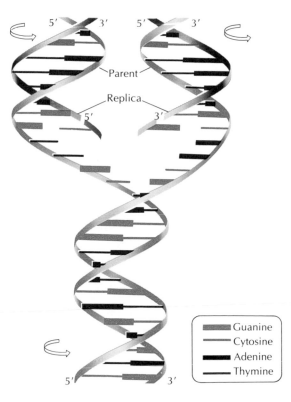

Fig. 8-9. Model of DNA in process of replication. As the new strands are laid down causing the parent strand to unwind, the daughter strands must also rotate. The model assumes that there is simultaneous enzymatic activity adding to both the 3'- and 5'-ends of the growing daughter strands. So far no enzyme which can attach a nucleotide to the 5'-end has been discovered.

If this picture of DNA duplication involving the laying down of complementary strands and thus bringing about strand separation is correct, then it should be possible to obtain direct evidence for the existence of Y-shaped growing points in duplicating DNA. Such evidence has, in fact, recently been obtained (Fig. 8-10). By using the technique of autoradiography, Cairns was able to show that the circular molecule of DNA in the bacterium *E. coli* has two Y-shaped regions, one representing the "growing point" and the other the "initiation point" of a cycle of DNA replication.

The formation of hydrogen bonds is a spontaneous process not requiring catalysis by an enzyme. The selection of the proper complementary bases, therefore, does not require catalysis, but hooking together the nucleotides with phosphodiester bonds is a covalent reaction and does require enzymatic catalysis.

(See Chapter 14.)

The second biological function of DNA, as we have already said, is the transmitting of information eventually to be used for the synthesis of proteins. In later chapters we shall discuss this phenomenon in detail. Suffice it to point out here that the linear array of the four bases can be thought of as "information" based on a language of four letters, which finally becomes transcribed into a protein language of 20 letters. Thus, according to Crick, if one were to imagine the pairs of bases corresponding to the dots and dashes of the Morse code, there is enough DNA in the human cell to encode 1000

(See Chapters 14 and 15.)

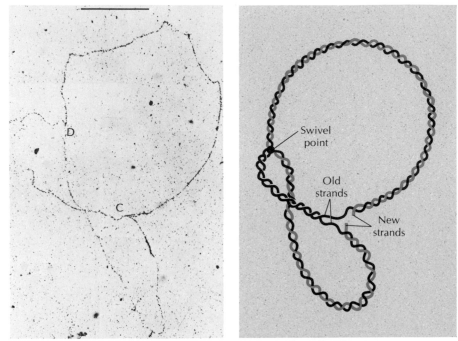

Fig. 8-10. Autoradiograph showing circular chromosome of *E. coli* cell replicating. "C" is the Y-shaped growing point, and "D" is the initiation or swivel point of the replicating circular chromosome. The DNA is rendered visible in the electron microscope by growing the cells in tritium (H^3)-labeled thymidine. When the radioactive DNA is placed against a photographic emulsion it produces the blackened grains, which are then seen in the electron microscope. (Courtesy of Cairns, 1963.)

large textbooks. We have suggested in Chapter 2 that a signal characteristic of the living machine is its capability for "microminiaturization." Here indeed is a striking example of this cellular property: the ability of all the hereditary information responsible for the development of a complex individual to be stored in 10^{-15} grams of material. The explanation for this remarkable phenomenon is that the "bits of information" of the language of the nucleic acids are the nucleotide residues, tiny pieces of a molecule, which are many times smaller than anything man has invented for storing information. It staggers the imagination to realize that the fertilized eggs that developed into all the people now living, and that possessed all the hereditary information available to mankind, contained no more DNA than would fit on the head of a pin.

Recent Evidence Supporting the Double Helix

In the years following 1953 a great deal of evidence has been accumulated that helped establish the Watson-Crick hypothesis on the structure of DNA as one of the most useful and secure generalizations in cell biology. Some of the more important lines of evidence are as follows.

1. The technique of X-ray diffraction is limited by the fact that it is difficult to derive directly by a deductive process a three-dimensional structure from the data one obtains by using this method. In the case of large molecules it is in fact impossible to do so. Moreover, the DNA fiber used in these experiments is oriented along one axis but does not form a three-dimensional crystal. However, once the experimenter has a hypothesis as to the three-dimensional structure of the molecule he is investigating, he is able to compute the diffraction pattern such a structure would produce and determine whether these predictions really appear in the experimental results. This is, in fact, what happened with DNA. Once there was a hypothesis as to its structure, subsequent workers were able to predict certain X-ray scattering properties based on spacings such as 20 A (width), 3.4 A (distance between bases), and 34 A (distance for a complete turn of the double helix). These spacings were indeed confirmed in the X-ray diagrams and so were a number of others, all of which corroborated the double-helix hypothesis. In fact recent X-ray studies by Wilkins and his co-workers were even able to refine the original DNA model by showing that the planes of the bases were not exactly perpendicular to the helix axis.

2. Direct visualization and precise measurements of DNA with the electron microscope confirmed the dimensions predicted by the double-helix model. Thus, if one knows the molecular weight of a certain species of DNA and assumes the width predicted from the model, one can predict the length of the molecule. In a number of studies the measured and predicted lengths agreed extremely well.

3. Numerous physical-chemical studies of the properties of DNA in solution such as light scattering, viscosity, sedimentation, and diffusion have yielded results that are consistent with the predictions of the model, that DNA is a somewhat stiff, unbranched rod and not a highly flexible polymer.

4. However, the properties of a flexible polymer are assumed by DNA when "denatured" by a variety of agents such as low pH or heat. Thus, Doty and his co-workers found that when solutions of DNA are heated, there is a marked decrease in viscosity occurring over a narrow pH or temperature range (Fig. 8-11). These phenomena, reminiscent of the melting point of a crystal, are due to the fact that at a given temperature the hydrogen bonds and hydrophobic interactions holding the complementary strands together are broken, thus allowing each strand to separate and form a flexible—and hence, on the average, a much more compact—molecule with a much lower viscosity. The decrease in viscosity is correlated with other properties such as an increase in ultraviolet absorption and a decrease in optical rotation. The change in ultraviolet absorption (called the hyperchromic effect) is the most convenient property to measure. The abruptness of the change in these properties of DNA is explained by the "cooperative" nature of the effect. By this we mean that as some bonds are broken, neighboring bases are affected as well, leading to a progressive melting in large regions of the molecule. Similar transitions occur when solutions of DNA are exposed to pH extremes, certain nonaqueous solvents, or a number of ionic species, as well as upon heating.

Modern studies of the melting phenomenon have shown that the double-helical structure of DNA is stabilized not only by the very specific hydrogen bonds between

Fig. 8-11A. Change in viscosity of DNA over narrow pH range explained by "cooperative" nature of helix-coil transition. Stiff double helix collapses over narrow pH range because each separation of base pairs facilitates the breaking of the remaining base pairs. (From Doty and Thomas, 1957.)

Fig. 8-11B. The increase in ultraviolet absorption of DNA over narrow temperature range (hyperchromic effect). The DNA of different species have different melting points (see Fig. 8-12 for explanation). (From Marmur and Doty, 1962.)

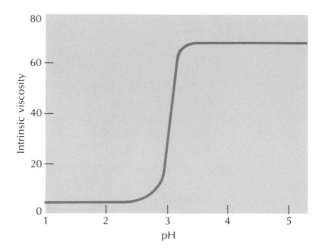

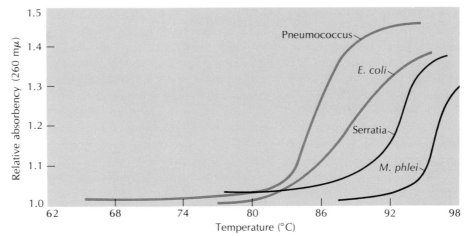

the complementary bases, but also by so-called "stacking interactions" between adjacent pairs of bases up and down the molecule, giving rise to the highly cooperative effects between base pairs.

Doty and his co-workers have noted that the melting point of DNAs from different sources varies according to their G + C content (Fig. 8-12). This can be explained both by the presence of three hydrogen bonds between G and C (only two between A and T) and by the stronger stacking interactions between base pairs involving G and C.

By very slow cooling of "melted" solutions of DNA obtained from viruses and bacteria, it has been possible to "anneal" it, that is, to re-establish the double-helical structure (Fig. 8-13). Skeptics have often questioned whether such melting and annealing experiments actually do re-establish the original molecule or whether base-pairing occurs on a more random basis. Furthermore, it has been questioned whether upon

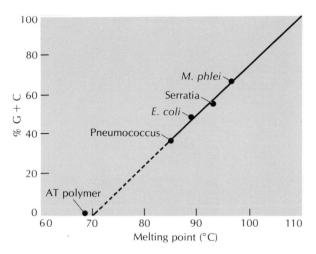

Fig. 8-12. Relationship between melting point and G + C content of DNA preparations obtained from a number of bacteria. The curve can be extrapolated to zero G + C content, which predicts a melting point corresponding quite closely with that of a synthetic polynucleotide consisting only of A and T nucleotides (AT polymer). (From Marmur and Doty, 1962.)

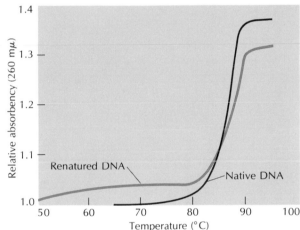

Fig. 8-13. Melting curves of native and renatured DNA. The latter is obtained by slowly cooling (annealing) denatured DNA. The fact that the two curves are not identical shows that some of the DNA does not return to the native state. The denatured DNA can be removed by treatment with an enzyme (phosphodiesterase), which does not hydrolyze native DNA, after which one obtains a renatured DNA preparation that has properties just like the native material. [From Marmur, J., Schildkraut, C. L., and Doty, P.: Biological and Physical Chemical Aspects of Reversible Denaturation of Deoxyribonucleic Acids. In *The Molecular Basis of Neoplasia* (A Collection of Papers Presented at the Fifteenth Annual Symposium on Fundamental Cancer Research, 1961.) Austin, Texas, University of Texas Press, 1962.]

melting the strands actually do separate or whether they remain stuck in one or more places, thus helping the annealing process. It now appears that it is possible to pair two strands of DNA that had been completely separated to form very precise, perfectly aligned helices. To demonstrate this, Ris and Westmoreland took the melted DNAs of two slightly different lambda phages and annealed a mixture of them. Genetic analysis had previously shown that one of the phages had a deletion (that is, a piece of hereditary material missing), and in addition another deletion into which a shorter piece of hereditary material had been substituted. The substituted piece was entirely different from the corresponding segment of DNA of the other phage. From these facts one could predict that should specific pairing occur between two strands of the different phages, then the effect of the simple deletion would be a loop, and the effect of the "nonhomologous" piece would be an open, unpaired region with one strand shorter than the

Fig. 8-14. DNA molecule produced by melting and annealing two lambda phage strains. The DNA of one of the strains has a piece missing (a deletion) in one portion of the molecule, as well as another deletion into which a shorter length of extraneous (nonhomologous) DNA has been substituted. After the two strands have formed a duplex, one can detect the region of the simple deletion by a loop and the nonhomologous piece by an open, unpaired region with one strand shorter than the other. (Courtesy of Ris and Westmoreland.)

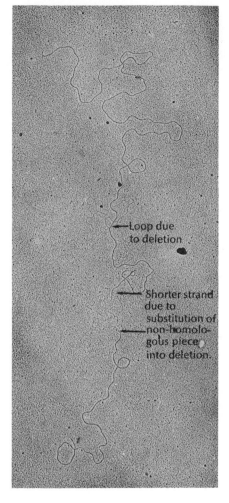

Loop due to deletion

Shorter strand due to substitution of non-homologous piece into deletion.

other. Figure 8-14 shows that this is indeed what occurs. This experiment constitutes a magnificent demonstration of the specificity of interaction between two strands of DNA even in an *in vitro* separation and annealing experiment.

It has been possible to show that when the DNA of two closely related species is mixed and melted, the double-helical content upon cooling is high; whereas, when the DNAs of unrelated species are used, then the double-helical content is low. The explanation of these observations, of course, is that the greater the similarity, the more readily do complementary strands form "hybrid" double-stranded molecules. Such DNA "hybridization" experiments have become a useful experimental tool in studying taxonomic relationships between different species of organisms (Fig. 8-15).

5. The spectacular work by Kornberg and his group in which they purified an enzyme (DNA polymerase) and accomplished the enzymatic synthesis of DNA in the

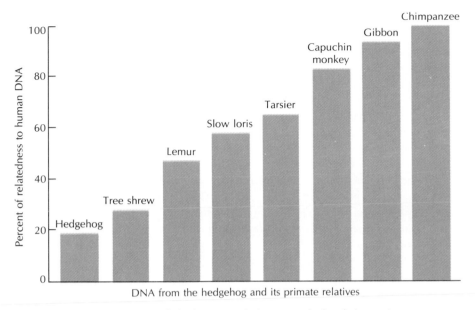

Fig. 8-15. The use of the DNA hybridization technique to study the phylogenetic relationships between organisms in the Primate order. The technique employed in these experiments was to use melted human DNA in agar and incubate it with fragments of melted ^{14}C-labeled human DNA and fragments of DNA from another species. The fragments of labeled human DNA hybridize with the DNA trapped in the agar, an effect which is reduced if related DNA from a different species is present. Interestingly, this method cannot distinguish between human and chimpanzee DNA since the latter DNA inhibits hybridization between the DNA-agar and the human-labeled DNA fragments as extensively as does unlabeled human DNA fragments. However, the gibbon, which on morphological grounds is more distantly related to man than the chimpanzee, appears to be only 94 percent related on the above scale. The degree of relatedness of the other species in this diagram also parallels the relatedness established on morphological grounds. The hedgehog, which showed the least relatedness to man, is not a primate but is thought to be derived from a more generalized insectivore ancestral to the primates. (Redrawn from Hoyer and Roberts.)

test tube also yielded a number of results which proved to be consistent with the Watson-Crick hypothesis. Thus, when the DNA polymerase and the necessary building blocks were present, no DNA synthesis normally occurred. For DNA synthesis to proceed, it was necessary to add some DNA as a template or "primer." This observation proved to be an elegant *in vitro* demonstration of the self-duplicative properties of DNA.

(See Chapter 14.)

Kornberg also developed an ingenious analytical tool for the characterization of the newly synthesized DNA. This method, called "nearest neighbor analysis," can determine the frequency with which any two bases are found next to each other in a given sample of DNA. As we shall see later this method demonstrated very clearly that the two polynucleotide strands run in opposite directions.

(See Chapter 14.)

6. Not only is the melting point of DNA related to its G + C content, but so also is its density, a fact which again points to the tightness of the G≡C bonding and of the stacking interactions of adjoining G≡C base pairs. These facts are consistent with the double-helix model.

Meselson and Stahl developed an ingenious and very important technique first devised by Vinograd whereby the density of DNA molecules can be measured and molecules of different density can be physically separated. They utilized this method to demonstrate that the *in vivo* duplication of DNA is consistent with the double-helical model. This important experiment will be described later. We shall therefore limit our-selves here to a discussion of the "density-gradient method" because it has become one of the major tools of research in molecular biology.

(See Chapter 14.)

When a concentrated salt solution, cesium chloride (CsCl) for example, is centri-fuged at high speed in an ultracentrifuge, its molecules will redistribute themselves slightly and, like the molecules in our atmosphere, will come to an equilibrium in which the concentration will vary in proportion with the strength of the gravitational field acting on the molecules. Thus, the CsCl concentration will be lowest at the meniscus and highest at the bottom of the tube. If DNA molecules are also present in the centri-fuge tube, they will come to equilibrium by "banding" in certain regions of the tube where the density of the DNA molecules will be identical to the density of the cesium chloride (Fig. 8-16). Thus, the formation of a band of DNA in a CsCl gradient is the result of two opposing forces: the high-gravitational field in the ultracentrifuge tube causing the molecules to sediment and the buoyancy of the DNA molecule in the concentrated cesium chloride solution exerting a force in the opposite direction. The width of the band thus formed is determined by thermal diffusion of the DNA mole-cules. The result is a gaussian distribution of concentration of molecules about the point where their density is equal to that of the salt. When a mixture of different DNA molecules is present in the tube, they will band in different positions. The exact

Fig. 8-16. A. The "banding" of three types of DNA in a density gradient of cesium chloride established by centrifuging the material at 45,000 rpm for 20 hours. The high centrifugal field causes a redistribution of the CsCl, which brings about the banding of the DNA molecules at places where their density is equivalent to that of the salt solution. B. Absorbence measurements obtained from fractions removed sequentially from the bottom of the centrifuge tube. Another way of obtaining such curves is to photograph the bands and perform densitometer tracings of the photographs.

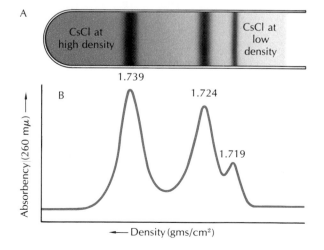

position of the band can be determined in a variety of ways. In an analytical ultracentrifuge, it is possible for an optical system to "look" at the rotating centrifuge tube and determine where the bands occur. More frequently, however, a preparative ultracentrifuge is used. After equilibrium is reached, the tube is removed from the centrifuge, and a small hole is punched into the bottom of the tube. As the tube empties from the bottom, fractions are collected. By that we mean that small volumes are collected until the contents of the centrifuge tube are transferred sequentially into a series of test tubes. The absorbency at 260 mμ, for example, of each fraction can be measured in a spectrophotometer, and the corresponding density of the salt in each fraction can be assayed from the refractive index. The results are usually presented graphically (Fig. 8-16). Furthermore, it is possible to calculate the percent G + C from the density, and it can be shown that this value agrees very well with the value obtained from direct chemical analysis. Density-gradient centrifugation has proved to be of tremendous power and versatility. Thus, Fig. 13-14 shows this resolving power of the CsCl method; a separation of nuclear and mitochondrial DNAs can easily be obtained, illustrating in this instance that the mitochondria of a cell have DNA of a different base composition from that of the nucleus of the same cell.

The density-gradient centrifugation method has also been used to separate helical (native) and nonhelical (denatured) DNA. Native DNA is lighter because the double-stranded structure excludes more water than does denatured DNA. Figure 8-17 shows how the method was used in a study of the denaturation and renaturation of DNA. In this particular study, the enzyme *E. coli* phosphodiesterase was used to remove the portion of the DNA that remained denatured, demonstrating that it is possible to anneal denatured DNA so that its density distribution returns almost exactly to that of the original native material. This study also points to an interesting phenomenon, that of the much greater resistance to enzyme degradation of native DNA. This property of DNA may prove to be one of the important biological functions of its double-helical structure. After all, DNA is the hereditary material, and it is eminently reasonable that it should have evolved a structure that is relatively resistant to enzymatic attack. As we shall see below, DNA need not be double-helical in order to function as hereditary material.

7. Exceptions can often be thought of as helping to prove the rule. A number of small bacterial viruses in which the rule A/T = G/C does not hold (see Table 8-1) have recently been discovered. Physical studies such as melting point and hyperchromicity studies demonstrated that the DNA packaged in these viruses is a single-stranded form. The interesting question now posed itself as to whether single-stranded DNA can enzymatically replicate itself. One might imagine that the strand that is packaged in the virus (the "+" strand) might, after infecting a bacterium, synthesize the complementary "−" strand, which then in turn might again give rise to the "+" strand. In fact, this exception to the general mechanism of DNA duplication does not occur. What does happen is that when the "+" strand infects a bacterium, it acts as a template for the synthesis of a "−" strand and thus forms the conventional double-helix structure; the so called *replicative* form. This double-stranded form can replicate itself to produce

(See Chapter 14.)

Fig. 8-17. Use of the density-gradient centrifugation method to show that a substantial portion of heat-denatured DNA can be renatured to give DNA of the same density as native DNA. The enzyme used to hydrolyze denatured DNA is phosphodiesterase obtained from the bacterium *E. coli*. This experiment also provided excellent evidence that heat denaturation of DNA is indeed reversible. (From Schildkraut, Marmur, and Doty, 1962.)

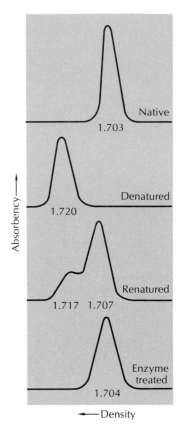

more double-helical structures. Eventually, though, and this *is* an exception to the general rule, the replicative form will begin to produce only "+" strands, which, after becoming packaged into a protein, emerge from the bacterium as new virus particles (Fig. 8-18).

The requirement for a double-helical structure in the replication of even single-stranded DNA is a dramatic confirmation of the generality of the double-stranded system. The mechanism whereby the DNA in this particular case is able to replicate selectively only a "+" strand is as yet not understood.

From these data we can conclude that the double-helical model of DNA is one of the most thoroughly validated hypotheses in biology and as we shall see from the remaining portion of this book, conceptually one of the most useful ones.

Some Overall Properties of DNA Molecules

As you will see in the next chapter, it is possible to purify protein molecules until they form a homogeneous preparation. By this we mean that it is possible to make a cell

A

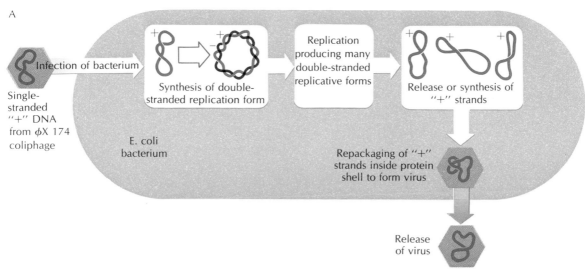

Single-
stranded
"+" DNA
from φX 174
coliphage

Infection of bacterium

E. coli
bacterium

Synthesis of double-
stranded replication form

Replication
producing many
double-stranded
replicative forms

Release or synthesis of
"+" strands

Repackaging of "+"
strands inside protein
shell to form virus

Release
of virus

·B

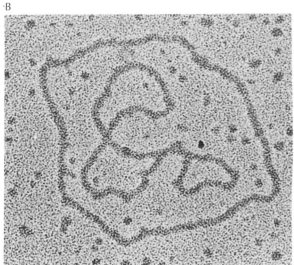

Fig. 8-18. A. Replication cycle of φX 174 coliphage which packages its DNA in a single-stranded form. However, the DNA replication still occurs in the double-stranded form. The release of "+" single strands from the duplexes is not as yet understood. B. Electron micrograph of partially synthetic φX 174 DNA (× 310,000). In a recent dramatic experiment, Kornberg and his group were able to take single-stranded DNA from the phage φX 174 and using the enzyme DNA polymerase, to synthesize a complementary strand, thus producing the double-helical structure (shown above) normally observed for the replicative form of this virus. This double-stranded form proved to be infective, thus providing the bacterium with all the information necessary for the synthesis of new virus particles. One duplex happens to lie inside the other. (Courtesy of Kornberg.)

extract and by a series of "fractionation" steps eliminate all the protein molecules except the one that we are interested in selecting. For many years this was not possible with DNA and RNA. Thus, although it was possible to remove other substances such as proteins, lipids, and RNA from DNA preparations, these solutions represented a complex, heterogeneous mixture of different DNA molecules. During the last decade, however, work with microbial systems such as viruses and bacteria finally led to the preparation of homogeneous molecules of DNA.

This work began when the attempt was made to determine the molecular weight of the DNA in the bacterial virus T2. Different workers obtained different values ranging from hundreds of thousands to several millions. It was soon discovered that the more carefully one handled the DNA solutions, the higher the molecular weight values would turn out to be. This led finally to the remarkable discovery that very large DNA molecules would break physically into two approximately equal fragments when subjected to the mild "shear" of flowing liquid. It was found that very large DNA molecules would break in half if their solutions were even pipetted.

One T2 bacteriophage contains sufficient DNA to form a molecule with a molecular weight of 1.2×10^8. Was it possible that all of the DNA in T2 was present in a single giant molecule? It soon turned out that conventional physical-chemical methods of measuring molecular weight were not suited for the handling of such giant molecules, but at that point the electron microscopists came to our aid! Cairns worked out an autoradiographic technique consisting of growing *E. coli* infected with T2 phage in the presence of tritium-labeled thymidine. Thus, T2 DNA was synthesized containing tritium label in the thymine residues. He then purified the phage particles and ruptured them delicately by placing them in a drop of distilled water on a copper grid. The grids were then placed against a photographic emulsion that was sensitive to the beta emissions of tritium and were left in position for weeks or even months. The beta emissions of tritium are "soft," that is, they penetrate only short distances into the photographic emulsion, causing the blackening of grains over only short distances and thus providing for reasonably high-resolution (1000 A) reproduction of the radioactive source. Using this method, Cairns was able to show that all the DNA of T2 was indeed present in a single giant molecule of 1.2×10^8 molecular weight.

Kleinschmidt even succeeded in rupturing T2 phages on electron microscope grids and obtaining pictures of the DNA itself. The pictures he got had such dramatic impact on the biological community that one particular example has been reproduced in dozens of textbooks and has become a symbol for modern molecular biology. We cannot resist reproducing it also (Fig. 8-19). Looking at the photograph, one wonders about the nature of the molecular process that packages this gigantic molecule into the head of the phage particle.

The bacterium *E. coli* has, as we have seen in Chapter 1, one or more nuclei depending on the growth conditions. The amount of DNA in each nucleus would give us a DNA molecule with a weight of about 2×10^9 daltons. Can it be that even this huge package of DNA exists as a single chromosome? Utilizing his autoradiographic technique, Cairns showed that in *E. coli* one really finds a single DNA molecule of 2×10^9 daltons in weight, its extended length being almost 1 mm! Furthermore, the DNA molecule frequently turned out to be circular. Other investigations of the DNA of bacteria and viruses, principally by genetic means, showed that a number of other systems also had circular DNA (Fig. 8-20). The DNA molecule of the tiny pleuropneumonialike organism (PPLO) *Mycoplasma hominis H39* has also been rendered visible in the electron microscope by a modification of the Kleinschmidt technique. This cell was found to contain a single circular chromosome 262μ in length and 5.1×10^8 daltons in weight. The question arises whether circularity of the DNA molecule at least at a certain stage of its replicative cycle is not a widespread phenomenon in procaryotic systems.

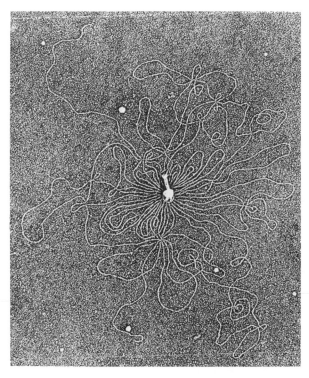

Fig. 8-19. A single molecule of DNA released from one T2 bacteriophage particle. This molecule has a weight of 1.2×10^8 daltons, is 52μ $(5.2 \times 10^5 A)$ long and consists of 10^5 base pairs ($\times$ 75,000). (From Kleinschmidt *et al.*)

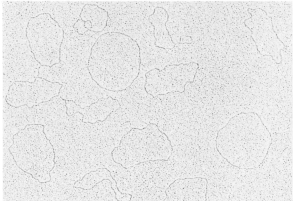

Fig. 8-20. Autoradiograph of circular single-stranded DNA molecule (chromosome) of ϕX 174 coliphage. (Courtesy of A. K. Kleinschmidt.)

In eucaryotic cells the DNA found in the mitochondria appears to be circular (Fig. 13-15), a fact interpreted by some workers to mean that the evolutionary origin of mitochondria was an "infection" by a procaryote of another cell, leading to a symbiotic relationship that eventually resulted in the development of the eucaryotic cell.

The organization of DNA in the eucaryotic chromosome has for a long time been a matter of conjecture. Recently, a number of workers have utilized the Kleinschmidt technique developed for viruses and bacteria and have succeeded in spreading mitotic

Fig. 8-21. Experiments on the organization of DNA in the eucaryotic chromosome. Upon spreading of the nuclei from erythrocytes of the salamander *Triturus viridescens* on an air-water interface, one obtains long branched fibers of nucleohistone which can be seen clearly in the electron microscope. (Courtesy of Ris.) A. Nucleohistone fibers 250 A thick. Single arrows indicate branching, and double arrows show contact between tips of branches and fibers causing netlike appearance of these structures. B. Nucleohistone fibers briefly treated with 5 mM sodium citrate. Notice that fibers are composed of two 100 A fibrils, the branching appearing to be due to the formation of loops of one of the two strands. C. Prolonged treatment with 5mM sodium citrate brings about the disassociation of the two strands giving a 100 A fibril composed of thin regions (arrows) and knobs. D. Treatment of the 100 A fibril with the proteolytic enzyme mixture pronase causes them to unravel and, produce 25 A threads which are DNAse sensitive and therefore are likely to be extremely long double-stranded DNA molecules. In many places (arrows) one can see short segments folded into small loops which might correspond with the knobs seen on the 100 A fibrils. The bar on each micrograph represents 0.1μ.

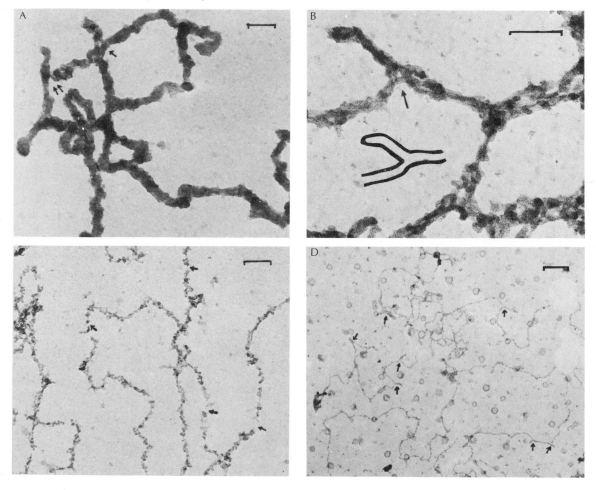

chromosomes or interphase nuclei on an air-water interface. Ris and his associates have been able to demonstrate the presence of branched nucleohistone fibers 250 A in diameter, which appear to be composed of two 100 A fibers. The branches appear to be caused by loops in one of the fibers (Fig. 8-21 A, B, and C). Treatment with the proteolytic enzyme mixture *pronase* removes the histones and leaves behind some DNase-sensitive threads 25 A in diameter (Fig. 8-21 D). These threads are very likely to be gigantically long DNA duplexes. It would therefore appear that the histones are involved in the folding and packaging of DNA in the eucaryotic chromosome in such a way as to permit orderly transcription of DNA to RNA, replication of DNA, and separation of chromosomes. The details whereby the interaction of histones and DNA brings about the highly ordered and specific structure of chromosomes and renders them capable of coiling and uncoiling during the mitotic cycle is a problem that will require intensive study in the near future.

The discovery that the DNA of many microorganisms is present as a single giant molecule enables us to prepare homogeneous samples of DNA. The problem of avoiding fragmentation of these huge molecules is still a serious one, but by selecting relatively small molecules like the DNA of the TF phage, which has a molecular weight of only 3.2×10^7, it is now possible to obtain relatively homogeneous DNA preparations.

If DNA is an "information tape" on which the hereditary message is encoded in the form of the sequence of bases, it would be of interest to be able to determine this sequence for a particular DNA molecule. Unfortunately, even a "small" molecule like that of the TF phage DNA is at present much too large to be sequenced by our present methods. In the next decade, no doubt, methods will be developed whereby we will be able to read the "message" directly from the DNA molecule.

We have studied the molecular properties of DNA in great detail. In a later chapter you will have the opportunity to use these new insights by applying them to the functioning of the hereditary material in the replication and transcription of information.

(See Chapter 14.)

THE DIFFERENT KINDS OF RNA

DNA can be considered a single kind of substance, in almost all cases having the function of storing information and of providing information for transcription to RNA. Not so with RNA, which can be divided into several categories in terms of both functions and physical properties. We shall discuss the biological function of RNA in some detail in later chapters and will therefore limit ourselves mainly to structural considerations with only brief references to function. The student, however, should realize that although we now assign different functions to these categories of RNA, we first obtained evidence for their existence from differences in their properties.

(See Chapters 14 and 15.)

The known categories of RNA are as follows:

1. viral RNA—normally wrapped in a protein coat;
2. ribosomal RNA (rRNA)—sedimented at high speed in the ultracentrifuge;
3. transfer RNA (tRNA)—remaining in the supernatant after ribosomes are sedimented and formerly called soluble RNA (sRNA);

4. messenger RNA (mRNA)—incorporates labeled uracil before other RNA.

RNA is also found in association with chromosomes, in the nucleolus, in the nuclear sap, and in chloroplasts and mitochondria. Whether these RNA forms all belong to one or another of the above categories or whether some additional categories of RNA exist is at present uncertain.

There are a number of structural generalizations regarding all forms of RNA that we shall discuss before treating each category separately.

The presence in RNA of ribose rather than deoxyribose and of uracil rather than thymine we have already discussed. We have also said that all forms of RNA, just like DNA, are unbranched polynucleotides with the nucleotide residues being linked by 3', 5'-phosphodiester bands. Because the 2'-OH is free, RNA is more susceptible to mild alkali hydrolysis than DNA, and this has been used for the removal of traces of RNA from DNA preparations as well as for the estimation of the amounts of DNA and RNA in cells.

The base composition of the different forms of RNA does not in general conform to the $A/U = G/C = 1$ rule observed by Chargaff for DNA. Some general regularities in the percent of $G + C$ among the various categories of RNA have been observed, but their significance, if any, is not yet understood. Thus, one could predict that if there is base-pairing between complementary bases, it can only involve a fraction of the bases. This prediction is borne out by hyperchromic melting-point studies of various forms of RNA (Fig. 8-22), which show lower hyperchromicity and less abrupt transitions than DNA. Since most RNA has been demonstrated to be single stranded (the RNA of reoviruses is an exception), the above hyperchromicity as well as X-ray diffraction and optical-rotation studies suggest that RNA can form an antiparallel intramolecular "hairpin loop" structure (Fig. 8-23). Because base-pairing is intramolecular, the melting of RNA is very readily reversed. In the case of DNA the annealing process must be slow so that complementary strands have time to "find" each other and come into proper juxtaposition.

Let us now examine each kind of RNA in greater detail.

Viral RNA

Single-stranded RNA is the hereditary material of most plant viruses and of a number of bacterial and animal viruses. In these systems RNA replaces the function of DNA (namely that of storage of information and of initiation of transcription), and it is entirely possible that it represents evolutionarily an earlier stage of information transfer. It seems plausible that the earliest system of protein synthesis involved a self-duplicatory form of RNA and that the permanently double-stranded DNA, with its added advantage of stability and later separation in a nucleus from the rest of the cell, evolved at a later stage.

(See Chapter 10.)

As we shall see later in greater detail, the RNA of viruses is usually encased in a shell composed of protein subunits. In some cases, such as tobacco mosaic virus (TMV), one can show that the RNA interacts with the protein particles of the virus but does not interact intramolecularly with itself.

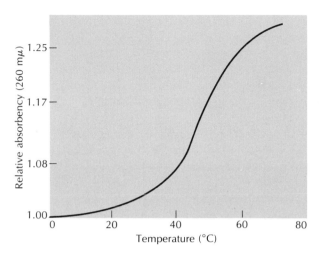

Fig. 8-22. Melting curve of tobacco mosaic virus (TMV) RNA. Notice that the transition is less abrupt, and the change in relative absorbance less extensive than in the case of DNA as shown in Fig. 8-13. This is because there is less base-pairing in RNA than in DNA. (From Doty *et al*, 1959.)

Fig. 8-23. Model of base-pairing that might occur in a single strand of RNA. The model consists of unpaired (nonhelical) regions and hairpin loops forming incompletely-paired double helices. The unpaired loops are due to the absence of complementary bases on the opposite strand. The precise structure of virus and ribosomal RNA is probably in great part due to interactions with proteins.

Viral RNA was the first form of RNA that was isolated in homogeneous preparations. This is because the virus infection and duplication provide a convenient biological system for the manufacture of large amounts of a particular form of RNA, while at the same time, by packaging the RNA in a distinctive protein coat, it is possible to

(See Chapter 10.)

separate the virus from other components of the cell. Thus, it has been possible for some time to prepare TMV free of other tobacco leaf components, isolate the RNA, and demonstrate that each virus particle of particle weight 4×10^7 daltons contains a single strand of RNA of molecular weight 2.2×10^6. TMV RNA has therefore become one of the most intensely studied and best understood molecules of RNA, having been used not only for physical studies, but also for the study of (1) virus assembly, (2) protein synthesis, (3) mechanism of mutagenesis, and (4) infectivity. This is an excellent example of the principle we have stated in the opening chapter, namely that the success of modern molecular biology lies in the utilization of nature's high degree of specificity for the study of the various manifestations of biological specificity that are central to the functioning of the machinery of the cell.

Studies of other RNA viruses have shown that the molecular weight of the RNA enclosed in the virus can vary greatly. Generally, the larger the virus, the larger the RNA molecule it contains, although the precise geometry and architecture do play a role. Thus, as is shown in Table 8-2, TMV is unusually large for its RNA content because of its cylindrical structure, and human influenza Type A is gigantic in relation to the amount of RNA it contains because it has two protein coats. All RNA viruses studied so far contain a single molecule of RNA, although a few such as the reovirus are double-stranded.

Table 8-2 MOLECULAR PROPERTIES OF SOME VIRUSES

RNA VIRUS	PARTICLE WEIGHT $\times 10^6$ daltons	MOLECULAR WEIGHT OF RNA $\times 10^6$	PERCENT OF RNA	SHAPE AND DIMENSIONS IN A
Tomato ring spot	1.5	0.66	44	polyhedral
Tomato bush stunt	10.6	1.65	15	polyhedral, d = 280
Poliomyelitis	16.7	4.8	22–30	polyhedral, d = 300
Tobacco mosaic	40	2.2	5–6	cylindrical, 150 × 3000
Human influenza type A	280	22.2	0.8	polyhedral, d = 750

From Routh, A. W., *Biophysical Journal* 5:527, 1965.

Messenger RNA

(Compare Chapter 15.)

When in the late 1950s it became clear that ribosomes are involved in protein synthesis, the question arose as to whether ribosomes are informational, that is whether each specific type of protein is synthesized by a special ribosome carrying the information required for the synthesis of the protein. There were a number of reasons even then why this seemed unlikely, most of these reasons having to do with the remarkably rapid versatility of the protein synthesizing system of the bacterium *E. coli*. Thus, for instance, it was observed that within 1 min after infection of an *E. coli* cell by a bacteriophage, the bacterium ceases to synthesize *E. coli* proteins and begins to make proteins required for viral replication. It seems unlikely that in such short time enough ribosomes could be synthesized to account for the massive synthesis of protein involved in viral replication.

Because of this and other reasons Jacob and Monod in 1961 postulated that DNA synthesizes a special messenger RNA that attaches to the ribosome, thus endowing it with the required specificity to synthesize a particular protein. The first critical experiment confirming this postulate was performed that very same year by Brenner, Jacob, and Meselson, who demonstrated by using radioactive labeling techniques that after phage infection no new ribosomes are synthesized, but instead a very short-lived form of RNA is synthesized, which becomes attached to the ribosomes.

The details of this important discovery and its crucial aftermath will be discussed later. Here, we shall limit ourselves to a discussion of the molecular properties of mRNA, which unfortunately are not as yet well understood. (See Chapters 14, 15, and 18.)

Messenger RNA is normally attached to ribosomes. This attachment appears to depend on the Mg^{2+} concentration, $10^{-2}M$ promoting binding and $10^{-4}M$ promoting dissociation. In recent years electron microscope and ultracentrifuge studies have shown that ribosomes that are actively engaged in protein synthesis are strung loosely together like beads on a string (polyribosomes). It is thought, but not yet proven, that the "string," which can sometimes be resolved in the electron microscope, is the mRNA (Fig. 15-2 and 15-5).

One of the problems involved in the preparation of pure mRNA is that in most bacterial systems for which the techniques of isolation have been worked out, it proved to be an unstable molecule. It now appears that higher forms seem to have mRNA that is very much more stable, and there is hope that a convenient system will be selected that will yield stable mRNA capable of being studied in detail.

Molecular weight determinations of mRNA show that it is very polydisperse, which means that it ranges from 5×10^4 to 5×10^6 daltons. This is not surprising because we know that (1) protein chains for which the messenger codes vary greatly in size, (2) one messenger RNA molecule codes for more than one kind of protein chain, and (3) it is likely that messenger is degraded during the isolation procedures.

Ribosomal RNA

As we shall discuss in greater detail later, ribosomal particles are made of two subunits unequal in size. The larger subunit is composed of one large RNA molecule, one small RNA molecule, and a number of protein chains, while the smaller subunit consists of just one RNA molecule and several protein chains. (See Chapter 15.)

The G/C and A/U ratios in ribosomal RNA of different sources vary considerably. The one regularity that has been observed is that ribosomal RNA from all sources has a G + C content of more than 50 percent. Physical studies of ribosomal RNA such as hyperchromicity-melting point and viscosity studies suggest that a great deal of base-pairing occurs within the isolated RNA molecule, but it is not clear how regularly periodic it is or whether it has any relevance to the structure of the RNA in the ribosome.

Recently, Sanger and his colleagues have determined the sequence of one of the 5S RNAs from the 50S ribosomal subunit of *E. coli* (Fig. 8-24). Sanger's method for sequencing RNA is a very ingenious recent development and will be described at the

Fig. 8-24. The base sequence of 5S-RNA of *E. coli.* found in the 50S subunit of ribosomes (From Brownlee and Sanger.)

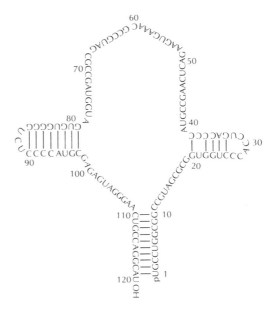

end of this chapter. The biological significance of 5S RNA is as yet unknown, but the fact that the 5S RNA of *E. coli* is a simple homogeneous molecule does suggest that it plays an important role in the mechanism of ribosome action.

Transfer RNA

When a cell homogenate containing $10^{-2}M$ of Mg^{2+} is centrifuged at high speed (100,000 × g for 120 min), the high molecular weight RNA bound to ribosomes is sedimented, and one is left with a supernatant containing a fraction of RNA, which used to be called "soluble" RNA. We now know that most, possibly all the RNA in this supernatant fraction, has a "transfer" function, that is of acting as "adaptors" (or connectors) between the amino acids and the mRNA.

(See Chapter 15.)

Transfer RNA's are a family of small molecules (M ≃ 25,000). For reasons that will become apparent later, the cell probably contains no more than 60 different kinds of tRNA molecules. Each tRNA molecule has a specificity for a given amino acid to which it can attach by a covalent linkage. Because of their small size it has been possible to use the countercurrent distribution method for separating them from each other. This method used the difference in solubility in two immiscible solvents of the solute molecules to be purified. The apparatus consists of a long row of tubes acting like separatory funnels hooked to each other in a continuous series. By attaching the appropriate amino acids to the various tRNA molecules it has been possible to enhance the differences in solubility properties of the different kinds of tRNA molecules, and thus, for the first time, the purification of small RNA molecules has been achieved. Thus, several tRNA fractions with specificity toward a single amino acid have recently been purified.

(See Chapter 15.)

The emergence of tRNA as a convenient material for precise molecular study is one of the exciting recent events in molecular biology. Holley and his co-workers after several years of intensive work succeeded in determining the sequence of bases of three different tRNA molecules. More recently, Sanger and his group have greatly simplified the methods of analysis so that henceforth the determination of sequence of small RNA molecules might well become a routine matter.

It has been known for some time that all tRNA molecules have a sequence of CCA at the 3' end, which is the end to which the amino acid is attached.

(See Chapter 15.)

It has also been known for some time that the A/U and G/C ratios approach equality so that there is a possibility for considerable base-pairing within the molecule.

It is with these general results as a background that Holley and his co-workers began their structural studies of tRNA in yeast. One of their early observations, which proved of value in their studies, was that tRNA contains a number of "unusual bases." Most of these bases differ from A, G, C, and U in that they have one or more methyl groups substituted in various positions of their ring structures. The method that Holley utilized in his structural studies was similar in principle to the method employed by Sanger for the determination of the structure of the protein insulin. A purified tRNA

(See Chapter 9.)

(alanine tRNA was the first one studied) was hydrolyzed under a variety of conditions by a number of hydrolytic enzymes (RNA nucleases) of varying specificity. They discovered that by carrying out the enzyme digestion at a low temperature, they would obtain a few (in one case only two) large fragments whereas at a higher temperature they obtained larger numbers of smaller fragments. By utilizing enormously long ion exchange columns and a solvent containing urea (which interferes with hydrogen bonding and thus decreases interaction between fragments), they were able to purify the fragments from each other. Each purified fragment was again subjected to enzyme or acid hydrolysis, the resulting fragments again purified, and so on. By this procedure the sequence of the small nucleotides could be determined, and since the methods of hydrolysis they utilized yielded overlapping sequences, it was possible to arrange the order of the small nucleotides within the larger nucleotides. This was considerably aided by the presence of the unusual bases. Finally, some overlap among the larger nucleotides permitted the determination of their order (Fig. 8-25) so that the complete sequence of yeast alanine tRNA could be determined.

As this book goes to press some five additional tRNA molecules have been sequenced, and it is therefore only a matter of time before we know the general as well as specific features of the molecule. Nevertheless, it is possible, even now, to make some general conclusions regarding the structure and function of the tRNA molecule

(Fig. 8-26).

1. The sequence of bases is such that in each case the molecule can form a hydrogen bonded "cloverleaf" configuration (Fig. 8-26) in which CCA bases at the 3' end stick out. One of the striking facts about this overall structure is that the topological similarity in structure is achieved even though there are very few places in which the same nucleotide appears in all three structures.

2. The structural similarities among the cloverleaf configurations are legion. They cannot all be enumerated here, but the student can amuse himself by making up a list of them.

3. One of the most important structural similarities among the cloverleaf patterns of the various tRNA molecules is that the overall distance from the CCA_{OH} at one end to the anticodon at the other end appears to be constant, the difference in nucleotide number in the different molecules being compensated for by the size of the little loop or "extra arm" located between the right hand and bottom limbs.

(See Chapter 15.)

4. Another interesting aspect of the cloverleaf structure is that the unusual bases such as DiMeG, DiMeA, MeI, and MeG are located in regions not forming hydrogen bonds. This suggests that their role is to help determine the three-dimensional structure of tRNA, and as will become evident later, the three-dimensional configuration of tRNA is of crucial importance. The molecule must have both some general features that are common to all tRNA molecules, and it must have some specific properties that distinguish one type of tRNA molecule from another. Thus, the common features must include (1) at least one, and probably two, binding sites for the ribosome, and (2) overall dimensions that are such that the position of the various amino acids are the same when the tRNA is attached to the ribosome. The features that differ from one tRNA to another must include (1) a binding site for the specific enzyme catalyzing the attachment of the

Fig. 8-25. The determination of the primary structure of alanine tRNA. Notice how the overlapping of the large oligonucleotide fragments permitted the establishment of a unique sequence for the tRNA molecule. The special bases are: 1-methylguanosine 3'-phosphate (MeG), 5, 6-dihydrouridine 3'-phosphate (DiHU), N²-dimethylguanosine 3'-phosphate (DiMeG), 1-methylinosine 3'-phosphate (MeI), inosine (I) pseudouridine (γ), ribothymidine 3'-phosphate (T). These specific bases proved to be helpful in the sequence determination. (From Holley *et al.*)

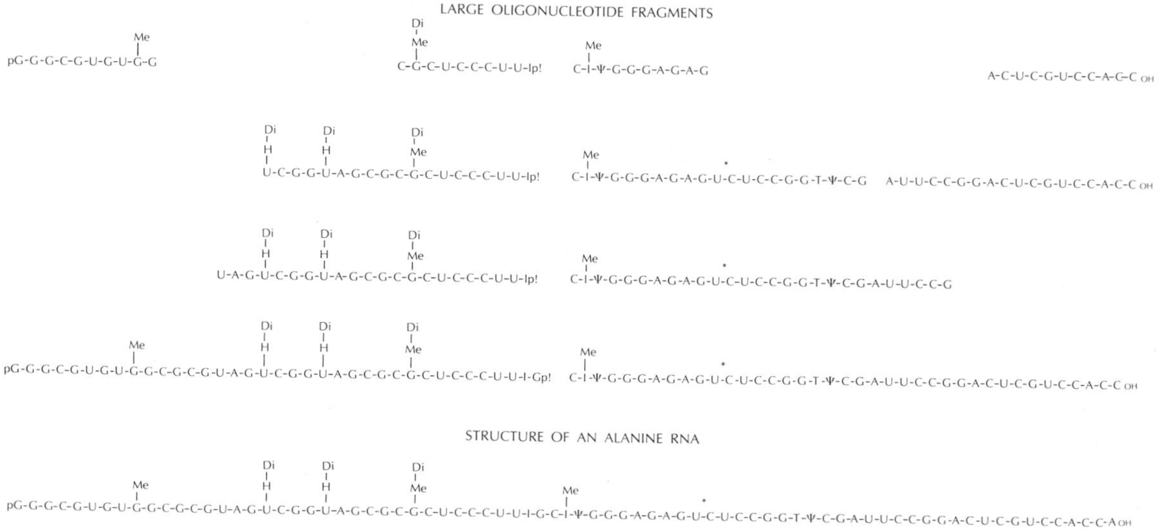

LARGE OLIGONUCLEOTIDE FRAGMENTS

STRUCTURE OF AN ALANINE RNA

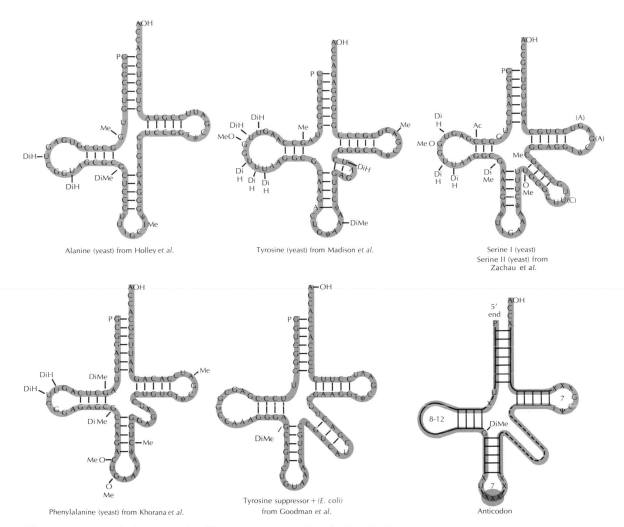

Alanine (yeast) from Holley *et al.*

Tyrosine (yeast) from Madison *et al.*

Serine I (yeast)
Serine II (yeast) from
Zachau *et al.*

Phenylalanine (yeast) from Khorana *et al.*

Tyrosine suppressor + (*E. coli*)
from Goodman *et al.*

Anticodon

Fig. 8-26. ''Cloverleaf'' pattern of 5 different tRNA molecules. The fact that the same secondary structure can be generated from widely differing base sequence is a dramatic confirmation of the cloverleaf pattern first proposed by Holley. The anticodon triplet with which the tRNA ''reads'' the messenger RNA is shaded. The student is encouraged to look for other structural similarities among these 5 different tRNA molecules and then compare them with the generalized cloverleaf structure shown above (bottom right).

amino acid to the tRNA, and (2) an ''anticodon'' site, which is the site of base-pairing between the tRNA and the mRNA.   (See Chapter 15.)

Thus, tRNA, unlike other forms of RNA, must have a highly specific three-dimensional structure, and in this it really resembles the proteins. It would appear that the variable of four different bases in the nucleic acids can provide for pairing, double-helix formation, but a finite 3-dimensional structure only when proteins are involved as in viruses, ribosomes, and eucaryotic chromosomes. Transfer RNA is unique in that it appears to be able to form a finite, highly specific three-dimensional structure by itself, and it is likely that the unusual bases are, at least in part, responsible for this. Thus, Holley and co-workers have noted that ribonuclease T1 at 0° and 0.02M $MgCl_2$ will cleave only the bottom loop of alanine tRNA (in the G position). This suggests that the two lateral loops are not susceptible to cleavage, and since there is a possibility for base-pairing between them, it is conceivable that the three-dimensional organization of tRNA is like a T (see Fig. 8-27). While no other information for the "T-structure" is available, the cloverleaf model is supported by some detailed physical chemical experiments, which show that at a higher temperature ribonuclease will remove as much as 35–40 percent of the nucleotides without greatly influencing overall structure. Figure 8-27 shows how the three loops might be digested at a high temperature while leaving the "stems" of the cloverleaf intact.

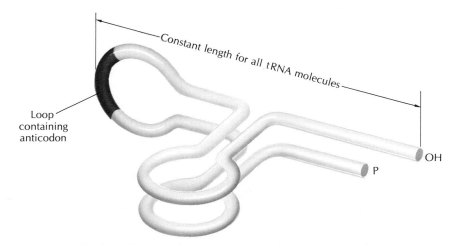

Fig. 8-27. Possible three-dimensional (tertiary) structure of tRNA. The secondary (cloverleaf) structure allows for a number of folding possibilities, the above being only one of them. Recent X-ray scattering measurements suggest models in which three arms are folded up tightly against each other with the fourth arm extended in the opposite direction, or two arms folded in one direction and two arms in the other. Though tRNA molecules vary a great deal in molecular weight, the cloverleaf pattern they form insures a constant length between the "anticodon" and the end to which the amino acid is attached.

5. There are three regions common to the tRNAs studied so far: (1) the diMeG in the corner between the left and bottom loop, (2) the T—ψ—C—G— region in the right loop, and (3) the A—C—C—A_{OH} regions at the amino acid acceptor end of the molecule. It is possible that these are three binding sites interacting with the ribosomes, which might contribute to the precise positioning of the tRNA molecule, one end making contact with the mRNA and the other end bringing the amino acid into juxtaposition with the growing polypeptide chain.

6. The middle loop of the cloverleaf contains a triplet of unpaired bases, the "anticodon" site, which presumably is complementary to the "codon" on the messenger RNA. The fact that every anticodon so far fits one of the codons assigned to the (See Chapter 15, Table 15-5.) respective amino acid is a most dramatic confirmation, not only of the tRNA structures proposed, but of our overall ideas regarding the mechanism of protein synthesis. (See Chapters 14 and 15.)

Finally, a word about the recent contribution of Sanger's group. As we shall see, (See Chapter 9.) Sanger's work on the structure of insulin ushered in the era of modern molecular biology. Recently Sanger's group has developed new methods for the structural analysis of nucleic acids, which will undoubtedly have very important repercussions in the next few years. While there are a variety of ingenious aspects to Sanger's new methods, the most important feature is the utilization of a two-dimensional ionophoresis-ion exchange system in which the ^{32}P-labeled enzyme partial digest of RNA is purified. The important innovation of this system is that in the case of small molecules of RNA such as the 5S rRNA and tRNA, it is possible to separate all the smaller nucleotides (from mono- to pentanucleotides) in one single operation (Fig. 8-28). The radioactivity helps to locate the nucleotides and they can be eluted and determined quantitatively.

Fig. 8-28. Two-dimensional chromatography ("fingerprint") of T1 ribonuclease digest of 5S RNA. A. Position of various nucleotides showing how one can determine composition of nucleotide from the position on the grid. B. Autoradiograph of a T1 ribonuclease digest of 5S RNA which has been labeled by growing *E. coli* cells in a ^{32}p medium. The student might try to identify the nucleotide content of the polynucleotides by using the standard grid in Part A and then comparing their results with Part C. (From Brownlee and Sanger.)

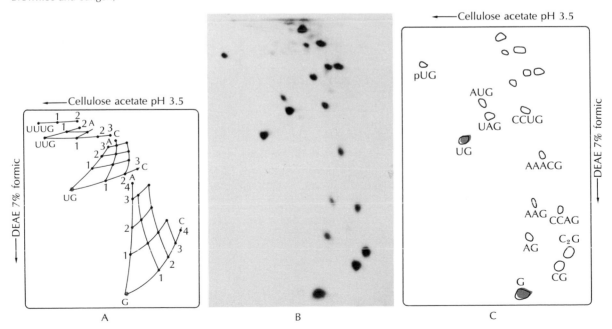

By growing the bacterium *E. coli* in ^{32}P of very high specific activity, Sanger and his group could work with very small amounts of nucleic acid, which they identified by their ability to blacken photographic plates. This method therefore makes possible structural studies of the tRNA species which occur in especially low amounts. The tyrosine tRNA (Fig. 8-26) that they have sequenced is, in fact, one of the so-called "minor" tRNA species resulting from a "suppressor" mutation in *E. coli*. This is indeed a remarkable achievement.

(See Chapters 14 and 15.)

We have studied the molecular properties of the nucleic acids in considerable detail because these properties bear important relations to the role the nucleic acids play in the cell. As we shall learn, the nucleic acids store, transcribe, and translate information, the final act of translation resulting in a new language—the language of the proteins. This new language brings about the three-dimensional configurations that are required for *biological activity*. It is therefore necessary that we now study the molecular properties of proteins in similar detail.

SUGGESTED READING LIST

Offprints

Beadle, G., 1948. "The Genes of Men and Molds." *Scientific American* offprints. San Francisco: W. H. Freeman & Co.

Beadle, G. and Tatum, E. L., 1963. "Genetic Control of Biochemical Reactions in Neurospora." *PNAS*, pp. 499–506. Bobbs-Merrill Reprint Series. Indianapolis: Howard W. Sams & Co.

Cairns, J., 1966. "The Bacterial Chromosome." *Scientific American* offprints. San Francisco: W. H. Freeman & Co.

Crick, F. H. C., 1954. "The Structure of the Hereditary Material." *Scientific American* offprints. San Francisco: W. H. Freeman & Co.

Crick, F. H. C., 1957. "Nucleic Acids." *Scientific American* offprints. San Francisco: W. H. Freeman & Co.

Fraenkel-Conrat, H. and Williams, R. C., 1955. "Reconstitution of Active Tobacco Mosaic Virus from Its Inactive Protein and Nucleic Acid Components." *PNAS*, pp. 690–698. Bobbs-Merrill Reprint Series. Indianapolis: Howard W. Sams & Co.

Hall, C. E., 1961. "Electron Microscopy of Nucleic Acids and Proteins." *Journal of Applied Physics*, p. 1640. Bobbs-Merrill Reprint Series. Indianapolis: Howard W. Sams & Co.

Hanawalt, P. C. and Haynes, R. H., 1967. "The Repair of DNA." *Scientific American* offprints. San Francisco: W. H. Freeman & Co.

Hershey, A. D. and Chase, M., 1951. "Genetic Recombination and Heterozygosis in Bacteriophage." *CSHSQB*, pp. 471–479. Bobbs-Merrill Reprint Series. Indianapolis: Howard W. Sams & Co.

Holley, R. W., 1966. "The Nucleotide Sequence of a Nucleic Acid." *Scientific American* offprints. San Francisco: W. H. Freeman & Co.

Hotchkiss, R. D., 1954. "Cyclical Behavior in Pneumococcal Growth and Trans-formability Occasioned by Environmental Changes." *PNAS*, pp. 49–55. Bobbs-Merrill Reprint Series. Indianapolis: Howard W. Sams & Co.

Hotchkiss, R. D. and Weiss, E., 1956. "Transformed Bacteria." *Scientific American* offprints. San Francisco: W. H. Freeman & Co.

Hurwitz, J., Furth, J. J., Anders, M., Ortiz, P. J. and August, J. T., 1961. "The Enzymatic Incorporation of Ribonucleotides into RNA and the Role of DNA." *CSHSQB*, pp. 91–100. Bobbs-Merrill Reprint Series. Indianapolis: Howard W. Sams & Co.

Meselson, M. and Stahl, F., 1958. "The Replication of DNA in *Escherichia coli*." *PNAS*, pp. 857–868. Bobbs-Merrill Reprint Series. Indianapolis: Howard W. Sams & Co.

Mirsky, A. E., 1968. "The Discovery of DNA." *Scientific American,* July, 1968.

Nagata, T., 1963. "The Molecular Synchrony and Sequential Replication of DNA in *Escherichia coli*." *PNAS,* pp. 551–559. Bobbs-Merrill Reprint Series. Indianapolis: Howard W. Sams & Co.

Sinsheimer, R. L., 1962. "Single-stranded DNA." *Scientific American* offprints. San Francisco: W. H. Freeman & Co.

Spiegelman, S., 1964. "Hybrid Nucleic Acids." *Scientific American* offprints. San Francisco: W. H. Freeman & Co.

Watson, J. D. and Crick, F. H. C., 1953. "Genetic Implications of the Structure of Deoxyribonucleic Acid." *Nature 171,* pp. 964–967. Bound with Watson, J. D. and Crick, F. H. C., 1953. "Molecular Structure of Nucleic Acids: A Structure for Deoxyribose Nucleic Acid." *Nature 171,* pp. 737–738. Bobbs-Merrill Reprint Series. Indianapolis: Howard W. Sams & Co.

Articles, Chapters and Reviews

Thomas, C. A., Jr., Ritchie, D. A. and MacHattie, L. A., 1967. "The Natural History of Viruses as Suggested by the Structure of Their DNA Molecules" in *The Molecular Biology of Viruses,* Colter, J. S. and Paranchych, W., eds. New York: Academic Press, Inc.

Books

Chargaff, E. and Davidson, J. N., eds., 1960. *The Nucleic Acids* (Vol. I–III). New York: Academic Press, Inc.

Davidson, J. N., 1960. *The Biochemistry of Nucleic Acids,* 4th ed. New York: John Wiley & Sons, Inc.

Jordan, D. O., 1960. *Chemistry of the Nucleic Acids*. Washington: Butterworths.

Kornberg, A., 1962. *Enzymatic Synthesis of DNA*. New York: John Wiley & Sons, Inc.

Watson, J. D., 1968. *The Double Helix*. New York: Atheneum.

Chapter 9

Proteins, Agents of Biological Specificity

Proteins, as the derivation of the word implies, are of primary importance to the life of the cell. As Table 9-1 shows, proteins constitute the major component of the dry weight of an actively growing cell. What is so remarkable about proteins is that they are not only the main building material of the cell, but are also the regulators of all the activities carried out by the living machine. To perform their regulatory function, proteins are endowed with specificity, the ability to distinguish among different molecules. This property, more than any other, is characteristic of the phenomenon of life itself, and the specificity of proteins is believed not only to permit the regulation of the multitude of cellular processes, but also to constitute the molecular basis of the differences that exist between individuals and between species.

As we have stressed repeatedly, it is an abiding law of nature that structure and function are intimately related. Thus, we believe that the key to understanding how proteins behave is to know in detail how they are put together. In the last 20 years an astounding series of developments has provided us with detailed insights into the exact structure of proteins. We shall attempt here to give an account of these developments in order to build in the student's mind a vivid picture of the architecture of the protein molecule.

PURIFICATION OF PROTEINS

The number of different proteins in a given cell is extremely large. In the smallest and simplest of bacteria there may be as few as one or two thousand, but human cells may contain as many as 100,000 different proteins. In order to study a given type of protein

Table 9-1 TYPICAL ANALYTIC RESULTS OBTAINED UPON
FRACTIONATION OF RAPIDLY GROWING CELLS

MATERIAL	CRITERIA USED FOR FRACTION	PERCENT DRY WEIGHT
Small molecules "acid-soluble fraction"	Solubility in 5 percent trichloroacetic acid in the cold	2–3
Lipids "organic solvent-soluble fraction"	Solubility in alcohol-ether at 50°C	10–15
Nucleic acids "hot acid-soluble fraction"	Soluble after 30 min treatment in 5 percent trichloroacetic acid at 90°C	10–20
Proteins "hot acid-insoluble fraction"	Insoluble after 30 min treatment in 5 percent trichloroacetic acid at 90°C	55–85

These data are characteristic for cells lacking polysaccharide cell walls or large amounts of other structural and storage materials.

molecule, however, one must prepare it in pure form, that is, one must discard all other proteins and increase the concentration of the particular protein one is interested in. The procedures used for the purification of proteins (fractionation) are a highly refined and rapidly developing art, which we shall be able to describe only briefly.

The first notable success in the purification of proteins was achieved by James Sumner (1926), who crystallized the protein *urease* from the tissue of the jack bean. This important achievement spelled the end of an era during which biologists had come to regard the protein with an awe that precluded the utilization of straightforward chemical approaches to the study of these complicated compounds. Indeed, many biologists gave Sumner's discovery little credence for a number of years. By now some 100 different proteins have been crystallized, and a much larger number have been prepared in highly purified state.

The reasons why proteins can be separated from each other are the very same reasons why proteins are biologically active in the first place. Proteins have a highly specific structure that endows them not only with their various catalytic or structural properties, but also with very specific physical properties that determine their solubility in various solutions or their binding at various interfaces. It is this difference in solubility and binding properties under a large variety of conditions that has been used to fractionate proteins.

There are three basic procedures that must be employed in the fractionation of a protein.

1. One must be able to distinguish between the protein one wants to purify and all the other proteins to be discarded, and one should be able to express this distinction in quantitative terms.

If the protein one is purifying is an enzyme (as most of them usually are), it is possible to devise a test that utilizes the rate of action of the enzyme on a specific

(See Chapter 10.) substrate as a criterion of enzyme concentration. The rate of action of an enzyme on a substrate under defined conditions is called the enzyme activity and is related to the enzyme concentration.

There are numerous tests that measure the total amount of protein. The tests usually involve the elimination of small molecules, lipids, and nucleic acids by washing with alcohol-ether and heating in 5 percent trichloroacetic acid (TCA) at 90°C for 30 min. This leaves the proteins in precipitated form, after which the amount of protein can be determined by drying and weighing, by nitrogen determination after digestion in acid (Kjeldahl test), or by a variety of colorimetric procedures of which the famous "Lowry test" is the most frequently used.

If one expresses the enzyme activity in terms of the amount of total protein present, one obtains the specific activity of the particular enzyme, which is of course related to its state of purity. The higher the specific activity reached during a protein purification of a given enzyme, the purer it is.

If the protein to be purified is not an enzyme, some other test endowed with sufficient specificity to distinguish it from other proteins must be utilized. In the case of a protein hormone a biological test capable of measuring its hormonal activity can be used. With structural proteins that do not possess any specific biological activity, it is possible to use immunological methods of assay. These consist of eliciting the production of specific *antibodies* by injecting the protein into a rabbit, after which the antibodies produced can be used in a variety of ways to determine the amount of that protein (*antigen*) in a given fraction.

Thus, because of the biological specificity inherent in protein structure, we have a variety of methods for determining the specific activity of a given protein in a complex mixture consisting of numerous other proteins.

2. In order to purify a given protein, it is also necessary to use some physical method to increase the concentration of the protein with respect to the other proteins ("impurities") in the mixture.

The most widely used method to achieve this is to change the medium in which the protein mixture is dissolved so as to bring about the precipitation of some of the proteins. If this is done correctly, it leaves most of the desired protein either in solution (supernatant) or in the precipitate. Which fraction the desired protein is in can be determined by measuring the specific activity of the enzyme in the supernatant and the precipitate. The fraction (supernatant or precipitate) with the higher specific activity is then kept, and the other one is discarded, after which the "active" fraction is subjected to another precipitation. By varying the conditions of precipitation each time and retaining the active fraction, it is possible to eliminate systematically more and more of the unwanted proteins.

The conditions of precipitation can be varied in a great variety of different ways; the most frequently used are the following.

a. One method is called "salting out," which means that large concentrations of a salt such as ammonium sulfate are added. The parameter that affects the solubility of a protein is the ionic strength of the solution as given by

$$\mu = \frac{1}{2} \sum c_i Z_i^2$$

That is, the ionic strength (μ) is equal to one-half the sum of the concentration of each ion (c_i) multiplied by the square of its charge (Z_i).

  b. Some proteins called "globulins" become insoluble at low ionic strength; therefore, it is possible to separate them from the water-soluble "albumins" by adding distilled water to a solution of proteins.

  c. The solubility of proteins is profoundly affected by pH, their solubility being lowest at their particular isoelectric point. As we shall see later, the isoelectric point is the pH at which the net charge on a particular protein is zero. Since the isoelectric point differs with each protein molecule, the use of pH to precipitate proteins selectively is very effective.

  d. It is also possible to precipitate proteins by adding a variety of organic solvents such as ethanol, acetone, or ether. Usually, precipitations with organic solvents are carried out at low temperatures (0 to $-5°C$) in order to prevent the irreversible effects that organic solvents have on many proteins at room temperature.

  These and other methods for precipitating proteins are useful for the early stages of purification when the protein that is desired might be outnumbered by the protein impurities by a factor of 100 or even 1000. As the specific activity of the protein increases, it is often desirable to use a chromatographic method, which distinguishes among proteins by the differences in their ability to bind at a given solid-liquid interface. Thus, a material such as diethylaminoethyl cellulose (DEAE cellulose), when packed into a column, will bind different proteins to different extents. By pouring the protein solution into the top of the column, *eluting* with different buffers, and collecting "fractions" (sequential samples), it is possible to separate many of the proteins from each other (Fig. 9-1). Here again, we measure the proteins by a general test such as the frequently used optical density (absorbance) at 280 mμ, whereas a specific test such as enzyme activity or immunological specificity is used to measure the protein to be purified.

  There are a number of other principles that have been used successfully for the purification of proteins. They include *electrophoresis,* which utilizes electric charge differences among proteins; molecular sieving, which discriminates among differences in size; sucrose gradient centrifugation, which depends on size, shape, and density differences among proteins; and finally crystallization.

  Table 9-2 illustrates the results of a typical fractionation of the enzyme plasma transglutaminase in which a variety of different methods such as salting out, heat precipitation of impurities, and chromatography have been used. The effectiveness of the fractionation procedure is determined by maintaining a quantitative account of the progress of the fractionation. A successful fractionation maximizes purification and yield while maintaining the number of steps at a minimum.

  3. When, by varying conditions of fractionation, no increase in specific activity

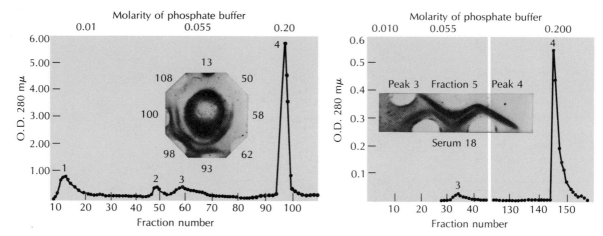

(See Chapter 11 for a
description of the role
of this enzyme.)

Fig. 9-1. A. DEAE cellulose chromatogram of a fraction of human blood plasma containing plasma transglutaminase. Peak 4 is the only peak containing the enzyme activity. The immunological test in the center of the diagram (Ouchterlony gel diffusion), however, shows that peak 4 contains an impurity that is homologous with the protein in peak 3. (For an explanation of gel diffusion methods, see Fig. 9-4.) Notice that peak 3 fractions 58 and 60 show an antigen-antibody precipitation zone that is continuous with a zone from peak 4 fractions 93, 98, and 100. These fractions in peak 4 also contain another protein, which is the enzymatically active plasma transglutaminase. Other experiments showed that the peak 3 protein is an inactive subunit of the peak 4 active enzyme. Thus, the subunit of the enzyme can be distinguished from the active enzyme by chromatographic as well as immunological tests. (From Loewy *et al.,* 1961.) B. Rechromatography of peak 4 shows that some peak 3 material can again be purified from peak 4. This is an unusual situation, which in this instance was explained by the fact that the peak 3 protein is an inactive subunit of the active enzyme in peak 4. Normally, rechromatography under different conditions constitutes an excellent criterion for purity.

is obtained, it is necessary to determine the purity of the protein by any one of a variety of procedures. Crystallinity is not a good criterion of purity in a protein because proteins form loose crystals that can include in them considerable amounts of impurities. There are numerous other methods, however, which can often be used on a preparative scale to purify proteins and on an analytical scale to determine their purity. These include:

1. *column chromatography* (Fig. 9-1) repeated under a variety of conditions;
2. *electrophoresis* (Fig. 9-2) at a number of pH values and in several supporting media;
3. *ultracentrifuge studies* (Fig. 9-3) at a variety of protein concentrations, ionic strengths, and pH values;
4. *immunological gel-diffusion* studies (Fig. 9-4); and
5. *solubility studies,* which surprisingly are very sensitive to small amounts of impurity (Fig. 9-5).

Thus, in summary, by using a specific test for the desired protein and by employing a variety of methods of fractionation, it is possible to purify a protein and establish its

Table 9-2 FRACTIONATION OF PLASMA TRANSGLUTAMINASE*

FRACTION	CONDITIONS OF FRACTIONATION	YIELD, PERCENT	SPECIFIC ACTIVITY	PURIFICATION
Plasma			2.1	1
1	Precipitate: 20% ammonium sulfate, pH 7.0	100	87	42
2	Precipitate: 16% ammonium sulfate, pH 5.4	80	190	91
3	Precipitate: 16% ammonium sulfate, pH 7.0	80	290	138
4	Supernatant: 56°C for 3 min	72	4,600	2200
5	Precipitate: 36% ammonium sulfate, pH 7.0	70	13,000	6200
6	Peak 4: DEAE cellulose chromatography eluted at 0.20 M phosphate buffer, pH 7.0	67	16,800	8000

*(Recalculated from Loewy et al, 1961.) The success of this purification depends on the fact that plasma transglutaminase is attached to fibrinogen and thus follows fibrinogen in fractions 1, 2, and 3. The heat treatment that follows denatures and precipitates fibrinogen, leaving active plasma transglutaminase in the supernatant. The enzyme now in the absence of fibrinogen shows entirely different solubility characteristics so that the impurities not removed in the earlier fractionations can be eliminated. This allows the high degree of purification while maintaining a high yield.

purity by a number of criteria. It must be remembered, however, that proteins are relatively labile molecules capable of losing their specific biological properties (denaturation) often under relatively mild conditions. A successful purification of a protein is one that involves a minimum of denaturation and satisfies a variety of criteria of homogeneity.

positive electrode

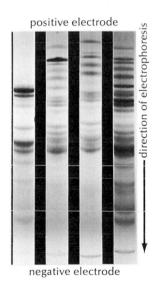

direction of electrophoresis

negative electrode

Fig. 9-2. The use of electrophoresis for the separation of proteins. The above method, referred to as *disc electrophoresis*, is carried out in narrow tubes filled with polyacrilamide gel. Numerous other supporting media such as paper, starch, and cellulose acetate can be used, but disc electrophoresis is the most sensitive method and is capable of unusually high resolution. The above samples are preparations of the protein components of 50 S subunits of ribosomes from *E. coli*. The tube at the extreme right contains all the proteins found in the 50 S subunits. Notice that at least 20 bands are visible of the 30 to 40 different proteins believed to be present in the 50 S subunits. The other three tubes show that these proteins can be separated from each other, in this instance by the chromatographic purification of ribonucleoprotein fragments of the 50 S subunit. The band at the bottom of the tubes is a "marker dye" used to standardize each gel run. The experiment is carried out at an acid pH so that the net charge on the proteins is positive. (From unpublished results by M. Santer.)

Fig. 9-3. Sedimentation in the ultracentrifuge as a criterion of purity. A solution of plasma transglutaminase (see Fig. 9-1) was centrifuged at 59,780 rpm for 36 min. The centrifugal force caused the protein molecules to sediment. An optical system in the instrument is capable of displaying the change of protein concentration (concentration gradient) in the form of one or more peaks which, with the passage of time, move from the meniscus to the bottom of the tube. In the above picture, the large peak represents the enzyme plasma transglutaminase whereas the small peak to the left is caused by the subunit of the enzyme.

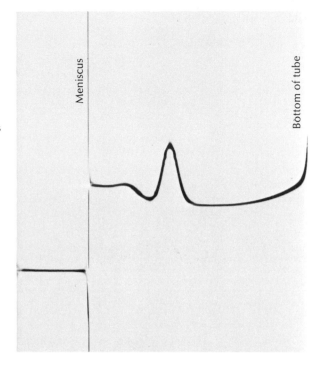

After a protein has been purified, it is possible to study it in a systematic manner. We shall begin by showing how one can study the overall size and shape of the molecule; then we shall "focus" into the molecule by beginning at the simplest level of structure; and slowly, by moving through various levels of structure, we shall build a detailed picture of the complete molecule.

THE SIZE AND SHAPE OF PROTEIN MOLECULES

The very large size and relative lability of protein molecules present the structural chemist with a number of problems that cannot be solved with the methods applied to small molecules. Consequently, a variety of methods have been developed and refined during the last 20 years that have made it possible to determine with considerable precision the molecular weight and the shape of macromolecules.

Methods Yielding Molecular Weight Directly

The measurement of osmotic pressure is perhaps the only method which, though used with small molecules, can also be adapted to macromolecules. The apparatus consists of a membrane that is not permeable to the protein but is permeable to water and buffer. The equipment must be capable of measuring very small pressures since proteins are large molecules and, therefore, cannot form solutions of high molar concentration.

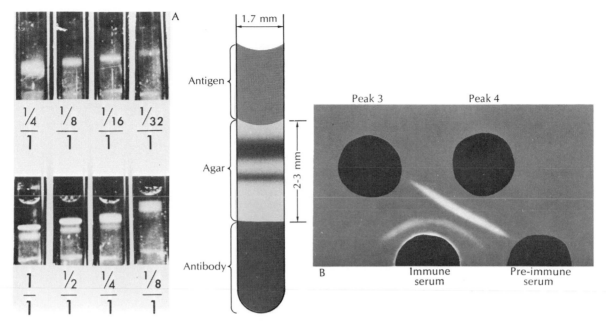

Fig. 9-4. The use of immunodiffusion methods to determine the purity of a protein preparation. A rabbit was immunized with a preparation of plasma transglutaminase (fraction 5). After immunization the serum contained antibodies against the enzyme as well as against impurities. Protein preparations (antigen) can then be diffused through agar against serum (antibody), and where they meet, a *precipitation zone* or *band* is produced. By varying relative concentrations of antigen and antibody, one obtains changes in the position and sharpness of the bands, and hence one can determine the number of antigens in the preparation. A. Double diffusion in tiny tubes (1.7 mm diameter) shows the presence of plasma transglutaminase (upper band) and its subunit (lower band). This method has high sensitivity (10 μgrams protein/milliliter) and precision. B. The Ouchterlony method of gel diffusion permits the identification of related and identical proteins in different preparations. Circular wells are cut into a thin layer of agar on the surface of a slide and are filled with antigen and antibody. This experiment shows that peak 4 contains two proteins, and peak 3 contains only one protein, which is homologous with one of the proteins in peak 4. Precipitation bands of proteins that are not homologous cross each other rather than fuse.

One of the potentially most useful methods, one which will without a doubt be used with increasing frequency, is the "sedimentation-equilibrium" method in the ultracentrifuge. The ultracentrifuge is a device that can rotate a protein solution at velocities as high as 100,000 rpm, thereby increasing the gravitational force on the macromolecules to as much as 200,000 × g.* In the analytical ultracentrifuge it is possible to observe the sedimentation of the macromolecules by a variety of optical devices, which in effect are capable of measuring the change in concentration of the

* g is the acceleration due to gravity, measured in cm/sec.²

Fig. 9-5. Solubility as a criterion of purity. A protein preparation is added in small increments to a salt solution, and each time after equilibrium is reached the protein in solution is measured. At first, a straight line at 45° is obtained. If the protein is pure, one obtains a sharp transition to a horizontal line as soon as the limit of solubility of the protein is reached; however, in the presence of even a small amount of impurity, one sees a gradual transition.

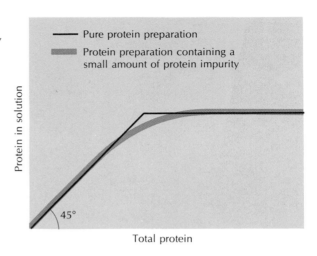

Fig. 9-6. The determination of molecular weight using the sedimentation-equilibrium method. (From Schachman, 1957.) The curve shows how the concentration of a solution of macromolecules is distributed in a moderately low centrifugal field after equilibrium is reached. The measurement of concentration in relation to distance is performed most accurately by using an optical system creating Rayleigh interference fringes, which can be photographed while the rotor is in motion. The data in the figure at right are then calculated by precise measurements of the interference fringes.

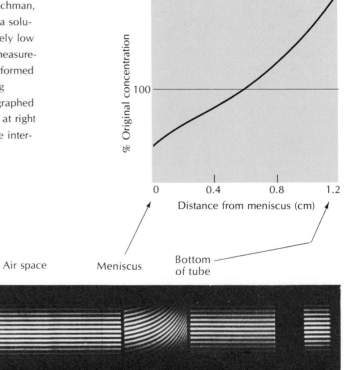

Rayleigh interference fringes

macromolecules with time at various positions in the centrifuge cell between the meniscus and the bottom of the cell. If a solution containing a purified protein is rotated at moderate speed for sufficient time, an equilibrium will develop in which the tendency for the molecules to sediment will be opposed by the tendency for the molecules to diffuse from the region of high concentration in the lower part of the tube to the region of low concentration in the upper part of the tube (Fig. 9-6). The position of the equilibrium will be dependent on the molecular weight and on the density of the protein. The density of the protein can be measured independently, and, therefore, it is possible to use this method to calculate the molecular weight with considerable precision and theoretical rigor.

Methods Sensitive to Shape and Yielding Molecular Weight Indirectly

By centrifuging a protein solution at high speeds, it is possible to sediment it to the bottom of the tube (Fig. 9-7). Thus, one can measure the rate of this sedimentation, and by converting it to standard conditions, one can obtain an S value, which we have already encountered in our discussion of nucleic acids. In recent years this method, which originally required the use of an analytical ultracentrifuge, has been adapted to the simpler and cheaper preparative ultracentrifuge. The method consists of establishing a sucrose gradient in the centrifuge tube and then layering the protein at the top of this solution. During centrifugation this narrow layer or "band" of protein will start sedimenting toward the bottom of the tube. Centrifugation is stopped before the protein has reached the bottom of the tube, a small hole is punched into the tube, and small fractions are collected. It is possible then to measure the total protein in each fraction as well as the activity of the protein that is being studied (Fig. 9-8). This method therefore is extremely useful because it provides a criterion of purity (specific activity can be determined for each fraction), and in the case of an impure protein preparation it allows purification as well as estimation of the sedimentation rate.

It is not possible to determine molecular weight directly from the sedimentation rate. By making another determination such as diffusion rate or viscosity, it is possible to calculate molecular weight from any two of these three parameters.

The rate of diffusion of protein molecules can be measured in a number of different pieces of apparatus including the ultracentrifuge. The diffusion coefficient (D) is frequently used in conjunction with S to calculate the molecular weight.

The virtue of viscosity determinations is that they are simple to perform. They consist essentially of measuring the rate of flow of a protein solution through a capillary tube (Fig. 9-9).

Each of the above three parameters depends not only on the molecular weight of the protein, but also on its shape and hydration. By making certain assumptions about hydration, it is possible to use these parameters to make some estimates as to shape. Before the advent of the electron microscope and the perfection of the light-scattering method, protein chemists had to rely on diffusion or viscosity measurements to calculate the "asymmetry," the degree of elongation of the macromolecules.

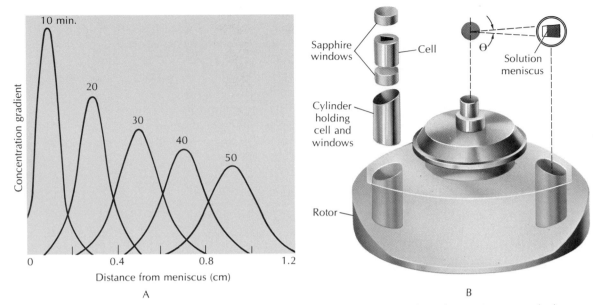

A

B

Fig. 9-7. A. The determination of molecular weight using the sedimentation-rate method. (From Schachman, 1957.) A high centrifugal force is used to create a "boundary" that moves through the solution at a rate that can be measured. In the above diagram a Schlieren optical system is used that displays the concentration gradient in the form of a gaussian curve. As the boundary leaves the meniscus, it is sharp; but as it moves to the bottom of the tube, the boundary widens because of the diffusion of the macromolecules. The sedimentation rate under standard conditions depends on the size and shape of the macromolecules, and lies between 0.25 and 500×10^{-13} seconds. S (the Svedberg unit) has been defined as 10^{-13} seconds, so that sedimentation rates are expressed as so-called "S-values." The value for plasma transglutaminase, for instance, is 9.9 S (see Fig. 9–3). B. Appearance of a centrifuge cell and of a rotor in the analytical ultracentrifuge. Light passes from a light source below the rotor, through the protein solution in the cell, and into an appropriate optical system that measures the concentration of protein at different points in the centrifuge cell.

Methods Yielding Size and Shape Independently

There are a number of optical methods that can provide us with independent estimates of the size and shape of macromolecules. Direct visualization in the electron microscope has come into increasing use, especially with large molecules. By using special techniques such as shadow casting (Fig. 9-10), it is possible not only to see the shape of the molecule, but also to calculate its dimensions. From the dimensions an approximate molecular weight can be calculated, but molecular weight can be determined more accurately by a counting technique. One uses for this purpose a preparation of polystyrene beads, which is homogeneous and in which the exact dimension of the beads is known. By spraying a known mixture of polystyrene and protein on a grid and counting the number of protein particles and beads in given fields, it is possible to

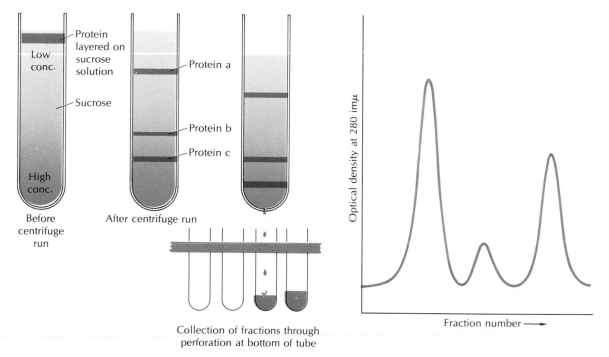

Fig. 9-8. Determination of the sedimentation rate and of the purity of a protein in a sucrose gradient centrifugation experiment. After centrifugation a hole is punched in the bottom of the centrifuge tube, and drops are collected in separate fractions. One can measure both total protein (absorbency at 280 mμ) and enzyme activity in each fraction. Since no complex optical system for measuring the concentration of protein during centrifugation is required, it is possible to use the simpler and cheaper preparative ultracentrifuge.

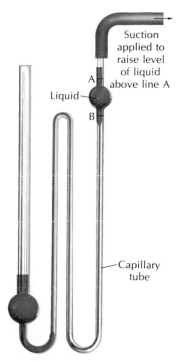

Fig. 9-9. Apparatus for measuring the viscosity of a solution of macromolecules. The solution is drawn above line A and then allowed to flow back by gravity. The time required for the meniscus to move from line A to line B is measured at constant temperature. If t_p is the time taken for the protein solution to flow through, and t_s the time taken for the salt solution used for dissolving the protein, then the relative viscosity of the protein is $\dfrac{t_p}{t_s}$.

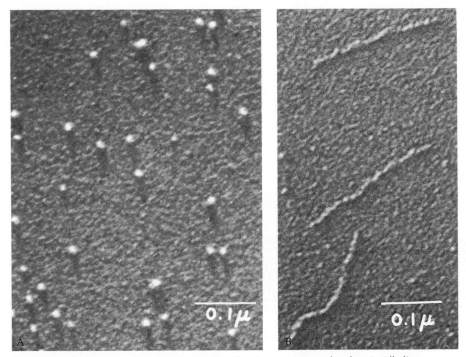

Fig. 9-10. Shadow-cast electron micrographs of various protein molecules. A. Alkaline phosphatase molecules (× 163,000). (Courtesy of Hall.) B. Collagen molecules (× 166,000). (Courtesy of Hall and Doty.) Shadow casting is done by evaporating a metal such as gold on the preparation at a given angle. The metal will coat one side of the molecule and be absent on the other, thereby casting a "shadow" and bringing about a three-dimensional effect. From the length of the shadow and the angle from which it was cast, one can calculate the height of the molecule. This permits one to determine the molecular weight. A more precise way of determining molecular weight is to count the number of particles in a known amount of protein. Since the total amount of protein in an electron microscope field cannot be measured directly, it can be determined indirectly by including in the protein preparation some polystyrene beads of uniform and known dimensions.

calculate the number of protein molecules per weight of protein in the field (as estimated by counting the polystyrene particles) and hence the molecular weight.

Solutions of macromolecules scatter light quite strongly. The degree of scattering at right angles to the incident light depends in part on the molecular weight of the macromolecule and in part on the "polarizability" of the molecules, which can be determined from refractive index measurements. Furthermore, from the angular dependence of the scattering (that is, the degree of scattering at a variety of angles to the incident light), it is possible to calculate the asymmetry of the macromolecules.

X-ray analysis of protein crystals yields very accurate molecular weights and, as we shall see later, very precise, high-resolution pictures of molecular shape. Although by far the most powerful method available to the protein chemist, it is also a most laborious and time-consuming method and has so far been applied successfully to only a very few crystalline proteins.

Methods for Measuring Minimum Molecular Weight

As we shall see later, a number of protein molecules are composed of two or more subunits, either identical or different. When a protein is composed not only of amino acids but also of one or more metal atoms, it is possible to calculate the total weight of protein per metal atom very accurately and obtain a "minimum" molecular weight. By dividing this accurate minimum molecular weight into a less accurate actual molecular weight obtained by some other method, one can obtain the number of subunits; after this one can calculate an accurate molecular weight by multiplying the accurate minimum molecular weight by the number of subunits.

Recently, there have been written computer programs with which one can calculate minimum molecular weights directly from the amino acid analysis of the protein. Although most effective with small protein molecules, this method may achieve wider application as the precision of our techniques of amino acid determination improves.

Table 9-3 shows some of the molecular weights of a number of proteins and nucleoproteins which have been determined with a variety of methods. It shows that, for the most part, the values obtained agree very well.

Table 9-3 MOLECULAR WEIGHTS OF A NUMBER OF PROTEINS AND NUCLEOPROTEINS, DETERMINED BY SEVERAL DIFFERENT METHODS

PROTEIN	OSMOTIC PRESSURE	LIGHT SCATTERING	SEDIMENTATION RATE AND DIFFUSION	SEDI-MENTATION EQUILIBRIUM	CHEMICAL METHODS
Insulin	12,000	12,000			6000*
Ribonuclease			12,700	13,000	13,683†
Pepsin	36,000	37,000	35,500	39,000	
Ovalbumin	46,000	38,000	44,000	43,500	
Hemoglobin	67,000		63,000		66,800
Bovine serum albumin	69,000	70,000	65,400	68,000	
Hemocyanin (Polynurus)		461,000	450,000	453,000	
Tomato bushy stunt virus	9,000,000		10,600,000	7,600,000	
Tobacco mosaic virus	40,000,000		40,700,000		

*The discrepancy in the case of insulin is due to insulin's tendency to dimerize.
†We now know this value with an accuracy of 5 significant figures because the complete amino acid sequence of ribonuclease has been determined.

We can conclude that molecular weights and overall molecular dimensions of proteins can be determined if they are available in pure state. We have emerged with the following picture of the overall structure of the protein molecule.

1. Molecular weights of proteins vary over a wide range (insulin 6000, snail hemocyanin 6,700,000). However, large proteins are generally composed of subunits, and it now looks unlikely that there are many instances of polypeptide chains of molecular weight larger than 100,000 occurring in nature.

2. Shapes vary considerably from near-spherical to highly elongated, but the latter appear to be polymers of near-spherical subunits.

3. Unlike many synthetic polymers, proteins appear to be rigid particles of finite shape. However, as we shall see later in this chapter, certain subtle but important "conformational" changes do occur in proteins, but these are finite changes of a finite structure. In fact, as we shall see when we study the use of X-ray diffraction, proteins are highly uniform molecules with an extremely precise and definite architecture down to atomic dimensions.

By starting with the amino acid building blocks and moving slowly through the various levels of structure, we shall attempt, in the remainder of this chapter, to build a precise picture of how proteins are put together.

PRIMARY STRUCTURE — THE SEQUENCE OF AMINO ACIDS

When a protein is heated in 6N hydrochloric acid at 115°C for several hours, it is broken or hydrolyzed into its constituent building blocks, called amino acids. There are some 20 different amino acids in the proteins of all organisms although certain proteins may contain fewer of them and certain organisms may contain special amino acids that represent secondary modifications of the 20 principal ones. Figure 9-11 shows the basic structure of all but one of the 20 amino acids. It shows that at neutral pH the amino acid is a "zwitterion," containing simultaneously a negative and a positive group. It also shows that the α-carbon atom is asymmetric since it has four different groups attached to it. All amino acids derived from proteins are *l*-amino acids (except for glycine, which is not optically active), a fact that strongly suggests the common origin of all living matter on earth.

The absolute configuration in space has recently been established for the isomers of tartaric acid by X-ray diffraction. This has enabled us also to determine the configuration of the amino acids (Fig. 9-12). The *d*-isomers of amino acids have on occasion been found in nature, but as far as we know they do not become incorporated into the structure of proteins.

The symbol *R* in Fig. 9-11 represents the variable portion of the molecule that differentiates one amino acid from the other. Figure 9-13 is a list of the 20 most common amino acids. Note that the last one, proline, is the only atypical member of the series in the sense that the α-amino group is not free but part of a ring structure. The amino acids are the alphabet of protein structure and are ultimately responsible for the specificity and variability of living matter. The student of biology is well advised to

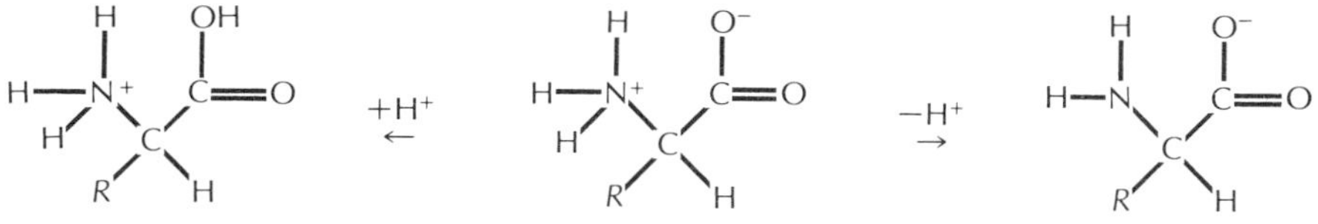

Fig. 9-11. Zwitterion forms of amino acids. At intermediate pH values, amino acids have a negatively charged α-carboxyl group and a positively charged α-amino group. Lowering the pH suppresses the dissociation of the carboxyl group and causes the molecule to become positively charged. Raising the pH brings about the dissociation of the bound proton of the amino group, causing the molecule to be negatively charged. The pH at which a given amino acid has equal positive and negative charge (that is, its net charge is zero) is called the *isoelectric point* of the amino acid. At this pH the amino acid will not migrate in an electric field.

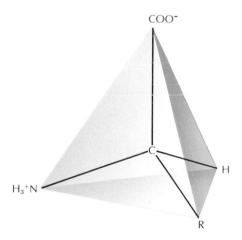

Fig. 9-12. The absolute configuration in space of *l*-amino acids. All naturally-occurring proteins are composed exclusively of *l*-amino acids. The absolute configuration has been determined by X-ray diffraction.

familiarize himself with these important compounds, for the next few decades will witness the elucidation of the precise roles played by the *R* groups of amino acids in the specific molecular interactions of the cell.

The separation of the 20 amino acids from each other was at first a formidable task to which the famous chemist Emil Fischer (1852–1919) devoted many years of his life. In 1941 Martin and Synge proposed a new approach to the problem of purification and estimation of compounds such as amino acids that closely resemble each other. This approach involved dissolving the amino acid mixture in a pair of solvents and then "percolating" this mixture along filter paper strips or through columns. By

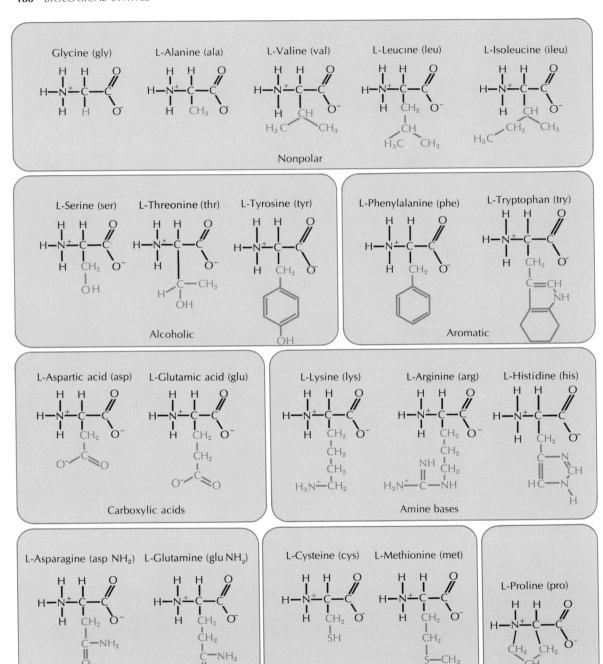

Glycine (gly) L-Alanine (ala) L-Valine (val) L-Leucine (leu) L-Isoleucine (ileu)

Nonpolar

L-Serine (ser) L-Threonine (thr) L-Tyrosine (tyr)

Alcoholic

L-Phenylalanine (phe) L-Tryptophan (try)

Aromatic

L-Aspartic acid (asp) L-Glutamic acid (glu)

Carboxylic acids

L-Lysine (lys) L-Arginine (arg) L-Histidine (his)

Amine bases

L-Asparagine (asp NH₂) L-Glutamine (glu NH₂)

Amides

L-Cysteine (cys) L-Methionine (met)

Sulfur containing

L-Proline (pro)

Imino

using a second pair of solvents and turning the paper through 90 degrees, it is possible to separate amino acids two-dimensionally over a large sheet of filter paper (Fig. 9-14). Because these methods bear some resemblance to a technique employed many years earlier by Tswett to the separation of leaf pigments, Martin and Synge named the procedure *chromatography*. In the last 20 years, chromatographic methods have been extended and perfected to such an extent that they are now the most widely used techniques for analysis and purification. The ultimate in amino acid analysis has been achieved in the laboratory of Moore and Stein, who have perfected column chromatography of amino acids and built an automatic machine for their determination. This device was capable of taking the hydrolysate of 1 mg of protein and estimating the concentration of each one of its component amino acids within an accuracy of a few percent. Moore and Stein utilized a column of ion-exchange resins made of sulfonated polystyrene. The amino acid hydrolysate was placed on the column in a buffer at low pH and low ionic strength, which are conditions for maximum binding of the positively charged amino acids with the $-SO_3^{2-}$ groups on the resin. Buffer was then run through the column, and both the pH and the temperature of the column were raised. This process, called elution, causes the different amino acids to percolate down the column at different rates, eventually separating into separate bands. The *eluate* emanating at the bottom of the column was then processed by an automatic machine that mixes the eluate with *ninhydrin* and heats it to 95°C to produce a blue color which then is measured in a recording spectrophotometer.

Numerous modifications and improvements of the original "Moore, Stein, and Spackman machine" have been made. Today, it is possible to hydrolyze 0.1 mg. of protein and measure 0.02 μmoles of each amino acid with an error of five percent and a total recovery of the original sample of 100 $\pm$ 3 percent, and even more sensitive techniques are being worked out.

Figure 9-15 is an amino acid chromatogram produced by a "Technicon" amino acid analyzer. This instrument can run 40 samples consecutively at the rate of 12 chromatograms per day.

The automation of amino acid and peptide analysis is by no means a mere "labor-saving gimmick." Our experience of the last ten years shows that the elucidation of the structure of a number of proteins would not have been feasible without this automatic device. Like computer methods in the field of physics, automated analytical methods have made, in the field of biology, a qualitative difference as to the nature of the problems that can be attacked. Further strides in amino acid analysis can be expected when techniques such as vapor-phase chromatography and mass spectroscopy are applied to the measurement of these compounds.

Once protein has been hydrolyzed into its building blocks, it is logical to examine how these building blocks fit together in the protein molecule. Fischer was able to

Fig. 9-13. (Opposite page.) The structure of the 20 commonly occurring amino acids. The student should memorize this basic 20-lettered alphabet with which the language for biological variability and specificity is built.

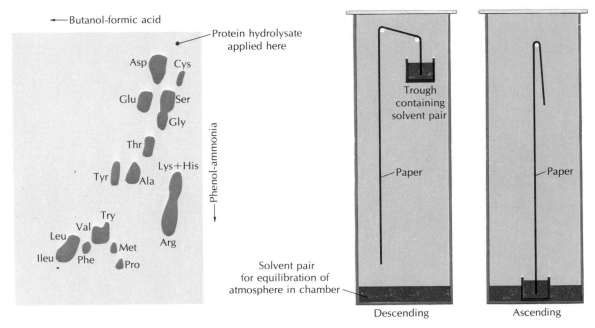

← Butanol-formic acid

Protein hydrolysate applied here

Asp Cys
Glu Ser
Gly
Thr
Tyr Lys+His
Ala
Try
Val
Leu
Ileu Phe Met
Pro
Arg

Phenol-ammonia →

Solvent pair for equilibration of atmosphere in chamber

Two-dimensional chromatogram

Trough containing solvent pair

Paper

Paper

Descending   Ascending

Two types of chromatography apparatus

Fig. 9-14. Two-dimensional paper chromatography of amino acids. The amino acid mixture is applied in one corner, the paper is inserted in the chromatography chamber, and a given pair of miscible solvents either ascends by capillary action or descends by siphon action. As the solvents move through the filter paper, the amino acids are partially separated along one dimension. Further separation is achieved by turning the paper through 90° and repeating the operation with a different solvent pair. The position of each amino acid is then determined by spraying the filter paper with ninhydrin, which produces a purple color with amino acids. The position of radioactive amino acids can be determined by placing the chromatogram on a sheet of X-ray photographic paper, which after development will show the amino acids as darkened spots. Glutamine and asparagine are deaminated by acid hydrolysis to glutamic and aspartic acids, respectively, and therefore do not appear on the chromatogram.

show that upon hydrolysis of a protein an equal number of amino and carboxyl groups are released. To explain this result he suggested that the amino acids were linked to each other by a peptide bond (Fig. 9-16). The equilibrium of this reaction is far on the hydrolysis side, and as we shall see later, the cell synthesizes the peptide bonds of the protein molecule by an entirely different mechanism. Fischer's peptide bond theory for protein structure has since been validated by many separate lines of evidence. Proteolytic enzymes, for instance, which are known to break the peptide bond of synthetic peptides, are also capable of hydrolyzing proteins.

Figure 9-17 is a diagram of a hypothetical chain of amino acids, termed a polypeptide. Notice that except for the terminal carboxyl (C-terminal) and the terminal amino (N-terminal) groups, all the remaining α-carboxyl and α-amino groups are involved in

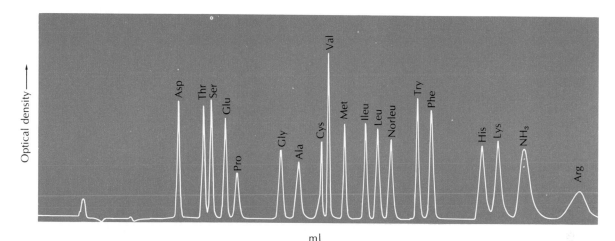

Fig. 9-15. Chromatogram of a mixture of amino acids obtained by using an automated amino acid analyzer. (Courtesy of Technicon Corporation.) This run took 2 hours and utilized 0.05 μmole of each amino acid. Norleucine is not a naturally occurring amino acid but is used here as an "internal standard." Asparagine and glutamine are missing because they are deaminated during hydrolysis of the protein with hydrochloric acid.

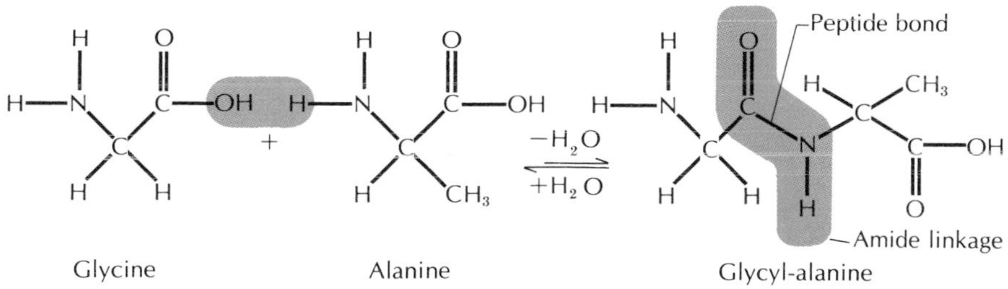

Fig. 9-16. The peptide bond. Glycyl-alanine is a *dipeptide* made of two *residues*, the glycyl and alanyl residues.

the amide linkage. This linkage is not capable of releasing or accepting protons and therefore does not contribute to the acid-base properties of the polypeptide. The electrochemical properties at physiological pH of the polypeptide (and hence of the protein) are determined by the *R* groups of the acidic amino acids, aspartic acid and glutamic acid; the basic amino acids, lysine and arginine; and by histidine. The latter is important because it is the only amino acid, with the exception of the N-terminal amino group, with a pK in the region of physiological pH, and thus it is the only amino acid that could contribute to a change in charge on the protein caused by small changes in pH under physiological conditions. The role of this amino acid side chain in mechanisms of cellular regulation has yet to be discovered.

The sequence of amino acids in polypeptides is generally written as in Fig. 9-17, that is, with the α-amino or N-terminal group on the left and the α-carboxyl or

glycyl-aspartyl-lysyl-glutamyl-arginyl-histidyl-alanine

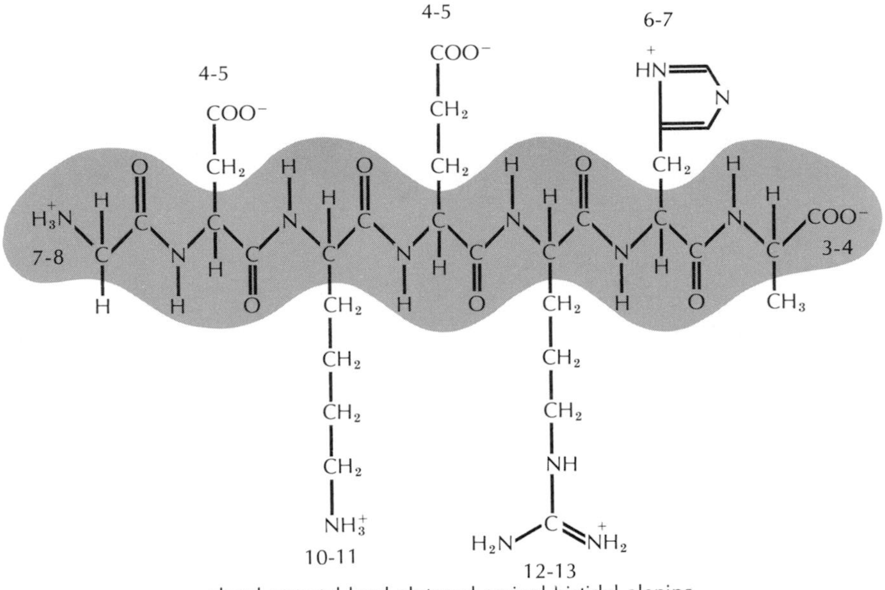

Fig. 9-17.   A hypothetical polypeptide containing all the groups that normally contribute positive and negative charges to proteins. The numbers represent the pK range of each dissociating group. Notice that the charged groups consist of (1) an "N-terminal" and a "C-terminal" group; (2) negatively charged aspartyl and glutamyl side chains; and (3) positively charged histidyl, lysyl, and arginyl side chains. In a large polypeptide chain, the charge contributions of the N-terminal and C-terminal groups are, of course, small compared with those of the side chains.

C-terminal group on the right. One generally uses the abbreviations, so that this particular peptide would be written

Gly.Asp.Lys.Glu.Arg.His.Ala.

The term "residue" is used to denote the amino acid moiety (minus the one molecule of water) found in the polypeptide chain. Amino acid residues are denoted with a "yl" ending (for example, one refers to the alanyl residue in the peptide chain).

The hypothetical polypeptide in Fig. 9-17 shows all the R groups that contribute to the charge on a protein at physiological pH. It shows by extrapolation that a protein is a polyvalent ion containing a number of positive and negative charges. From the dissociation properties of the groups one can predict what should happen when acid or base is added to a protein. Thus, when acid is added, —COO⁻ will change to —COOH and therefore leave the protein with a net positive charge; when base is added, —NH₃⁺ will change to —NH₂, leaving the protein with a net negative charge. At a certain intermediate pH called the *isoelectric point* the number of positive charges

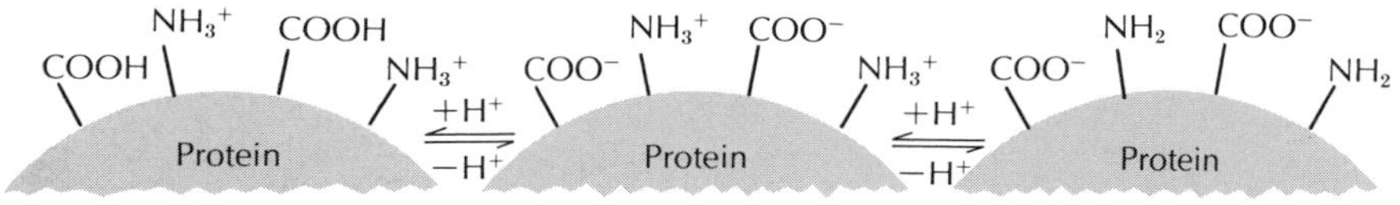

Acid pH	Isoelectric pH	Alkaline pH
Net charge is +	Net charge is zero	Net charge is −
Migration toward cathode	No migration in an electric field	Migration toward anode

Fig. 9-18. The effect of pH on the charge of proteins. At the isoelectric pH, the net charge is zero although the total number of positive and negative charges is maximum. At its isoelectric point the protein is at its minimum solubility and can be precipitated by salt or organic solvents more easily than at other pH values.

equals the number of negative charges (Fig. 9-18). Since most proteins have isoelectric points on the acid side, they will carry a net negative charge at physiological pH. However, the protamines and histones, which are proteins found in the chromosomes, contain large amounts of lysine and arginine, and they therefore carry a net positive charge under normal conditions. Without a doubt this property plays an important role in the molecular structure of the chromosome, which in part is composed of the positively charged histones interacting with negatively charged DNA.

We have shown that it is possible to determine the amino acid composition of a polypeptide. Is it possible, however, to determine the specific sequence of amino acids in the polypeptide? Sanger was the first to show that this indeed could be done. As is usually the case with a first contribution in a given field, his achievement is notable not only for the very important technical contribution made in answering this question, nor just for the far-reaching theoretical conclusions deriving from the results, but also for the insight and faith involved in asking the particular question. The early traditions of protein chemistry are derived from colloid chemistry, which regarded its materials as chemically indefinite entities to be characterized by statistical rather than precise chemical parameters. Sanger challenged this tradition by asking a question that as recently as 1945 was considered preposterous by most protein chemists. He asked: "Do proteins have a specific and finite chemical composition down to the *sequence* of the 20 building blocks?" To ask such a question and to be willing to invest 10 years of one's life in answering it implies a belief in the *absolute* accuracy by which biological macromolecules are replicated and synthesized. The extent to which we take this for granted today is a tribute to the magnitude of Sanger's contribution. When Sanger began his epoch-making study on the protein *insulin,* the method of paper chromatography for the separation of amino acids had just been developed by Martin and Synge. To this Sanger added a technique of his own, the "labeling" of the amino end group (N-terminal group) of a peptide by combining it with the yellow compound 2,4-dinitrofluorobenzene (DNFB) to give a dinitrophenyl (DNP) peptide.

2,4-dinitrofluoro-
benzene
(DNFB)

N-terminal end
of peptide

dinitrophenyl
peptide
(DNP)

This yellow compound is stable during the acid hydrolysis of the peptide, and it is possible by the use of chromatographic separation after hydrolysis to identify the particular amino acid to which the DNP is attached. By determining the N-terminal amino acid, it was possible for Sanger to orient the peptide, that is, to distinguish one end from the other. Sanger also developed a number of methods for partially hydrolyzing the insulin molecule into smaller peptides of various lengths. He developed chromatographic procedures for separating these peptides from each other, which permitted him to determine their amino acid composition and the identity of the N-terminal amino acid. Some of the longer peptides had to be hydrolyzed for a second time, chromatographed, and again analyzed for end groups. As the data accumulated, more and more of the sequence became uniquely defined. Eventually all of the sequence had become uniquely defined, after which time all the new data simply verified the existing structure without providing any evidence against it. Figure 9-19 is a summary of the results that Sanger used in determining the amino acid sequence of one of the two polypeptide chains of the insulin molecule.

Using the above approach, Sanger was able to show that only one unique sequence would satisfy the data he had obtained. It was thus shown for the first time that despite the complexity of the protein molecule, the cell is able to synthesize it in a reproducible and chemically precise manner. It is interesting to note that although insulin is one of the smallest proteins we know, it is not the simplest in structure since it is composed of two polypeptides.

Figure 9-20 shows the sequence or so-called primary structure of a much larger protein (18,000 molecular weight) that is composed of a single chain. The determination of sequence of this larger polypeptide proved a much more difficult problem requiring the precision of the automated column chromatographic method of Moore, Stein, and Spackman.

The "sequencing" of this and other larger polypeptide chains in protein molecules also depended on the technique of specific enzyme hydrolysis. The proteolytic enzyme trypsin, for instance, cleaves only those peptide bonds that have the positively charged arginyl or lysyl residues on the carboxyl side of the bond. This procedure breaks the large polypeptide chain into smaller specific "peptides." These can be separated by column chromatography, and each of their structures can be determined separately.

Since such specific enzyme hydrolysis produces "nonoverlapping" peptide sequences, it is necessary then to determine the order of the tryptic peptides by an additional technique. This usually consists of using a second enzyme with a different specificity, which produces a different set of peptides and provides the necessary information for determining the order of the tryptic peptides. Chymotrypsin is the enzyme that is often used for this purpose. Its specificity is not as great as that of trypsin, although it hydrolyzes more rapidly the peptide bonds of the aromatic amino acid residues (tyrosyl, tryptophanyl, and phenylalanyl). Other enzymes of great value for amino acid sequence determination are *leucine aminopeptidase* and the *carboxypeptidases*. The former hydrolyzes the peptide bond closest to the N-terminal end of the peptide, and the latter hydrolyze the peptide bond closest to the C-terminal end. Since, therefore, these enzymes digest peptides sequentially, they can often be used for establishing the sequence of a few amino acids from either end of a chain. In Fig. 9-21 the amino acid sequences of peptides of the protein glucagon following various types of digestion have been provided. As an exercise in sequence determination of proteins, we ask the student to write these sequences on separate cards and then line them up with respect to each other (each digestion procedure in a separate row) until a unique sequence is obtained such as the one illustrated in Figs. 9-19 and 9-20. In a unique sequence a place is found for every peptide, and every vertical row contains only one particular amino acid. Figure 9-21 provides more results than are really needed, which shows that if the sequence is correct, all additional results are merely redundant rather than contradictory.

The primary structure, that is, the sequence of amino acids forming the one or more peptides of a protein, has now been determined for a considerable number of proteins. Each year advances are made that simplify the process and permit the study of larger peptide chains. So far a large chain, such as the 68,000 molecular weight chain in serum albumin, has not been sequenced. With the application of new approaches such as the use of automated sequencing procedures and of the mass spectrometer, however, it can be expected that the primary structure of most proteins will soon become susceptible to analysis.

There are two good reasons why an understanding of the primary structure of proteins is of importance. Firstly, as we shall see later, the primary structure of proteins is determined uniquely from the nucleotide sequence of the respective DNA by a process of transcription and translation. It is therefore necessary to know the primary structure of proteins if one is to understand the overall phenomenon whereby the hereditary materials control the synthesis of proteins. Secondly, as we shall see later in this chapter, it is the primary structure of a protein that controls the final three-dimensional structure of the molecule and hence the physiological role it plays. The primary structure is the fundamental link between the genetic material on one hand and the physiological function on the other. We should therefore not think of the considerable efforts invested in sequence determination as a pedantic exercise in analytical chemistry but rather as a crucial methodological contribution to the solution of important biological problems. The following "stories" are illustrations of this.

(See Chapters 14 and 15.)

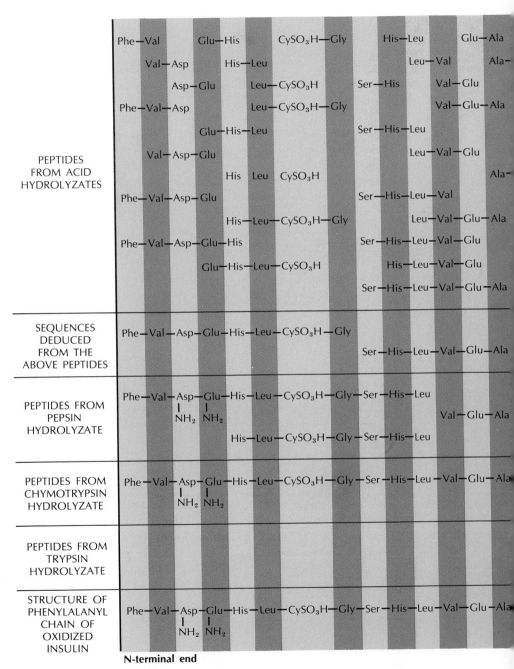

N-terminal end

Fig. 9-19. Compilation of Sanger's analytical results showing how the final sequence of one of the polypeptide chains of insulin was deduced. Notice that acid hydrolysis permitted the determination of the sequence of five short peptides. Enzyme hydrolysis provided the overlap that established the overall sequence. The student can work through

C-terminal end

the logic of this procedure by putting the sequences on separate cards and deducing the overall sequence from them. (From "The Insulin Molecule," by E. O. P. Thompson, in *Scientific American.* Copyright © 1955 by Scientific American, Inc. All rights reserved.)

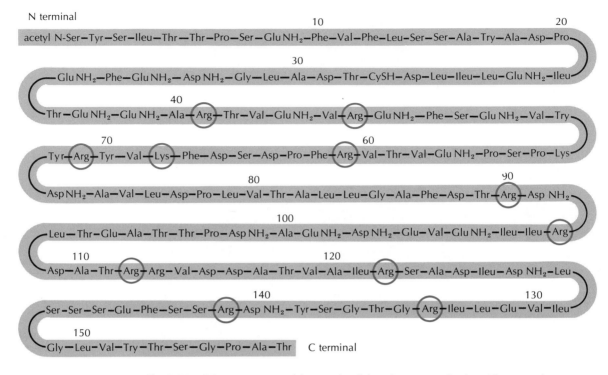

Fig. 9-20. Primary structure of the protein of the tobacco mosaic virus. (Courtesy of Virus Laboratory, University of California.) Circles are placed around the lysine and arginine residues that are on the N-terminal side of the peptide bond cleaved by trypsin. The sequence of this protein has been determined independently by two laboratories. Notice that at the N-terminal end the amino group is not free but occurs in an acylated form.

The first discovery relating a genetic mutation to a known molecular change in protein structure was made by Ingram, not in a microorganism, but of all things, in human hemoglobin! There is a disease in humans known as sickle-cell anemia, which is inherited in a simple Mendelian fashion, the gene producing a form of hemoglobin which Pauling and Itano showed has a lower electrophoretic mobility than normal hemoglobin. Ingram developed a technique, the peptide map (or fingerprint) method, which consists of hydrolyzing the hemoglobin with trypsin and separating the resulting peptides from each other by moving them on paper "two-dimensionally," that is, chromatographically in one direction and electrophoretically in the other. Figure 9-22 shows a comparison of maps of the normal and abnormal hemoglobin. It shows that of all the 26 spots, only 1 (colored) appeared in a different position. Analysis of the amino acid content of this peptide indicated that a glutamic acid residue was missing and a valine residue had been substituted in the mutant form. Human hemoglobin is made of four polypeptide chains, two α-chains and two β-chains. Ingram was able to show that the abnormal gene was responsible for the synthesis of a hemoglobin molecule in which the β-chains had a valyl residue substituted for the glutamyl residue in the sixth position from the N-terminal end. A number of other abnormal hemoglobin molecules have since been discovered (Fig. 9-23); and the study of amino acid substitutions resulting from genetic mutations, especially in microorganisms,

Fig. 9-21. Sequences of peptides obtained from the protein glucagon with a number of different enzyme and acid hydrolyses. The student should write these sequences on individual cards and try to deduce a unique sequence for the glucagon chain. It is important to demonstrate to your own satisfaction that all the data are consistent with a single sequence. The order of the amino acids within the bracket was not determined, and therefore the amino acids are listed alphabetically.

Trypsin (2.25 hr.)	His (Asp, Glu, Gly, Lys, Phe, Ser, Ser, Ser, Thr, Thr, Tyr) Tyr (Arg, Asp, Leu, Ser) Arg (Ala, Asp, Asp, Glu, Glu, Leu, Met, Phe, Thr, Try, Val) Ala (Asp, Asp, Glu, Glu, Leu, Met, Phe, Thr, Try, Val) Arg
Chymotrypsin	Val (GluNH$_2$, Try) Ser (Lys, Tyr) Thr (Asp, Ser, Tyr) Leu (Asp, Met, Thr) His (GluNH$_2$, Gly, Phe, Ser, Thr) Leu (Ala, Arg, Arg, Asp, Asp, Glu, Phe, Ser)
Carboxypeptidase action	Ala (Asp, GluNH$_2$, GluNH$_2$, Leu, Phe, Try, Val) Met, AspNH$_2$, Thr
Trypsin (50 hr.)	Arg Leu (Asp, Met, Thr) Tyr (Arg, Asp, Leu, Ser) His (Glu, Gly, Phe, Ser, Thr) Thr (Asp, Lys, Ser, Ser, Tyr) Ala (Asp, Glu, Glu, Phe, Try, Val) Ala (Asp, Asp, Glu, Glu, Leu, Met, Phe, Thr, Try, Val)
Subtilisin	Arg (Ala, GluNH$_2$) AspNH$_2$ · Thr Lys · Tyr Asp · Phe His (Glu, Ser) Leu · Met Thr · Ser Leu (Arg, Asp, Ser) Val (Glu, Try) Gly (Phe, Thr) Asp (Ser, Tyr)
Acid degradation products	Thr His Asp · Ser Tyr Leu · Asp Asp · Tyr Glu · Gly Thr · Phe Ser · Lys Ser (Asp, Tyr) Ser · Asp Ser (Glu, Gly) Ser · Arg Tyr · Leu Tyr (Asp, Leu)

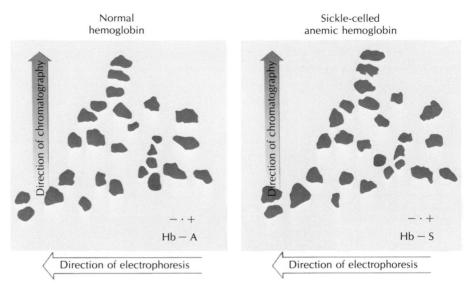

Normal
hemoglobin

Sickle-celled
anemic hemoglobin

Direction of chromatography

Direction of chromatography

Hb − A

Hb − S

Direction of electrophoresis

Direction of electrophoresis

Fig. 9-22. Peptide maps (or "fingerprints") of normal hemoglobin (Hb A) and abnormal (sickle-cell anemia) hemoglobin (Hb S). The position of one peptide identified by color has changed because the mutation has caused the substitution of one valine residue for a glutamic acid residue in this peptide. (From Baglioni, 1961.)

(See Chapters 14 and 15.)

has become a major approach to the study of the "coding problem." Thus began one of the most important discoveries in molecular biology, namely that a genetic mutation can bring about a specific amino acid substitution in a protein.

Another interesting biological result of the study of primary structure is the discovery that proteins with similar functions have similar primary structures even if they

Fig. 9-23. The sequence of the β-peptide chain from the N-terminal end of normal hemoglobin and of three mutants. The sickle-cell anemia mutation causes the disease by substituting a nonpolar amino acid (valine) for a negatively charged one (glutamic acid). This change causes a considerable change in the solubility of the hemoglobin molecule so that in human red cells hemoglobin forms large aggregates that deform the cells and exhibits an impaired oxygen-exchange capacity. By now, numerous mutations in both the α- and β-chains of human hemoglobin have been identified as amino acid substitutions at certain points. The arrows represent one point at which the proteolytic enzyme trypsin breaks the β-chain. (From Ingram, *The Biosynthesis of Macromolecules.* New York: Benjamin, 1966.)

Normal hemoglobin	(HbA)	NH₃-Val-His-Leu-Thr-Pro-Glu-Glu-Lys · . . .
Sickle celled anemic hemoglobin	(HbS)	NH₃-Val-His-Leu-Thr-Pro-Val-Glu-Lys · . . .
Mutant hemoglobin	(HbC)	NH₃-Val-His-Leu-Thr-Pro-Lys-Glu-Lys · . . .
Mutant hemoglobin	(HbG)	NH₃-Val-His-Leu-Thr-Pro-Glu-Gly-Lys · . . .

occur in quite unrelated organisms. Thus, the important protein cytochrome c, which occurs in almost all cells, exhibits only minor modifications in its primary structure as we proceed from the primitive yeast cell to much more recent species such as mammals (Fig. 9-24). Thus, for instance, of the 104 amino acids in the cytochrome of

(See Chapter 13.)

Fig. 9-24. A comparison of the number of amino acid differences between the enzyme cytochrome c of man and of other organisms. Notice that with an increase in phylogenetic distance, the number of amino acid differences also increases. By a detailed study of the pattern of substitutions, it is possible to deduce a phylogenetic tree resembling the phylogenetic relationships that are based on generally accepted morphological grounds. (From the *Atlas of Protein Sequence and Structure*, Dayhoff and Eck, 1968.)

	Human	Monkey	Pig, Bovine, sheep	Horse	Dog	Rabbit	Kangaroo	Chicken, turkey	Duck	Rattlesnake	Turtle	Tuna fish	Moth	Neurospora	Candida	Yeast
Human	0															
Monkey	1	0														
Pig, bovine, sheep	10	9	0													
Horse	12	11	3	0												
Dog	11	10	3	6	0											
Rabbit	9	8	4	6	5	0										
Kangaroo	10	11	6	7	7	6	0									
Chicken, turkey	13	12	9	11	10	8	12	0								
Duck	11	10	8	10	8	6	10	3	0							
Rattlesnake	14	15	20	22	21	18	21	19	17	0						
Turtle	15	14	9	11	9	9	11	8	7	22	0					
Tuna fish	21	21	17	19	18	17	18	17	17	26	18	0				
Moth	31	30	27	29	25	26	28	28	27	31	28	32	0			
Neurospora	48	47	46	46	46	46	49	47	46	47	49	48	47	0		
Candida	51	51	50	51	49	50	51	51	51	51	53	48	47	42	0	
Yeast	45	45	45	46	45	45	46	46	46	47	49	47	47	41	27	0

the rattlesnake and man there are only 14 amino acid differences. This and other discoveries have provided a new avenue of insight into the phenomenon of evolution. By studying the primary structure of proteins, one is looking at the earliest phenotypic effect of the mutations in the hereditary material. Thus, by following the evolution of the primary structure of proteins, one is studying morphology at a very fundamental level.

SECONDARY STRUCTURE—THE RELATIONSHIP OF NEAREST NEIGHBORS

As we have seen, physical studies of protein molecules have shown that they are compact, rigid, not overly elongated molecules. The conclusion must be that the polypeptide is folded in some manner to achieve this compact structure, but the precise geometry of folding has been a subject of many years of speculation. In 1951 Pauling and Corey provided the first clue. They had started their work with a meticulous X-ray analysis of the structure of a number of simple peptides, which yielded a precise description of the amide linkage of the polypeptide chain (Fig. 9-25A). They discovered that the six atoms of the amide group (CCONHC) are coplanar, lying within a fraction of an Angstrom unit in a common plane. Furthermore, they found that the C—N bond of the amide group was unusually short (1.32 A). Pauling and Corey concluded that the C—N bond had a "partial double-bond character" (Fig. 9-25B), thus explaining both the planar configuration and the shortness of the bond. Furthermore, they found that the amide group was arranged in the *trans*-configuration, that is, the two asymmetric carbon atoms lie in the opposite corners of the group. Finally, Pauling and his associates

Fig. 9-25. The amide linkage of the polypeptide. A. Exact structure of the amide group on a polypeptide. Colored atoms lie in one plane. The distance of 1.32 A between C and N is unusually low for a typical C—N bond. This is due to the "partial double-bond" character of this carbonyl carbon-to-nitrogen bond, as shown in B. The partial double-bond character is responsible for the coplanarity of the amide linkage of the polypeptide chain. Notice also that the amide linkage is in the transconfiguration, with the two asymmetric carbons lying in opposite corners of the amide group.

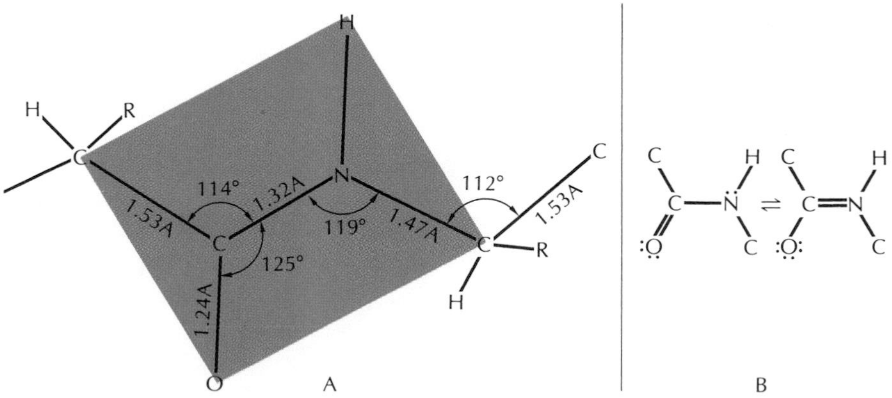

pointed out that the bonds to the corner carbons of the amide group are single chemical bonds and therefore capable of rotation. They argued that rotation around this bond would determine the configuration of the polypeptide. By building scale models, they soon discovered that a large number of different configurations could be generated if one merely accepted the above restrictions. It is precisely here, in deciding which of these many configurations is the most likely to occur, that Pauling made the crucial intuitive leap. Having had a great deal of experience with the importance of hydrogen bonds in the crystal structure of model compounds, Pauling predicted that the preferred configuration of the polypeptide would be the one that favored maximum hydrogen bonding between the —C=O and —NH groups of the amide linkage. He very wisely,

Fig. 9-26. Rotation around an asymmetric carbon atom, which puts each successive amide group into a new plane, thus generating a helical configuration. The H atom and the *R* group on the asymmetric carbon atom do not lie in the same plane as either of the amide groups.

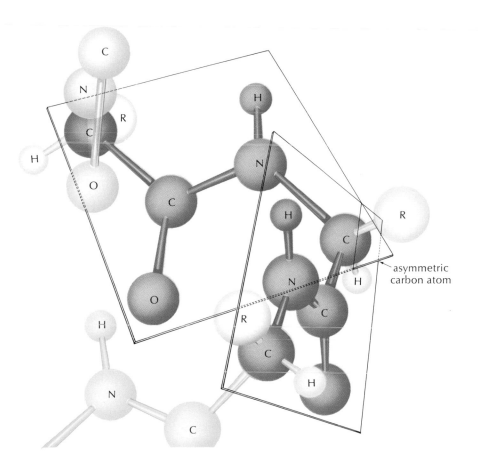

as a first approximation, decided to ignore the possible effects of the *R* groups. He then discovered that by subjecting each amide group along the polypeptide to a small and equivalent rotation (Fig. 9-26), a helix was generated that had the property of forming hydrogen bonds between the turns of the helix. The so-called α-helix (Fig. 9-27) is capable of forming hydrogen bonds between all its amide groups, a fact that on a priori grounds would give it the greatest degree of stability. The characteristics of the α-helix can be described by stating the number of amino acid residues per turn (3.6), the pitch of the helix (5.4 A), the diameter including the side chains (10.5 A), and the number of amino acid residues in one hydrogen-bonded loop (3). The formula for the hydrogen-bonded loop of the α-helix is

Subsequent investigations of the structural possibilities of helix formation showed that a number of other helices are possible, five of them with low deformation energies (less than 3 kcal/mole of residues). But the original α-helix of Pauling and Corey turned out to be the only one that requires no deformation and at the same time forms hydrogen bonds between all the —C=O and —NH groups.

The dimensions of the α-helix were looked for in a number of proteins and synthetic polypeptides by the method of X-ray analysis, and a great deal of evidence soon accumulated to bear out Pauling's brilliant hypothesis. As we shall see later, there is now a considerable body of direct evidence for the existence of the α-helix in proteins.

Finally, it is necessary to consider the role played by the *R* groups in the formation of the α-helix. It is of course possible to generate two kinds of helices, a right-handed and a left-handed one (the one in Fig. 9-27 is right-handed). These would be equivalent to each other if it were not for the presence of the asymmetric carbon atom of the *l*-amino acid residues. In a left-handed α-helix generated by *l*-amino acids the *R* groups come too close to the neighboring —C=O groups, and one would therefore predict that *l*-amino acids would normally give rise to right-handed helices. As we shall see, this prediction has been validated by direct evidence.

The hypothesized α-helix, as recent studies show, is a stabilizing structure between amino acid residues that are close to each other on the polypeptide chain (nearest neighbors). However, as studies of the shorter polypeptides in aqueous solution indicate, it is not sufficiently stable by itself to maintain itself in water, which, after all, competes with the protein in the formation of hydrogen bonds. During the last decade it has become clear that other interactions, often between side chains of amino acid residues that are at considerable distance from each other on the poly-

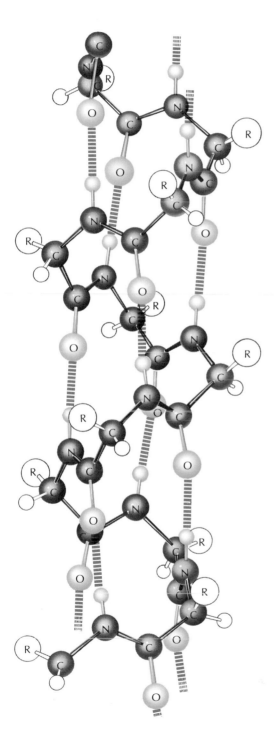

Fig. 9-27. Model of a right-handed α-helix proposed by Pauling and Corey. There are three amino acids in one hydrogen-bonded loop, 3.6 amino acids per turn, a pitch of 5.4 A, and a diameter of 10.5 A. Notice that all —C=O and —NH groups form hydrogen bonds. In the right-handed α-helix, all R groups point away from the helix. (Redrawn from Low and Edsall, 1956.)

peptide backbone, play an important role in stabilizing α-helix formation and other aspects of protein structure.

TERTIARY STRUCTURE—THE CONFORMATION OF THE POLYPEPTIDE

As we have already said, most proteins are compact, rigid molecules. Although the formation of an α-helix would shorten the extended polypeptide chain, this would by no means give a compact, near-spherical molecule. Thus, the dimensions of an α-helix of the protein of tobacco mosaic virus would be 10 × 240 A when in actual fact they are approximately 25 × 75 A. We therefore must imagine that the polypeptide, besides forming α-helices, must also bend and fold in such a way as to generate a more compact structure. Furthermore, from the high specificity of enzymes and antigen-antibody interactions it is clear that this folding is highly determinate and forms a uniquely characteristic structure for each given protein. The recent dramatic X-ray diffraction studies of globular proteins have amply confirmed these predictions. However, before we discuss these important results, we must first examine the kinds of interactions that we would expect to play a role in stabilizing the conformation of the polypeptide in the protein molecule.

We have already discussed hydrogen-bond formation between the —C=O and —NH groups of the amide linkage of the peptide bond. But another source of interaction, one which we must consider now in some detail, comes from the 20 different *R* groups, which we have so far ignored. These can interact with each other as well as with the amide groups of the peptide bond to form the following types of covalent and secondary interactions.

Covalent Interactions

Disulfide bonds are the most frequently encountered covalent bonds between side chains of the protein molecule. We have known for a long time that such bonds occur in proteins, but it is only recently, when the structure of insulin was elucidated by Sanger and that of ribonuclease by Moore, Stein, and their co-workers, that the precise structural chemistry of intrachain disulfide bonds was documented in detail (Fig. 9-28). Disulfide bonds can form spontaneously when two —SH groups of two cysteinyl residues are brought close to each other:

$$2\text{—SH} + 1/2\,O_2 = \text{—S—S—} + H_2O$$

The protein structural chemist can break these bonds by a variety of reducing agents such as cysteine, mercaptoethanol, or dithioerithritol; by oxidizing agents such as perchloric acid; or by sulfitolysis with sodium sulfite. Indeed, the disulfide bonds of a protein such as ribonuclease must be broken before its primary structure can be determined. Once this is done, it is possible to determine the exact location of the disulfide bonds by leaving them intact, digesting the protein with a variety of enzymes,

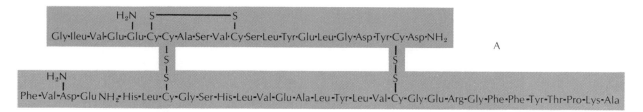

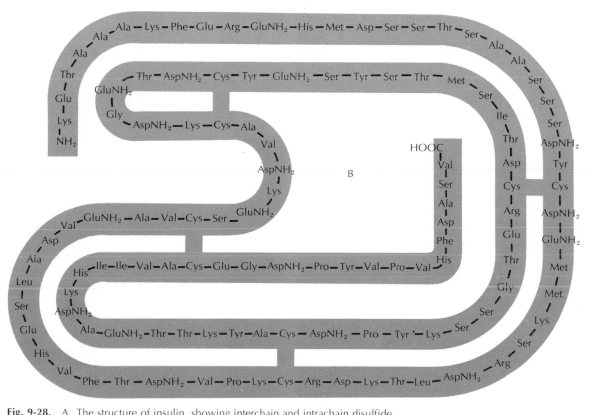

Fig. 9-28. A. The structure of insulin, showing interchain and intrachain disulfide bonds. Recent work shows that the cell synthesizes a precursor of insulin. This precursor consists of a single chain from which the active protein (hormone) is formed by the action of a proteolytic enzyme. The enzyme "chews out" a length of peptide from the chain, thus producing the two-chained structure shown above. B. The primary structure of the enzyme ribonuclease, showing disulfide bonds cross-linking the peptide chain in four places. (From Smyth, Stein, and Moore, 1963.)

and identifying the amino acids that are connected with the cystine (that is, disulfide containing) peptides.

A considerable number of proteins do not contain disulfide bonds (the proteins of the bacterium *E. coli* have very little or none) so that one cannot claim that they are of universal importance in maintaining protein structure. On the other hand, it has been possible to show in a number of proteins that do contain disulfide bonds that when they are reduced, the protein loses its specificity or becomes considerably more unstable. Thus, disulfides contribute to the stability of those proteins containing them.

(See Chapter 11.)

Until very recently there has been very little, if any, direct evidence for other covalent bonds involved in the maintenance of protein structure. As will be seen, it has been demonstrated recently that an enzyme found in the plasma of blood can cross-link fibrin molecules by forming "isopeptide" bonds between the γ-carboxyl groups of glutaminyl residues and the ε-amino groups of lysyl residues. Since similar transglutaminase enzymes are also found in cells, it is entirely possible that this type of bond is more widespread than formerly suspected, not only between molecules, but also within protein molecules.

Apolar or Hydrophobic Interactions

In recent years a great deal of evidence has accumulated that *apolar* or *hydrophobic* interactions involving nonpolar amino acid side chains (such as valine, leucine, isoleucine, phenylalanine, and tryptophan) occur. These interactions bring about an overall molecular conformation that brings the nonpolar amino acid side chains into the interior of the molecule, leaving the polar and ionic groups at the surface where they can interact with water. The driving force for this effect is believed by Kauzmann to be entropic. It can be shown that hydrophobic groups in water tend to orient the water molecules and restrict their rotational and translational freedom so that by withdrawing these groups from the water, the water becomes more disorganized, and entropy is increased. Hydrophobic bonds are increasingly thought to be important for the organization of macromolecules and even larger structures such as membranes. As we have indicated in the previous chapter, they are responsible for the "stacking interactions" between nitrogen bases in the DNA double helix.

Polar Interactions

There are a number of polar interactions that can occur within a polypeptide chain. We have already considered the hydrogen bonds formed between amide groups of close neighbors as being responsible for helix formation. Similar hydrogen bonds can form between amides of residues brought close to each other by peptide chains running next to each other in the same direction (*parallel pleated sheet*), or in opposite directions (*antiparallel pleated sheet*) (Fig. 9-29). Pauling and Corey first suggested this "pleated sheet" arrangement in relation to the structure of certain fibrous (insoluble) proteins, but as we shall see, this interaction also occurs in globular proteins.

Hydrogen bonds can also occur between a number of side chains, as, for example, between tyrosyl and histidyl and between seryl and aspartyl residues. A number of physical studies have definitely identified such interactions, especially in the case of tyrosyl in which there is a small shift of the ultraviolet absorption spectrum when the hydroxyl groups of the phenolic side chain ionize. The hydrogen-bond interactions

N-terminal

C-terminal

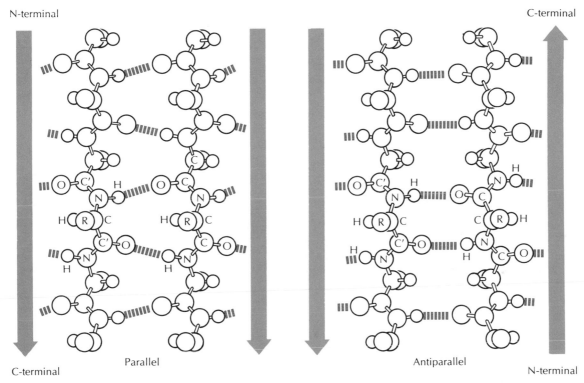

C-terminal

Parallel

Antiparallel

N-terminal

Fig. 9-29. Parallel and antiparallel pleated sheet arrangements of polypeptide chains proposed by Pauling and Corey. In the parallel case the polypeptide chains run in the same direction; in the antiparallel arrangement, neighboring chains run in opposite directions. A short length of antiparallel configuration has been discovered in lysozyme by Phillips (see Fig. 9-36) and of both configurations in carboxypeptidase A by Lipscomb (see Fig. 9-37). (Redrawn from Pauling and Corey, 1951.)

tend to shift the pK of the tyrosyl groups, a fact that can be observed by studying the effect of pH on the absorption spectra of proteins at 280 mμ.

Another form of a polar interaction is the ionic bonds that may occur between oppositely charged side chains such as the positively charged lysyl, arginyl, or histidyl side chains and the negatively charged glutamyl or aspartyl side chains.

The ambiguity about polar interactions is that proteins are dissolved in aqueous salt solution in which the water can compete with interpeptide hydrogen bonds and the ions with interpeptide ionic bonds. And indeed as we shall see, the majority of polar groups do interact with the solution on the surface of the protein molecule. However, a number of polar interactions have been shown to occur, probably because of nonpolar interactions that might contribute to their stability or even shield them from the action of the aqueous solvent and from ions. This latter possibility is entirely likely since X-ray studies have recently shown that protein molecules are extremely compact, and it allows for the presence of very few water molecules inside them. An ionic bond is an extremely strong form of interaction if other ions are not available to compete with it.

Figure 9-30 is a summary of the types of interactions that may stabilize the highly specific conformation of the polypeptide chain in the protein molecule (tertiary structure). This is as much as we knew until the dramatic work started by Kendrew and Perutz on the X-ray crystallographic analysis of the structure of globular proteins confirmed and extended the ideas regarding the nature of the forces stabilizing the conformation of the polypeptide in the protein molecule.

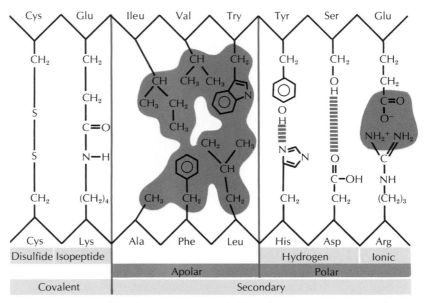

Fig. 9-30. Variety of bonds or interactions stabilizing the tertiary structure (conformation) of protein molecules. It is important to realize that these bonds do not act in a simple additive manner. Although most charged side chains will be oriented toward the aqueous solvent at the surface of the molecule, some polar bonds might remain inside the molecule, presumably "shielded" from the water and salt by nonpolar groups.

It is not possible here to discuss in any detail the method of X-ray crystallography employed by Kendrew and Perutz, and the student is advised to read carefully the *Scientific American* articles by Kendrew and by Phillips in the Suggested Reading List. Instead, we would like to discuss in very general terms the kind of problems encountered in the X-ray crystallography of proteins and how they were solved.

It has been known for a long time, largely through the efforts of Perutz, that X-ray diffraction diagrams of protein crystals exhibit regularity and detail down to dimensions less than 2 A. Figure 9-31 shows a "precession diagram" of myoglobin obtained by recording X-ray diffraction patterns while moving a protein crystal and the photo-

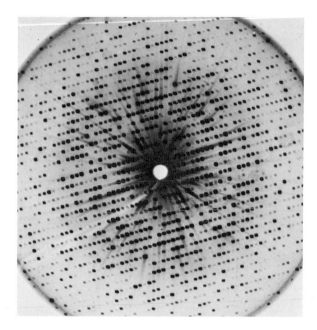

Fig. 9-31. Precession X-ray diagram of myoglobin. (From Crick and Kendrew, 1957.) This pattern is obtained when the crystal and the camera are moved simultaneously. In this type of diagram the scattering maxima near the center represent long spacings in the crystal whereas the scattering maxima at the outer edge of the diagram represent short spacings in the crystal. Kendrew and his associates were able to study myoglobin at 1.5 A resolution, which involved the measurement of some 10,000 intensities and the calculation of an equal number of Fourier transforms.

graphic plate simultaneously. It shows numerous scattering maxima; the ones close to the center represent large regularities in the protein crystal, whereas the ones at the periphery represent regularities within the protein molecule at resolutions as small as 1–2 A. Thus, we already knew in the early 1940s that there was enough data in these diagrams to work out the detailed structure of protein molecules. There was, however, a problem here which seemed unsurmountable at that time. The information obtained from the intensities of the scattering maxima does not permit the direct calculation of the "phase angles" needed for solving the Fourier equations. This is a mathematical way of saying that when the electron densities of atoms act as scatterers and the scattered X-rays are photographed, then the traces they leave on the photographs do not provide sufficient information to enable us to calculate backwards and obtain the structure that originally scattered the X-rays. With simpler structures one employed the procedure of making an educated guess about the possible structure, calculating what kind of scattering maxima one would obtain, and comparing it with the experimental maxima; then if need be, improving one's guess and seeing if it could be brought closer to the experimental data. Thus, by a series of convergent guesses one hoped to arrive at a structure from which one could calculate the same scattering maxima obtained experimentally. Protein molecules are far too complex to permit the use of such a procedure. There is, however, a procedure that permits one to calculate the structure directly from the scattering maxima. This involves comparing the scattering of the normal crystals of a given compound with crystals in which a strongly scattering atom or groups of atoms (such as heavy metals) have been substituted at a specific place in the molecule. This creates a change in the intensities (Fig. 9-32) of the scattering maxima, and from the change in intensities it is possible to calculate the exact position of the heavy atom; once that is done it is possible to calculate the phase angles and solve the Fourier equations. This procedure, referred to as the *isomorphous replacement method,* had worked well for complicated organic compounds such as penicillin, but until the middle 1950s no one had succeeded in preparing isomorphous replacements of protein crystals.

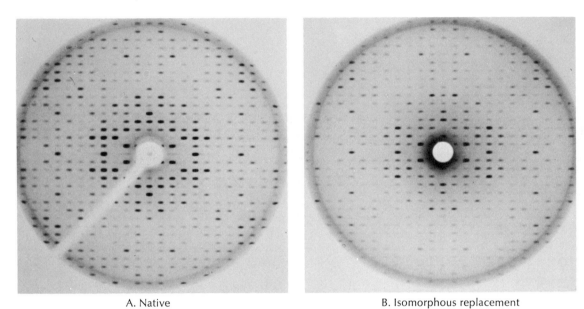

A. Native   B. Isomorphous replacement

Fig. 9-32. Comparison of X-ray diagram obtained from a normal horse heart cytochrome c crystal and that from an isomorphous replacement crystal obtained by diffusing $PtCl_4^=$ into it. Numerous differences in intensity of the scattering maxima can be seen. The measurement of intensity differences permits the determination of phase angles and from these the three-dimensional structure can be deduced. (Courtesy of Dickerson.)

(Fig. 10-19)

But in 1956 Dintzis and Bodo, while working with Kendrew, succeeded in preparing several heavy-metal derivatives of myoglobin, and when X-ray diffraction measurements of the crystals showed that they were indeed isomorphous replacements, a new era in the structural chemistry of proteins had begun. By now, detailed structural analyses of myoglobin, hemoglobin, lysozyme, carboxypeptidase A, and ribonuclease have been made, and our understanding of protein structure has been solidified enormously. We shall not dwell on how Kendrew and his associates measured thousands of scattering intensities and, utilizing special high-speed electronic computers, traced electron-density maps (Fig. 9-33) at various levels, superimposed them (Fig. 9-34), and built models of the myoglobin molecule first at 6 A resolution (Fig. 9-35) and then at 1.5 A resolution.

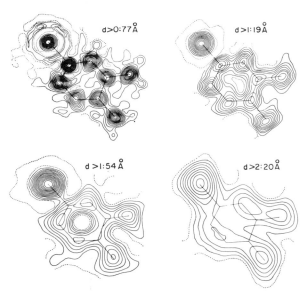

Fig. 9-33. An example of an electron-density map (Fourier projection). In this instance 4,5-diamino-2-chloropyrimidine was represented at four different resolutions. Notice how the six-membered ring structure becomes more clearly defined as the resolution increases. (From Dorothy Hodgkin.)

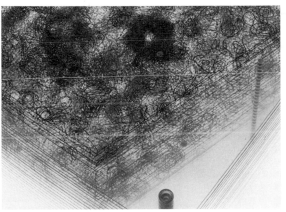

Fig. 9-34. A three-dimensional electron-density map of myoglobin. Electron densities at various levels are plotted on lucite sheets, and then the sheets are stacked over each other so that the three-dimensional structure of the molecule can be determined. (From Kendrew.)

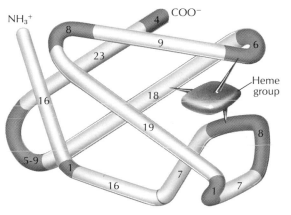

Fig. 9-35. The conformation of the myoglobin chain at 6 A resolution. The heme group is the shaded flat section at upper right. The eight straight portions are α-helical regions interrupted by seven nonhelical corners of different degrees of tightness. Black numbers represent the number of amino acids in a given α-helical region. Numbers in shaded regions represent the number of amino acids in a change of direction (corner) that are not involved in a α-helical configuration. (After Kendrew.) (For a detailed representation of myoglobin at 1.5 A resolution, see the *Scientific American* article by Kendrew.)

The 6 A resolution model that Kendrew built immediately provided direct evidence for the Pauling and Corey α-helix. Figure 9-35 shows that the myoglobin molecule is composed of 8 α-helical segments that are twisted around each other, interrupted by "corners" consisting of extended polypeptide chains (random coils) of different lengths.

It is the 1.5 A resolution diagram that provided detailed structural information at all levels of protein structure. First of all, at the level of primary structure it was possible to identify directly a large number of the amino acid residues. By combining these data with the chemical sequencing work being done at the time, it was soon possible to work out the complete amino acid sequence of the myoglobin polypeptide. This permitted the building of detailed three-dimensional structures in which the exact conformation of the polypeptide, both helical and nonhelical portions, was specified. When this had been done, Kendrew was able to analyze visually the interactions that presumably stabilized the myoglobin molecule. A number of generalizations emerged that, on the whole, confirmed the earlier speculations of the physical chemists regarding protein structure. Thus,

1. myoglobin is a compact molecule containing fewer than four molecules of water in its interior;
2. most of the polar side chains are on the surface of the molecule interacting with some water molecules (bound water);
3. the nonpolar side chains are found in the interior of the molecule; and
4. a number of polar interactions do appear to occur between various side chains and sections of the polypeptide chain as evidenced by the closeness (2–3.1 A) of these groups to each other (Table 9-4).

This magnificent piece of work by Kendrew and his co-workers supplied the capstone of the structural work on proteins initiated by Sanger. The latter's hunch that the primary structure of proteins is a highly specific entity for each type of protein molecule proved to be true also for the secondary and tertiary structure of proteins. And indeed, looking at it in retrospect, how could it have been otherwise? How else could one have accounted for the high degree of specificity exhibited by protein molecules such as enzymes, antibodies, and perhaps even structural proteins?

Recently, the results of a structural study of the enzyme lysozyme has been published by Phillips and his co-workers (Fig. 9-36). Its structure exhibits the same principles as those observed for myoglobin such as helical and nonhelical regions (although the helical content in lysozyme is less), as well as the presence of hydrophilic side chains outside and hydrophobic side chains inside the molecule. However, there are also some differences. The helices do not seem to be perfect α-helices but appear to be somewhat distorted. On the other hand, there appears to be another conformation of the polypeptide chain predicted by Pauling and Corey, the antiparallel "pleated sheet" arrangement formed by a polypeptide chain that bends on itself.

Even more recently, Lipscomb and his co-workers worked out the structure of carboxypeptidase A at 2 A resolution (Fig. 9-37). This molecule contains four major and four minor α-helical regions, constituting 30 percent of the molecule, whereas 20

Table 9-4 POLAR INTERACTIONS IN MYOGLOBIN*

RESIDUE	TOTAL NUMBER	NUMBER ON SURFACE INTERACTING WITH SOLVENT	NUMBER "BURIED" INSIDE MOLECULE	NUMBER INVOLVED IN STRONG INTRAMOLECULAR INTERACTIONS	PARTNERS
Lys	19	19	0	3	Glu
Arg	4	4	0	1	Asp
Glu or GluNH$_2$	19	19	0		Lys, Try Chain NH
Asp or AspNH$_2$	8	8	0	4	Arg, His
Ser	6	5	1	4	Chain CO Chain NH
Thr	5	3	1–2	3	Chain CO Chain NH
His	11	7–8	3	3	Chain CO Fe^{++}(H$_2$O) Asp
Try	2	2	0	1	Glu
Tyr	3	3	0	1	Chain CO

* From F. M. Richards.

percent of the molecule consists of eight extended chains forming a pleated sheet of antiparallel as well as parallel chains. The pleated sheet of carboxypeptidase A is a most astounding structure: it runs down the center of the molecule, changing its plane so that the top chain is rotated 120 degrees with respect to the bottom chain. The stereoscopic view presented in Fig. 9-37 is included here because the authors believe that this device will play an increasingly important role in the presentation of three-dimensional structure of biological macromolecules.

Although the fine structure of only a few protein molecules has been worked out so far, a number of general conclusions seem to emerge. Perhaps the most important conclusion is that unlike DNA structure no single architectural principle seems to dominate protein structure. To be sure, we expect to find hydrophilic groups on the outside of the molecule and hydrophobic groups in one or more regions ("oil droplets") in the interior, but we can also expect exceptions to this rule. We expect to see α-helices but also other kinds of helices. It is likely that pleated sheets will be found quite frequently with chains running in antiparallel and also parallel sense. Finally, we expect to see extended chains performing a variety of functions such as allowing for changes in direction of chains or even endowing the molecule with a certain amount of conformational flexibility. In short, the 20-lettered alphabet of proteins is used to elaborate a complex structure utilizing a variety of structural principles. This is to be expected, for, as we shall see, proteins play a large number of highly specific functions—be they

Key to 9-36A: Explanation of symbols for amino acids.

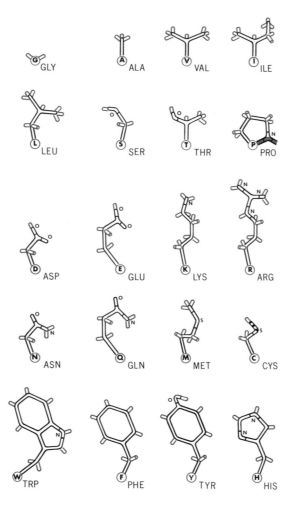

Fig. 9-36A. Three-dimensional model of the structure of lysozyme. (From Phillips as drawn by Irving Geis for the *Atlas of Protein Sequence and Structure.*) The molecule consists of a single chain of 129 amino acids. There are some α-helical regions in the lower right-hand corner (amino acid residues 25 to 35), at the center of the molecule (amino acid residues 89 to 95), and a short antiparallel pleated sheet region on the lower left-hand side. The polypeptide chain in the molecule is folded so as to form a cleft running from upper left to lower right. This cleft is the "active site" and interacts with the polysaccharide chain that is the substrate of this enzyme. A detailed study of this model reveals numerous examples of the interactions (Fig. 9-30) that had been thought to stabilize the folding of the polypeptide chain before high resolution techniques such as X-ray diffraction in fact demonstrated that they occur. The folding pattern of the first 56 amino acid residues shows the tight (mainly α-helical) configuration of the first 36 residues as well as the antiparallel pleated sheet folding pattern in the hairpin loop formed by residues 41 through 54. It is likely that this structure forms as the polypeptide chain is being synthesized and "comes off" the ribosome. It is therefore possible that the final structure of the enzyme is influenced by the "kinetics of release" of the polypeptide chain. (For a more detailed treatment of the three-dimensional aspects of protein structure see *The Structure and Action of Proteins* by Dickerson and Geis, Harper & Row, 1969.)

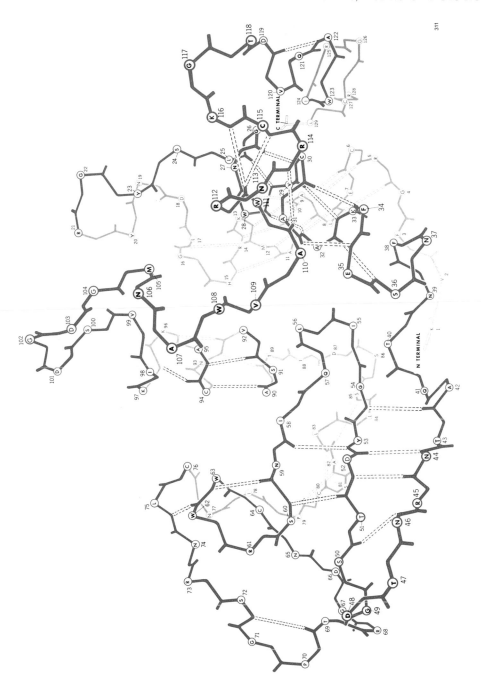

Fine Structure of the Lysozyme Molecule

Fig. 9-36B. Stereoscopic view of lysozyme molecule. (Courtesy of Phillips and Browne.) The terminal amino end (*bottom left rear*) and carboxyl end (*bottom right rear*) are marked by two gray hemispheres; every fifth residue is numbered; the oxygens of Asp52 and GluNH$_2$35, which may be involved in catalysis, are shown by dark hemispheres; the disulfide bridge sulfurs are represented by white spheres. (This view may be seen in three dimensions by using a stereoscopic viewer such as Model CF-8 from the Abrams Instrument Corporation, Lansing, Michigan.)

structural, catalytic, or regulational—and therefore must rely on a large variety of structural principles to achieve this.

The high-resolution work of Kendrew, Phillips, Lipscomb and others not only gives us a static picture of the structure of a protein molecule, it also helps us to develop important insights on (1) how such structures become generated, and (2) how they might function in their particular biological role. Let us deal with these in turn.

(See Figs. 10–18 and 10–19.)

Even before the three-dimensional structure of ribonuclease was worked out in detail, Anfinsen and his co-workers began an interesting series of experiments with the enzyme ribonuclease. Anfinsen asked the following question: Can the three-dimensional structure of ribonuclease be generated purely from the sequence of amino acids in the polypeptide, or is some other structural entity or template necessary to contribute to the formation of the specific structure?

Anfinsen and his co-workers treated ribonuclease with mercaptoethanol and 8M urea, the former to break the four disulfide bonds and the latter to bring about the unfolding of the molecule. He found that by *dialyzing* the urea away slowly in the presence of mercaptoethanol, which is then also slowly removed, a ribonuclease molecule is formed which appears to have the same physical properties as the original molecule, the same enzymatic properties, and what is more, the same four pairs of disulfide bonds. In fact, one can calculate that by a random process only 1 percent of the molecules would have formed the four correct pairs of disulfide bonds when actually 95 percent of the ribonuclease molecules turned out to be correct. Thus, Anfinsen and his group showed that ribonuclease, when it is unfolded, can refold spontaneously to produce essentially the same ordered structure as before. And, we can conclude that at least in the case of ribonuclease, its primary structure, which originally was determined by the genetic material, is sufficient for the determination of the final, biologically specific structure of the enzyme. This would suggest that the "native" structure of a protein is also thermodynamically the most stable form. Similar

Fig. 9-37. Stereoscopic view of the conformation of the polypeptide chain of carboxy-peptidase A. (Courtesy of Lipscomb.) By using a stereo viewer one can see quite clearly various structural features such as the four major α-helices and the eight chains that form a pleated sheet structure. Note that the pleated sheet changes its direction, the uppermost chain pointing east-west and the lowest being at an angle of 120° to it. Symbols are: 1 = N-terminus (Ala 1); 12 = C-terminus (AspNH$_2$ 307); white sphere = zinc atom necessary for the enzyme's activity; 2, 4, 8 = residues binding zinc atom (His 69, Glu 72, Lys 196); 6 = binding site of COO$^-$ terminus of substrate; 10, 11 = side chains involved in enzyme action (Tyr 248, Glu 270); 9 = Tyr 198 side chain which moves 8 A upon binding of substrate (this observation is so far the best evidence for Koshland's Theory of adaptive fit); 5, 7 = disulfide bond (Cys 138, 161). (This model may be seen in three dimensions by using a stereoscopic viewer such as Model CF-8 from the Abram Instrument Corporation, Lansing, Michigan.)

(See Chapter 10.)

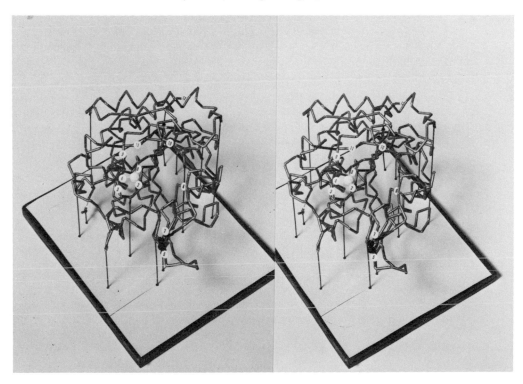

(See Chapter 15.)

experiments have been repeated with a number of enzymes, in some cases with considerable success, in other cases with less. As will be discussed later, the release of the peptide chain from ribosomes during peptide synthesis occurs from the N-terminal end and may be gradual. Thus, the kinetics of the release may in part play a role in the determination of the final structure. In the case of some enzymes this final, biological structure may be close to the thermodynamically most stable configuration but not necessarily the most stable. In the case of lysozyme, Phillips and his co-workers found that the first 40 amino acid residues form a compact structure, which may act as a kind of "nucleation device" or substructure around which the rest of the structure folds. This process may bring about a structure that is not the thermodynamically most stable one, but possibly it is the kinetically most feasible way of producing a structure endowed with a certain kind of specificity. The fact that enzymes, even when stored in the cold, lose their structure with varying degrees of rapidity also suggests that not all enzymes have the thermodynamically most stable structure, although admittedly other interpretations of this observation are possible. Eventually, this important problem might be resolved by calculating (with the aid of electronic computers) stable structures from given amino acid sequences and determining whether nature uses these structures or some other less stable one. It is our hunch that since stability is only one criterion for the evolutionary selection of a given enzyme structure (range of specificity, catalytic efficiency, and ability to be regulated by feedback inhibition are others), nature might well have to make compromises in the primary structure of proteins and occasionally might have had to sacrifice thermodynamic stability in favor of biological function.

Another consequence of understanding the detailed structure of proteins is the elucidation of the mechanism of their specific biological role. Here again the study of lysozyme has made important contributions. As we shall see, the mechanism of the specific enzymatic activity of lysozyme can now be understood in terms of the geometry of the molecule and, interestingly, in terms of changes in geometry. Phillips was able to show that upon binding of the polysaccharide substrate, small but definite conformational changes occurred in the enzyme. This and other data we shall presently consider indicate that superimposed on the basic structural rigidity of proteins is a possibility of limited but highly specific conformational changes, which may ultimately prove to play a very important role in the understanding of mechanisms of biological specificity.

QUATERNARY STRUCTURE—WHEN POLYPEPTIDES GET TOGETHER

So far we have discussed proteins such as tobacco mosaic virus protein, myoglobin, ribonuclease, and lysozyme, which are composed of single polypeptide chains. Numerous proteins, however (Table 9-5), contain more than one polypeptide chain. Furthermore, as Table 9-5 demonstrates, we know of a number of instances in which a protein molecule is constructed of two or more different chains. In fact, the variability is even greater than indicated in Table 9-5, for some enzymes such as glutamic dehydrogenase exist as large complexes, which upon dilution dissociate into smaller ones that are still enzymatically active. These subunits are then in turn composed of

Table 9-5 QUATERNARY STRUCTURE OF PROTEINS*

PROTEIN	MOLECULAR WEIGHT	NUMBER OF CHAINS†	NUMBER OF —S—S—BONDS
Insulin	5798	1 + 1	3
Ribonuclease	13,683	1	4
Lysozyme	14,400	1	5
Myoglobin	17,000	1	0
Papain	20,900	1	3
Trypsin	23,800	1	6
Chymotrypsin	24,500	3	5
Carboxypeptidase	34,300	1	0
Hexokinase	45,000	2	0
Taka-amylase	52,000	1	4
Thiogalactoside transacetylase	65,000	2	—
Thioredoxin reductase	65,800	—	—
Bovine serum albumin	66,500	1	17
Yeast enolase	67,000	2	0
Hemoglobin	68,000	2 + 2	0
Liver alcohol dehydrogenase	78,000	2	0
Alkaline phosphatase	80,000	2	4
Hemerythrin	107,000	8	0
Glyceraldehyde-3P-dehydrogenase	140,000	4	0
Lactic dehydrogenase	140,000	4	0
γ-Globulin	140,000	2 + 2	25
Yeast alcohol dehydrogenase	150,000	4	0
Tryptophan synthetase	159,000	2 + 2	—
Aldolase	160,000	4(?)	0
Phosphorylase-b	185,000	2	—
Threonine deaminase (Salmonella)	194,000	4	—
Fumarase	200,000	4	0
Tryptophanase	220,000	8	4
Formyltetrahydrofolate synthetase	230,000	4	—
Aspartate transcarbamylase	310,000	4 + 4	0
Glutamic dehydrogenase	316,000	6	0
Fibrinogen	330,000	2 + 2 + 2	—
Phosphorylase-a	370,000	4	—
Myosin	500,000	2 + 3	0
β-Galactosidase	540,000	4	—
Ribulose diphosphate carboxylase	557,000	24	—

*(Courtesy of Schachman.)
†Wherever we know that the chains in a molecule are dissimilar we have indicated it by noting the number of dissimilar chains found. Thus, for example, hemoglobin, which has 2 α-chains and 2 β-chains, is here characterized by the notation "2 + 2."

several chains. In other instances such as plasma transglutaminase, activity is lost when the enzyme breaks down into subunits upon dilution. This raises the difficult question as to the applicability of the term "molecule" in relation to proteins. Where does a given entity cease being a molecule and begin to be an aggregate of molecules? The usual chemical criterion of covalent interactions is not useful in relation to biological macromolecules—it would make two molecules out of every DNA duplex and four molecules out of every hemoglobin unit. Biologists prefer a functional criterion for the definition of a molecule, one which takes into account the fact that the DNA duplex or the hemoglobin unit has a certain biological function that the individual subunits do not possess or at least cannot duplicate as efficiently. In certain instances such as glutamic dehydrogenase in which the subunit still possesses the enzymatic activity, it nevertheless loses the ability to be regulated by feedback inhibition. Finally, as we shall see later, we now know of the formation of multi-enzyme complexes whose association appears to have a metabolic function. Wherever possible we shall use the following terminology in relation to enzymes:

(See Chapters 11 and 18.)

chain—single polypeptide
subunit—two or more chains lacking the full range of catalytic and regulatory function, found in the molecule as a whole
molecule—unit of full catalytic and regulatory activity
complex—more than one unit of similar or dissimilar catalytic and regulatory activities.

One interesting aspect of quaternary structure is the formation of *isozymes*. A number of instances are known such as chicken lactic dehydrogenase in which a range of different forms of a given enzyme exists. Markert and Kaplan have shown that lactic dehydrogenase is made of four chains, and they suggested that there are two different genes capable of making slightly different chains. Thus, chicken heart is capable of making a great deal of an enzyme we can denote as BBBB, and chicken muscle is capable of making one we shall call AAAA. This would suggest that three other enzymes could in principle occur in the tissues of chicken (BAAA, BBAA, and BBBA), and in fact Markert and Kaplan found three additional lactic dehydrogenase enzymes to occur. Electrophoretic studies capable of separating the two chain types of the five different enzymes confirmed beautifully this analysis: two distinct bands appeared in the proper amounts to fit the stoichiometry of Markert and Kaplan's predictions (Fig. 9-38).

The biological implications of the isozyme story are of considerable interest. The cell and, more importantly, the multicellular organism are able to manufacture five enzymes with slightly different physical properties by utilizing in a variety of combinations only two polypeptide chains. The importance of this in cellular regulation and in the differentiation of multicellular systems is being actively studied at the moment.

Finally, we should consider briefly the mechanisms whereby the association of chains and subunits in protein molecules is brought about. Numerous studies have by now been performed showing that enzymes can be dissociated into subunits or even into separate chains, after which the process can be reversed and an active enzyme *reconstituted*. A variety of techniques have been used for the dissociation of proteins. Reagents such as urea, guanidine hydrochloride, detergents, and (if disulfide bonds are to be split) mercaptoethanol or dithioerithritol must be used. The simplest explanation for the association of subunits is to implicate the same covalent, apolar, and polar interactions that are responsible for tertiary protein structure, thus ultimately relating the formation of a multichain protein molecule to the primary structure of its polypep-

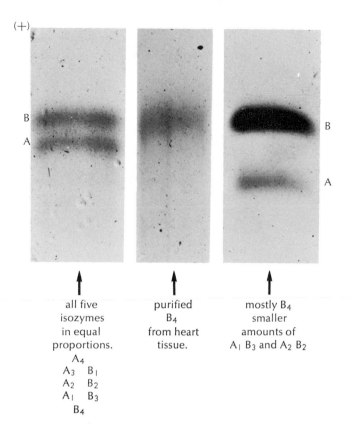

(+)

B
A

all five
isozymes
in equal
proportions.
A_4
A_3 B_1
A_2 B_2
A_1 B_3
B_4

purified
B_4
from heart
tissue.

mostly B_4
smaller
amounts of
$A_1 B_3$ and $A_2 B_2$

B

A

Fig. 9-38. Electrophoretic evidence for chain composition of isozymes of lactate dehydrogenase (LDH). Enzymes from different sources were dissociated into their component chains with urea. The chains were separated from each other by electrophoresis on urea starch gels. Notice that in the channel on the left all 5 isozymes give only two bands. (Courtesy of Markert.)

tides. The most detailed support for this view comes from the X-ray crystallographic studies and association studies of hemoglobin.

Hemoglobin has a molecular weight of 68,000 and consists of two α-chains and two β-chains with slightly different primary structures. Perutz and his co-workers have used a number of isomorphous replacements and worked out the three-dimensional structure to 5.5 A resolution. Interestingly, although the α- and β-chains of hemoglobin differ somewhat from the single chain of myoglobin, the conformation achieved by each of these chains is strikingly similar (Fig. 9-39). A detailed comparison between amino acid sequences of the three chains may eventually contribute to a better understanding of the factors responsible for tertiary folding. The conformation of the hemoglobin chains is such that the four chains fit perfectly with each other to form a near-spherical structure (Fig. 9-41).

The interaction of the four chains in hemoglobin is of considerable importance in its oxygen-carrying function. Hemoglobin contains four heme groups, one associated with each chain. It is a remarkably efficient oxygen carrier as it is able to absorb four

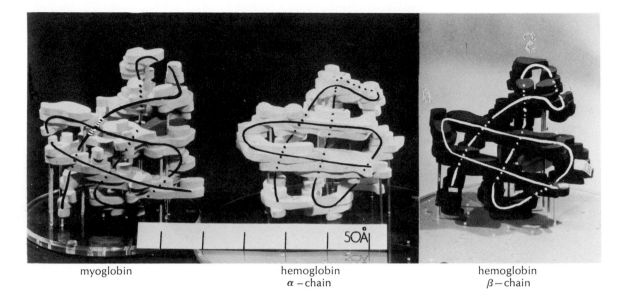

myoglobin hemoglobin hemoglobin
α – chain   β – chain

Fig. 9-39. Comparison of the conformation of the polypeptide chain of myoglobin with that of the α- and β-chains of hemoglobin. These three chains show very similar conformations in spite of a number of differences in amino acid content. (From, Perutz *et al.*, 1960.)

oxygen molecules, the binding of each actually raising the affinity of the hemoglobin for the succeeding oxygen molecule (Fig. 9-40). Since the iron atoms of the four heme groups are at least 25 A apart, one must conclude that the effect these four oxygen-absorbing centers have on each other must be exerted through conformational changes in the peptide chains. This indeed was suspected for a long time since it was known that

Fig. 9-40. Comparison of the oxygen-dissociation curves of myoglobin and hemoglobin. Myoglobin with its single heme group exhibits the typical hyperbolic dissociation curve normally encountered in such phenomena. In the case of hemoglobin one obtains an "S-shaped" or sigmoid curve, which shows that the binding of the first oxygen molecule facilitates the binding of the second one; similarly, the second molecule facilitates the binding of the third, and the third, the fourth. This cooperative effect makes the hemoglobin molecule a very effective oxygen carrier because in the intermediate range of oxygenation it is able to pick up and release oxygen with considerable ease.

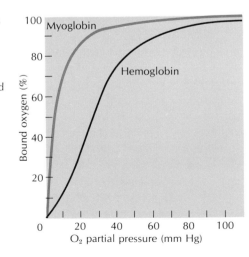

reduced horse hemoglobin crystals (hexagonal plates) change their crystal structure to needles when they become oxygenated. And, indeed, recent detailed studies by Perutz and his co-workers show considerable conformational difference between reduced and oxidized hemoglobin (Fig. 9-41).

Fig. 9-41. The effect of oxygenation on the relationships between the subunits of hemoglobin. (From Perutz *et al.*, 1964.) Upon oxygenation the subunits can be shown by X-ray analysis to turn with respect to each other so that the distances between the iron atoms of the β-chains decrease by 6.5 A and those between the α-chains increase by 1.0 A. In the above photographs the 5.5 A resolution model of hemoglobin is shown as well as the axes of rotation of all four subunits. The exact mechanism whereby these conformational changes are brought about is being worked out by Perutz and his collaborators, and we can expect a more detailed understanding of this interesting phenomenon in the near future.

Hemoglobin can easily be dissociated and associated in solution, a clear demonstration, it would seem, that the final structure of the molecule is determined by the amino acid sequence of the two types of chains of which it is composed. However, the cell must have some tricks up its sleeve since it apparently does place some restrictions on the association of the chains. A human heterozygote for the sickle cell anemia gene makes two kinds of β-chains, a normal β and an abnormal β_s. Interestingly, such an individual makes two kinds of molecules $\alpha\alpha\beta\beta$ and $\alpha\alpha\beta_s\beta_s$. He does not make hybrid molecules $\alpha\alpha\beta\beta_s$, although such molecules can be made *in vitro* by dissociating and reassociating a mixture of normal and abnormal molecules. Whether this observation is of importance for the understanding of *in vivo* assembly of protein molecules from their chains and subunits remains to be seen.

We have discussed in considerable detail the structure of protein molecules because we hope not only to convey to the student the importance of this subject for the understanding of the cellular machinery but also to provide sufficient background for an understanding of the developments of the future. In the next chapter we shall utilize the knowledge we have gained to examine the mechanism whereby enzymes speed up the rate of chemical reactions.

SUGGESTED READING LIST

Offprints

Gray, G. W., 1951. "Electrophoresis." *Scientific American* offprints. San Francisco: W. H. Freeman & Co.

Gray, G. W., 1961. "The Ultracentrifuge." *Scientific American* offprints. San Francisco: W. H. Freeman & Co.

Hall, C. E., 1961. "Electron Microscopy of Nucleic Acids and Proteins." *Journal of Applied Physics,* p. 1640. Bobbs-Merrill Reprint Series. Indianapolis: Howard W. Sams & Co.

Ingram, V. M., 1958. "How Do Genes Act?" *Scientific American* offprints. San Francisco: W. H. Freeman & Co.

Ingram, V. M., 1962. "The Evolution of a Protein." *Fed. Proc.,* pp. 1053–1057. Bobbs-Merrill Reprint Series. Indianapolis: Howard W. Sams & Co.

Kendrew, J. C., 1961. "The Three-dimensional Structure of a Molecule." *Scientific American* offprints. San Francisco: W. H. Freeman & Co.

Kendrew, J. C., 1963. "Myoglobin and the Structure of Protein." *Science*, pp. 1259–1266. Bobbs-Merrill Reprint Series. Indianapolis: Howard W. Sams & Co.

Levinthal, C., 1966. "Molecular Model-building by Computer." *Scientific American* offprints. San Francisco: W. H. Freeman & Co.

Perutz, M. G., 1964. "The Hemoglobin Molecule." *Scientific American* offprints. San Francisco: W. H. Freeman & Co.

Phillips, D. C., 1966. "Three-dimensional Structure of an Enzyme Molecule." *Scientific American* offprints. San Francisco: W. H. Freeman & Co.

Porter, R. R., 1967. "The Structure of Antibodies." *Scientific American* offprints. San Francisco: W. H. Freeman & Co.

Richards, F. M., and Vithayathil, P. J., 1959. "The Preparation of Subtilisin-modified Ribonuclease and the Separation of the Peptide and Protein Components." *Jour. Biol. Chem.,* pp. 1459–1465. Bobbs-Merrill Reprint Series. Indianapolis: Howard W. Sams & Co.

Sanger, F., and Smith, L. F., 1957. "The Structure of Insulin." *Endeavour*, pp. 48–53. Bobbs-Merrill Reprint Series. Indianapolis: Howard W. Sams & Co.

Stein, W. H., and Moore, S., 1951. "Chromatography." *Scientific American* offprints. San Francisco: W. H. Freeman & Co.

Thompson, E. O. P., 1955. "The Insulin Molecule." *Scientific American* offprints. San Francisco: W. H. Freeman & Co.

Williams, C. A., Jr., 1960. "Immunoelectrophoresis." *Scientific American* offprints. San Francisco: W. H. Freeman & Co.

Zuckerkandl, E., 1965. "The Evolution of Hemoglobin." *Scientific American* offprints. San Francisco: W. H. Freeman & Co.

Articles, Chapters and Reviews

Crick, F. H. C., and Kendrew, J. C., 1957. "X-ray Analysis and Protein Structure." In: *Advances in Protein Chemistry*. New York: Academic Press, Inc.

Edsall, J. T., and Wyman, J., 1958. "Problems of Protein Structure." Chapter 3 in *Biophysical Chemistry*, Vol. I. New York: Academic Press, Inc.

Books

Bailey, J. L., 1967. *Techniques in Protein Chemistry.* New York: Elsevier Publishing Co.

Dickerson, R. E., and Geis, I., 1969. *The Structure and Action of Proteins.* New York: Harper & Row Publishers, Inc.

Haurowitz, F., 1963. *Chemistry and Biology of Proteins.* New York: Academic Press, Inc.

Neurath, H. (ed.) 1964. *The Proteins,* Vols. I-IV. New York: Academic Press, Inc.

Scheraga, H. A., 1961. *Protein Structure.* New York: Academic Press, Inc.

Schroeder, W. A., 1968. *The Primary Structure of Proteins.* New York: Harper & Row Publishers, Inc.

Steiner, R. F., 1965. *The Chemical Foundations of Molecular Biology.* Princeton, N. J.: D. Van Nostrand Co., Inc.

BIOLOGICAL DYNAMICS

Part FOUR

In this part we shall attempt to show how the molecules
we have described in Part Three interact to produce
the cellular properties we have discussed in Parts One and
Two. Although no complete molecular description of
cell structure and function is as yet possible, modern cell
biology has made sufficient progress to indicate the
direction in which our understanding will take us. It is
therefore important, in this dynamically oriented portion
of the book, to provide the experimental bases on
which our conclusions rest. Ultimately, it will be the
success of this molecular approach to the phenomena of
life that will justify our basic assumption that living
matter and physical matter are part of the same continuum
and subject to the same natural laws.

Chapter 10

We pointed out in Chapter 7 that most organic compounds of the cell are remarkably stable at physiological temperatures, pressures, and hydrogen-ion concentrations. If urea, for instance, is dissolved in water, it will not react with its solvent at an appreciable rate even though to do so would release a considerable amount of energy.

$$\begin{array}{c} H_2N \\ \diagdown \\ C{=}O + H_2O \longrightarrow CO_2 + 2NH_3 + 13{,}800 \text{ cal} \\ \diagup \\ H_2N \end{array}$$

Eyring's explanation of this is that the urea and water do not react at an appreciable rate because the reaction has to pass through an "activated complex," the formation of which takes a great deal of energy (Fig. 10-1). Thus, in order to form the activated complex, water and urea molecules must collide with a certain minimum amount of energy. As Fig. 10-2 shows, only an infinitesimal number of urea and water molecules have this minimum amount of energy at room temperature. By raising the temperature, a larger proportion of molecules will achieve this minimal energy, and the rate of the reaction will increase correspondingly. This is what the laboratory organic chemist does, although he also at times varies the pressure and utilizes extremes of pH in order to speed up the rate of organic reactions. The cell, however, carries out its reactions at mild temperatures, low pressures, and pH values close to neutrality although the reactions often proceed at considerable rates.

These mysterious powers of the cell are of course due to the specific biological catalysis carried out by enzymes. The speeding up of the chemical reactions in the cell,

Fig. 10-1. Energy diagram of the hydrolysis of urea showing that in order to react, urea and water must have sufficient energy to form an activated complex. Since at room temperature very few molecules have sufficient energy to form the acitivated complex, the rate of the reaction is exceedingly slow even though the formation of products is very favorable energetically.

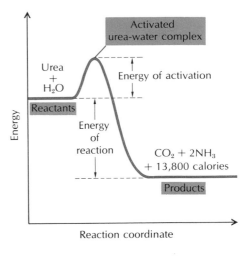

Fig. 10-2. Energy distribution of urea in water at two temperatures in relation to activation energy necessary for hydrolysis of urea. At 100° C a much higher proportion of molecules have energies equal to or greater than the activation energy; this explains the considerable effect of temperature on the rate of a chemical reaction.

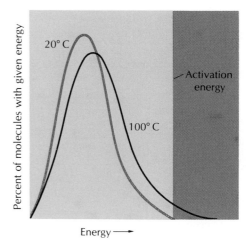

however, is only one side of the catalytic coin. The other side is the regulation or control of the reactions. Thus, in a sense, it is fortunate that uncatalyzed organic reactions are so terribly sluggish; if they were not so, it is hard to see how the cell could regulate them. Hence, since organic reactions are slow, enzymes can both speed them up and control their rate.

The concept of catalysis was proposed by the Swedish chemist Berzelius in 1853 to explain the speeding up of chemical reactions not only by acids, but also by extracts obtained from living tissues. He compared, for instance, the much greater effect of the enzyme *diastase* on the hydrolysis of starch with the effect of sulfuric acid. Berzelius actually suggested that the so-called "vital force" of organisms resided in the unusual property of biological catalysts that could greatly speed up chemical reactions, a concept to which many of his contemporaries, including the great chemist Liebig, objected violently. The precise measurements of the rates of reactions by Van't Hoff and Ostwald finally convinced most chemists that extracts from tissues could indeed speed up the rate of certain chemical reactions; these works finally led to the definition of catalysis as the speeding up of a reaction without affecting the overall results of the reaction. This notion at first was thought to imply that the catalyst does not enter into the reaction, but subsequent work has made it clear that the catalyst does react with the *substrate* and is regenerated at the end in its original form. This enables the catalyst to go through many cycles of reaction, and explains why it is effective in such very low concentrations. The substrate-catalyst complex has an energy of formation lower than that of the "activated" intermediates of the uncatalyzed reaction, thus permitting a larger number of molecules to react at a given temperature (Fig. 10-3). It should be noted that since the formation of the catalyst-substrate intermediate is temporary, having nothing to do with the initial and final states, it has absolutely no effect on the equilibrium of the reaction. Catalysts are not thermodynamic genii capable of pushing reactions "uphill." They do not shift equilibria, but merely increase the rate at which the equilibrium is reached.

In the case of urease the rate of the enzymatic reaction is speeded up by a factor of 10^7 over the uncatalyzed reaction, giving an absolute rate (turnover number) of 10^4 sec^{-1}. This means that one enzyme molecule will catalyze the hydrolysis of 10,000 urea molecules per second.

Fig. 10-3. A. Energy diagram of the hydrolysis of urea showing that catalysis lowers the energy of formation of the activated complex. Enzyme catalysis is much more effective than H^+ catalysis because the energy of activation in the enzymatic reaction is much lower.   B. Energy distribution of urea in water showing the effect of lowering the activation energy on the percentage of molecules having sufficient energy to form the activated complex and therefore to react. Notice that the dramatic increase in the proportion of molecules having sufficient energy to react when the activation evergy is lowered by enzyme catalysis.

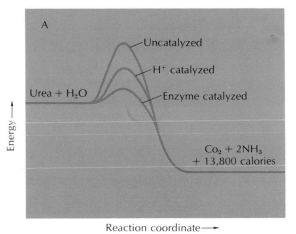
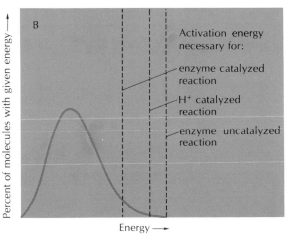

Biological catalysis by enzymes obeys the same general rules observed for nonenzymatic catalysis. It differs from other forms of catalysis chiefly in the greater effectiveness of lowering the energy of activation barrier (see Fig. 10-3) and in the extraordinary degree of specificity exhibited. Herein lies a clue to the delicate control of cell processes that enzymes exert. This control is possible because enzymes can act at very low concentration and with a high degree of specificity. By regulating the supply and activity of tiny amounts of active enzymes, the cell can control the metabolic flux of the compounds within it. In order to understand their regulatory capacity we must first study the general properties of enzymes.

THE ENZYME ASSAY

Modern biochemistry owes its success to a biological phenomenon that living systems have utilized from their inception, namely the specificity of enzyme action. This in practice permits us to mix a purified substrate with enzyme preparations that might contain 1000 times as much impurity and obtain a quantitative assay of the amount of enzyme present. The reason for this, as we have indicated, is that the enzyme can ignore all the extraneous compounds and catalyze specifically the reactions of its substrate. It is possible, for instance, to grind up some jack beans and add this crude preparation to purified urea. The hydrolysis of the urea can then be followed by measuring the production of ammonia with an appropriate chemical test (Fig. 10-4). The practice usually followed is to draw a tangent to the curve with its origin at zero time. The slope of this line is called the initial rate of the reaction and is related to the enzyme concentration.

As we have described in Chapter 9, to estimate the purity of an enzyme preparation, the initial rate is divided by the total amount of protein present. This measure is called the specific activity, and it increases as successive steps of purification provide a purer

Fig. 10-4. A typical enzyme assay. The rate of production of ammonia by the action of urease on urea and water. The initial rate is measured by drawing a tangent to the curve starting at zero time.

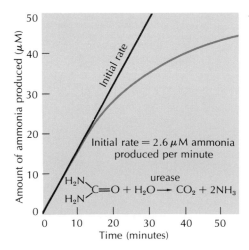

enzyme preparation. When the specific activity cannot be increased beyond a certain limit, even though a wide range of purification procedures have been employed, it is generally assumed that the enzyme is pure. This may or may not be the case, depending on the stringency of the criteria used for judging purity.

(See Chapter 9.)

An enzyme assay is not an absolute test. The initial rate depends on a number of parameters such as substrate concentration, pH, ionic strength, the kind of buffer used, and the temperature. Care must always be taken that all of these are specified. Furthermore, care should normally be taken that the substrate concentration is sufficiently high or enzyme concentration sufficiently low so that the enzyme is fully "saturated." Thus, if one plots initial velocity against enzyme concentration at a given substrate concentration, one obtains a curve that at first is a straight line (Fig. 10-5). It is in this linear region, when the initial rate is proportional to enzyme concentration, that the enzyme assay should be carried out.

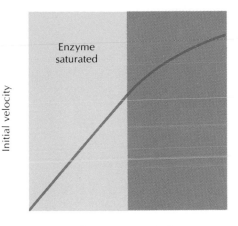

Enzyme saturated

Initial velocity

Enzyme concentration

Fig. 10-5. Relationship between initial velocity and enzyme concentration. In the early, linear portion of the curve, the enzyme is "saturated" by the substrate so that the rate of the reaction is proportional to the enzyme concentration. Enzyme assays should be carried out in this region. At higher enzyme concentrations the enzyme may no longer be saturated by the substrate so that the relationship between velocity and enzyme concentration is no longer linear.

THE PROTEIN NATURE OF ENZYMES

The general procedure we have just outlined was employed by Sumner, who in 1926 was the first to crystallize an enzyme, urease. This achievement was especially remarkable because it took place at a time when an atmosphere of mystery and awe surrounded the phenomenon of biological catalysis. The crystals of urease that Sumner obtained proved to be made of protein, but owing to the prestige of the great chemist Willstätter, who opposed the notion that enzymes were "mere" proteins, it took a great deal of further experimentation before the protein nature of enzymes became generally accepted.

By 1966 some 100 different enzymes had been crystallized, and a great many more had been purified; all of them were found to contain protein. Perhaps the most convincing proof for the protein nature of enzymes is the often-repeated experiment whereby a given enzyme is treated with a protein-digesting enzyme (proteinase). It can be shown that the loss of enzyme activity parallels the disappearance of protein. It is possible also to demonstrate a parallelism between protein denaturation, as caused by a number of agents, and loss of enzyme activity.

In the last few years occasional claims have been made for enzymatic activity of preparations containing no protein. None of these claims has ever been fully repeated or substantiated, however, and it must therefore be concluded that the protein nature of enzymes is fully established.

The fact that enzymes are proteins has a number of consequences related to their properties. One of these is the effect of temperature on enzyme action. Figure 10-6 shows that this effect has two components: a curve with a positive slope, which is caused by the effect of temperature on a chemical reaction (see Fig. 10-2), and a curve with a very steep downward slope, which is due to the effect of temperature on the denaturation of the enzyme. The marked effect of temperature on denaturation is a unique and characteristic property of proteins and nucleic acids, in part because their highly ordered structure "melts" or becomes disordered by a cooperative phenomenon.

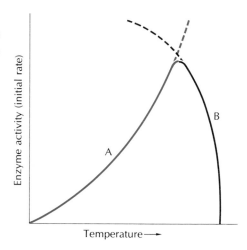

Fig. 10-6. The effect of temperature on enzyme activity. The curve is a composite of two processes: *A* is the effect of temperature on an enzyme catalyzed reaction; *B* is the effect of temperature on the rate of denaturation of the enzyme. The abruptness of this latter curve is due to the very high temperature coefficient of protein denaturation.

The entropy term of a protein denaturation reaction caused by heat is consequently very large. It is therefore important that enzyme assays be normally carried out at temperatures at which no appreciable irreversible inactivation of the enzyme occurs.

Enzymes, being proteins, are polyvalent ions, their net charge being dependent on pH. Enzyme activity is therefore highly pH-dependent, and the pH maxima vary considerably from one enzyme to another (Fig. 10-7). The pH dependence of enzyme activity is due to a number of effects such as (1) the dissociation of the substrate, (2) the dissociation of side chains on the enzyme directly interacting with the substrate during catalysis, and (3) the effect of pH on the overall conformation of the enzyme molecule.

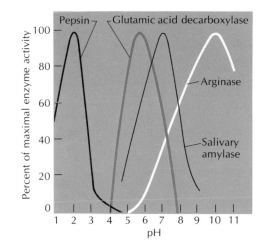

Fig. 10-7. Effect of pH on the activity of a number of enzymes. (Redrawn with permission from J. S. Fruton and S. Simmonds. *General Biochemistry,* New York: John Wiley & Sons, Inc., 1953.) The effect of pH is a complex phenomenon being often exerted on the substrate as well as the enzyme.

For a number of different reasons enzymes are not always found in an active state. Some enzymes, among them the proteolytic or protein-digesting variety, are often found in the cell as inactive precursors. *Trypsinogen,* for instance, is inactive but can be activated to *trypsin* by a number of enzymes, including trypsin itself. The activation seems to involve the removal of a short peptide, followed by some structural changes in the protein molecule that finally yield the active trypsin enzyme. Other enzymes, as we have already pointed out in Chapter 4, require for activity certain divalent cations like Ca^{2+}, Zn^{2+}, Mn^{2+}, Mg^{2+}, or Co^{2+}. Another category of enzymes requires as a "coenzyme" a complicated organic compound. Both the divalent cation and the coenzyme can usually be separated from the enzyme by dialysis. This procedure involves putting the protein into a cellophane bag, with pores too small to allow the enzyme to go through, and surrounding it with large volumes of water or salt solution. The coenzyme is able to pass through the pores of the dialysis bag and become diluted in the large external volume of solution. The coenzyme is usually stable in boiling water, while the *apoenzyme,* or protein portion of the active enzyme, is inactivated by heat. Dialyzability and heat stability are, then, the usual tests for a coenzyme. Some enzymes, however, contain a nonprotein portion that is sufficiently firmly attached so that it cannot be removed by dialysis. This nonprotein portion, called the *prosthetic group,* plays an important role in the interaction between enzyme and substrate.

THE ENZYME-SUBSTRATE COMPLEX

If the activity of an enzyme is measured at a number of different substrate concentrations (Fig. 10-8A) and if the rate of reaction is plotted against substrate concentration, a curve depicted in Fig. 10-8B is obtained. As the substrate concentration is increased, the value of v approaches a limiting velocity, V. At this velocity enzyme activity is proportional to enzyme concentration. To explain these relationships, Henri (1902)

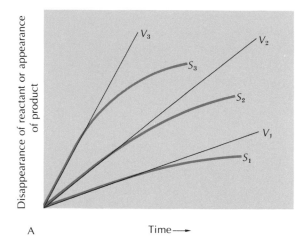

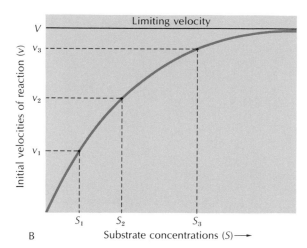

Fig. 10-8. The kinetic properties of most enzyme reactions. A. The initial rates of reactions (v_1, v_2, v_3) at three substrate concentrations (S_1, S_2, S_3). B. Curve obtained when the initial velocities measured at different substrate concentrations are plotted against substrate concentrations. The limiting velocity (V) is the initial velocity at "infinite" substrate concentration. The curve has the shape of a rectangular hyperbola.

first suggested that the enzyme and substrate combine with each other. Thus, it can be seen on qualitative grounds that only when the substrate concentration is high enough to keep the enzyme fully occupied with the catalytic process does the proportionality between enzyme activity and enzyme concentration occur. Michaelis and Menten (1913) were able to validate Henri's hunch in quantitative terms.

Michaelis and Menten began their formulation by assuming that the law of mass action applied to the formation of an enzyme-substrate complex

$$E + S \underset{k_2}{\overset{k_1}{\rightleftharpoons}} ES \overset{k_3}{\rightarrow} E + P$$

where E is the free enzyme; S, the substrate; ES, the enzyme-substrate complex; P, one or more products; and k_1, k_2, and k_3, the velocity constants of the reactions. Furthermore, the following additional assumptions were made:

1. Only one substrate is involved, or if more than one are involved the others are held at constant concentration. This is a plausible assumption in a hydrolysis reaction in which the concentration of water is constant.
2. The reverse reaction $E + P \rightarrow ES$ is negligible.
3. The ES complex has reached its limiting concentration so that the amount of free enzyme E is negligible.
4. Only a negligible amount of substrate has been converted to product, that is, the treatment only holds for the initial velocity of the reaction.

5. All variables such as temperature, pH, ionic strength, and type of buffer are defined and maintained constant.

The above conditions are easy to attain in practice so that as we shall see the relationship developed by Michaelis and Menten has considerable practical usefulness in the study of enzyme reactions.

We shall not dwell on the mathematical techniques of derivation that Michaelis and Menten used. Since then, a number of other derivations have been worked out, some simpler and others more rigorous than the original one. Whatever the method of derivation, the final statement is the same. It is

$$v = \frac{V\,[S]}{K_m + [S]}$$

where

v = velocity of the enzymatic reaction

V = the limiting or maximal velocity at "infinite" substrate concentration

$[S]$ = the substrate concentration in moles per liter of solution

K_m = the Michaelis constant in moles per liter of solution

This equation states that given the fact that the above restrictions are obeyed, it is possible to describe the velocity of an enzymatic reaction in terms of just two parameters that are characteristic for a given enzyme-substrate system. They are the limiting velocity (V) and the Michaelis constant K_m.

If one plots the Michaelis-Menten equation, one obtains a curve (Fig. 10-9) that has the same shape—rectangular hyperbola—as the curve one obtains experimentally (Fig. 10-8B). Since the derivation of this relationship is based on the assumption of the formation of an *ES* complex, one can consider the congruence between the experi-

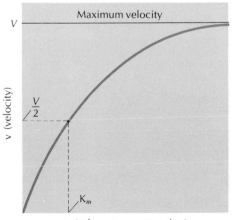

Fig. 10-9. Curve obtained by plotting the Michaelis-Menten equation. The shape of this curve is a rectangular hyperbola of the same general properties as the curve obtained experimentally in Fig. 10-8B. By substituting $V/2$ for v into the Michaelis-Menten equation, one obtains K_m. Notice that the steeper the curve, that is the more rapidly the limiting velocity is reached with respect to substrate concentrations, the smaller is K_m.

mental and theoretical curves to constitute good, if indirect, evidence for the formation of an enzyme-substrate complex.

For colored enzymes more direct evidence of an enzyme-substrate complex can be obtained by studying changes in their absorption spectra when the enzymes interact with their substrates. Chance and his co-workers have perfected a rapid mixing and flow technique (Fig. 10-10) that enables them to measure the absorption spectra of colored enzymes such as catalase or peroxidase even though the duration of the *ES* complexes was in the millisecond range. They found in a number of instances that more than one *ES* intermediate occurred; in the case of peroxidase for instance, the brown enzyme first turns green because of the formation of ES_1 and then turns red

Fig. 10-10. Continuous flow apparatus developed by Theorell and Chance for following rapid reactions. By using enzymes that absorb visible light, it is possible to show that a number of transient changes in the absorption spectrum of the enzyme occur. These can be explained by the formation of a number of intermediate enzyme substrate complexes.

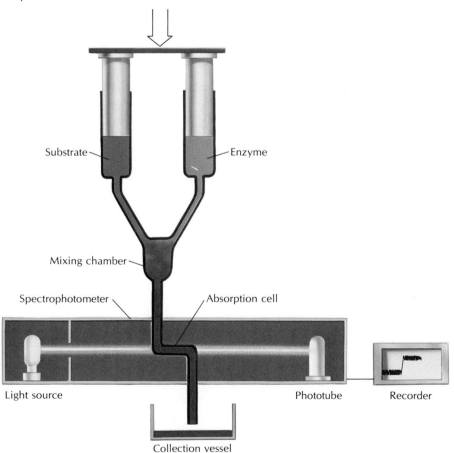

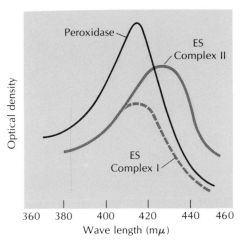

Fig. 10-11. Direct demonstration of the formation of enzyme-substrate intermediates by using the continuous flow apparatus. The enzyme, horseradish peroxidase, is brown in the absence of substrate because of the presence of a heme group. The first *ES* complex is green; the second one is red. (After Chance, 1949.)

from the formation of ES_2 (Fig. 10-11). Thus, one can write

$$E + S \; \rightleftharpoons \; ES_1 \; \rightleftharpoons \; ES_2 \; \rightarrow \; E + P$$

and so on for additional intermediates if necessary. It turns out that even when more than one intermediate is formed, Michaelis-Menten kinetics apply, probably because only one of the steps is rate-limiting.

Recently an ingenious method using very rapid temperature "jumps" has been introduced by Eigen and his collaborators. This method is able to measure effects as short as 10 microseconds, and, furthermore, its use is not restricted to colored enzymes. Temperature-jump methods are adding much new information on the nature of the *ES* complexes, on the kinetics of enzyme reactions in general, and on the structure of macromolecules.

Let us examine the Michaelis-Menten equation and the curve it generates (Fig. 10-9) a little more closely. If one takes an initial velocity (v), which is equal to one-half the limiting velocity (V), and substitutes this into the equation, one obtains the Michaelis constant (K_m). The limiting velocity, however, given the above formulation, is hard to determine experimentally. Lineweaver and Burk suggested an ingenious reciprocal plot, which gives a straight line and allows the determination of V and K_m by extrapolation (Fig. 10-12). Another advantage of using the reciprocal plot is that one can determine immediately whether the experimental points of $1/v$ versus $1/[S]$ fall on a straight line, thus obeying the Michaelis-Menten formulation. As it turns out, enzyme systems that we now know have far more complex enzyme-substrate interactions than assumed by the Michaelis-Menten treatment nevertheless obey the kinetics predicted by that treatment. Thus, the determination of V and K_m have turned out, in practice, to be useful parameters for characterizing a given enzyme-substrate system. However, it should be remembered that K_m is not a simple affinity constant between enzyme and substrate and that its precise physical meaning depends on the reaction mechanism of a given enzyme-substrate reaction that must usually be worked out by other means.

Fig. 10-12. The "Lineweaver-Burk plot" of experimental values of v and $[S]$ giving a straight line, which upon extrapolation allows the determination of K_m and V. The advantage of a straight line is that each experimental point contributes to the determination of K_m and V.

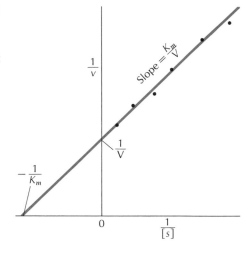

ENZYME INHIBITION

One considerable value of the Michaelis-Menten treatment of enzyme-substrate kinetics is that it also helps to describe the phenomenon of enzyme inhibition. As we pointed out in Chapter 5, we can expect that virulent poisons exert their effect by inhibiting enzymes. The inhibition of catalysis is a plausible explanation of the drastic effects minute quantities of poisons have on living systems.

Enzyme inhibition can occur in a variety of ways, and its study can reveal a great deal about the nature of the enzyme and of the catalytic reaction. Some inhibitors have an irreversible effect on enzymes by reacting with them to form an irreversible derivative. Thus, diisopropylfluorophosphate (DFP) reacts with enzymes such as trypsin, chymotrypsin, and thrombin to form an inactive product. ^{32}P-labeled DFP has been used to study this reaction, and as we shall see, considerable information has been obtained about the mechanism of action of trypsin and related enzymes by these means. This type of inhibition is independent of substrate concentrations, so that Michaelis-Menten kinetics are irrelevant to a description of their mechanism of action.

There are, however, forms of enzyme inhibition in which the substrate concentration does play a decisive role. If one uses the Lineweaver-Burk plot for determining the effect of constant inhibitor concentration on the initial velocity of the reaction at different substrate concentrations, one can obtain a variety of curves, the extremes of which are represented in Fig. 10-13A and B.

At one end of the spectrum we have fully competitive inhibition in which V is not affected by the presence of the inhibitor but K_m is increased. In this case the mechanism of action is thought to be competition between the substrate and the inhibitor for the same portion of the enzyme (active site) involved in the catalytic process. In Fig. 10-13A we have a situation in which when $1/[S] = 0$ (that is, substrate concentration is infinite), inhibition has been reduced to zero.

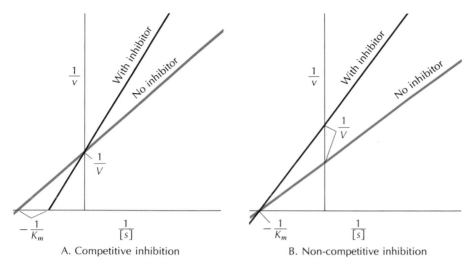

Fig. 10-13. Lineweaver-Burk plots of two forms of enzyme inhibition. A. Competitive inhibition in which the inhibitor competes with the substrate for the "active site" of the enzyme. Notice that when $1/[s] = 0$ (infinite substrate concentration), the maximal velocity of the reaction is the same in the presence and absence of inhibitor. The K_m, however, is different. B. A form of noncompetitive inhibition in which different maximal velocities but the same K_m are obtained in the presence and absence of inhibitors. Types of inhibition that are intermediate between these examples are known.

The effect of malonic acid as a competitive inhibitor of the oxidation of succinic acid is a classic example of competitive inhibition.

(See Chapter 13.)

COOH
|
CH₂
|
CH₂
|
COOH

succinic
acid

catalyzed by
succinic dehydrogenase
⟶
inhibited
competitively by

COOH
|
CH₂
|
COOH

malonic acid

COOH
|
CH
‖
CH
|
COOH

fumaric acid

The competitive inhibitor is sufficiently similar to the real substrate to form an inhibitor-enzyme complex, yet its difference from the substrate prevents the enzyme from acting catalytically on it. In the presence of an inhibitor some of the enzyme is "tied up" and cannot act catalytically, an effect that can be decreased by increasing the substrate concentration.

Another well-known example of this is the effect of sulfanilamide (a sulfa drug), which "fools" the enzyme normally acting on *p*-aminobenzoic acid, an important growth factor in many bacteria.

p-aminobenzoic acid   sulfanilamide

Note again the general structural similarities between the two compounds. Numerous competitive inhibitors are known, their use having contributed greatly to our understanding of the mechanism of enzyme action as well as to the study of metabolic sequences.

At the other end of the spectrum of reversible enzyme inhibition we have fully noncompetitive inhibition in which K_m is not affected by the presence of the inhibitor, but V is decreased. This type of inhibitor is thought to interfere with the dissociation of the *ES* complex, that is, it reduces k_3. Only when the reduction of k_3 does not affect K_m, that is, when K_m is a true equilibrium constant, does one obtain fully noncompetitive inhibition.

Other cases of enzyme inhibition are known exhibiting kinetic properties that are intermediate between the above extremes. The important point about the formulation is that it allows us to distinguish between various mechanisms of inhibition.

The last and most important form of enzyme inhibition we should mention is the "natural" one occurring as a regulatory mechanism in cells. We refer to the *allosteric effect* brought about by feedback inhibition. However, in order to do this subject justice, we should first discuss the topics of *enzyme specificity* and *mechanisms of enzyme action,* and then we shall return to a discussion of allosteric enzyme inhibition at the end of this chapter.

SPECIFICITY OF ENZYME ACTION

Specificity is perhaps the most characteristic property of the living machine, and the enzyme-substrate interaction perhaps the most dramatic example of biological specificity. By now some 700–800 different enzymes, each with its own catalytic properties and range of specificities, have been isolated from various tissues. It is not necessary here to review how these large numbers of catalytic functions have been classified; in fact, a special Commission on Enzymes has been established by the International Union of Biochemistry to classify enzymes and propose a proper nomenclature.

In addition to classifying enzymes according to a variety of catalytic functions it is also possible to divide them into a number of categories differing in their degree of specificity. Thus we have the following categories.

1. *Absolute specificity* is found in an enzyme such as *urease,* which can utilize only urea as a substrate and will fail to act on even the slightest chemical modification of urea.

2. *Absolute group specificity* is found in an enzyme such as *alcohol dehydrogenase* that can act only on alcohols.

$$CH_3CH_2OH + \begin{matrix} \text{oxidized} \\ \text{coenzyme} \end{matrix} \quad \underset{\text{alcohol dehydrogenase}}{\xrightleftharpoons{\hspace{2cm}}} \quad CH_3CHO + \begin{matrix} \text{reduced} \\ \text{coenzyme} \end{matrix}$$

In such cases the rate of enzyme activity will depend on the nature of the remaining portions of the molecule. In the case of alcohol dehydrogenase, the enzyme "prefers" ethanol but can act at decreasing rates on straight-chain alcohols of increasing lengths.

3. *Relative group specificity* is found in an enzyme such as *trypsin* that can hydrolyze a peptide bond as well as act as an esterase but requires a positively charged side chain of lysine or arginine on the carbonyl (C=O) side of the peptide or ester bond. This high degree of proteolytic specificity of trypsin has been used to great advantage in the sequence studies described in Chapter 9.

arginine residue — peptidase activity of trypsin

benzoyl-L-arginine methyl ester — esterase activity of trypsin — benzoyl-L-arginine + methanol

4. *Stereochemical specificity* is a property of most enzymes and is the most extreme example of the ability of an enzyme to discriminate between related substrates. Although in other respects it may fit into one of the three previous classes, an enzyme may also be able to discriminate between a given substrate and its optical isomer. An *l*-amino acid oxidase, for instance, will not act on *d*-amino acids, and vice versa.

$$+H_3N \diagdown \quad COO^- \diagup$$
$$C + O_2 + H_2O \xrightarrow[\text{oxidase}]{\text{l-amino acid}} O = C \diagdown + NH_3 + H_2O_2$$
$$H \diagup \quad \diagdown R \qquad\qquad R$$

l-amino acid   *α*-keto acid

MECHANISM OF ENZYME ACTION

Kinetic studies of enzyme inhibition and of degrees of enzyme specificity as well as protein structural studies discussed in Chapter 9 have gradually given us an insight into mechanisms of enzyme action. Although no overall theory of enzyme action can as yet be enunciated, we can at least state the general problems which a theory of enzyme catalysis must elucidate.

1. As we have stated at the beginning of the chapter, the enzymes increase the rate of a reaction by decreasing the energy of activation a substance must acquire before it can react. Thus, for example, the uncatalyzed decomposition of hydrogen peroxide occurs slowly with an energy of activation of 18 kcal/mole. Traces of iron catalyze this reaction, bringing down the energy of activation to 13 kcal, whereas the enzyme *catalase* speeds up the reaction tremendously by lowering the energy of activation to 4 kcal (Table 10.1).

2. As we have learned, the enzyme does not only speed up the rate of a chemical reaction, but in addition does so specifically, for it can often distinguish be-

Table 10-1. THE EFFECT OF ENZYMES ON THE ENERGIES OF ACTIVATION OF TWO REACTIONS.

REACTION	CATALYST	ACTIVATION ENERGY* kcal/mole
$2H_2O_2 \rightarrow 2H_2O + O_2$	none	18
	traces of iron	13
	enzyme catalase	4
$H_2N \diagdown$ $C = O + H_2O$ $H_2N \diagup$ $\downarrow$ $CO_2 + 2NH_3$	hydrogen ion	25
	enzyme urease	13

* These activation energies are frequently measured as ΔH of activation rather than as ΔF of activation but the error involved is not great since the ΔS of activation is usually small.

tween very small differences in the structure of the substrate.

3. As we shall see toward the end of this chapter, the rate of enzymatic catalysis is capable of being regulated under biological conditions.

A general theory of enzyme action must therefore be able to explain in molecular terms (1) the enzyme-induced lowering of the energy of activation, (2) the high degree of specificity, and (3) the ability of the catalytic process to be regulated under biological conditions.

During the last decade, a concept has gradually emerged that, although it will require considerable sharpening in its molecular details, has gone far in providing a unifying theme for enzyme action. We refer to the idea of the *active site*.

We have already discussed the evidence that the enzyme and substrate form a complex and, furthermore, that the interaction must be highly specific. Stereochemical specificity suggests that the interaction between enzyme and substrate must include at least three of the four substituents of the optically active carbon atom (Fig. 10-14). A close relationship between enzyme and substrate is also suggested, even if somewhat indirectly, by the phenomenon of competitive inhibition.

According to the concept of the active site, there are one or more regions on the enzyme at which catalysis occurs through a close (multipoint) attachment between

Fig. 10-14. Demonstration that optical specificity must be due to minimum of three-point contact. Two-point contact cannot distinguish between the optical isomers *A* and *B*.

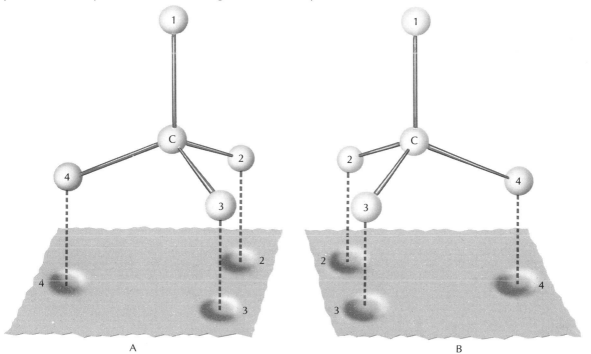

the enzyme and a specific substrate. We think of the active site as being composed of a few reactive groups consisting of some of the 20 different side chains of the amino acid residues and possibly also of amide groups of the polypeptide backbone. In the case of enzymes that require prosthetic group for their catalytic activity, we expect that the prosthetic group, at least in part, supplies one or more groupings for the active site. Recently, the concept of the active site has been enlarged to include the entire region of contact between enzyme and substrate, including reactive forms as well as groups involved in binding of substrate to enzyme.

It has generally been thought that implicit in the active site concept is the notion that the various functional groupings acting as points of attachment for the substrate are held rigidly on the enzyme molecule. Thus, it was thought that their position is fixed in space and that when the substrate becomes attached to the enzyme, the former becomes distorted so as to render a given bond or grouping more reactive. Once a reaction has taken place, the structure of the product is now sufficiently changed that its affinity for the active site has presumably vanished or at least been lowered. However, some cases are known in which the product does behave as a competitive inhibitor of the substrate.

Thus, the concept of a specific arrangement and rigid bracing of functional groups in the active site has been used to explain not only the lowering of the free energy of activation, but also the specificity of enzyme action. Furthermore, the induction of subtle changes in the conformation of the active site would explain how enzyme activity can be regulated; this will be discussed later.

While a full and detailed documentation of the active site concept has not yet been achieved, a number of studies have brought to bear considerable evidence in its favor. There are the so-called conjugated enzymes in which "prosthetic groups" are necessary for enzyme activity. Since we can expect the prosthetic group to be at least part of the active site of the enzyme, we have the opportunity to study the structural chemistry of this region. Wang carried out an interesting study on *catalase,* the enzyme catalyzing the decomposition of hydrogen peroxide into oxygen and water. Catalase has a prosthetic group composed of an iron atom and a *heme group* (hematin). It is possible to measure the catalytic effects of the iron alone (weak) and of hematin (less weak); the conclusion, as might be expected, is that the protein portion of the enzyme does play a crucial role in catalysis (Fig. 10-15). It is very likely that the hematin group is located on the enzyme in such a manner that the substrate will interact with portions of the protein molecule as well as with the prosthetic group. An alternative hypothesis assumes that the interaction of the protein and the hematin modifies the properties of the latter in such a way as to increase its catalytic activity. Wang was able to synthesize a compound (ferric triethylenetetraamine) that is a much more effective catalyst than hematin but is still far less effective than the enzyme (Fig. 10-15). By placing the NH_2 groups close to the iron, this compound possibly simulates the situation of the iron in the hematin-protein complex of catalase. Such studies of the catalytic activity of enzyme analogues are likely to be of increasing importance in the immediate future.

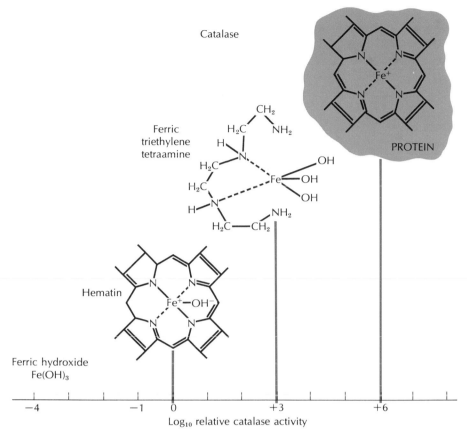

Fig. 10-15. Catalytic activity exerted by a number of compounds on the decomposition of hydrogen peroxide. (Courtesy of J. H. Wang.) Notice that iron catalysis is ten billion times slower than the enzyme catalyzed system, whereas the hematin group without the protein is only one million times slower. Ferric triethylenetetraamine appears to be a good model of the enzyme, being only a thousand times slower. This is probably due to the two charged amino groups, which in the case of the catalase molecule are probably provided by the protein portion of the enzyme. To account for the thousand-fold greater activity, the enzyme must have some additional aspects about which we do not as yet know.

Another approach that has recently yielded some interesting results is to remove portions of the enzyme molecule by selectively splitting specific peptide bonds with certain proteolytic (protein-splitting) enzymes. F. M. Richards, using the action of the enzyme subtilisin on the enzyme ribonuclease, was able to split off a 20-residue peptide. It was possible to abolish ribonuclease activity by removing this peptide and to regenerate enzyme activity by adding it back although the peptide does not thereby become attached by a peptide bond but appears to be held to the enzyme by secondary interaction. Studies such as these will provide an increasingly precise picture of the portion of the enzyme required for catalytic activity.

An approach that has come back into use in recent years is the specific blocking of side chains (R groups) on the enzyme molecule. It has been known for a long time that certain enzymes require for their activity the presence of one or more specific sulfhydryl groups. When such groups become tied up by a mercurial reagent like p-chloromercuribenzoate or an alkylating reagent like iodoacetamide, then enzyme activity is lost.

$$\text{enzyme—SH} + \text{I—}\underset{\underset{H}{|}}{\overset{\overset{H}{|}}{C}}\text{—}C\!\!\!\overset{NH_2}{\underset{O}{\diagdown\!\!\diagup}} \rightarrow \overset{\text{inhibited}}{\underset{\text{enzyme}}{}}\text{—S—}\underset{\underset{H}{|}}{\overset{\overset{H}{|}}{C}}\text{—}C\!\!\!\overset{NH_2}{\underset{O}{\diagdown\!\!\diagup}} + \text{HI}$$

iodoacetamide

Recent work on a number of enzymes has implicated some additional groups like histidine, aspartic acid, tyrosine, and serine; and we can expect this list to grow rapidly in the immediate future.

However, one difficulty with this approach is that it does not provide unambiguous evidence that such a group is involved in the active site. One can imagine that the blocking of a group at a distance from the active site might well bring about a conformational change in the enzyme molecule, which only secondarily causes an alteration in the active site and hence a loss of enzyme activity. To avoid such ambiguities, recent work in this area has sought to identify R groups which were unusually reactive and hence likely to play a role in the active site.

(Fig. 9-28B.)

In the case of ribonuclease, Barnard and Stein discovered that a histidine R group located in the 119 position is unusually reactive toward bromoacetic acid. This reagent does not react as readily with the other three histidines in the molecule nor does it react readily with histidine in solution. Furthermore—and this is most significant—the special reactivity of histidine 119 is lost if ribonuclease is denatured since it is dependent on the specific configuration of the native enzyme. Further kinetic studies of ribonuclease using a variety of inhibitors have also implicated histidine 12 and lysines 7 and 41 in the active site of the enzyme. These studies were verified in a dramatic manner when Harker and his co-workers succeeded in working out the three-dimensional structure of ribonuclease to 3.5 A resolution using the X-ray diffraction method. According to this structure histidines 119 and 12 and lysines 7 and 41 are all in the same vicinity. It would appear that it is only a matter of time before the details of the mechanism of action of ribonuclease will be worked out.

Another interesting example of blocking techniques has been the use of diisopropylfluorophosphate (DFP), which reacts stoichiometrically with a number of esterase and protease enzymes to form a diisopropylfluorophosphoryl derivative. Hydrolysis of the enzyme derivative has shown that in each case the hydroxyl group of a serine side chain is involved in this reaction.

Seryl residue   Diisopropylfluorophosphate
(DFP)

Partial hydrolysis of a number of these enzymes and recovery of the DFP peptides has shown that in many cases glycyl or alanyl residues are placed next to the reactive serine (Table 10.2). The fact that other serine side chains in the enzyme molecule are not equally reactive and that in all of these enzymes only one serine reacts rapidly and stoichiometrically constitutes strong presumptive evidence that this particular serine is part of the active site. However, the involvement of this serine hydroxyl group in such a wide range of enzymatic reactions suggests that it plays a role in the bond-breaking aspect of the catalytic reaction rather than in that aspect of the enzyme-substrate interaction responsible for its specificity. One can therefore imagine that for each of the enzymes listed in Table 10-2, additional side chains in the active site help determine the specificity of the enzyme for a particular substrate.

Table 10-2. AMINO ACID SEQUENCE AROUND THE DFP-
REACTIVE SERINE RESIDUE OF SOME ESTERASES
AND PROTEASES*

Sequence	Enzyme
Gly Val Ser Ser Cys Met Gly Asp *Ser* Gly Gly Pro Leu Val Cys Lys	Chymotrypsin
$\overset{\displaystyle NH_2}{\underset{\displaystyle \mid}{}}$ Asp Ser Cys Glu Gly Glu Asp *Ser* Gly Pro Val Cys Ser Gly Lys	Trypsin
Asp *Ser* Gly	Thrombin
Asp *Ser* Gly	Elastase
Phe Gly Glu *Ser* Ala Gly (Ala, Ala, Ser)	Butyryl cholinesterase
Glu *Ser* Ala	(Eel)acetylcholinesterase
Gly Glu *Ser* Ala Gly Gly	Liver aliesterase (horse)
$\overset{\displaystyle NH_2}{\underset{\displaystyle \mid}{}}$ Asp Gly Thr *Ser* Met Ala Ser Pro His	Subtilisin (*B. subtilis*)
Thr *Ser* Met Ala	Mold protease (*Aspergillus oryzae*)
Thr Ala *Ser* His Asp	Phosphoglucomutase
$\overset{\displaystyle NH_2}{\underset{\displaystyle \mid}{}}$ Lys Glu Ile *Ser* Val Arg	Phosphorylase
Thr Gly Lys Pro Asp Tyr Val Thr Asp *Ser* Ala Ala Ser Ala	Alkaline phosphatase (*E. coli*)

* Reprinted with permission from R. A. Oosterbaan and J. A. Cohen, in "Structure and Activity of Enzymes," Goodwin, Hartley, and Harris (eds.), Academic Press, New York, 1964.

One of the most extensively studied enzymes listed in Table 10-2 is chymotrypsin. Studies of the effect of pH on the rate of chymotrypsin catalysis and of the inhibiting effects of alkylating reagents suggest that a histidyl side chain is also involved in the catalytic process. Furthermore, the use of a number of reagents such as N-acetyl-tryptophan, which forms an acyl-enzyme intermediate, suggests that in the case of the normal enzyme-substrate interaction a covalent acyl-enzyme intermediate also takes place (Fig. 10-16).

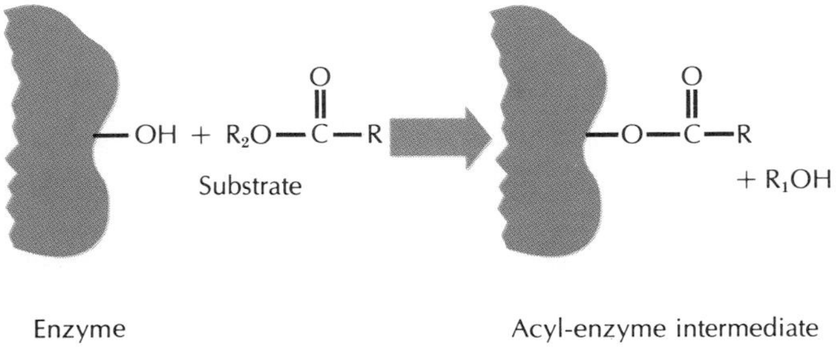

Fig. 10-16. Formation of covalent acyl-enzyme intermediate. Evidence has accumulated recently that a number of enzymes form covalent acyl intermediates. This fact shows that the enzyme-substrate complex need not always be held together by secondary forces but can at times be joined by chemical bonds even if only temporarily.

Enzyme   Acyl-enzyme intermediate

Similar covalent enzyme-substrate intermediates have by now been demonstrated in a number of other enzymes.

Using a variety of such approaches a tentative mechanism for the action of chymotrypsin has been worked out (Fig. 10-17A) and has been confirmed in a most remarkable manner by the recent X-ray crystallographic analysis at 2 A resolution of chymotrypsin by a Cambridge group working at the Laboratory of Molecular Biology (Fig. 10-17B). The model shows that in a region that has been identified as being the active site by reacting the enzyme with a specific inhibitor, there are indeed a seryl (195) and a histidyl (57) residue. Furthermore, in the vicinity of His-57 there is a terminal —NH_3^+ group apparently in contact with a —COO^- group of Asp-194—a fact that will probably have to be taken into account when the mechanism of chymotrypsin is subjected to more detailed scrutiny.

So far we have discussed the active site as if it were a completely rigid and unchangeable structure irrespective of the presence or absence of the substrate. In recent years Koshland and his co-workers have presented evidence that suggests that, at least in the case of a few enzymes, the presence of the substrate induces a change in the configuration of the active site. Koshland based his arguments on a number of observations, some of which could possibly be explained without resorting to an induced fit theory but which when taken together make such a theory extremely plausible. Thus, for instance, the observation that in certain instances the presence of substrate increased the reactivity of a given sulfhydryl group in an enzyme could best be explained by assuming that the substrate brought about conformational changes in the enzyme that caused a sulfhydryl group to become more reactive.

Koshland's induced fit theory has recently been confirmed by Lipscomb's observation that in carboxypeptidase A (Fig. 9-37), a tyrosyl residue moves as much as 8 A

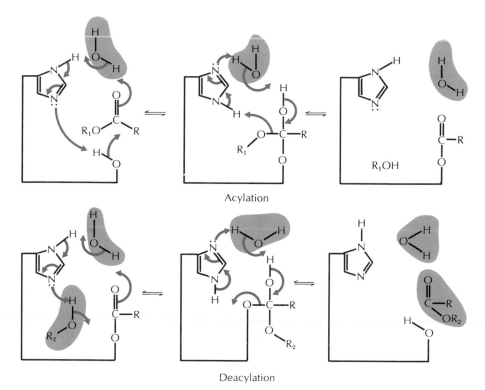

Acylation

Deacylation

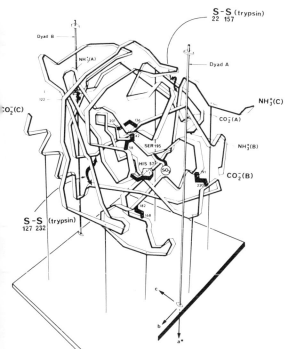

Fig. 10-17. A. The structure and function of chymotrypsin. The attack of the serine hydroxyl group on the carbonyl carbon atom of the substrate is pictured as being aided by both general acid and general base catalysis by an imidazole moiety of a histidine residue. A molecule of water serves as proton transfer agent for the general acid-catalyzed portion of the reaction. In a symmetrical fashion, the decomposition of the tetrahedral intermediate is also considered to proceed via a general acid-base catalyzed route, a molecule of water serving as proton-transfer agent for the general base catalysis. The deacylation part of the overall reaction is considered to be essentially the microscopic reverse of the acylation part. (Reprinted with permission from Mahler and Cordes, *Biological Chemistry*, New York: Harper & Row, 1966.) B. The structure of chymotrypsin at 2 A resolution. This enzyme is derived from the inactive form chymotrypsinogen by the enzymatic cleavage of four specific peptide bonds, leaving three chains cross-linked by two of the five disulfide bridges. Notice that most of the chains are in extended form, there being only one short helical region. The stability of the molecule is probably derived from interchain interactions (disulfide bonds and hydrogen bonds) since chains can be seen to run next to each other in parallel or antiparallel arrangements. The close juxtaposition of Ser-195 and His-57 can be seen readily. (Courtesy of David M. Blow, Medical Research Council, Cambridge, England.)

upon binding of the substrate, and by the beautiful X-ray analysis of lysozyme carried out by Phillips and his co-workers. For details of this latter piece of work the student is referred to the *Scientific American* article by Phillips listed in the suggested readings. The following is but a brief summary.

Lysozyme is an enzyme that can bring about the rupture of bacteria by hydrolyzing a specific linkage in the polysaccharide chains of bacterial cell walls. The lysozyme molecule is composed of a single chain of 129 amino acids linked in four places by disulfide bonds (Fig. 10-18), the amino acid sequence and position of disulfide pairs having been worked out independently by Jollès and Canfield and their respective co-workers. In 1960 the first isomorphous replacements were prepared, and in 1962 a low-resolution image of the molecule was worked out. In 1965 a high-resolution (2 A) image was completed, which gave us a detailed picture of the conformation of the polypeptide chain (Fig. 9-36). The helical content of lysozyme is less than one-half of the 75 percent helicity observed in myoglobin, and some of the helices are different from the classical α-helix of Pauling and Corey: the C=O and N—H groups are not as close as they are in the α-helix. Furthermore, the lysozyme molecule contains a structure, the "antiparallel pleated sheet," which Pauling and Corey had predicted to occur in certain fibrous proteins. The remainder of the structure does not appear to have any obvious regularity except that, just as in myoglobin, the polar side chains are on the outside of the molecule interacting with the water, whereas the nonpolar side chains are buried in the interior. One other interesting feature of the molecule is the fact that the first 40 residues from the amino-terminal end form a compact helical structure that might well represent a "substructure" around which the rest of the molecule is folded as it is being synthesized on the ribosomes.

The most dramatic aspect of the structure of lysozyme is the fact that a cleft runs down the middle of the molecule (Fig. 9-36). It was tempting to assume that this cleft represents the binding site for the polysaccharide chain that is the normal substrate of the enzyme. This assumption was validated by a brilliant series of experiments carried out by Phillips and his group, who used X-ray crystallography to study the structure of lysozyme in the presence of a number of competitive inhibitors closely resembling the normal substrates. These studies showed the precise location of the competitive inhibitors on the binding site (Fig. 10-19) and helped formulate the exact interactions between enzyme and the polysaccharide chain of the substrate. Some six hydrogen bonds and a number of nonpolar interactions seem to be involved. Furthermore, it appears that upon binding of the substrate a small change in conformation occurs in such a way as to cause the cleft to narrow and deepen slightly. This then is the first direct confirmation of the induced fit theory of Koshland although, admittedly, the extent of the conformational change induced by the substrate is very slight.

It is not possible to sketch out in detail the evidence which led Phillips to suggest the mechanism for the enzymatic action of lysozyme, but the student is urged to study the mechanism in detail in the above-mentioned article. It suffices to point out that

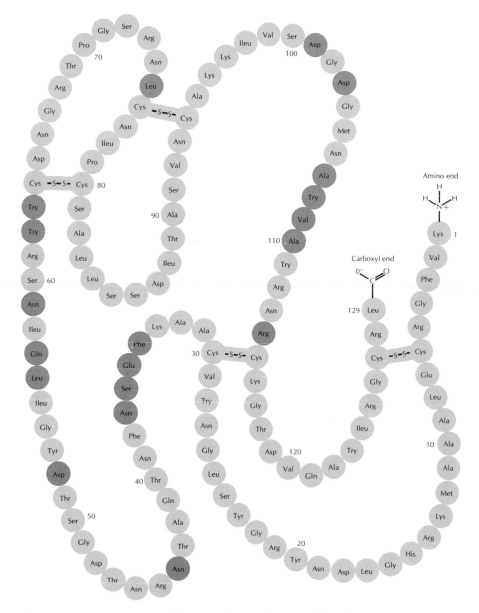

Fig. 10-18.   Primary structure of lysozyme showing location of four disulfide bonds. The amino acids forming the active site of the molecule are in color. The three-dimensional folding of the molecule has the role of bringing these amino acids into the proper juxtaposition. (From J. Jolles, J. Jarequi-Adell and P. Jolles, 1963.)

the mechanism involves (Fig. 10-20) (1) distortion of the stable "chair" configuration of the six-membered ring of the appropriate sugar residue, thus rendering the C—O

313

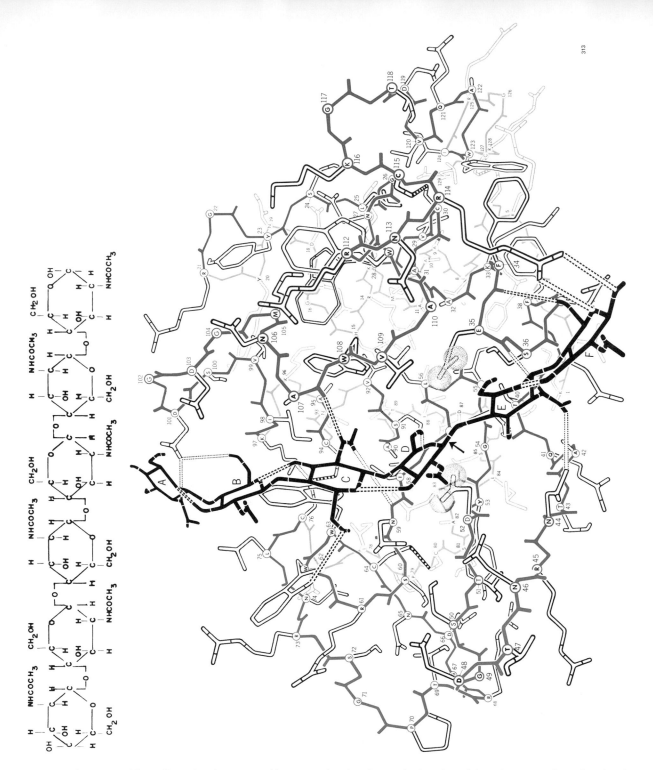

Fig. 10-19. Three-dimensional structure of lysozyme showing the precise location of the substrate on the active site of the enzyme. The amino acid side chains involved in the interaction of the substrate with the enzyme are shown as well as the position of the four disulfide bonds. The four spheres represent the oxygen atoms belonging to the carboxyl groups of a glutamyl and an aspartyl residue, which Phillips believes are involved in the immediate mechanism of the hydrolytic reaction catalyzed by lysozyme. (Reprinted with permission from the *Atlas of Protein Sequence and Structure, 1967–68.* Drawn by Irving Geis. Appeared originally in *Scientific American*.)

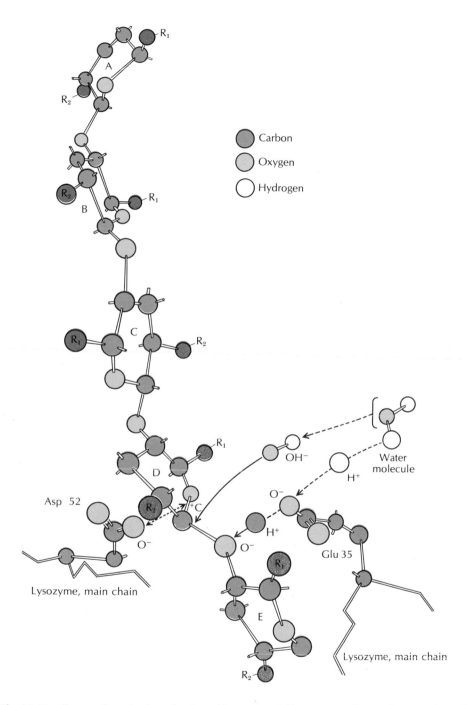

Labels within figure:
- R₁ (A ring, top)
- A
- R₂
- Carbon
- Oxygen
- Hydrogen
- R₂
- B
- R₁
- C
- R₁
- R₂
- D
- R₁
- OH⁻
- Water molecule
- H⁺
- Asp 52
- R₂
- +C
- O⁻
- O⁻
- H⁺
- O⁻
- Glu 35
- Lysozyme, main chain
- R₁
- E
- R₂
- Lysozyme, main chain

Fig. 10-20. Proposed mechanism of action of lysozyme. Phillips suggests that residue 35 (Glu) and residue 52 (Asp) are directly involved in the splitting of the C—O bond in the polysaccharide chain. It is proposed that a hydrogen ion (H⁺) becomes detached from the —OH group of residue 35 and attaches itself to the oxygen atom that joins rings D and E, thus breaking the bond between the two rings. This leaves carbon atom 1 of the D-ring with a positive charge, in which form it is known as a carbonium ion. The atom is stabilized in this condition by the negatively charged side chain of residue 52. The surrounding water supplies an OH⁻ ion to combine with the carbonium ion and an H⁺ ion to replace the one lost by residue 35. The two parts of the substrate then fall away, leaving the enzyme free to cleave another polysaccharide chain. (From Phillips, 1966. Reprinted with permission, Scientific American Inc., W. H. Freeman and Co.)

bond more reactive; (2) the presence of a carboxyl group belonging to a glutamic acid residue that donates a hydrogen ion to the glycosidic oxygen of the C—O bond to be split; and (3) the presence of a carboxyl group belonging to an aspartic acid side chain that temporarily interacts with the positively-charged carbonium ion formed.

Although the details of the mechanism of action of lysozyme have yet to be worked out, we nevertheless have here the first relatively complete, high-resolution model of an enzyme-catalyzed reaction. Interestingly, most of the concepts that had been formulated prior to this important study received some independent and direct support. These include (1) the binding of the substrate to an active site on the enzyme, (2) the induction by the substrate of a conformational charge in the active site, (3) the binding of the substrate to the enzyme, bringing about a distortion of some bonds in the substrate, and (4) the presence of some specially reactive side chains in the neighborhood of the bond to be split.

The scientific harvest from the study of lysozyme is rich indeed, and it is clear that X-ray analysis of other enzymes is helping to confirm the general results the study has yielded.

It is impossible in this brief treatment to do justice to the many mechanistic studies of enzyme action that are at present in progress. Suffice it to say in summary that by combining kinetic studies in the presence of modified substrates or inhibitors with data obtained by high-resolution X-ray analysis, it has been possible to identify the active site of a number of enzymes and the specific groups it contains. So far the groups that have been most frequently found to be directly involved in the catalytic process are the side chains of the following residues: Asp, Cys, Glu, His, Lys, Met, Ser, and Thr, as well as the C- and N-terminal groups of the polypeptide chains. Whether these groups are held rigidly in place on the enzyme molecule or "close in" on the substrate or whether a little bit of both happens must yet be determined. But what seems clear is that the mechanisms of enzymatic catalysis involve simple processes of *nucleophilic* and *electrophilic* or *general acid-base catalysis*, which are familiar to the modern organic chemist, but which we cannot discuss here in any detail. For an excellent treatment of this subject the student is referred to the book by Bernhard entitled *The Structure and Function of Enzymes.*

ALLOSTERIC CONTROL OF ENZYME ACTIVITY

(See Chapter 18.)

We have said at the beginning of this chapter that enzymes not only speed up reactions involving organic compounds, but also help regulate them with respect to each other. As will be discussed later at greater length, recent studies have shown that the activity of a number of enzymes can be enhanced or inhibited by certain small molecules that usually are intermediates in the metabolic pathways in which the particular enzyme is involved.

One of the earliest such regulatory phenomena was discovered by Umbarger and Brown in 1958 and is depicted in Fig. 10-21. Note that isoleucine, which is located far along the metabolic pathway, acts as an inhibitor of L-threonine deaminase.

From a number of subsequent studies the following has become clear.

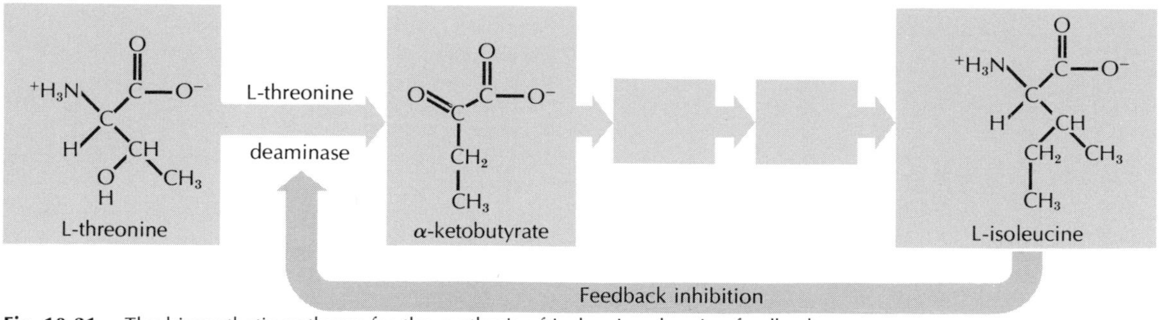

Fig. 10-21. The biosynthetic pathway for the synthesis of isoleucine showing feedback inhibition. Isoleucine synthesized at the end of a chain of four enzymatic steps was shown to regulate the rate of the first step by inhibiting the enzyme threonine deaminase. (From Umbarger and Brown, 1958). The enzyme thus controlled is called a regulatory enzyme and often occurs in the step following a branching of the metabolic network.

1. The inhibitor is not structurally similar to the substrate of the enzyme. Hence, Monod, Changeux, and Jacob have coined the term *allosteric* to emphasize this important fact.

2. The regulatory enzyme is usually located after a branching of the metabolic pathway, and the allosteric inhibitor is usually at the end of the particular pathway. The regulatory function of such an arrangement will be discussed at greater length.   (See Chapter 18.)

Since the allosteric inhibitor is not structurally similar to the substrate, one must conclude that the inhibitor cannot interact with the enzyme at the same active site and thus behave in a purely competitive manner. This has been confirmed by kinetic studies. Thus two other possibilities remain: (1) that the active site for the substrate and the active site of the inhibitor overlap; or (2) that the respective active sites are located on entirely different portions of the enzyme.

Monod, Changeux, and Jacob, who in 1963 wrote a truly prophetic paper in which they developed a general theory of allosteric regulation of enzyme action, felt that the balance of evidence favored a model in which the two sites were located in different places on the enzyme. Furthermore, they suggested that a third site, one involving the action of activators, was also operative in certain regulatory enzymes. Thus, in the threonine deaminase reaction they concluded that there are two *allosteric effector* sites: one to react with isoleucine to cause inhibition, and the other to react with valine to bring about activation.

At the time at which Monod, Changeux, and Jacob proposed these ideas, the evidence supporting them was fragmentary and inconclusive. Recently, Gerhart and Schachman in a beautiful series of studies provided direct confirmation for the idea that the catalytic and regulatory sites are located on different portions of the enzyme. They also showed that binding of the substrate induces in the enzyme a conformational change that is reversed by binding of the inhibitor.

The enzyme they studied is aspartate transcarbamylase (ATCase), which catalyzes the first step of the pathway responsible for pyrimidine biosynthesis (Fig. 10-22).

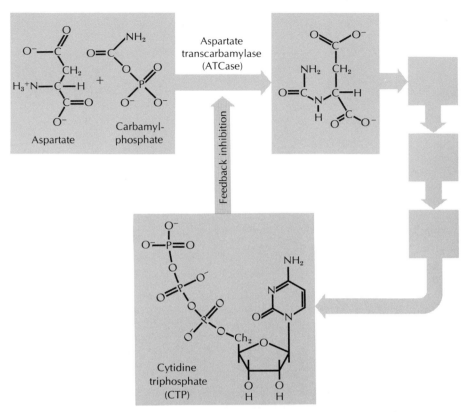

Fig. 10-22. The biosynthetic pathway for the synthesis of cytidine triphosphate showing feedback inhibition. Cytidine triphosphate at the end of a chain of five enzymatic steps can be shown to regulate the rate of the first step by inhibiting the enzyme aspartate transcarbamylase. Cytidine triphosphate bears no similarity to the reactants or the product of the reaction catalyzed by ATCase, and therefore it is unlikely that CTP acts as a competitive inhibitor. (From Gerhart and Pardee, 1962.)

The product of this pathway, cytidine triphosphate (CTP), had been shown by Gerhart and Pardee to act as an inhibitor of ATCase. ATP, on the other hand, appears to be an activator.

Gerhart and Schachman showed the following.

1. The molecular weight of ATCase is 310,000.
2. One molecule of ATCase binds eight molecules of CTP.
3. When the enzyme is treated with *p*-mercuribenzoate (PMB), it dissociates into two distinct types of subunits with molecular weights 96,000 and 30,000 (Fig. 10-23). The original enzyme is presumably composed of two large and four small subunits, and recent work shows that each large subunit is in fact composed of two polypeptide chains. Thus, this protein is composed of eight chains, four weighing 48,000 each and four weighing 30,000 each.

Dissociation of ATCase by Mercurials

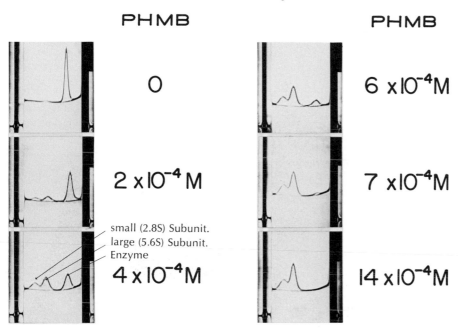

Fig. 10-23. The effect of *p*-mercuribenzoate (PMB) on the dissociation of aspartyl transcarbamylase (ATCase) into subunits. The above six figures are ultracentrifuge sedimentation velocity runs. Sedimentation is from the meniscus on the left to the bottom of the tube on the right. The intact enzyme sedimenting most rapidly is at the right and is followed by the large subunits (5.6S) which in turn is trailed by the small subunits (2.8S). Notice that at 14×10^{-4} M PMB the enzyme appears to be fully dissociated into subunits. The above Schlieren photographs of the sedimenting proteins were taken 50 minutes after centrifuging at 60,000 rpm. (From Gerhart and Schachman, 1965.)

4. Each of the two large subunits is still enzymatically active although CTP can no longer inhibit the activity (Fig. 10-24). Each large subunit has two binding sites for the substrate — a fact that fits the finding that each subunit is composed of two chains.
5. The small subunits bind CTP, but the large ones do not.
6. By mixing the two types of subunits, one can regenerate a molecule that is the same size as native ATCase and that again has both the catalytic and the regulatory functions (Fig. 10-24).
7. The activating effect of ATP competes with the inhibiting effect of CTP, and therefore these two effects probably share a single site on the regulatory subunit.
8. When very precise sedimentation studies of the enzyme are performed in the ultracentrifuge, one can see that the sedimentation coefficient obtained

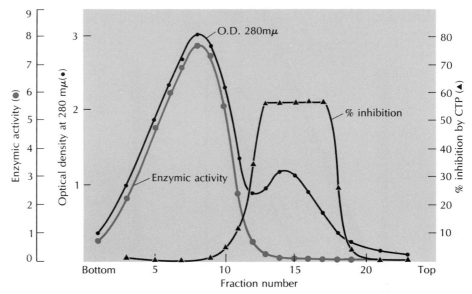

Fig. 10-24. Experiment demonstrating catalytic role of large subunits and regulatory role of small subunits of the enzyme aspartyl transcarbamylase. After dissociating the enzyme with *p*-mercuribenzoate (PMB), the two kinds of subunits were separated by sucrose gradient centrifrigation. The larger subunit appearing in the earlier fractions contained all the enzyme activity whereas the smaller subunits retained the property of binding the inhibitor (CTP). When material from the peak on the left (catalytic subunit) was added to all the fractions and the product tested for the ability to be inhibited with CTP, then only the fractions in the right-hand peak were shown to have that function. This proves that the subunits in the right-hand peak have the regulatory role. (From Gerhart and Schachman, 1965.)

in the presence of succinate (a substrate analog) is slightly different from the one obtained without succinate (Fig. 10-25). This result confirms the hypothesis that the substrate brings about a conformational change in the enzyme although it is too early to say whether the effect is due to swelling or to a change in asymmetry of the molecule.

9. Precise kinetic studies indicate that when only 15 percent of the substrate sites on the enzyme are filled, 50 percent of the conformational change has already proceeded to completion.

This series of studies by Gerhart and Schachman not only brought direct evidence to bear on the theory by Monod, Changeux, and Jacob (that the catalytic and inhibition sites are on separate portions of the enzyme), but it also confirmed their prediction that these sites would be located on distinct subunits of the enzyme molecule. And this leads us to the important question as to how *allosteric effectors* bring about activation or inhibition of enzyme activity.

It had been observed by a number of workers that regulatory enzymes give an "S-shaped" curve when the velocity of the reaction is plotted against the concen-

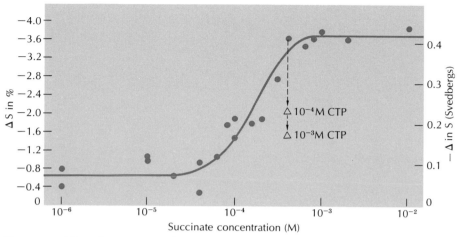

Fig. 10-25. Effect of succinate (a substrate analogue) on the sedimentation coefficient (S) of aspartate transcarbamylase (ATCase) and the antagonistic effect of the allosteric inhibitor cytidine triphosphate (CTP). In the presence of both succinate and carbamyl phosphate there is a 3.6 percent reduction in the sedimentation coefficient. Notice also that CTP opposes this effect. This change in sedimentation coefficient parallels the enhanced reactivity of sulfhydryl groups also observed when the substrates are added to the enzyme. One concludes that the substrates cause the enzyme to swell or become more asymmetric, an effect that is reversed by the addition of CTP, the allosteric inhibitor. (From Gerhart and Schachman, 1968.)

tration of substrate. This S-shaped curve changes to a classical Michaelis-Menten curve when the inhibitory effect is "uncoupled" by heat treatment (Fig. 10-26) or enzyme poisons such as mercurials or by the effect of certain mutations. The S-shaped curve obtained with regulatory enzymes is reminiscent of the one obtained in the binding of oxygen by hemoglobin (Fig. 9-40) and suggests a "cooperative effect" among a number of subunits. And indeed such curves can be generated from simple theoretical models. The best-known of these models, which are currently championed, attempt to explain the cooperativity and other features of allosteric effects in somewhat different ways.

Koshland proposed a general model (Fig. 10-27) in which an induced fit conformational change can be brought about in each subunit. The effect this has on an adjacent subunit, Koshland argues, will vary from protein to protein. When the "coupling" effect is slight or nonexistent, we will have Michaelis-Menten kinetics; whereas, when the binding of substrate in one subunit helps produce a conformational change in the adjacent subunit to enhance reactivity, a positive cooperative effect will be produced. The reverse, a negative cooperative effect, will be produced if the binding of substrate produces a conformational change in a neighboring subunit to slow the reaction. Koshland's model has the advantage of being catholic, for it includes every conceivable type of interaction between subunits of protein molecules. Whether nature indeed uses so much regulational diversity remains to be seen.

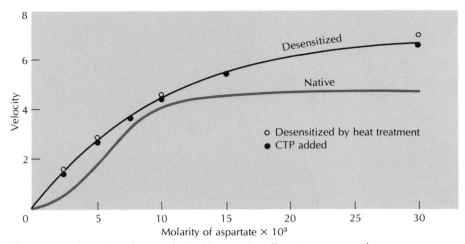

Fig. 10-26. The S-shaped curve obtained by a native allosteric enzyme such as aspartate transcarbamylase. The enzyme can be desensitized by heat treatment, whereupon its kinetics become the usual Michaelis-Menten variety (open circles). The desensitized enzyme can also be shown to become insensitive to CTP, the feedback inhibitor (closed circles). Thus, there appears to be a correlation between the loss of the "cooperative" S-shaped kinetics and the inability to be inhibited by the feedback inhibitor. (From Gerhart and Pardee, 1962.)

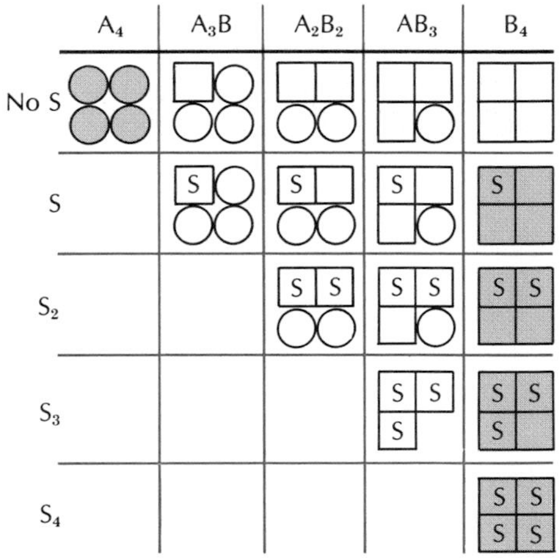

Fig. 10-27. A general model by Koshland of an allosteric enzyme involving four subunits, exclusive binding of substrate and two possible conformations per subunit, showing the number of different states in which an allosteric enzyme might find itself. According to this general model the subunits can be in two alternate forms—circles (A) or squares (B). Binding of substrate (S) shifts the equilibrium from the A form of the subunit to the B form. A variety of pathways going from $A_4 S_0$ to $B_4 S_4$ can be imagined, and all of them can be shown to act cooperatively and generate S-shaped curves. Whether most or even a number of these pathways are used by living systems remains to be seen. The four colored states ($A_4 S_0$ and $B_4 S_{1-4}$) are the only ones to occur according to the theory of Monod, Wyman, and Changeux. (Modified from Haber and Koshland, 1967.)

Monod, Wyman, and Changeux, on the other hand, proposed that when substrate binds on one of the subunits, it brings about a conformational change in *all* the other subunits, thus enhancing the reactivity of all the sites of the enzyme. The reverse happens when the inhibitor binds at one site, for then all the other sites also reverse their conformation and become unreactive to the substrate. This model, often referred to as the *symmetry model* (Fig. 10-28), suggests that all the subunits of a regulatory enzyme are in either one state or the other—a magnificently simple and therefore appealing approach to allosteric control.

It is so far too early to say whether nature uses exclusively the symmetry model or whether nature is more complex and uses the greater versatility offered by Koshland's model. In the case of ATCase (which at present is the system we understand best), the data as we have seen seem to fit the symmetry model satisfactorily (see point 8 on page 261). On the other hand, it now seems likely that hemoglobin does *not* fit the symmetry model, and there may be one or two other enzymes, which although not as well understood, do not fit the ATCase story.

Whether all allosteric regulatory effects necessarily require that the enzyme consist of more than one subunit remains to be seen. Certainly, the study of lysozyme shows that an induced fit can be brought about in a protein consisting of a single chain, and one can imagine that allosteric inhibition could also occur via an induced fit at a special inhibition site, thus bringing about a conformation change and thereby a change in the catalytic site of a single-chain enzyme. It is entirely possible that the multisubunit nature of regulatory enzymes is a consequence of their evolutionary history and that the symmetry principle is a result of evolutionary selection. Thus, it may well be that the first stage in the evolution of metabolic regulation was the formation of two separate proteins, one with catalytic function and the other acting as a specific inhibitor of the catalyst. This might have been followed by a coming together of these two proteins, their interaction being then modulated by an appropriate small molecule. The cooperativity brought about by subunit interaction may be a further step in the evolution of these enzymes.

The role of small molecules in bringing about conformational changes in proteins will without doubt become one of the most fruitful areas of research in the near future. In Chapter 18 we will note how allosteric interactions can be used by the cell to regulate the functioning of metabolic pathways. It is likely that other important physiologi-

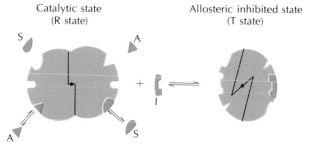

Catalytic state
(R state)

Allosteric inhibited state
(T state)

Binding of substrate (S) and of effector (A) stabilizes catalytic state

Binding of allosteric inhibitor (I) stabilizes inhibited state

"Symmetry principle" excludes a mixed T-R state

Fig. 10-28. The "symmetry model" for allosteric regulation of enzyme action. (Modified from Changeux, 1964). This model assumes that there is a symmetry principle that regulates the subunit interaction and requires that the conformation of all the subunits change simultaneously. This enzyme is composed of two "protomers." When the allosteric inhibitor binds on one protomer, it produces a conformational change that enhances the binding of a second inhibitor molecule, while at the same time bringing about a change in the active site for the substrate to prevent catalytic activity (T state). This effect can be reversed by the binding of the allosteric activator and of the substrate (R state). The theory excludes the possibility of a mixed *T-R* state.

cal phenomena such as mechanochemical, bioelectric, and osmotic transductions involve conformational changes in proteins brought about by their interaction with small molecules. Unfortunately, a large number of theories, based on very little experimental evidence but attempting to explain many physiological phenomena in terms of allosteric effects, are being enunciated; and it will indeed be difficult in the near future to keep this subject within the limits set by the experimental evidence.

The foregoing discussion of enzyme action shows that we are well on the way toward understanding the general features of the mechanism whereby enzymes increase and regulate the rate of chemical reactions. At present we are in position to "explain" how a certain reaction is catalyzed by a series of interactions with certain functional groups in the active site of an enzyme. We do not have a rigorous understanding of the mechanism, however, to account for the *magnitude* of the rate, which can be as high as 10^8 molecules of substrate per molecule of enzyme per second. Undoubtedly, the solution of this problem will be achieved through a detailed understanding of the interaction between the substrate and the *polyfunctional* macromolecular catalysts. This interaction brings about a series of *intramolecular* states, each with a low energy barrier, so that the overall energy barrier is much lower than that which obstructs the *intermolecular* processes catalyzing the same reaction in the test tube of the organic chemist. But then the enzyme may yet have a few surprises in store for us.

SUGGESTED READING LIST

Offprints

> Gale, E. F., 1963. "Mechanisms of Antibiotic Action." *Pharmacological Reviews*, pp. 481–530. Bobbs-Merrill Reprint Series. Indianapolis: Howard W. Sams & Co.
> Koshland, D. E., Jr., 1963. "Correlation of Structure and Function in Enzyme Action." *Science*, pp. 1533–1541. Bobbs-Merrill Reprint Series. Indianapolis: Howard W. Sams & Co.
> Neurath, H., 1964. "Protein-digesting enzymes." *Scientific American* offprints. San Francisco: W. H. Freeman & Co.
> Phillips, D. C., 1966. "The Three-dimensional Structure of an Enzyme Molecule." *Scientific American* offprints. San Francisco: W. H. Freeman & Co.

Articles, Chapters and Reviews

> Baldwin, E., 1957. "Enzymes," Part I in *Dynamic Aspects of Biochemistry*, 3d ed. Cambridge: Cambridge University Press.
> Eigen, M., and Hammes, G. G., 1963. "Elementary Steps in Enzyme Reactions." In: *Advances in Enzymology*. New York: Interscience Publishers, Inc.
> Kirtley, M. E., and Koshland, D. E., 1967. "Models for Cooperative Effects in Proteins Containing Subunits." In: *Jour. Biol. Chem. 242*, pp. 4192–4205.

Singer, S. J., 1967. "Covalent Labelling of Active Sites." In: *Advances in Protein Chemistry 22*, pp. 1–51. New York: Academic Press, Inc.

Stadtman, E. R., 1966. "Allosteric Regulation of Enzyme Activity." In: *Advances in Enzymology*. New York: Interscience Publishers, Inc.

Books

Bernhard, S., 1968. *The Structure and Function of Enzymes.* New York: W. A. Benjamin, Inc.

Boyer, P. D., Lardy, H., and Myrbäck, K., 1959. *The Enzymes.* New York: Academic Press, Inc.

Colowick, S. P., and Kaplan, N. O., 1955. *Methods in Enzymology.* New York: Academic Press, Inc.

Dixon, M., and Webb, E. C., 1964. *Enzymes*, 2d ed. New York: Academic Press, Inc.

Klotz, I. M., 1957. *Some Principles of Energetics in Biochemical Reactions.* New York: Academic Press, Inc.

Neilands, J. B., and Stumpf, P. K., 1958. *Outlines of Enzyme Chemistry*, 2d ed. New York: John Wiley & Sons, Inc.

Chapter 11   From Molecules To
Biological Structures

We have seen in previous chapters that protein molecules differ in complexity, consisting in the simplest case of one polypeptide chain and in the most complex of two or more subunits each in turn composed of one or more polypeptide chains. We have suggested that the highly specific structure of each type of protein molecule can in certain cases be shown to be determined by the primary structure of the polypeptide chains only, although the possibility cannot be excluded that in other cases the pathway of folding may produce a native protein that does not necessarily correspond to the most stable, lowest energy configuration. However, although the issue of structure determination has not been entirely settled for the formation of all protein molecules, we nevertheless feel that it is not premature to ask similar questions about the mechanism of assembly of more complex structures such as enzyme complexes, viruses, cytoplasmic fibrils, and beyond those, complex structures of the cytoplasm such as cortical regions and plastids. As we shall see, we know as yet very little about the molecular mechanism of assembly of these structures, particularly about those at the more complex end of the spectrum. Nevertheless, techniques and approaches are being developed that are beginning to yield results, and it is important to indicate the direction in which this work shall take us in the near future.

Let us begin by drawing a formal distinction between *self-assembly, aided assembly,* and *directed assembly.* In self-assembly the only information needed for the formation of the structure is contained in the macromolecules themselves, which form the assembled product. In its purest form no other macromolecular component is necessary for the assembly process. In an *aided assembly* process other macromolecules that do not enter into the final product are needed for the assembly process. Thus, one can imagine that certain enzymes may be needed that modify the macromolecules in such a way as to prepare them for the assembly process. In *directed assembly* a certain structure is needed in order to organize the components to form more of that structure. It is possible that we shall encounter two fundamentally different directed assembly processes. One of these with which we are familiar is the *template-* (or primer-) controlled process in which detailed information is faithfully transferred from one macromolecule to another. The other process is more analogous to the seeding of a crystal in which the presence of a given structure acts as an *initiator* of an otherwise prohibitively slow assembly process. Whether the distinction between template and initiator will stand the test of time or whether intermediate situations will appear that will reduce the distinction we are drawing remains to be seen.

THE MULTIENZYME COMPLEX—AN EXAMPLE OF SELF-ASSEMBLY

The biochemists of the 1930s often treated the cell as if it were a "bag of enzymes in solution." We have known for some time, however, that many enzymes do not occur freely in solution but are associated with structures such as membranes or with other enzymes in multienzyme complexes. One such system of enzymes that has been studied lately in considerable detail is the *pyruvate dehydrogenase complex* (PDC) in pig heart and *E. coli*. These huge complexes, when looked at with the electron microscope, exhibit a regular polyhedral appearance (Fig. 11-1A).

Table 11-1. THE COMPOSITION OF THE PYRUVATE DEHYDROGENASE COMPLEX OF *E. COLI* (AFTER REED *ET AL.*, 1964.)

ENZYME	NUMBER OF MOLECULES	NUMBER OF SUBUNITS	MOLECULAR WEIGHT
Pyruvate decarboxylase	16		183,000
Lipoic reductase-transacetylase	1	64	1,600,000
Dihydrolipoate dehydrogenase	8		112,000

It is also possible to dissociate the complex into its component enzymes, and as Table 11-1 shows, the approximate stoichiometry of its composition has been worked out. Furthermore, the PDC can be reconstituted from its components to form a fully active and morphologically comparable structure (Fig. 11-1B). It would thus appear that in the case of the PDC we have a straightforward case of self-assembly in which only the molecules forming part of the structure are involved. The biological importance of having enzymes involved in metabolically related processes that form structures in which they are also in geometrical proximity will be discussed later. (See Chapters 13, 14, and 18.)

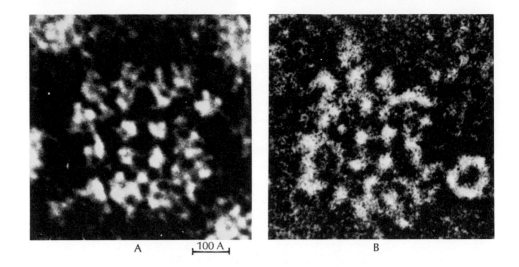

A   100 A   B

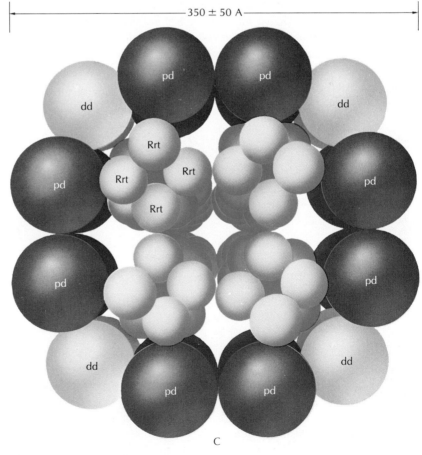

350 ± 50 A

C

16 pd = pyruvate decarboxylase
64 Rrt = lipoic reductase transacetylase
8 dd = dihydrolipoate dehydrogenase

THE ASSEMBLY OF VIRUSES

The study of viruses has proved to be a fertile area of research in the mechanism of the assembly process of biological structures. The reasons for this are obvious: viruses can be grown in large quantities, their infectivity is a convenient biological assay to evaluate the success of the assembly, they are generally large enough to be seen clearly under the electron microscope, and they provide us with a broad range of structural complexity from the relatively simple tobacco mosaic virus to the complex "T-even" coliphages.

Tobacco mosaic virus, the first virus to be purified (Stanley 1935), is a cylindrical structure with a molecular weight of 40 million and dimensions of 160×3000 A (Fig. 11-2). Extensive structural studies of the virus—which include X-ray diffraction, electron microscopy at various stages of dissociation, solution properties (sedimentation, diffusion, light-scattering), as well as studies of the separate protein and RNA components—have yielded the following remarkably detailed and precise model of the virus particle.

Fig. 11-2. *A tobacco mosaic virus particle.* This virus has a particle weight of 40 million. It is a cylinder with dimensions of 160×3000 A. The shadow-casting method used above permits the calculation of the height of the particle. The small piece is a fragment of the virus. (Courtesy of Williams, Virus Laboratory, University of California.)

◄ **Fig. 11-1.** The pyruvate dehydrogenase (PDC) complex—dissociation and reassembly ($\times$ 700,000). A. Electron micrograph of native PDC. B. PDC dissociated and reassembled (reconstituted). C. Model of PDC. After some practice the student will be able to identify the three types of enzymatic components in the electron micrographs of the native and reconstituted molecules. (Courtesy of Fernandez-Moran, 1964.)

A   STRUCTURE OF TOBACCO MOSAIC VIRUS

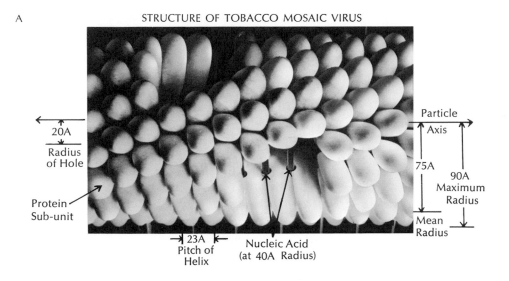

20A
Radius
of Hole

Protein
Sub-unit

23A
Pitch of
Helix

Nucleic Acid
(at 40A Radius)

Particle
Axis

75A

90A
Maximum
Radius

Mean
Radius

B

Fig. 11-3. A. Model of structure of tobacco mosaic virus. (Franklin et al, 1959.) A close up view of a short length of a model of TMV. The complete particle contains 2130 identical protein subunits forming a cylindrical structure composed of protein particles in which the RNA strand is embedded in the form of a gigantic helix. B. A plausible picture of the assembly process of TMV. As we shall later see, it is possible to dissociate and reassemble the virus *in vitro.* (Courtesy of Stanley, Virus Laboratory, University of California.)

The protein coat of the virus consists of 2130 identical protein subunits, each with a molecular weight of 18,000. The subunit is composed of a single chain of 158 amino acids of known sequence; the folding of the chain produces an ellipsoid with an axial ratio of 3:1. The RNA of the virus is a single strand of molecular weight 2.4 million, consisting of 6500 nucleotides. The RNA strand forms a helix with a radius of 40 A that lies imbedded in a helix (radius 85 A) of identical pitch (23 A) formed by the protein subunits (Fig. 11-3). Down the middle of the molecule runs a cylindrical hole with a radius of 20 A that can be seen in the electron microscope if a special staining technique with phosphotungstate is used. An alternative approach is to dissociate the virus and reassemble it partially, forming little "doughnut-shaped" structures showing the holes at the center if viewed end on (Fig. 11-4). It is possible, by using hot detergent, to dissociate TMV in stages and to take EM photographs of the dissociation process. Figure 11-5, showing TMV particles at various stages of dissociation, with the strand of RNA extending from them, was taken after the structure of TMV had been worked out from more indirect evidence. It constitutes a dramatic visual confirmation of the early ideas regarding the structure of TMV.

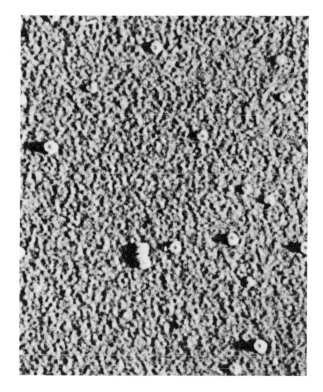

Fig. 11-4. Demonstration of cylindrical structure of TMV with hole at center. TMV can be dissociated into protein and RNA. The protein can be reassembled by itself, and when this is done partially, one can obtain little doughnut-shaped particles which, if viewed end on, show clearly the circular cross section of the particle and the hole at the center. (Courtesy of Williams and Smith.)

Fractionation procedures of TMV have been worked out whereby it is possible to prepare native protein free of RNA and, conversely, undegraded RNA free of protein. Interestingly, the protein-free RNA has infectivity (though reduced to 1 or 2 percent of that of the normal virus), showing that the protein is not necessary for virus replication but that it is useful for the infection process.

Fig. 11-5. Stepwise dissociation of TMV showing strands of RNA sticking out from one end of the particle. The virus was partially dissociated with a solution of hot detergent. The circular object is a polystyrene bead of known dimensions that is used for purposes of standardization of measurements. From the position and the length of the RNA strand one must conclude that it is wound in a helical conformation inside the virus particle, a conclusion that was first suggested by Franklin on the basis of her X-ray studies. (From Hart, Virus Laboratory, University of California.)

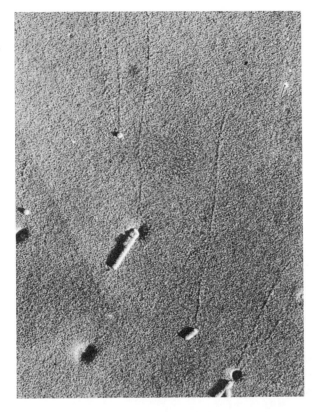

Under conditions of acid pH and low ionic strength differing from the natural conditions found in the tobacco leaf cell, the protein subunits can aggregate to form rods that are identical in diameter to the TMV virus but which are of indeterminate lengths, frequently much longer than the normal virus.

Titration studies suggest that the assembly of the protein subunits is in part controlled by two sets of hydrogen-bonded carboxyl groups, probably of glutamic acid.

$$\underset{\displaystyle -\overset{\textstyle O}{\overset{\|}{C}}-O}{} \cdots \cdots \underset{\displaystyle H-O-\overset{\textstyle O}{\overset{\|}{C}}-}{}$$

The hydrogen bonding appears to raise the pK of these groups to 7.5. It is thought that these bonds are stable in an aqueous medium because they are constrained to be close to each other by other interactions in the assembled structure. Since the pK of these groups is 7.5, the assembly of the protein coat alone will not occur in neutral or slightly alkaline solutions. If, however, the virus RNA is mixed with the protein, then assembly will occur in neutral or slightly alkaline solutions, and, furthermore, the assembly process will terminate when the entire length of RNA is used up. The infectivity of the reconstituted virus is almost as high as that of the normal virus.

These and other results suggest that the RNA does interact with the protein and has some guiding influence in the assembly process. Precise kinetic studies of the dissociation and of the assembly process show that the interaction between the RNA and the protein is not equally strong all along the length of the rod. There appear to be "strong points" at each end of the molecule and some in between. This suggests that some bases or base sequences interact more strongly with the coat protein than others and that some features of the assembly process have been coded into the base sequence of the RNA.

A very recent study utilizing a reagent that reacts with ε-amino groups of lysine shows that one of the two lysines (lysine-53) in the TMV coat protein seems to be involved in the stabilization of the intact virus. It was found that lysine-53 is unreactive in the intact virus but reacts readily in the isolated protein, and it is therefore likely that this positively charged side chain may be involved in an ion-pair within the virus surface.

It is possible, as we indicated in Chapter 8, to create hybrid viruses by reconstituting a virus made up of the coat protein of one strain and the RNA of another strain, and vice versa. These hybrid viruses, however, will replicate the virus from which the RNA has been derived (that is, hybrid virus with the RNA of strain A will produce strain A, and similarly, the hybrid virus with the RNA of strain B will produce strain B).

We have here, in the assembly of the TMV particle, a clear example of self-assembly since the molecules incorporated in the structure are sufficient to guide the assembly process. The exact molecular properties involved in this assembly process are being actively studied so that we can expect to know a great deal more about this dramatic phenomenon in the near future.

There is another group of viruses that from the point of view of their geometrical structure appear to be constructed in a simple way. They have *icosahedral symmetry;* are composed of a shell of apparently identical, morphologically distinct subunits made of protein (capsomeres); and contain single (often circular) molecules of RNA or DNA. They vary in size from the tiny bacterial virus ϕX 174 (12 capsomeres) to structures as large as the adenovirus (252 capsomeres). Figure 11-6 shows electron micrographs of ϕX 174 and adenovirus as well as the respective icosahedral structures to which these viruses seem to conform. Caspar and Klug formulated an ingenious theory, which states that the icosahedral symmetry "represents the optimum design for a state of minimum energy of a closed shell built of regularly bonded, identical subunits." Their theory predicted a number of icosahedral arrangements that have been subsequently discovered in a variety of viruses.

A most ingenious demonstration of icosahedral symmetry was made by Horne with *Tripula iridescent* virus in which he used a technique of double-shadow casting in the electron microscope and compared the shape of the shadows with those cast by an icosahedral model (Fig. 11-7).

Until very recently attempts to bring about the *in vitro* self-assembly of viruses of cubic symmetry had proven fruitless. It was thought that in closing a curved surface

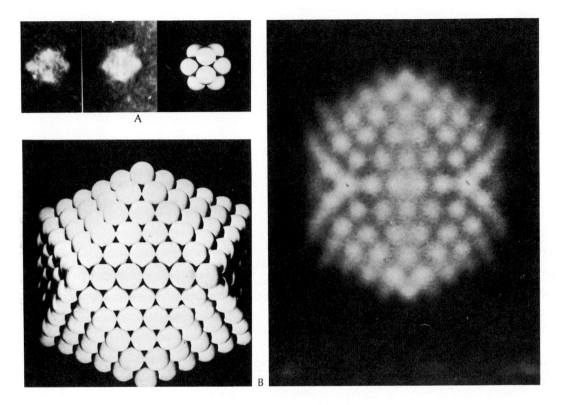

Fig. 11-6. Two icosahedral viruses and their respective models. The resolution of the individual capsomeres is facilitated by the use of a "negative-staining" technique with phosphotungstate. This technique causes stain to arrange itself around structures, leaving them less electron charge. (Courtesy of Horne.) A. ϕX 174 (× 400,000) and its ping-pong ball model composed of 12 spheres. B. Adenovirus (× 850,000) and its model composed of 252 spheres.

Fig. 11-7. Demonstration of the icosahedral structure of *Tipula iridescent* virus. A. EM photograph of virus particles shadowed from two different directions. B. Cardboard model of respective icosahedron shadowed by two light sources. Notice two types of shadows: one blunt, the other pointed, which is characteristic of the icosahedral geometry. Notice also that the same shapes appear in the EM photograph of the virus particle. (Courtesy of Horne.)

more bonds are formed when the last unit is added than for any of the previous ones, and thus some template, extraneous to the virus, might be needed for stabilizing the structure until the last unit has been introduced. An analogy to such a phenomenon would be the scaffolding that is used in building an arch before the keystone has been set in place. However, recently a claim has been made of an *in vitro* assembly of an RNA phage with cubical symmetry, and it is possible that even members of this group of viruses will prove to be capable of self-assembly.

Perhaps the most complex assembly process being studied at present is that of the bacteriophages (Fig. 11-8). These, although they are the most complex of viruses,

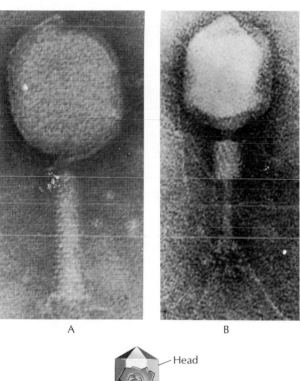

Fig. 11-8. T2 bacteriophage in its normal and contracted forms. This virus is one of the most complex and differentiated forms known, being made of a number of structures each built of one or more structural proteins. The contraction of the tail contributes to the process whereby the phage DNA is injected into the *E. coli* cell. The tailplate appears to contain some lysozyme that has the function of "chewing" a hole into the mucopolysaccharide wall of the bacterium. A small amount of enzymatic proteins is injected into the cell along with the DNA. (Courtesy of Brenner, Horne, *et al.*) A. Normal intact form (× 600,000). B. Contracted (triggered) form (× 680,000). C. Diagrams of intact and "triggered" T2 phage.

A   B

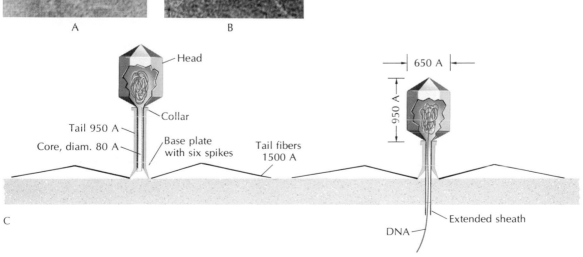

C

are nevertheless extremely suitable for experimental study because of their susceptibility to genetic manipulation and to electron microscopic visualization.

Although it is not clear whether icosahedral viruses require a factor other than
their protein subunits and their nucleic acid strand to contribute to their assembly,
the formation of the head (capsid) of the "T-even" bacteriophages almost certainly
requires one or more additional factors. Thus, in T4 there are 1000–2000 subunits per
capsid, and it is difficult to see how the properties intrinsic to the subunits alone can
explain the formation of the biprolate icosahedron characteristic of these viruses.
The discovery of mutants (Fig. 11-9) that are responsible for the formation of short-

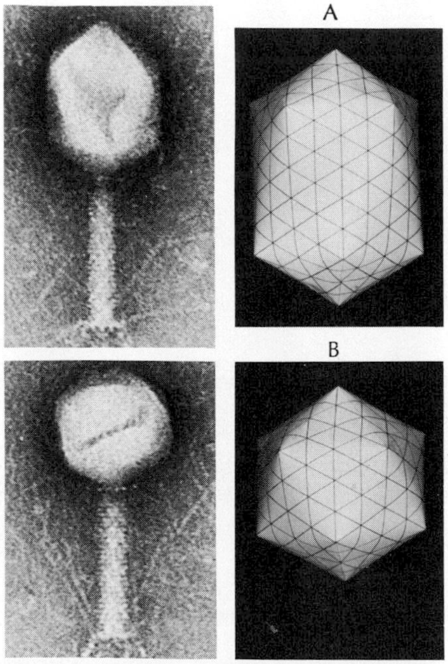

Fig. 11-9. Mutants of T4 bacteriophage resulting in the
formation of short-headed and tubular forms. A. Normal T4 bacteriophage and biprolate icosahedral model.
B. Mutant short-headed form and icosahedral model.
C. Mutant tubular (polyhead) form and models with different angular pitches. It is hoped that a detailed study of
mutants will provide insights into the mechanism of
assembly of these complex viruses. (Courtesy of Kellenberger, Eiserling, and Boydelatour.)

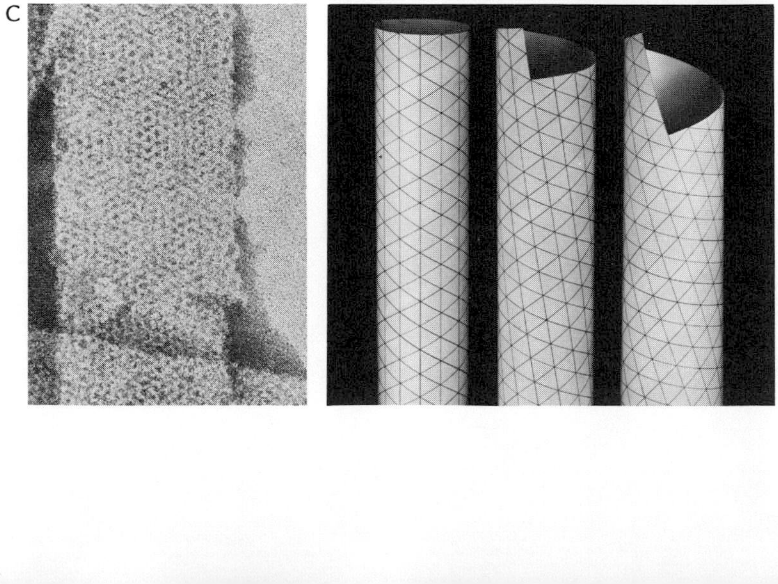

headed (icosahedral) forms and of tubular forms (polyheads) may contribute to the identification of the "morphopoietic factors," believed to be involved in the assembly of the virus head. Whether the morphopoietic factors are part of a "core," as Kellenberger suggests, around which the head is constructed, or arch materials not even incorporated into the virus remains to be seen. The student is referred to the excellent *Scientific American* articles by Kellenberger and by Edgar for a detailed discussion of this fascinating subject.

THE PARTIAL DEGRADATION AND REASSEMBLY OF RIBOSOMES

We have seen that in the case of viruses the assembly process varies from the simplest self-assembly, involving one molecule of RNA and numerous protein molecules of one given kind (TMV), to complex assembly processes, involving a single DNA duplex and a number of different kinds of proteins (T2 bacteriophage). In the case of ribosomes we have a situation that resembles the more complicated viruses. As we have stated in Chapter 8, ribosomes from procaryotic cells are composed of two subunits, 50S and 30S, which together form the 70S ribosomal particle. The 50S subunit is composed of one 23S and one 5S RNA chain as well as some 40 different proteins. The 30S subunit is composed of one 16S RNA chain and also some 20 different proteins, many of which differ from the proteins of the 50S subunit. As we have already indicated, the 70S ribosome can be dissociated into 50S and 30S subunits by lowering the Mg^{2+} concentration from 10^{-2} M to 10^{-4} M. Furthermore, by treating the subunits with very high concentrations of cesium chloride, it is possible to strip some of the proteins from the ribosomal subunits, lowering their sedimentation rate to 43S and 28S. Needless to say, these "core particles" cannot be reassociated at 10^{-2} M Mg^{2+}, nor do they retain their ability to carry out protein synthesis. When, however, the core particles are incubated in the presence of 10^{-2} M Mg^{2+} with the proteins that had been stripped off, Nomura has shown that it is possible to regenerate 50S and 30S particles, which can again be reassembled into 70S ribosomes. But more importantly the reassembled ribosomes show some ability to carry out protein synthesis. Although nobody has as yet succeeded in degrading ribosomes down to their constituent RNA chains and proteins and then reassembling them *in vitro*, the above experiments are an important first step in that direction. The stepwise degradation and reassembly of ribosomes will not only help us understand the nature of assembly processes, but will also provide us with important information about the structure and function of ribosomes as such.

THE AIDED ASSEMBLY OF FIBRIN

So far we have discussed examples of molecular assembly in which we have no clear evidence for intercession by external agents. In the formation of fibrin we have the best understood and perhaps the simplest example of an aided assembly process that is controlled, in part, by external factors not incorporated into the final structure.

Fig. 11-10. Structure of the fibrin clot. A. Low-resolution picture showing the type of branching that occurs. B. High-resolution picture showing precise banding and fine structure. Although the dimensions of the bands have not yet been explained in terms of the dimensions of the fibrinogen molecule and the precise geometry of the polymerization process, it is nevertheless clear that fibrin polymerization is a highly specific and precise phenomenon comparable, for instance, with the assembly of virus particles. (Courtesy of Hall and Slayter.)

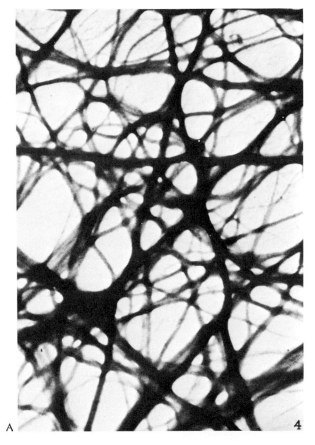

A   4

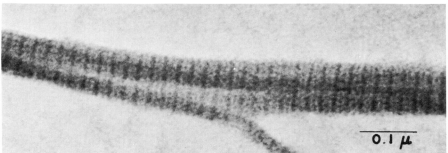

B   0.1 μ

Fibrin formation is the morphological basis for the clotting of blood. We now know that a complex series of events, resulting from the interaction of a number of factors, leads to the conversion of a protein *prothrombin* to an active proteolytic enzyme *thrombin*. This enzyme then acts on the protein *fibrinogen* to convert it to *fibrin*, a massive branched polymer responsible for the structural rigidity of the blood clot (Fig. 11-10). Fibrinogen is a protein (MW = 330,000) composed of six poly-peptide chains (two A, two B, and two C) arranged in a three-beaded structure as shown in Fig. 11-11. Thrombin is a proteolytic enzyme with a *trypsinlike* specificity, that is, it can hydrolyze peptide bonds in which arginyl or lysyl residues donate the carboxyl group. But interestingly, thrombin hydrolyzes only four bonds in the native fibrinogen molecule, all of which involve arginyl-glycyl residues, releasing four peptides (two A peptides and two B peptides) from the two A chains and the two B chains (Fig. 11-12). The release of the peptides changes the properties of fibrinogen in such a way that these molecules (now called fibrin monomers) aggregate in an

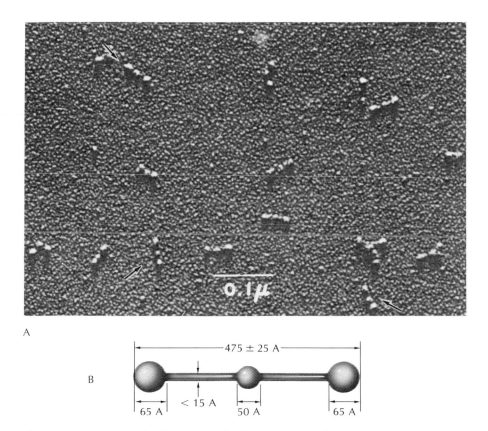

Fig. 11-11. Structure of the fibrinogen molecule. (From Hall and Slayter.) A. Electron micrograph of fibrinogen showing structure composed of three beads. The molecular weight of the molecule is 330,000, and it is composed of three pairs of chains (two A, two B and two C chains) of approximately 55,000 molecular weight each. The chains are held together by disulfide bonds. B. Model of fibrinogen showing dimensions of the molecule. The length of the molecule is greater than one complete repeat sequence of bands in the fibrin clot (see Fig. 11-10), which suggests that the polymerization of fibrin involves some sort of overlapping arrangement of the fibrin monomer units.

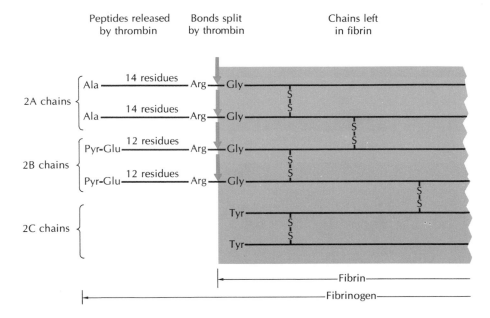

overlapping manner to produce the fibrin polymer. The forces involved in this aggregation are probably apolar and hydrogen bonds since this kind of fibrin is soluble in urea, detergents, and even in buffers of low or high pH. Yet clots formed from blood plasma are not soluble in solvents that do not break covalent bonds such as peptide or disulfide bonds. We now know that this conversion of *soluble* to *insoluble fibrin* is brought about by another enzyme (plasma transglutaminase), which in the presence of Ca^{2+} forms C—N (isopeptide) cross-links between fibrin monomers in a reaction involving the ϵ-amino groups of lysyl residues and the γ-carboxylamide groups of glutaminyl residues, ammonia being released in the process (Fig. 11-13). It is interesting to note that the thrombin is used not only for the formation of soluble fibrin, but also for the conversion of the inactive form of plasma transglutaminase (Factor XIII) to the active enzyme.

Thus, we have:

$$\text{Prothrombin} \xrightarrow[\text{factors} + Ca^{2+}]{\text{a number of}} \text{Thrombin}$$

$$\text{Fibrinogen} \xrightarrow{\text{thrombin}} \text{Soluble Fibrin}$$

$$\text{Factor XIII} \xrightarrow{\text{thrombin}} \text{Plasma Transglutaminase}$$

$$\text{Soluble Fibrin} \xrightarrow[Ca^{2+}]{\text{plasma tranglutaminase}} \text{Insoluble Fibrin}$$

◄ **Fig. 11-12.** Chain structure of human fibrinogen and fibrin. Fibrinogen is composed of three pairs of chains. Two A chains with alanyl residues as the N-terminal groups, two B chains with pyroglutamic acid as the N-terminal groups, and two C chains with tyrosyl residues as the N-terminal groups. Thrombin acts as a specific proteolytic enzyme cleaving two Arg-gly peptide bonds in the A chains and two Arg-gly bonds in the B chains, thus releasing four peptides. The resulting molecule, which we call the fibrin monomer, is transformed in such a way so as to cause it to polymerize to form the fibrin clot shown in Fig. 11-10. Such specific proteolytic actions by an enzyme on a protein substrate may also be involved in intracellular assembly processes.

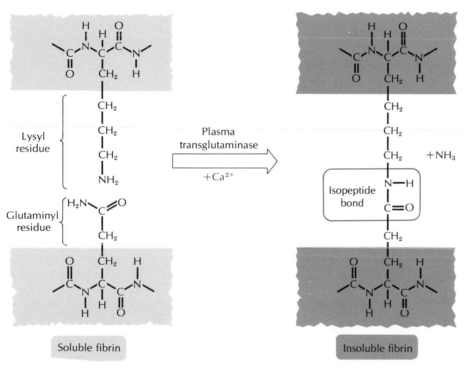

Soluble fibrin   Insoluble fibrin

Fig. 11-13. Molecular mechanism involved in the transformation of soluble fibrin into insoluble fibrin. The enzyme plasma transglutaminase catalyzes the formation as isopeptide bonds by reacting the γ-carboxyl group of glutaminyl residues with the ε-amino groups of lysyl residues, ammonia being released. Whether similar mechanism of cross-linking are found inside cells remains to be seen. (From Loewy, 1968.)

The fibrinogen coming off the ribosome does not have all the structural properties required for insoluble fibrin formation. The sequential intercession of two enzymes is needed before the final product, insoluble fibrin, can be made; thus, the process can be regulated in a highly precise manner.

It is possible that contemporary biology has become overly obsessed with the simplicity of replication of certain viruses such as TMV. One would expect, since replication is the virus's main business, that there has been a natural selection bias in the direction of replicative simplicity. Thus, for instance, the subunits of viruses appear not to be cross-linked by covalent bonds, which would require enzymatic intercession in the assembly process. Cells, on the other hand, might make considerable use of aided and directed processes of assembly. A considerable proportion of the proteins of a eucaryotic cell, for instance, are insoluble, probably because the molecules are cross-linked with covalent bonds, and it seems likely that this can occur only after soluble protein precursors have been synthesized by the ribosomes. As we shall see in the remainder of this chapter, we are beginning to encounter a number of phenomena in which aided or directed assembly processes must be involved.

We have dwelled on the above example because it is a clear-cut case of an aided molecular assembly process in which two enzymes participate but do not become incorporated into the final structure.

DIRECTED POLYMERIZATION OF BACTERIAL FLAGELLA

Recently, a very interesting assembly system has been described in which it is clear that an initiator mechanism is involved. The bacterium Salmonella has on its surface wavy projections called bacterial flagella. Upon heating, these structures break down into protein molecules (flagellin), which according to disk electrophoresis give a homogeneous preparation. Using this material, Asakura and his co-workers demon-

Fig. 11-14. Flagella reconstituted from the purified protein flagellin obtained from the bacterium *Salmonella* (× 24,000). In both the normal and the curly strains flagellar fragments are necessary for the assembly of flagella from the flagellin molecules. Curly flagella fragments tend to impose the curly pattern on flagellins from normal flagella whereas the reverse does not seem to be the case. (From Asakura *et al.* 1964.)

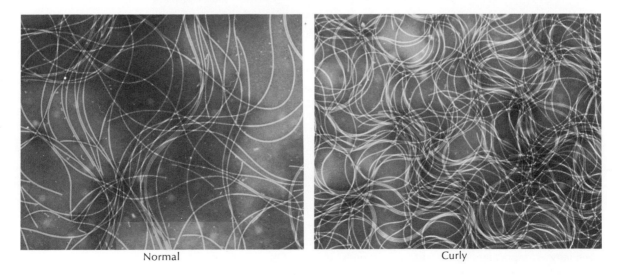

Normal   Curly

strated that flagellin molecules will not aggregate into flagella unless some flagellar fragments are added (Fig. 11-14). They also showed that polymerization occurs at the end of the flagellar fragments, which therefore seem to act as initiators for the polymerization reaction. By using flagellins of different strains of Salmonella that polymerize at different rates, Oosawa and Asakura showed that the flagellins determine mainly the rate of polymerization, whereas the fragments determine whether polymerization will occur.

Some strains of Salmonella have flagella with a tighter wave than others and are denoted as "curly." Curly flagella fragments tend to impose the curly pattern on flagellins from normal flagella whereas the reverse does not seem to be the case. This suggests that both the initiator and the monomer units are important for the morphology of the final product.

The above initiator-induced assembly process can be characterized as being self-replicative in nature. The initiator here is not different from or complementary to the structure being duplicated. Unlike DNA, in which replication of a complementary structure occurs, the flagellar system appears to utilize a structure that is capable of directing the formation of more of itself.

These experiments suggest that in the cell, a flagellar structure is necessary in order to manufacture more of it at a reasonable rate. Thus, the DNA specifies the information required for making flagellin molecules, but the latter do not contain sufficient information to polymerize rapidly into flagella. Some of that information must apparently come from the flagella themselves. As we shall see in the remaining portion of this chapter, it is likely that there are other structures in the cell that are similarly *self-reproductive* in nature. However, we shall not discuss such self-reproductive structures as mitochondria and chloroplasts since they are found to contain DNA and ribosomes and thus have an information transfer system that is analogous to that of the cell as a whole. This does not exclude the possibility that these organelles contain self-reproductive components which, like the bacterial flagella, are involved in assembly processes.

(See Chapter 13.)

ARE SOME CELLULAR STRUCTURES SELF-DUPLICATIVE?

The *microtubules* of the cytoplasm and the structural elements of the *mitotic spindle* are structures that may have similar properties to the bacterial flagella although no technique has yet been devised for the removal, dissociation, and assembly of these structures *in vitro*. Inoue has found it possible to dissociate the mitotic spindle inside the cell by lowering the temperature; and Marsland, by raising the pressure. Porter and his co-workers have been able to do the same with the microtubules in the axopods of *Actinosphaerium*. It is, of course, not possible to deduce from these experiments whether the assembly of microtubules is spontaneous, aided, or directed.

The mechanism of assembly of membranes is also as yet not well understood. Many biologists believe that the assembly of membranes is self-replicative in nature and that molecules or smaller assemblies of lipoproteins are introduced into preexisting membrane structures. Racker and his co-workers however have been able to disassemble the complex formed by the inner membranes of mitochondria and particles with ATPase activity and then to reconstitute them into structures that regain some of the enzymatic and structural properties of the original membranes. Such experiments on membranes will without question be one of the most important areas of research during the next decade.

So far the most convincing evidence for a cytoplasmic self-duplicative structure comes from Sonneborn's extensive study of the cortex of *Paramecium aurelia*.

This organism, like other protozoa, has a well-defined cortical layer of the cytoplasm located beneath the plasma membrane. Unlike the cytoplasmic interior it is not fluid but stiff or gelated and forms the matrix of a number of morphological features such as the contractile vacuole pore, the mouth, and the very precisely marked "kinety fields" composed of dots consisting of the points where the cilia and the basal bodies (kinetosomes) are joined. It has been known for some time that "doublet" animals, consisting of two sets of these surface structures, occur (Fig. 11-15). These animals are usually double only in the cortical region, the endoplasm in the interior being one region containing only one macronucleus.

Fig. 11-15. Normal and doublet forms of *Paramecium aurelia*. Observe the precise location of the kinety fields of the normal animal and how some of the kinety fields are duplicated in the doublet animal. Notice that the doublet animal has two mouths (m), two preoral sutures (S) and two anterior left kinety fields (AL). CVP = contractile vacuole pore. (Courtesy of Sonneborn and Dipple.)

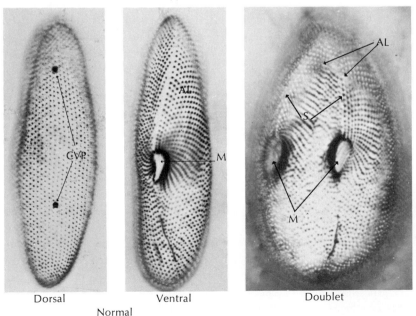

Dorsal   Ventral   Doublet

Normal

Paramecium aurelia is a useful animal for the investigation of this phenomenon because it is possible to put the animal through a variety of controlled exchanges. Thus, in an extensive series of experiments, Sonneborn and his co-workers were able to exchange the micronuclei, most of the macronuclear material, and varying amounts of the fluid cytoplasm (endoplasm). In all these cases the animals with a doublet cortex bred true, and animals with a singlet cortex also bred true. In other words, all exchanges of material below the cortical zone had no effect on the inheritance of the singlet or doublet character of the cortex. The inference one can draw from these experiments is that a number of structures in the cortical region are self-reproductive in character. This indeed Sonneborn was able to confirm by a series of positive experiments. It is not easy to remove a piece of cortex from one *Paramecium* and graft it on another, but, if one carries out numerous cytoplasmic exchanges, it is possible, on infrequent occasions, to find an animal that has taken with it a piece of the cortical region of another animal. When that happens, the new animal, which may have two mouths for instance, breeds true—at least this is what happened in one celebrated instance.

Sonneborn also found that when a row of kinety fields became inverted in one organism this abnormality also was transmitted to succeeding generations.

Detailed electron microscopic studies showed that when the cortical region enlarges during cell elongation, new kinetosomes are formed in the vicinity of existing ones, and although the mechanism of replication is not understood, it is clear that the orientation of the new kinetosome is determined by that of the "parent" kinetosome. Recent experiments have indicated that kinetosomes contain some DNA, and while it is possible that this DNA contains the information for kinetosome protein synthesis, it is unlikely that this DNA is responsible for the orientation and disposition of the kinetosomes. One is tempted to conclude that we have here an assembly process that as in the case of the bacterial flagella, requires the pre-existence of a structure.

It is clear that methods have not yet been developed to obtain unambiguous answers about the mechanism of assembly of these and other cell structures. We are accustomed to thinking of chromosomes, mitochondria, and chloroplasts as being self-reproductive, but to what extent some structures other than DNA in these organelles are required for their assembly process is not yet known.

The molecular mechanisms of problems such as these represent the greatest challenge to cell biology in the immediate future, and we can expect that new approaches to the mechanism of cellular assembly processes will emerge from our present efforts.

SUGGESTED READING LIST

Offprints

Burnet, Sir Macfarlane, 1951. "Viruses." *Scientific American* offprints. San Francisco: W. H. Freeman & Co.

Fraenkel-Conrat, H., 1956. "Rebuilding a Virus." *Scientific American* offprints. San Francisco: W. H. Freeman & Co.

Gross, J., 1961. "Collagen." *Scientific American* offprints. San Francisco: W. H. Freeman & Co.

Kellenberger, E., 1966. "The Genetic Control of the Shape of a Virus." *Scientific American* offprints. San Francisco: W. H. Freeman & Co.

Sonneborn, T. M., 1960. "Does Preformed Cell Structure Play an Essential Role in Cell Heredity?" *The Nature of Biological Diversity*, pp. 165–221. J. M. Allen, ed. McGraw-Hill Book Co., Inc. Bobbs-Merrill Reprint Series. Indianapolis: Howard W. Sams & Co.

Stent, G. S., 1953. "The Multiplication of Bacterial Viruses." *Scientific American* offprints. San Francisco: W. H. Freeman & Co.

Wood, W. B., and Edgar, R. S., 1967. "Building a Bacterial Virus." *Scientific American* offprints. San Francisco: W. H. Freeman & Co.

Articles, Chapters and Reviews

Henning, U., 1966. "Multi-enzyme Complexes." In: *Angew. Chim. Intern. Ed. Engl. 5*, pp. 785–790.

Weidel, W., and Pelzer, H., 1963. "Bag-shaped Macromolecules—A New Outlook on Bacterial Cell Walls." In: *Advances in Enzymology*. New York: Interscience Publishers, Inc.

Books

Allen, J. M. (ed.). 1967. *Molecular Organization and Biological Function*. New York: Harper & Row Publishers, Inc.

Fraenkel-Conrat, H., 1962. *Design and Function at the Threshold of Life: The Viruses*. New York: Academic Press, Inc.

Wolstenholme, G. E. W., and O'Connor, M., (eds.). 1966. *Principles of Biomolecular Organization*. Boston: Little, Brown & Co., Inc.

Metabolic Pathways

Chapter 12

Thus far we have discussed enzymes and how they act on substrates either to oxidize or reduce them, to split them with water or with phosphate, or to combine two smaller molecules to make a larger molecule. But what are these substrates, these compounds acted upon by enzymes? Primarily, they are the foodstuffs taken in by cells or the breakdown products of these foodstuffs. For example, whatever we take into our own bodies, be it carbohydrate, fat, protein, nucleic acids, or various other constituents, is broken down into smaller molecules by enzymes in the saliva and in the gastrointestinal tract. The breakdown products then enter the various cells of the body, via the blood stream, through mechanisms detailed later. In some cells breakdown proceeds still further; in some other cells the synthesis of larger molecules from the smaller ones occurs; in yet others the predominant mode of attack may be various transformations of these compounds. In most cases, the various cells of the body can perform all of these functions. All in all, the preponderant mode of utilization or metabolism of these compounds in the various cells of different organs varies according to the nature of the cells. Some, like the liver cell, are very active in providing a good deal of the circulating proteins of the blood and in mobilizing glucose for the use of the rest of the cells, mostly the muscle cells, of the body. Other cells, like the various secretory cells of the different glands, are active in synthesizing and secreting hormones, which, via the blood, are then deposited at their respective end-organ sites and are then active in some respects in regulating the overall metabolism of the body. In these tasks and in many others that will be outlined in later chapters, the cell does work, and in doing work it requires the expenditure of energy. In general it obtains energy by metabolizing a portion of the same breakdown products that are used in the synthesis of the larger compounds of the cell. Thus, some of the breakdown products are used to synthesize glucose or glycogen in liver and muscle cells, whereas some of the rest is metabolized to provide the energy for these syntheses.

HIGH-ENERGY PHOSPHATE COMPOUNDS

The cell obtains usable energy primarily by utilizing the high-energy phosphate compounds. The name high-energy phosphate was given to those phosphate esters whose hydrolysis leads to a uniquely high release of energy in the form of heat. Among these substances are some that will be mentioned later, such as adenosine triphosphate (ATP), the triphosphates of uridine, cytidine, and guanosine (UTP, CTP, GTP), phosphocreatine, phosphoenolpyruvic acid, amino acid adenylates, and uridine diphosphate glucose. For example, the hydrolysis of ATP leads to ΔH values of about 9000 calories per mole, whereas other compounds such as glucose-6-phosphate have ΔH values of about one-half of this. Actually, the high-energy content of the compound (See Chapter 2.) is more correctly given by the free energy of hydrolysis, the ΔF. The ΔF values are obtained from measurements of concentrations of the high-energy compounds in reactions in which they participate; they are a measure of differences between the free energies of reactants and of products. If the latter is higher, energy must be supplied for the reaction to proceed. The reasons for the "high-energy" nature of these compounds are not precisely known, but it is thought to be due to several factors that make these compounds unique: there are striking differences between the compounds and the products of their hydrolysis, as in their resonance stabilities, their ionizations,

Fig. 12-1. Properties of high-energy compounds.

1. Inorganic phosphate resonating forms:

2. Carboxylate-phosphate anhydride, as in acetyl phosphate:

3. Phosphate-phosphate anhydride, as in the nucleoside di- and triphosphates; ADP and ATP:

4. Basic nitrogen-phosphate link, as in phosphocreatine:

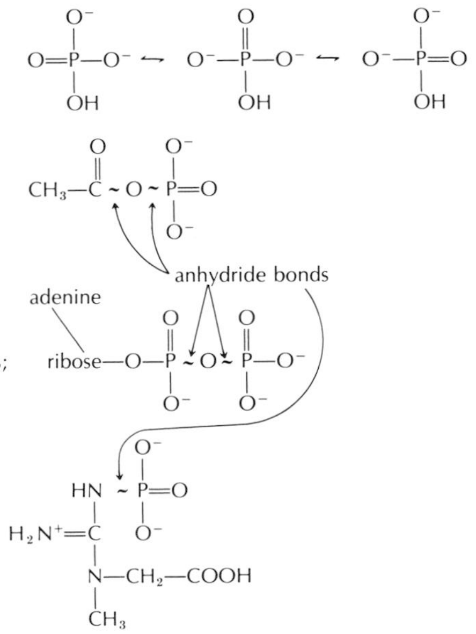

and the intramolecular electrostatic repulsions. It was noted by Lipmann and Kalckar in 1941 that a prominent feature of the high-energy compounds is that they are anhydrides of phosphoric acid with a second acid, such as a substituted phosphoric acid, to form the nucleoside di- and triphosphates; or with a carboxylic acid compound (acetic acid), to form acetyl phosphate; or even with an enol, to form phosphoenolpyruvic acid. In another case, however, phosphoric acid combines with a basic nitrogen compound, as in phosphocreatine. In all cases, the formation of the anhydride bond reduces the number of resonating groups in the molecule, and since the thermodynamic stability of a substance is increased by its ability to assume many resonating forms, the high-energy compounds are less stable, and their hydrolysis will yield a large amount of heat. Figure 12-1 shows some of these properties.

The most important high-energy compound is ATP. Its formula is shown in Fig. 12-2, and its modes of formation by the cell are given in Chapter 13. It can function in various ways: (1) as a phosphorylating agent, transferring inorganic phosphate to an acceptor compound, as in the hexokinase reaction to form glucose-6-phosphate

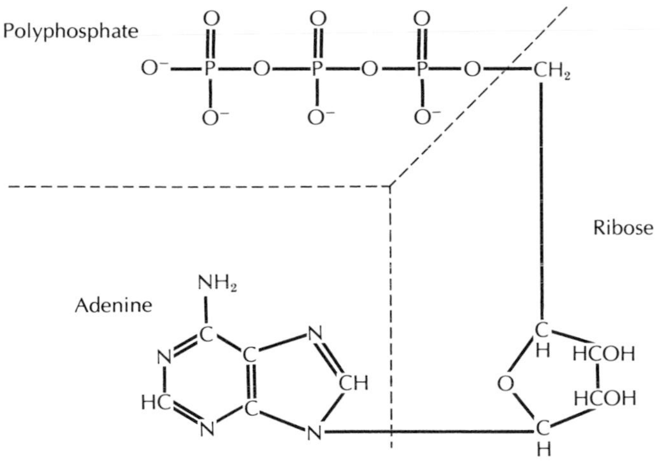

Fig. 12-2. Chemical formula of ATP.

(Fig. 12-4); (2) as a pyrophosphorylating agent, transferring inorganic pyrophosphate to an acceptor; or (3) as an adenylating agent, transferring the adenylate moiety to a suitable acceptor, as in the formation of the amino acid adenylates. In all these cases the product synthesized can be at a lower, at the same, or even at a higher energy level than is ATP. In the cell, ATP functions mostly as a phosphorylating agent in cases where energy is required to drive a reaction. ATP acts as a messenger between those reactions that supply energy (exergonic) and those that utilize energy (endergonic); it does so because it is common to both types of reactions.

(Fig. 15-6.)

Similarly, the other nucleoside diphosphates, guanylic, cytidylic, and uridylic, can be phosphorylated by ATP to give the corresponding triphosphates. These reactions

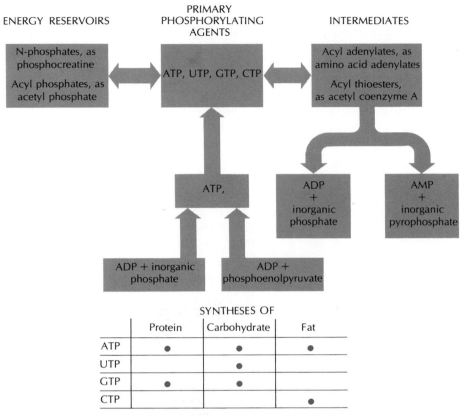

ENERGY RESERVOIRS — PRIMARY PHOSPHORYLATING AGENTS — INTERMEDIATES

Fig. 12-3. Interrelationships between high-energy compounds. The lower table gives the specifities for the various triphosphates in the general synthetic pathways.

	SYNTHESES OF		
	Protein	Carbohydrate	Fat
ATP	●	●	●
UTP		●	
GTP	●	●	
CTP			●

(See Chapter 15.)

(See Chapter 13.)

(See Chapter 16.)

are important in that all the nucleoside triphosphates are instrumental in synthetic reactions. ATP is involved in fatty acid and protein synthesis; CTP, in phospholipid synthesis; UTP, in glycogen synthesis; and GTP, in carbohydrate and protein synthesis. The ultimate energy requirement for the synthetic needs of the cell is ATP since only it is formed in glycolytic, photosynthetic, and oxidative phosphorylations. But from ATP the energy is funneled by countless transphosphorylation reactions to other "energy-donating" compounds. For example, glycogen is made from UDPG, a compound synthesized by the reaction of UTP with glucose-1-phosphate; the UTP is formed from ATP through the reaction $ATP + UDP \rightleftharpoons ADP + UTP$. In addition, ATP can phosphorylate creatine, forming the phosphocreatine utilized by muscle fibrils. It can also phosphorylate acetate, giving acetyl phosphate, or it can transfer its adenylate moiety to acetate, forming acetyl adenylate. Both these compounds can be transformed into acetyl coenzyme A, which is the immediate precursor of fatty acids. The interconversions between the various classes of phosphorus compounds are given in Fig. 12-3.

The problem presented by such a tabulation is how the cell can regulate, through the shunting of the high-energy groupings, the innumerable synthetic reactions that occur therein. In some cases, since the energy state of the phosphorylated acceptor compound is the same as or lower than that of ATP, it would appear that the availability of the acceptor compound, be it UDP or glucose, is used as a regulatory device. In other instances, as in the case of phosphocreatine or acetyl adenylate, the energy content is higher than that of ATP; hence, the reaction would not go in the direction of synthesis of the phosphorylated acceptor compound unless the concentration of the latter

Fig. 12-4. General pathway for glucose in liver cells.

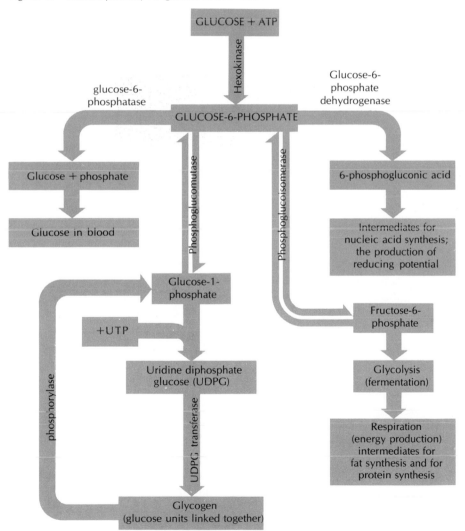

compound is reduced by being employed in yet another reaction, that of synthesis. In this way, the energy is used in the synthetic reaction; phosphate or pyrophosphate is split off, and thus the overall reaction, starting from ATP, will be in the direction of synthesis of the phosphorylated intermediate, and hence synthesis of glycogen or protein or whatever else the case may be.

METABOLIC NETWORKS: AN EXAMPLE OF THEIR MEANING

From the above bare outline it is clear that many of the substrates or metabolites are attacked not just in one way, by a single enzyme. In many cases, there is more than one enzyme attacking the same metabolite—one to oxidize it, one to reduce it, and one to attach some other compound to it. We can thus speak of metabolic pathways, that is, the various directions a metabolite can go. The same substance, for example, can be attacked by one enzyme to produce chemical energy and can also be put by another enzyme onto the pathway for the synthesis of some of the larger molecules of the cell such as proteins and nucleic acids. A good example is the fate of glucose-6-phosphate in the liver cell (Fig. 12-4). It can be formed in various enzymatic ways such as from glycogen or from smaller intermediates of fermentation, or from free glucose. Once it is formed, it can be acted upon by four different enzymes, that is, there are four vectors to its metabolism: (1) to replenish the supply of glucose in the blood; (2) to build up liver glycogen as a storehouse for blood and muscle cell glucose; (3) to provide intermediates for the syntheses of fats, proteins, and nucleic acids; and (4) to provide energy.

Exactly how all this is regulated by the liver cell we do not know, but it is almost certain that the relative rates of the enzymatic reactions and various hormones have a hand in this regulation, by determining what proportion of the glucose molecules travel along each route. Nature is conservative, in the sense that the same compound acts as a starting point for many different metabolic pathways. We can guess that the reason is that regulation of the overall metabolism—the physiology—of the cell is much easier if there are fewer key points that must be accommodated. The switch from one pathway to another becomes then more responsive to the needs of the cell. In the example given above the key compound might be glucose-6-phosphate. Also, although the example is the liver cell, other cells, as distant biologically as the yeast cell, have most of the same enzymes, have the same pathways, and are probably prone to the same sort of regulation.

However, even the simplified diagram of Fig. 12-4 does not give all the other known alternative pathways in this large subfield of metabolism. For example, the synthesis of nucleic acids is a necessary requisite of cell growth, as later chapters will point out. One of the constituents of these complicated compounds is the 5-carbon sugar, ribose. The immediate precursor of ribose, ribose-5-phosphate, is synthesized via many pathways. From the carbohydrate precursor, glucose-6-phosphate, there are actually three pathways leading to the formation of this compound (Fig. 12-5). One of these, the pentose phosphate pathway, has been known for a long time.

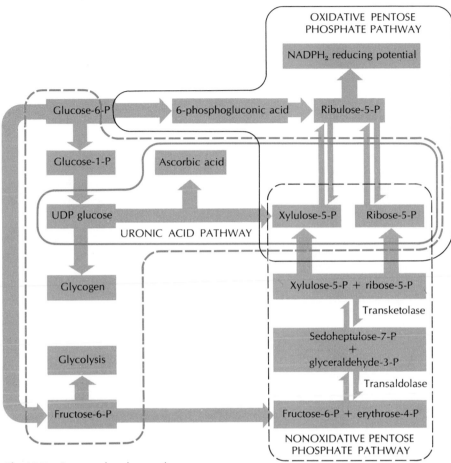

Fig. 12-5. Pentose phosphate pathways.

Actually, two pentose phosphate pathways are now recognized, the oxidative and nonoxidative. The former, leading to ribose-5-phosphate in four enzymatic steps, is a prime source of reducing potential in the form of $NADPH_2$; it is a reducing potential because this coenzyme, $NADPH_2$, is a necessary concomitant reactant for the syntheses of fatty acids and steroids. The nonoxidative pentose phosphate pathway begins with the conversion of glucose-6-phosphate to fructose-6-phosphate. Instead of continuing on this glycolytic road (see Fig. 12-6), however, the carbon atoms of fructose-6-phosphate go through a series of condensation and transfer reactions, leading, via the important enzymes transaldolase and transketolase, to xylulose-5-phosphate and ribose-5-phosphate. The former compound is also made via the oxidative pathway and via another road, called the uronic acid pathway, by way of uridine diphosphate glucose and 11 enzymatic reactions. Since xylulose-5-phosphate can be converted to ribose-5-phosphate via the intermediacy of ribulose-5-phosphate,

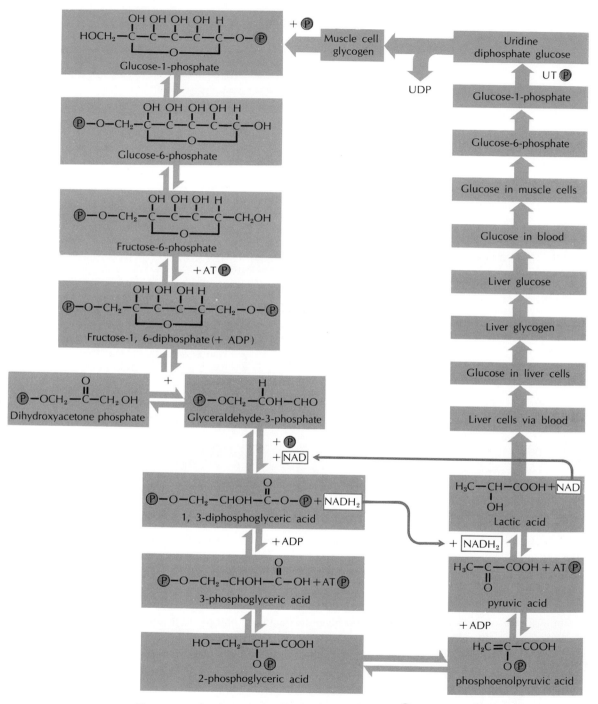

Fig. 12-6. Glycolysis in muscle cell. The colored circle ⓟ traces the pathway of phosphate, leading to high energy, AT ⓟ, production; the white blocks show the cycling of reducing potential within the glycolytic pathway.

the uronic acid pathway is also a source of the ribose moiety of nucleic acids. Another feature of the nonoxidative pathway is found in the use made of it by chloroplasts for the fixation of CO_2, in what has been called the carbon-reduction cycle.

(Fig. 13-13.)

Radioactive tracers indicate that all these pathways seem to be operating in the cell at the same time, although the traffic on the road leading to ribose via the uronic acids seems to be insignificant compared to that on the other two. In the liver and muscle cells at least, the uridine diphosphate glucose formed thereby is funneled off into glycogen synthesis. But why should these multiple sources of ribose exist? We are not sure, but we think that since nucleic acid synthesis is necessary for the cell to grow or even to exist, the cell has provided itself with more than one pathway for its synthesis in case the others are blocked off. This can be illustrated experimentally by making use of the observation that the vitamin thiamine, in the form of thiamine pyrophosphate, is a necessary cofactor in the functioning of the nonoxidative pathway enzymes transaldolase and transketolase. In normal animals it was demonstrated that of all the carbon atoms of glucose going to ribose, about 40 percent of the radioactive carbon atoms of administered glucose was converted to ribose carbon atoms by the oxidative pathway and 60 percent by the nonoxidative pathway. However, in animals rendered thiamine deficient, about 80 percent of the ribose carbons were formed via the oxidative pathway. Thus, the cell seems to have evolved multiple pathways to make certain that a necessary compound is formed no matter under what environmental conditions the cell might find itself. In the above case this compound is ribose-5-phosphate. Another example involving the same alternative pathways could be the need of the cell to make enough $NADPH_2$ for the synthesis of steroids and fats.

ENERGY SUPPLY: GLYCOLYSIS

Let us go back a bit and observe how glucose is used either for energy purposes immediately or stored in the form of glycogen to be used later by the cell when needed; this involves the glycolytic pathway (Fig. 12-6), based on the work of G. Cori and C. Cori and of Meyerhof and Embden in elucidating the energetic requirements of the muscle cell. Muscle glycogen breaks down to the 6-carbon compound, fructose-1, 6-diphosphate, via several intermediate enzymatic steps, including one that requires ATP. Fructose diphosphate is then split into two 3-carbon compounds that are in equilibrium with each other. One of these, glyceraldehyde-3-phosphate, is oxidized to 3-phosphoglyceric acid, generating ATP. This step, catalyzed by the important enzyme glyceraldehyde-3-phosphate dehydrogenase, is the sole oxidative step in the whole glycolytic pathway. Oxygen is not required for this oxidation; only the presence of NAD, which in the process is reduced to $NADH_2$, is required. Via several other enzymatic steps, pyruvic acid is formed, and at the same time another molecule of ATP is produced. Under anaerobic conditions, this pyruvic acid is reduced to lactic acid by the $NADH_2$ generated in the oxidative step of glycolysis, with NAD becoming re-formed, for further use in the oxidative step. All these enzymatic steps are performed by soluble enzymes located in the cell sap or matrix. Under aerobic conditions, pyruvic acid is oxidized via the Krebs cycle in the mitochondria, the formation of lactic acid being thus bypassed.

Since one ATP is needed to form fructose diphosphate and since four ATP molecules are produced during glycolysis (two 3-carbon glyceraldehyde-3-phosphate molecules being generated from each 6-carbon sugar phosphate), a net production of three moles of ATP is obtained for each mole of glucose split off from glycogen. When lactic acid accumulates under anaerobic conditions, it travels to the liver via the blood, and there it is built into liver glycogen. Muscle glycogen is rebuilt by a process that starts with liver glycogen being degraded and the liberated free glucose reaching the blood. From the blood it traverses the muscle cell membrane, is phosphorylated to glucose-6-phosphate, and is then transformed into glucose-1-phosphate. The glucose of this compound is then transferred into an "active" glucose compound, uridine diphosphate glucose, with the aid of UTP. This "active" glucose compound releases its glucose enzymatically to the end of the small amount of muscle glycogen molecules that are still present; in this way glycogen is built up of a very large number of glucose units. Since, by this pathway, one more ATP molecule is needed (by the enzyme hexokinase to form glucose-6-phosphate), the net result in glycolysis starting from free glucose is only two ATP molecules. Thus, either 20,000 or 30,000 cal of biologically usable energy in the form of ATP (about 10,000 cal per ATP) are obtained from a total heat-energy yield of about 58,000 cal in going from glucose to lactic acid; this is an efficiency of from 30 to 50 percent. In Chapter 16 will be detailed the use that the muscle cell makes of this energy derived from glycolysis.

ENERGY SUPPLY: THE KREBS CYCLE; ITS MEANING

The conservatism of nature is shown too in the fact that almost the same pathway exists for the synthesis of large molecules and for the production of energy. This is not entirely true for the entire length of the pathway, for it is now known that there are different portions of the pathway for the breakdown and synthesis of proteins, for the breakdown and synthesis of carbohydrates, and for the breakdown and synthesis of fats. But lower down on the metabolic ladder, where the intermediates for all these processes lie, the metabolic pathways are so arranged that a change in direction, so to speak, can lead either to the combustion of an intermediate to provide energy or to the coupling of an intermediate with another compound to give larger molecules. In this sense the biochemist speaks of cycles, of true wheels. Perhaps it is easiest to explain this by going into the details of the most famous and oldest of these cycles, the tricarboxylic acid (TCA) cycle, or Krebs cycle, named for its formulator. In doing so, we will also learn a bit of how a biochemist works and how these pathways came to be formulated. It is a fascinating story.

The cycle is shown in Fig. 12-7. Essentially, it is a scheme by which the two carbon atoms of acetic acid are oxidized to carbon dioxide and water, and at the same time a great deal of biological energy is produced. This 2-carbon compound is actually acetyl coenzyme A, and its production is a result of the previous breakdown of fat

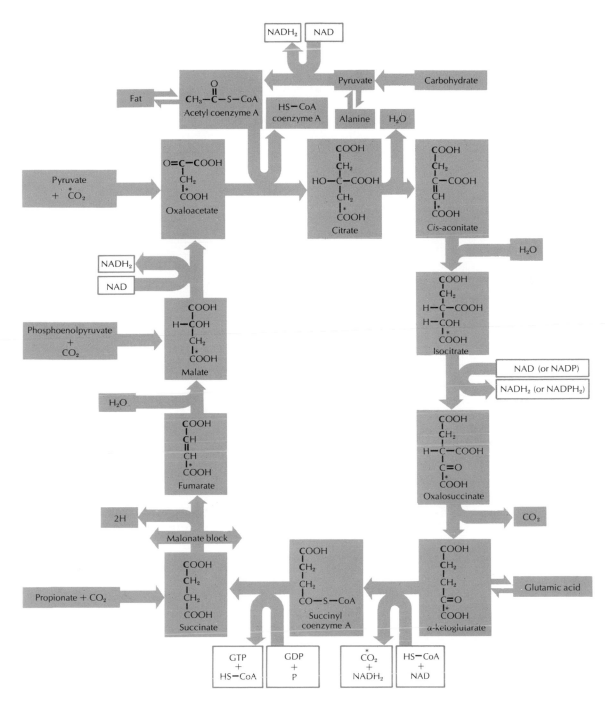

Fig. 12-7. Tricarboxylic acid (Krebs) cycle. The starred C (C*) shows the entrance and pathway of CO_2 in the cycle; the boldface two-carbon unit traces the pathway of the carbons of the initially condensed acetyl coenzyme A.

and carbohydrate foodstuffs. Carbohydrate, when taken into the body of a multicellular organism, is in the form of large polymers of glucose units that must first be broken down to smaller units. The breaking down into single glucose molecules is done by mainly extracellular enzymes, enzymes that are secreted by some cells. Glucose then enters the cells of the body by means that will be outlined later. In the case of single-celled organisms like bacteria and some fungi, the carbohydrate in the environment is broken down either by enzymes secreted into the medium by the cells or by enzymes situated on the surface of the cells. The glucose now inside is then broken down by a series of soluble enzymes known collectively as the glycolytic enzymes, as shown above. The end product of all this is usually pyruvic acid. This compound gets into the mitochondria and is oxidized to acetyl coenzyme A in a very complicated enzymatic reaction. The latter compound can either become further oxidized via the TCA cycle, or it can go through a series of reactions in which it successively condenses with itself or with a derivative of itself to form the higher fatty acids. These fatty acids are straight-chain compounds, 14–20 carbon atoms long. This is how carbohydrate becomes converted to fat, and how sugar eaten ends up as body fat.

(See Chapter 13.)

Why in some circumstances the acetyl coenzyme A gets oxidized to provide energy and in other instances it condenses to form longer fatty acids to provide fats, we do not know, but there are some possible explanations based on the circumstances of the TCA cycle. For example, note that in Fig. 12-7 the first step in the oxidation of acetyl coenzyme A is the condensation, with the aid of an enzyme called simply "condensing enzyme," of this compound with oxaloacetate, forming citrate. Now suppose there were very little oxaloacetate available for this step. The condensation would be much reduced in amount, and acetyl coenzyme A could not be further oxidized in large amounts; instead, a good deal of it would go onto an alternate pathway. The latter pathway is the one enabling acetyl coenzyme A to form long-chain fatty acids. The shunting at this point is one of the possible ways a common precursor is transformed into either carbohydrate or fat.

To get back to the cycle, the citrate formed in the above reaction is attacked by an enzyme, aconitase, that causes an equilibrium to be established between three compounds, citrate, *cis*-aconitate, and isocitrate. The direction of reaction is constantly being shifted toward isocitrate formation because there exists an enzyme that attacks isocitrate; by removing isocitrate from the reaction with aconitase, the overall equilibrium results in the formation of more isocitrate from *cis*-aconitate and citrate. This kind of an event, the shifting of one enzymatic equilibrium by means of another enzymatic attack on one of the compounds involved, is not a singular occurrence in the cell but goes on constantly. In this way many reactions that are thermodynamically reversible, in which very little change in free energy is involved, are constantly being rendered "enzymatically" irreversible, that is, caused to go most of the way in one direction. If the concentration of isocitrate is increased, as can be done experimentally, formation of citrate from isocitrate will result.

The enzyme that attacks isocitrate, called isocitrate dehydrogenase, uses nicotin-amide-adenine-dinucleotide (NAD) as a coenzyme, whereas a similar enzyme uses nicotinamide-adenine-dinucleotide phosphate (NADP). These two coenzymes were formerly known as diphosphopyridine nucleotide (DNP) and triphosphopyridine nucleotide (TPN). In the process, isocitrate is oxidized to oxalosuccinate, with the loss of two hydrogen atoms, giving $NADH_2$, or reduced NAD (and in the other case, $NADPH_2$). The same enzyme, probably at the same time, causes a decarboxylation of oxalosuccinate, giving α-ketoglutarate and carbon dioxide. The carbon dioxide comes from the oxidation of one of the carbon atoms of the original oxaloacetate. The formation of $NADH_2$ is notable in that its subsequent oxidation leads to a production of biological energy. How the latter process is accomplished will be discussed in the next chapter. It suffices here to point out this further function of the cycle, the originating of the production of most of the usable energy of the cell.

The next step, the oxidative decarboxylation of α-ketoglutarate to form succinate and carbon dioxide, by means of the enzyme α-ketoglutarate dehydrogenase, is also a very complicated reaction. Coenzyme A again takes part, and what is actually obtained is not succinate but succinyl coenzyme A. This dehydrogenation takes place again with the aid of NAD, forming $NADH_2$, and again this $NADH_2$ is further oxidized to provide energy. There is also, however, another step in this overall reaction that provides energy. This is the enzymatic removal of coenzyme A from succinyl coenzyme A, with the aid of guanosine diphosphate (GDP) and inorganic phosphate, to form guanosine triphosphate (GTP). This GTP is a high-energy compound, the properties of which were mentioned earlier. Thus, the whole sequence of reactions from α-ketoglutarate to succinate provides for energy production, in the formation of $NADH_2$ and GTP. This GTP can enzymatically react with adenosine diphosphate (ADP) to form GDP and adenosine triphosphate (ATP). It is the latter nucleotide, ATP, that has been called the energy coin of the biological realm. Notice, incidentally, that during the oxidation of α-ketoglutarate another molecule of carbon dioxide is removed from the original oxaloacetate.

The next reaction, succinic dehydrogenation, is a very powerful one in which two hydrogen atoms are removed from succinate and fumarate is formed. In the process the two hydrogen atoms, left dangling in the scheme, are transferred to cytochrome b, to form reduced cytochrome b. This cytochrome b is an electron carrier protein containing an iron heme as a prosthetic group; we will have more to say about this type of compound in the next chapter.

The fumarate resulting from the above reaction then takes on a molecule of water, with the aid of the enzyme fumarase, to form malate. Malate is oxidized by malic dehydrogenase and NAD, forming $NADH_2$ and oxaloacetate again. By following the two carbon atoms (boldface type) of acetate throughout the cycle, you will notice that they end up in the new oxaloacetate molecule. Thus, although oxaloacetate is back again, it is not precisely the same molecule since it has lost carbon atoms as carbon dioxide and has regained them from the acetyl coenzyme A. This oxaloacetate molecule can now accept another molecule of acetyl coenzyme A, and the cycle will be repeated. We can say, then, that the two carbon dioxide molecules result from the oxidation of acetate.

The cycle is the key to the oxidation of fats and carbohydrates, for we now know that most of the fat and a good deal of the carbohydrate is oxidized via the cycle. Since α-ketoglutaric, oxaloacetic, and pyruvic acids can be formed very easily from the amino acids, glutamic and aspartic acids, and alanine, the cycle is also responsible for the oxidation of some of the protein intermediates of the cell. This view is strengthened by the fact that many of the other amino acids can be broken down to intermediates such as acetyl coenzyme A, α-ketoglutarate, oxaloacetate, succinate, or fumarate. It can be estimated that the oxidation of the carbon atoms of most of the substrates taken in by the cell proceeds through this cycle. This happens in all kinds of cells, from most of the different kinds of cells in our body to plant cells to yeast and bacterial cells.

What about the reverse of this coin? The cycle is also involved in the syntheses of intermediates leading to the formation of larger molecules. For example, by amination (the putting in of ammonia in various ways) many amino acids are formed from the intermediates of the cycle. In turn, some of the amino acids are precursors of purines and pyrimidines, and thus of nucleic acids, and also are the precursors of porphyrins. Oxaloacetate, the initial condensant in the cycle, is also on the pathway for the syntheses of hexose and pentose, the latter being involved as the sugar moiety of the nucleic acids. In ancillary reactions to the cycle, as given in the scheme, carbon dioxide can be fixed into organic form, combining with phosphoenol pyruvate to form oxaloacetate, with pyruvate to form malate, and with propionate to form succinate. These reactions bring carbon dioxide into the scheme to form cellular substance. This use of carbon dioxide is found not only in photosynthesis, via different reactions, but also to some extent via the above reactions in our own bodies. Even from this very brief survey it can be seen that the cycle has a central role in the metabolism of foodstuffs by cells. Through it pass most of the chemical intermediates—most of the carbon atoms, to be exact—of the chemical constituents of the cell. It is the hub of a great many of the pathways involved in the breakdown and synthesis of cellular constituents. The reason this is so is that the intermediates of the cycle, particularly the 4-carbon ones, are very much involved as intermediates in other pathways of the chemical metabolism of the cell. Sooner or later the compounds involved in the latter pathways reach some part of the cycle. When this happens, their carbon atoms can either be oxidized to provide energy or can be shunted onto another pathway.

Although the framework of the cycle has been well established, very little is known of how carbon atoms are shunted about, of how the cycle "knows" which way to function—that is, whether as a biological energy source or as a mechanism for the syntheses of needed cell constituents, or as a means by which metabolism is shifted from one major pathway to another, say from carbohydrate to fat. From physiologic studies, we do know that these shifts can and do occur. Carbohydrate carbons do end up in fat; carbohydrate is made from smaller molecules and held in storage; carbohydrate is broken down to provide energy. Fat is synthesized or broken down; proteins are made or broken down. Carbohydrate, fat, and protein carbons are all interchangeable. All these events take place at one time or another. And because we now know most of the metabolic pathways along which these intermediates travel, we realize that the key place where the regulation of these pathways occurs is at the Krebs cycle.

What are the possible alternative pathways? A look at the cycle suggests several possibilities almost immediately, and experiments can be set up with minced tissues (homogenates) to test several of them. For example, pyruvate can be oxidized to carbon dioxide and water by these homogenates, via the Krebs cycle; for this to occur, however, it is necessary to add an adequate supply of 4-carbon acids such as oxaloacetic, malic, or fumaric acid because all are interconvertible. If any of these 4-carbon acids is omitted from the reaction mixture, pyruvate is not oxidized, but is converted via acetyl coenzyme A to acetoacetate, one of the intermediate products of fat metabolism. This result shows that the acetyl coenzyme A did not condense with oxaloacetate to form citrate but condensed with itself to form acetoacetate. One can also add ammonium chloride, which, too, largely prevents the oxidation of pyruvate and causes it to be converted to acetoacetate. The reason probably is that the added ammonium chloride drains away the intermediates of the Krebs cycle, like α-ketoglutarate and oxaloacetate, causing them to be converted to the amino acids glutamate and aspartate. Since all the intermediates are in equilibrium with each other, a decrease in some will be reflected in a decrease in all, and thus the levels of intermediates will be too low to perform effective condensation with the acetyl coenzyme A derived from the pyruvate. Thus again, two of the carbon atoms of the pyruvate will end up as fat carbons. More recently, it has been found that citrate, and indeed the dicarboxylic acids of the cycle, cause an increase in fat formation from acetyl CoA by activating acetyl CoA carboxylase, the first enzyme in the synthesis of fatty acids from acetyl CoA. This observation indicates that the levels of the intermediates of the Krebs cycle are strictly controlled, that is, perhaps a part of the acetyl CoA pool of the cell is oxidized via the cycle to keep the intermediates (including citrate) at a certain level and to provide energy so that another part of the acetyl CoA pool can be converted to fatty acids.

(See Chapter 18.)

KREBS CYCLE: ITS VERIFICATION

But how do we know that something like the cycle actually happens within the whole cell or within the animal? We do not know precisely, but we can make fairly good guesses. Radioactive tracers such as carbon-14 can be incorporated into certain chemical compounds such as pyruvic acid, and the labeled pyruvic acid then given to the whole organism, be it a yeast cell or a rat. Various chemical compounds — glucose, fat, and amino acids — are then isolated. By knowing which carbon atom of the pyruvic acid was labeled and by determining which carbon atoms of the isolated compounds are labeled, we can infer pretty well, knowing the individual biochemical reactions involved, what happened in the organism to the individual carbon atoms of the ingested pyruvic acid. On the basis of many such experiments, many investigators have come to the inescapable conclusions that the Krebs cycle is operative in the cell, as well as in the test tube; that there are alternative metabolic pathways for many of the intermediates of fat, carbohydrate, and amino acid metabolism; and that these intermediates intersect each other at the level of the Krebs cycle. The reactions of the Krebs cycle can explain fully the means whereby the carbon atoms of pyruvate end up in various other compounds. In science this verifiable predictability is considered to be proof of the present-day "truth" of the theory.

This brings us to the original evidence for the cycle and the grounds on which Krebs formulated the scheme. In minced tissue, like breast muscle or liver, it was very early noticed that the addition of only very small amounts of citrate catalyzed a large respiration. It was also found that citrate could be synthesized when oxaloacetate was added and that citrate, isocitrate, *cis*-aconitate, and α-ketoglutarate were all rapidly oxidized by these minced tissues. But the key finding in the early laboratory work involved the use of a particular enzyme inhibitor, malonic acid, which specifically inhibits the enzyme succinic dehydrogenase and prevents the interchange of hydrogens between succinate and fumarate. However, it was noticed that even in the presence of malonate, succinate could actually be synthesized when fumarate or oxaloacetate was added. By looking at the cycle, we can see that succinate could not have been formed by the direct reduction of fumarate, for this enzyme was blocked by malonate. Krebs deduced that there must be a "back" reaction by which succinate could be formed via the breakdown of α-ketoglutarate. Thus, a cycle was postulated, using the tricarboxylic acids, citrate, isocitrate, and *cis*-aconitate, and the dicarboxylic acids, succinate, fumarate, malate, and oxaloacetate. Furthermore, in liver cells, in which carbon dioxide fixation can take place (via the reaction of phosphopyruvate plus carbon dioxide giving oxaloacetate), citrate and α-ketoglutarate could be formed from pyruvate.

The next piece of evidence involved the use of radioactive carbon dioxide. When this compound was incubated with liver mince in the presence of pyruvate and malonate, it was found that the isolated succinate contained no radioactivity and that the radioactivity in the isolated α-ketoglutarate was in the carboxyl carbon next to the carbonyl group. In Fig. 12-7, the radioactive carbon dioxide is marked with an asterisk; note that the radioactive label is found in oxaloacetate. Because of the block by malonate, the cycle has to go around "clockwise," and thus the label is found in citrate, *cis*-aconitate, isocitrate, oxalosuccinate, and α-ketoglutarate; it is not in succinate, however, because the labeled carboxyl group of α-ketoglutarate was removed by α-ketoglutarate dehydrogenase. If you look closely at the cycle, you might ask whether radioactivity was found in fumarate and malate, coming from radioactive oxaloacetate. It was, and since no radioactivity was present in succinate, it is apparent that the succinate did not arise from fumarate but must have been formed via a back reaction. When malonate was omitted, all the dicarboxylic acids, including succinate, contained radioactivity, showing the interconvertibility of the compounds in question. Also, because malonate specifically blocks succinate oxidation, it can be used as a test system for any compound that will form succinate in the presence of the inhibitor since succinate actually accumulates under these conditions and can be extracted and its amount estimated. It was found that all of the intermediates of the Krebs cycle, as postulated, can form succinate under these conditions.

These pieces of evidence are explicable only by the reactions of the cycle. We could go on; for example, it was found that citrate could be formed from acetoacetate and oxaloacetate. Again, when radioactive acetate or acetoacetate was added to kidney minces, radioactive α-ketoglutarate was formed, and when radioactive acetate was injected into the whole animal, radioactive aspartic and glutamic acids were formed. Since the latter two amino acids are in equilibrium with oxaloacetate and α-ketoglutarate, we are sure that the cycle does function in the whole animal. A great deal of such evidence has been accumulated from both test tube and whole-body experiments. The inescapable conclusion is that there is indeed such a thing as the Krebs cycle in the many cells of the mammalian body as well as in most other cells of the plant and animal kingdoms. Perhaps the most telling bit of evidence has been the isolation and purification of all of the enzymes involved in the cycle, indicating that there are specific enzymes catalyzing these specific reactions. That the cycle is important for the economy of the cell is illustrated by the finding that the ingestion of fluoroacetate can kill an animal; this poison stops the cycle by inhibiting the enzyme aconitase, concomitantly piling up fluorocitrate in the cell.

The Krebs cycle, then, is a complex metabolic relationship between many of the intermediates of the large classes of foodstuffs taken in by the organism. It is a means whereby all these intermediates can be somewhat interchanged; whereby, for example, the energy in carbohydrate can be stored for future use in the form of fat, and the protein stores of the body, mostly in muscle, can be changed to fat and carbohydrate and used for energy. The cycle has another equally important function, namely, the transformation of chemical energy into the energy the cell needs to perform its various functions; this is the subject of Chapter 13.

ANOTHER CASE STUDY: GLYOXALATE CYCLE

The Krebs cycle is only one of a number of what we call intersecting pathways of metabolism. It is the rare chemical compound that is metabolized along only a single metabolic sequence. Even the Krebs cycle itself is only a central hub of impinging pathways. For example, it has been long noted that the growth of many bacteria and fungi can be sustained by the addition to the medium of simple two-carbon compounds like acetate. Acetate in these organisms, unlike the case in mammals, can give rise to a net increase in carbohydrate and protein via the glyoxalate cycle (Fig. 12-8). The cycle can be seen to intersect the Krebs cycle at the malic and isocitric acid levels. The bacteria and fungi have at this point two enzymes not found in animals: isocitrase, which splits isocitrate to glyoxalate and succinate; and malate synthetase, which condenses the glyoxalate with another molecule of acetate to form malate. Thus, over all, two moles of acetate are converted to one mole of succinate. The succinate can of course enter the Krebs cycle and thereby become a source of carbohydrate and protein carbon atoms. Thus, the difference between the Krebs cycle and what might be called its glyoxalate cycle bypath is that acetate carbons are oxidized by the former, whereas they are converted to oxaloacetate and hence to carbohydrate

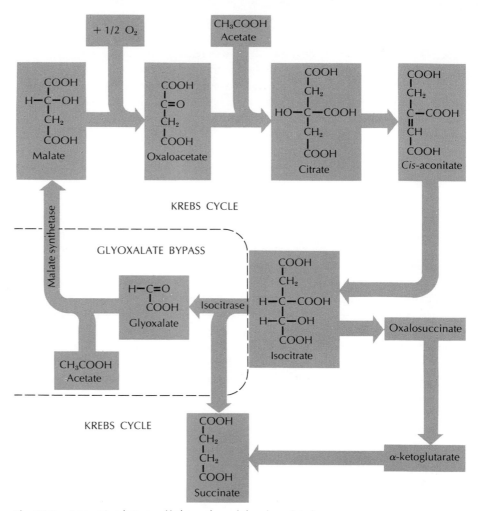

Fig. 12-8. Interaction between Krebs cycle and the glyoxalate bypass.

and protein by the latter. It is intriguing to speculate about the control mechanisms that determine whether isocitrase or isocitric dehydrogenase acts; this is the key point, the intersection between these two cycles. Another intriguing bit of speculation concerns the interactions between the products of the Krebs cycle in mitochondria and the products of the carbon-reduction cycle in chloroplasts. At present we can only speculate, for we have no answers to any of these questions.

(See Chapter 13.)

ANOTHER CASE STUDY: MULTIPLICITY

Finally, the case of methionine metabolism (Fig. 12-9) will illustrate once again the manifold roles in which a single molecule can masquerade. Methionine is an amino acid constituent of proteins. Also, because of its labile methyl group (CH_3), it acts as a methylating agent; it can methylate or transfer its methyl groups to methyl acceptors to form such compounds as choline and creatine. The former is important in that it is a constituent of the phospholipids forming the internal and external membranes of the cell, and it is also a key factor in lipid metabolism. Creatine, as phosphocreatine, is an energy source in muscle. These involvements of methionine can be illustrated by injecting ethionine into an animal. Various physiological disorders ensue, particularly to liver and pancreas; derangements in protein, nucleic acid, and phospholipid metabolism occur. Ethionine is a metabolic antagonist of methionine; it differs from methionine only in having a —S—CH$_2$CH$_3$ group instead of a —S—CH$_3$ group, and thus it can replace methionine in many metabolic reactions. It acts like methionine in that it becomes attached to other amino acids by peptide bonds and becomes incorporated into proteins; some of these proteins, because of this replacement, are thus rendered unsatisfactory for their metabolic roles. Ethionine also competes with methionine for various enzymatic reactions that require methionine. Methionine is "activated" for methyl group transfer by reacting with ATP, the high-energy compound, forming the "active" methionine compound, S-adenosyl methionine. Likewise, ethionine is attacked by the same enzyme but converted to S-adenosyl ethionine. However, the latter compound is not, like S-adenosyl methionine, a substrate for the various methyl-

(See Chapter 17.)
(See Chapter 16.)

Fig. 12-9. Ethionine, a metabolic antagonist.

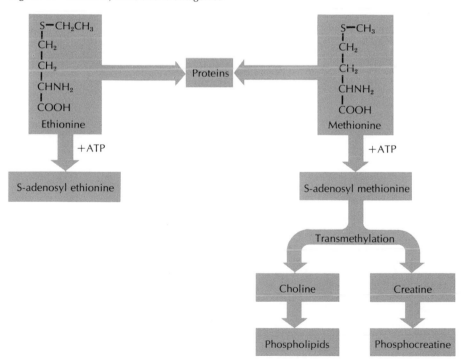

group transferring enzymes, and hence it is a dead end. If enough ethionine gets into a cell that has the methionine-activating enzyme, it can drain off the adenosyl moiety of ATP into its cul-de-sac; hence, both nucleic acid metabolism and energy metabolism are interfered with. All these multiple events take place when a cell ingests this simple, single metabolic antagonist. This observation is a fitting ending to the recital of the metabolic schemes of a cell in which one compound enjoys not one, but many functions. The cell is conservative, harboring its substance, hoarding its energy.

We now know of the existence of many enzymes, of many enzymatic pathways, involving the small molecules of the cell. Metabolic "maps" have even been published in book form, listing the actions and interactions between substrates and enzymes along these pathways, involving not only carbohydrate, but protein, fat, and nucleic acid precursors. Although this mapping has not been an easy task, requiring as it has years of work on the part of hundreds of individuals, it is still much easier than the formidable task still ahead: the understanding of the regulations along these pathways, of the means whereby a cell determines which roads are chosen in order that the final destinations will have some benefit to the economy of the cell as a whole. We must not forget that the cell is an integrated metabolic entity, not merely a collection of spatially and temporally separated enzymatic reactions. Some of the ideas that are currently held as to the nature of metabolic control mechanisms will be outlined later.

(See Chapter 18.)

SUGGESTED READING LIST

Offprints

> Green, D. E., 1949. "Enzymes in Teams." *Scientific American* offprints. San Francisco: W. H. Freeman & Co.
> Green, D. E., 1954. "The Metabolism of Fats." *Scientific American* offprints. San Francisco: W. H. Freeman & Co.
> Green, D. E., 1960. "The Synthesis of Fat." *Scientific American* offprints. San Francisco: W. H. Freeman & Co.
> Kamen, M. D., 1949. "Tracers." *Scientific American* offprints. San Francisco: W. H. Freeman & Co.
> Kornberg, H. L., and Krebs, H. A., 1957. "Synthesis of Cell Constituents from C_2-units by a Modified Tricarboxylic Acid Cycle." *Nature,* pp. 988–991. Bobbs-Merrill Reprint Series. Indianapolis: Howard W. Sams & Co.
> Stumpf, P. K., 1953. "ATP." *Scientific American* offprints. San Francisco: W. H. Freeman & Co.

Articles, Chapters and Reviews

> Axelrod, B., 1967. "Glycolysis." In: *Metabolic Pathways*, 3d ed., Vol. 1, D. M. Greenberg, ed. New York: Academic Press, Inc.

Caputto, R., Barra, H. S., and Cumar, F. A., Jr., 1967. "Carbohydrate Metabolism." In: *Ann. Rev. Biochem. 36*, pp. 211–246. Palo Alto, California: Annual Reviews, Inc.

Greenberg, D. M. (ed.). 1968. Chapters on fat and amino acid metabolism in *Metabolic pathways*, 3d ed., Vols. II and III. New York: Academic Press, Inc.

Ingraham, L. L., and Pardee, A. B., 1967. "Free Energy and Entropy in Metabolism." In: *Metabolic Pathways*, 3d. ed., Vol. I, D. M. Greenberg, ed. New York: Academic Press, Inc.

Lowenstein, J. M., 1967. "The Tricarboxylic Acid Cycle." In: *Metabolic Pathways*, 3d. ed. Vol. 1, D. M. Greenberg, ed. New York: Academic Press, Inc.

Potter, Van R., and Heidelberger, C., 1950. "Alternative Metabolic Pathways." *Physiological Reviews*, Vol. 30, p. 505.

Wood, H. G., 1955. "Significance of Alternate Pathways in Metabolism of Glucose." *Physiological Reviews*, Vol. 35, p. 841.

Books

Baldwin, E., 1957. *Dynamic Aspects of Biochemistry*, 3d ed. Cambridge, England: Cambridge University Press.

Cohen, G. N., 1967. *Biosynthesis of Small Molecules*. New York: Harper & Row Publishers, Inc.

Karlson, P., 1968. *Introduction to Modern Biochemistry*. New York: Academic Press, Inc.

Mahler, H. R., and Cordes, E. H., 1966. *Biological Chemistry*. New York: Harper & Row Publishers, Inc.

Chapter 13

Mitochondria, Chloroplasts, and the Fixation of Energy

Cells are energy converters; they are necessarily so because they need energy to perform their numerous tasks. To survive and to divide, to bring in substrates from the environment, to move about, to contract, and to expand—all the activities of various cells, be they mechanical work, osmotic work, or the concentrating of various solutes, require energy. Finally, the synthesizing of compounds in order to make new cells requires work. The importance of mitochondria in animal cells and of mitochondria and chloroplasts in plant cells lies precisely in the fact that they supply practically all of the necessary biological energy. Mitochondria do so primarily by oxidizing the substrates of the Krebs cycle; chloroplasts, by utilizing light energy. In this chapter we will describe what is known about these processes.

MITOCHONDRIA: HISTORY AND DESCRIPTION

In 1910 Warburg, an innovator in biochemistry, found that the oxidative reactions that take place in most tissues were concentrated in a small part of the cells. By grinding up the tissue, he made what we now call a homogenate of the tissue; we now make a homogenate with a revolving pestle that fits snugly inside a glass tube into which the tissue and a suitable medium have been placed. The homogenate can be spun in a centrifuge, and if this is done at various speeds, a fractionation or separation of the cell constituents will occur. When Warburg did this, he found that most of the enzymes responsible for the oxidation of the acids that we now know to be intermediates in the Krebs cycle were contained in a large granule fraction, so-called because it could be spun down at low speeds. Like many observations, this one lay dormant in the collective biochemical mind for many years although many biochemists did use the Warburg technique to determine some properties of these enzymes. But no one gave much thought to what was in the large granule fraction nor to what part of the cell contained all these enzymes. About 20 years ago, however, various biochemists became interested in just this problem, and, led by such men as Claude, Schneider and Hogeboom, Lehninger, and Green, they repeated, in essence, Warburg's experiments with techniques made more refined by the accumulation of biochemical experience. Since they were also interested in cytology, they discovered the biochemical importance of mitochondria.

These men, along with others, developed schemes for the fractionation of a cell homogenate into its constituent subcellular organelles. These fractionations are successful because the various organelles within a cell are, fortunately for the experimenter, greatly different in size and density. Thus, the nucleus is by far the largest organelle in the cell whereas nucleoli and ribosomes are the densest. The latter are also the smallest distinctive organelles whereas the mitochondria are intermediate to all these in size and density; the various membranes are the lightest because of their relatively high lipid content. Use is made of all these characteristics to separate one organelle from the rest. In general, there are two centrifugation methods one can use in these separations. One is called differential centrifugation, in which a homogenate is first spun at low speeds, forming a pellet and a supernatant; the supernatant is then spun at higher speeds, to sediment a second pellet and leave a second supernatant; and so on. The first pellet contains the nuclei, while the second contains mitochondria. Since both of these pellets are contaminated by other cellular constituents, they must be further purified, usually by washing and resedimenting the pellets. The homogenizing medium may be a salt solution, but it usually is a sucrose solution since sucrose is an uncharged molecule and is very soluble. Both of these characteristics are important since salts tend to harm organelles in various ways, while in many cases high densities are required for adequate separations. The other method is gradient centrifugation, in which a gradient of density is made in the centrifuge tube and the homogenate is then carefully layered on top of the gradient. In many cases impure fractions obtained from the first method are further purified by layering them on top of a suitable density gradient. Upon centrifugation, the denser particles spin down to lower levels in the gradient than do the less dense ones. If the centrifugation speed is high and the time of centrifugation prolonged, the particles will come to rest in zones equivalent to their own densities; hence this method has also been called zonal centrifugation. Sometimes a modification of this method is used, in which solutions of various densities are layered one on top of the other, with the densest on the bottom and with the homogenate or subcellular fraction in a solution of the lightest density on top. Upon completion of centrifugation, the various organelles will come to rest at the interfaces between the density layers; particles lighter than the density of the solution beneath them will not penetrate into that solution but will come to rest at the interface between the two solutions. In general, the differential centrifugation method separates organelles by virtue of differences in their size whereas density-gradient centrifugation separates them through their differences in density. Mitochondria are usually obtained by the first method in the form of a tan pellet that can be washed several times through resuspension and resedimentation.

Cytologists had known about the existence of mitochondria for a long time. They had seen these microscopically small bodies in most cells of the body; they could stain them with certain vital dyes; and they knew that the mitochondria contained enzymes that could react with these dyes. After a dozen years of work with these mitochondria, biochemists now know quite a bit about them: they are the important constituents of Warburg's large granule fractions; all the enzymes of the Krebs cycle are found in these bodies; and they are therefore responsible for most of the energy transformation of the cell.

What is this mitochondrion? It is a structure, bounded by a limiting membrane, found in all cells except the bacteria and the mature red blood cells of the multi-cellular organism; it is found in plant cells, in algae, and in protozoans. In all these cells its appearance is very similar: an outer membrane limits the structure, while just inside is another membrane that has folds or invaginations, called cristae, that reach deep inside the mitochondria, sometimes as far as the other side. This typical structure occurs again and again in cells so that it is easy to recognize a mitochondrion simply by its morphology. Figure 4-32 shows mitochondria *in situ* in a pancreas acinar cell. This is shown at a higher magnification in Fig. 13-1A; Fig. 13-1B shows an isolated mitochondrion from a rat kidney; and Fig. 13-1C shows, in negative staining, a broken preparation of heart mitochondria with the inner membrane particles lining the cristae membrane (these particles are destroyed by the positive-staining techniques used in A and B). Figure 13-2 is a diagram, derived from these electron micrographs.

Enzymes in Mitochondria

Even when isolated, the mitochondria retain their typical appearance, although the process of isolating them has modified their texture, damaged their membranes, and made their outlines less sharp. These particles are nowadays separated essentially by the same method used by Warburg, that is, by first homogenizing the tissue in some medium, isotonic or hypertonic sucrose solutions being commonly used. By differential centrifugation, the nuclei and unbroken cells are then spun down, and the supernatant from this spin is sedimented at a higher centrifugal force. The mito-chondria come down in the form of a packed tan pellet. By washing this pellet several times, a fairly clean mitochondrial fraction can be obtained and has been obtained from many kinds of animal and plant cells and even from some protozoans. Investigation has shown that the mitochondria from all these sources have almost the same enzymatic activities. In some cases they seem to be the sole depositories of certain enzymes in the cell. Table 13-1 shows the results of a typical experiment in which a liver homogenate was separated into its various subcellular fractions by differential centrifugation. Of the total succinoxidase activity and cytochrome oxidase

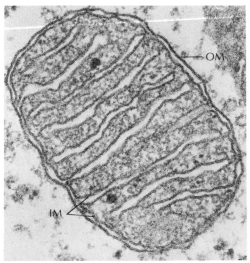

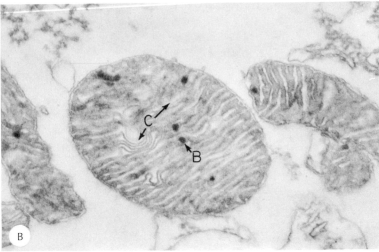

A

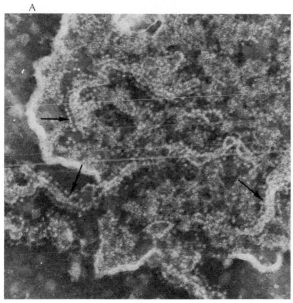

C

Fig. 13-1. A. Pancreas acinar cell mitochondrion showing outer membrane and inner membranes (× 230,000). (Courtesy of G. E. Palade, Rockefeller University.) B. Mitochondria isolated from rat kidney (× 56,000). (Courtesy of G. E. Palade and F. Miller, Rockefeller University.) C. Isolated and disrupted beef heart mitochondria, negatively stained with phosphotungstic acid (× 135,000). The picture clearly shows the inner membrane particles (arrows) attached to what we know from other pictures are the *cristae* (inner membranes) of the mitochondria. (Courtesy of S. Fleischer, B. Fleischer, and W. Stoeckenius, Rockefeller University.)

activity in the liver cell, some 60 percent in one case and some 80 percent in the other seems to reside in the mitochondria. The presence of these enzymes in the other cell fractions is thought to be due to contamination by whole or broken mitochondria. The fact that both these enzymes are concentrated in the mitochondria is shown by the observation that not only are most of their activities found there, but also that these activities are concentrated in this fraction; the amount of enzymatic activity per milligram of protein is greater in the mitochondria than in any other fraction or in the whole homogenate.

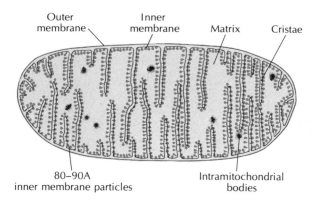

Outer membrane   Inner membrane   Matrix   Cristae

80–90A
inner membrane particles   Intramitochondrial bodies

Fig. 13-2. Diagram of a typical mitochondrion as seen from an electron micrograph.

In some cases we can infer that all of a particular enzyme is in one compartment of the cell, for the enzymatic activities in other cell fractions—like the cytochrome oxidase activity in the small particle and nuclear fractions—are there as a result of contamination by mitochondria. Sometimes there is a cofactor for the enzymatic activity in some other cell fraction, as in the supernatant fraction; thus, although little or no enzymatic activity resides in the supernatant fraction, there is something there that when added to the mitochondria increases their succinoxidase activity.

Experiments like these, in which recovery, activation, and recombination effects can be noted, have permitted us to determine the site of many enzymes and many cofactors in the cell. In just this way it was found that many of the enzymes of the Krebs cycle were localized in the mitochondria. This discovery, of course, could have been inferred from the early observation that mitochondria could oxidize pyruvate completely to carbon dioxide and water. Since we know that this can happen only via the Krebs cycle, we can say that all the enzymes of the cycle must reside in the mitochondria. In the case of some of these individual enzymes there is also a large portion of the total cellular activity that is extramitochondrial. What these enzymes—for example, malic dehydrogenase—are doing outside the mitochondria is not precisely known. Like all scientists, biochemists are willing to let awkward facts be laid aside for a while until some theory has caught up with them. For the moment we can say that the Krebs cycle is a mitochondrial cycle, and because it is an energy-transmutating cycle, we can call the mitochondria the energy transformers of the cell—the agents whereby the energy inherent in chemical compounds is modulated into the biological energy that is useful to the organism.

Table 13-1. DISTRIBUTION OF SUCCINOXIDASE AND CYTOCHROME OXIDASE ACTIVITIES IN MOUSE LIVER*

FRACTION	SUCCINOXIDASE		CYTOCHROME OXIDASE	
	ACTIVITY	ACTIVITY/MG PROTEIN	ACTIVITY	ACTIVITY/MG PROTEIN
Homogenate	4.25	1.34	6.86	2.06
Nuclei, cell debris	0.84	1.65	1.36	2.44
Mitochondria	2.40	3.18	5.39	6.46
Mitochondrial supernatant	0.18	0.09		
Small particles			0.29	0.35
Final supernatant			0.00	0.00
Mitochondria plus supernatant	3.15			
Mitochondria plus nuclei, cell debris plus mito-chondrial supernatant	4.18			

* Data from Hogeboom and Schneider.

MITOCHONDRIA AS ENERGY CONVERTERS

The problem of biological energy is closely related to inorganic phosphate metabolism. As long ago as 1907, the biochemist Harden found that inorganic phosphate disappeared during cell-free fermentation in yeast. He further found that enzymes in yeast esterify this phosphate into organic forms such as the hexose monophosphates and hexose diphosphates, which he later isolated from yeast cells. During the period 1930 to 1938 it was found that inorganic phosphate also disappears during the aerobic oxidation of carbohydrate to be converted also into an ester form, which in this case was 1,3-diphosphoglyceric acid (see Fig. 12-6). We now know that these enzymatic steps are part of the glycolytic scheme in which glycogen gets broken down anaerobically to lactate and during which some energy is made available. But respiration-dependent phosphorylation was not observed until 1930 by Engelhardt and 1937 by Kalckar.

It was in 1939 that the real start was made toward the determining of the role of phosphates in the oxidation of what we now call the Krebs cycle substrates. Many laboratories, almost simultaneously, observed that inorganic phosphate was necessary for the cellular oxidation of citrate, glutamate, fumarate, malate, and pyruvate, and that without the addition of phosphate to the test tube, very little oxidation was observed. Hence, phosphate disappearance and oxidation of substrate were coupled to each other. In muscle tissue the phosphate that disappeared was found to be esterified to creatine; and in kidney, liver, and muscle again, to be esterified to glucose, or fructose, or adenylic acid. The amount of phosphate that disappeared could be measured, and it was found that much more disappeared than could be accounted for by the then-known step of glycolytic phosphorylation—that is, the oxidation of glyceraldehyde-3-phosphate to 1,3-diphosphoglyceric acid (see Chapter 12). Hence, it was proposed that the remainder of the phosphorylations occurred during the transfer of electrons from substrate to oxygen, the substrates being the familiar ones of the Krebs cycle. The object of research in those days was to find out how many moles of inorganic phosphate were taken up into organic form per atom of oxygen consumed during the oxidation of substrate. This ratio is a measure of the efficiency of energy production, of how much substrate has to be oxidized to give so much energy. This ratio would also provide some idea of the number of individual enzyme steps involved. Later on, as the Krebs cycle was fully worked out, it became very clear that most of the phosphorylations associated with the oxidation of substrate occurred not directly with the oxidation of substrate but during the subsequent flow of electrons to the final electron and hydrogen acceptor, oxygen. For example, from Fig. 12-7 it can be seen that during the oxidation of α-ketoglutarate to succinate, one inorganic phosphate becomes esterified to form guanosine triphosphate. But when α-ketoglutarate was oxidized completely by a tissue preparation, not one, but up to four moles of phosphate were esterified for every atom of oxygen consumed or for every mole of α-ketoglutarate oxidized. Thus, three more phosphate esterifications had to be accounted for.

The Electron-Transport Chain

The electron-transport chain is the shorthand biochemical way of describing a reaction in which a donor transfers, with the aid of an enzyme, a pair of electrons to an acceptor. For example, pyruvate is oxidized by an enzyme that transfers a pair of electrons from pyruvate to an electron acceptor, this one being the coenzyme NAD, to form reduced NAD, or $NADH_2$. The electrons are then transferred via a series of coupled oxidations and reductions to the final hydrogen and electron acceptor, oxygen, forming water. These electron transfer couplets have been known for a long time. The best known, cytochrome c, is an established pigment in biochemistry. But we now know that cytochrome c is only one of a series of respiratory pigments. Figure 13-3 illustrates the current concept of the nature of electron transport from oxidizable substrate to reducible oxygen.

Many substrates are oxidized by enzymes known as dehydrogenases, so-called because they subtract two hydrogen ions and electrons from the substrate. Like most enzymes, dehydrogenases are specific for their particular substrate. In the Krebs cycle we speak of an α-ketoglutarate dehydrogenase, of a malate dehydrogenase, of a

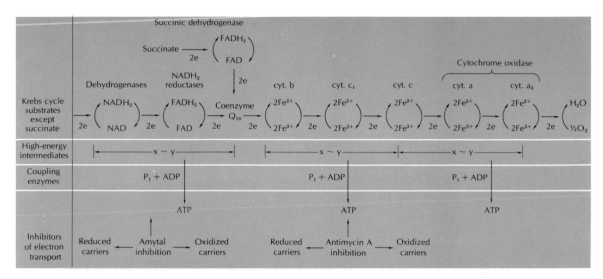

Fig. 13-3. Electron-transport chain of mitochondria, showing also sites of high-energy intermediate formation and of enzymes coupling these intermediates to phosphate esterification (sites of ATP formation), and sites of carrier inhibitions.

succinate dehydrogenase, and so on. In some cases the dehydrogenase is a flavoprotein, for it consists of a protein plus a prosthetic group, in this case a flavin, and both together constitute the active enzyme. The hydrogen acceptor for some dehydrogenases is NAD; others use NADP specifically; and still others can use either NAD or NADP. NAD and NADP are known as coenzymes because they participate in the reaction directly by accepting electrons, thus becoming reduced.

The next step in the electron-transport scheme is the oxidation of $NADH_2$ by an $NADH_2$ oxidase; this enzyme is a flavoprotein in which the prosthetic group, flavine-adenine-dinucleotide, or FAD, becomes reduced and then reoxidized and thus participates in the reaction. There seem to be two flavoproteins at this point in the electron-transport chain—one that couples or works with the succinic dehydrogenase, and one that works in conjunction with the remainder of the dehydrogenases of the Krebs cycle. In the former case the electrons from succinate are transferred directly to the flavoprotein, whereas in the latter case the electrons go first to NAD and then from $NADH_2$ to the FAD of the flavoprotein. These two reduced flavoproteins then give up their electrons to a common acceptor, cytochrome b, which then becomes reduced. In turn, cytochrome c_1, cytochrome c, cytochrome a, and cytochrome a_3 become alternately reduced and oxidized by the passage of electrons. Somewhere along this chain, probably between the reductases and cytochrome b, is involved a newly discovered carrier, coenzyme Q_{10}, a quinone chemically related to vitamin K.

Finally, the enzyme cytochrome oxidase, which seems to be the entity that we now more precisely call cytochrome a and cytochrome a_3, catalyzes the transfer of electrons and hydrogen to oxygen to form water. All these cytochromes are heme proteins in which the iron in the center of the heme becomes alternately oxidized and reduced. This shipment of electrons from the substrate to oxygen takes place because the energy potential of each of the intermediate carriers is such that the hydrogen or electrons go from a compound with a high-reducing potential to one with a lower-reducing potential. It is as if the electron-transport chain were made up of a series of "waterfalls": the water starts out at a high-energy level, at substrate, and ends up at a lower energy level, at oxygen; this is naturally a one-way flow. However, the flow of electrons can be reversed under certain conditions by putting energy into the system, just as water can be made to go uphill if energy is applied (see below).

Oxidative Phosphorylation

It is logical to ask why the cell employs this complicated means of capturing the energy in substrate. The reason seems to be that chemically it is more efficient to break up the large difference in energy potential between substrate and oxygen into many small steps, and this is precisely what the electron-transport chain does. At each waterfall the energy latent in the chemical configuration of the substrate is not only captured but transformed into a chemical form that the cell can use. We now know almost precisely where these energy-transforming waterfalls are located in the electron-transport chain; they are the places where ATP production takes place and are so designated in Fig. 13-3. So we can say that electron-transport phosphorylation is the coupling of the phosphorylation by inorganic phosphate of adenosine diphosphate (ADP) to give adenosine triphosphate (ATP), with the concomitant transfer of electrons from a donor to an acceptor. We have some idea how this occurs (see below). We know more precisely that it takes place during the oxidation of $NADH_2$, during the oxidation of reduced cytochrome *b,* and probably during the oxidation of reduced cytochrome *c.* Thus, for every substrate oxidized via NAD — that is, oxidized by an $NADH_2$-coupled dehydrogenase — there are three ATP molecules formed for every atom of oxygen that is reduced to form one molecule of water. In the case of succinate oxidation the NAD step is bypassed, and two ATP molecules are formed for every oxygen atom that is reduced. In some cases as in α-ketoglutarate oxidation, during the oxidation of α-keto-glutarate by NAD, there is also a phosphorylation step, a formation of GTP or ATP; this is an example of the so-called substrate phosphorylation, the production of energy in the form of ATP by the direct oxidation of substrate. Another example of this is the oxidation of glyceraldehyde-3-phosphate to 1,3-diphosphoglyceric acid and the conversion of the latter to phosphoglyceric acid and ATP (Fig. 12-6). In the case of α-ketoglutarate, its oxidation will therefore give four molecules of ATP for each atom of oxygen reduced. This ratio, moles of inorganic phosphate esterified to form ATP per atom of oxygen consumed during the oxidation, is called the P/O ratio. It is a measure of the energy yield from the oxidation of a substrate; in the case of α-keto-glutarate it is four, with succinate it is two, with malate it is three, and with $NADH_2$ it is three. This overall process, oxidative phosphorylation, is measured simply by adding substrate to mitochondria and observing the consumption of oxygen and the disappearance of inorganic phosphate. The P/O ratios that were observed during the oxidation of substrate showed that for each pair of electrons transferred to oxygen more than one phosphorylation step occurred. It is data such as these that produced the theory of electron-transport phosphorylation.

Concept of Coupled Phosphorylation and Control of Respiration

The term coupled phosphorylation needs explanation. In the working mitochondria no oxidation of many of the substrates of the Krebs cycle takes place without a concomitant phosphorylation. This can be shown very easily by studies with isolated mitochondria. It has been observed many times that mitochondria would not oxidize α-ketoglutarate or pyruvate in the absence of inorganic phosphate and ADP. In the presence of inorganic phosphate, but without ADP, there is no oxidation; as soon as ADP is added, the substrate is oxidized, inorganic phosphate disappears, and ATP is formed. This coupling of phosphorylation to oxidation is mandatory, that is, normally the electrons are not transported unless a synthesis of ATP also takes place. ADP is a necessary component in order for inorganic phosphate to be accepted, to form ATP; addition of only ATP has very little effect on the oxidation. However, if ATP is broken down to ADP, oxidation ensues, and the ADP is then rephosphorylated to ATP. This can easily be done by adding the enzyme hexokinase, which catalyzes the transfer of phosphate from ATP to glucose to form glucose-6-phosphate and ADP, the ADP necessary for oxidation to take place. Table 13-2 shows a typical experiment in which the addition of hexokinase, in the presence of ATP and glucose, causes an increase in the oxidation of pyruvate and fumarate. This effect of ADP is called the "phosphate acceptor" effect in increasing oxidation. Likewise, similar experiments have indicated the necessity of having inorganic phosphate present for oxidation to take place; the reason is precisely the same as in the case of ADP. Experiments such as these have firmly established the basis of the coupled oxidative phosphorylation of substrate in mitochondria. In Chapter 16 the possible meaning of this for the regulation of certain aspects of cellular metabolism will be discussed.

Table 13-2. EFFECT OF PHOSPHATE ACCEPTORS ON OXIDATIVE RATE

	OXIDATIVE RATE	
ADDITIONS	PYRUVATE AND FUMARATE AS SUBSTRATES	GLUTAMATE AS SUBSTRATE
ATP, inorganic phosphate	12	8
Inorganic phosphate	18	12
ADP, inorganic phosphate	75	101
ATP, inorganic phosphate, hexokinase, glucose	80	116

KINETICS OF ELECTRON TRANSPORT AND ATP FORMATION

In recent years brilliant work by Chance and his co-workers has verified the existence of the electron-transport chain in mitochondria, in isolated whole cells, and just recently, even in various tissues of the animal *in situ*. In addition, this group has elucidated the kinetics of electron transport. Through the use of micromethods, of oxygen electrodes to measure oxygen consumption, and of rapid-flow double-beam spectrophotometry, and by simultaneous measurement of the reduced and oxidized states of the electron carriers, the sequence of electron carriers along the chain has been verified. Instead of measuring respiration by its end product, oxygen consumption, it has been measured by the reduction, not of oxygen, but of the intermediate electron carriers. Since each of the electron carriers has a characteristic absorption spectrum in the oxidized and reduced states, measurement of the concentration at the absorption peaks provides an estimate of the percentage of the compound that is reduced or oxidized.

Furthermore, as a result of these techniques we now know precisely where ATP is formed along the chain. Previously, separate parts of the chain could be assayed for their P/O ratios; the oxidation of 1 mole of reduced cytochrome *c* gave 1 mole of ATP, the formation of which must have occurred between cytochrome *c* and oxygen; the oxidation of succinate gave two ATP's, the other occurring between the succinate-linked flavoprotein and cytochrome *c*; the oxidation of NAD gave three ATP's, the third one occurring between NAD and cytochrome *b*. The very interesting result of Chance's work is that he came to exactly the same conclusion using radically different methods. Simply put, his method consisted of measuring the states of oxidation and reduction of the various components of the chain under conditions where (1) no substrate was added, (2) adequate substrate but no ADP was added, and (3) both adequate substrate and ADP were added. For example, if the mitochondria are starved of substrate, all the electron carriers become oxidized; during substrate oxidation and consequent phosphorylation they become reduced. If, however, substrate and a small amount of ADP are added to mitochondria, respiration rapidly increases until all the ADP is phosphorylated to ATP, at which time the respiration rate begins to fall precipitously. By observing the changes in the oxidation-reduction states of the various components during the latter inhibited stage of respiration, points along the chain—the so-called "crossover" points—can be identified as sites of energy conversion to a form that can be made available for the phosphorylation of ADP. Under these conditions there occur reductions of NAD, flavoprotein, and cytochromes *b* and *c*, and an oxidation of cytochrome *a*. The "crossover" point is thus between cytochromes *a* and *c*, implicating this site as one of those where ATP formation could have taken place. These results were compared with those of experiments involving known electron-transport inhibitors such as antimycin A and amytal. These react with the components of the chain as shown in Fig. 13-3 and cause a reduction of some components and an oxidation of others, thus identifying their points of inhibition. Thus, conclusions are now firmly drawn as to the sites of ATP formation and even as to the possible mode of this formation.

Moreover, we can now calculate by this method the amount of each of the carriers in the chain. They seem to exist in distinct stoichiometric relationships to each other; all the cytochromes and the flavoproteins are present in about equal concentrations, with about ten times as much of the pyridine nucleotides. From these values and from the known figures for mitochondrial volume and protein content one can calculate the number of respiratory assemblies for each mitochondrion; this amounts to a figure somewhere between 5000 and 10,000 sets of electron-carrier chains.

Possible Mechanisms of Electron Transport Coupled to Phosphorylation

Another gratifying achievement of this method was to establish firmly the role of phosphate acceptors, like inorganic phosphate (P$_i$) and ADP, in the oxidation and reduction cycles of the carriers. The immediate changes in the reduction of these electron carriers that occur within seconds upon the addition of ADP strongly suggest a real chemical coupling between electron transport and phosphorylation. From the scheme given in Fig. 13-4 one can infer that not only is electron transport reversible, but the energy flow along the phosphorylation carriers can also be reversed. Recently, it has been found that when succinate and ATP are both added, the coenzyme NAD is reduced to NADH$_2$. This can be explained by the concept that upon the oxidation of succinate by its flavoprotein, the electrons flow first to cytochrome b; then, instead of electron flow to cytochrome c, the high energy of the added ATP causes a reversal, and the electrons go toward the NAD reductase, causing the reduction of NAD. ATP thus seems to be the energy source necessary to drive this reaction uphill, against the electrochemical potential. The tightness of the coupling between electron transport and phosphorylation is amply illustrated by the ease with which this reversible reaction takes place.

$$1. \quad AH_2 + B + X \;\rightleftharpoons\; A + BH_2 \sim X$$
$$2. \quad BH_2 \sim X + Y \;\rightleftharpoons\; X \sim Y + BH_2$$
$$3. \quad X \sim Y + P_i^{**} \;\rightleftharpoons\; Y \sim P_i^{**} + X$$
$$4. \quad Y \sim P^{**} + A^*DP \;\rightleftharpoons\; Y + A^*DP—P^{**} \; (ATP)$$

Fig. 13-4. Presently accepted hypothetical scheme for the coupling of electron transport to phosphorylation. A and B are electron carriers; X and Y are the hypothetical coupling compounds; P_i is inorganic phosphate; P_i^{**} is radioactive inorganic phosphate, and A^*DP is radioactive adenosine diphosphate.

We do not know how this coupling occurs, but we have a fair idea of the individual enzymatic steps that must be involved. The scheme given in Fig. 13-4 is evolved from the works and thoughts of biochemists like Lipmann, Slater, Lardy, Lehninger, and Hunter, and is the presently accepted one. The central idea is that the coupling between electron transport and the associated phosphorylation occurs via an intermediate common to both these types of reactions. This need not be the case, but it is the simplest hypothesis to entertain at the present time, and it seems to fit the known observations.

The experimental evidence that inorganic phosphate is not required for electron transport per se indicates that this common intermediate does not involve either P_i or ADP. Hence, it has been postulated that a reduced carrier (AH_2) reacts with the next carrier in line (B) in the presence of another substance (X), reducing B to BH_2. At the same time a high-energy bond (symbol $\sim$) of about the same energy content as is in ATP is formed between BH_2 and X, giving ($BH_2\sim X$). The latter compound then reacts with another hypothetical substance (Y) to form the high-energy intermediate ($X\sim Y$). At the following step P_i finally comes in, forming ($Y\sim P_i$) and, once again, free X. $Y\sim P_i$ then couples with ADP, forming, finally, ATP and free Y again. $X\sim Y$ is thus the intermediate common to both the electron carriers and the phosphorylation enzymes. Evidence for this scheme has been obtained in various ways. One could predict that reactions (3) and (4) should be reversible, involving as they do compounds of approximately equal energy content. One way to show this involves the use of the compound 2,4-dinitrophenol; this compound has for some time been known to be an uncoupler of oxidative phosphorylation, for in the presence of small amounts of the compound, substrate and ADP, oxidation proceeds at a fast rate, but no phosphorylation ensues. This compound is also an activator of a mitochondrial adenosine triphosphatase, splitting P_i from ATP. All these actions can be explained by postulating the dinitrophenol acts by causing $X\sim Y$ to be broken down to free X and free Y, and thus $X\sim Y$ cannot act as an acceptor for P_i to form ATP.

Two other partial reactions of the phosphorylation sequence have been described; both are inhibited by dinitrophenol and are thus involved in the coupling of electron transport to phosphorylation. One is the exchange reaction between P_i and the terminal phosphate of ATP, measured by mixing the mitochondria with radioactive P_i and nonradioactive ATP and measuring the radioactivity in the isolated ATP. Another is a similar exchange reaction between ADP and ATP, measured by incubating mitochondria with ADP labeled in the adenosine moiety and P_i, and then counting the labeled ATP. These are true exchange reactions in that there is no net synthesis of any of these compounds; no more ATP comes out than is put in. Reactions (3) plus (4) could explain the P_i-ATP exchange reaction, and reaction (4) could explain the ADP-ATP exchange. The identities of the hypothetical couplers, X and Y, are unknown. Both, or either, might be a phosphorylated enzyme. Recently, compounds such as an enzyme-bound "activated" NAD or an enzyme-bound phosphorylated histidine have all been suggested to function at some point in this scheme. Also, certain quinones, like vitamin K or ubiquinone, another recently identified mitochondrial electron carrier, have been thought to be involved, perhaps as being compound X. Finally, we do not know whether these postulated Y's and X's are the same in all the three sites of phosphorylation along the chain or if there are different couplers at every site. Since these partial phosphorylation reactions are tightly coupled to the electron-transport chain, one might expect that the state of reduction of the electron carriers has some effect on the rates of these partial reactions; and indeed, this is the case.

Recently, P. Mitchell has proposed a slightly different scheme to account for coupled phosphorylation. He has proposed the possibility that an electrical potential (a difference in charge) across the cristae membrane develops during respiration and that this difference powers the energy for ATP production. The positively charged hydrogen ions are moved to one side of the membrane and become separated from the negatively charged electrons on the other side. The separation of charges gives rise to an energy differential that, according to a complex mechanism envisaged by Mitchell, produces the high-energy intermediate, $X{\sim}Y$ (Fig. 13-4), which then powers the formation of ATP. This hypothesis is not very different from the currently accepted one given above. $X{\sim}Y$ is produced in both cases, but this intermediate is not formed, according to Mitchell, as is given in Fig. 13-4 but by means of an electrical membrane potential. Also, the second intermediate, $Y{\sim}P$ in Fig. 13-4 is nonexistent in Mitchell's scheme. At present, no clear-cut decision can be made between the two hypotheses, the subject being a very difficult one to evaluate experimentally. Later we take note of the use Mitchell has made of his scheme to account for ion transport in the mitochondria.

(See Fig. 17–11.)

Another aspect involving the phosphorylation sequence has recently been uncovered, and this has to do with the energy requirements of the mitochondria themselves. Later we mention that mitochondria have an active transport system for various ions, that is, they transport ions across a concentration gradient and as a consequence require energy to do this. Formerly, it was thought that the mitochondria-generated ATP was the only high-energy compound that was usable for this transport, but we now know this is not so. A high-energy intermediate seems to be involved, of the nature of the $X{\sim}Y$ compound in Fig. 13-4. The relevant experimental findings involved the use of an inhibitor, oligomycin, a compound that acts somewhat like dinitrophenol in uncoupling ATP synthesis from electron transport but at a different site from dinitrophenol, probably at step 3 in Fig. 13-4 so that $X{\sim}Y$ is unable to react with phosphate and ATP is not formed. It was previously known that the transport and accumulation within the mitochondria of ions such as Mg^{++}, Ca^{++}, or K^+ could be accomplished either by substrate addition and oxidation, generating ATP, or by the direct addition of ATP itself. It was now found that when oligomycin was added, ATP addition would not work for ion transport, but oxidation of substrate could still occur; furthermore, under the latter condition ion transport was not affected. Since it was known that ion accumulation required energy, it was inferred that under the conditions of oligomycin addition, where no ATP was formed, a high-energy intermediate could still be formed as a result of substrate oxidation, and it was this intermediate ($X{\sim}Y$) that was somehow responsible for the attendant accumulation of ions. Thus, it might be that for their own energy requirements, the mitochondria use the compound $X{\sim}Y$; however, under conditions where energy is required by the rest of the cell, this $X{\sim}Y$ can be phosphorylated to ATP (Fig. 13-4, steps 3 and 4,), and the ATP is thereby secreted to the rest of the cell.

(Chapter 17)

Submitochondrial Particles

The reason so much difficulty has been encountered in the solution of this problem of mechanism lies in the fact that the respiratory assemblies with their associated phosphorylations are a part of the insoluble lipoprotein membranes of the mitochondria. The laboratories of Lehninger, Green, and Racker have made large strides in attempting to fragment these complex multienzyme units into workable pieces. These membrane fragments, which can be isolated by breaking up mitochondria with detergents such as digitonin or cholate, turn out to be miniature mitochondria, for it appears that the respiratory assemblies seem to be fairly evenly spread over the intact mitochondrial membranes. A tentative idea that the cristae membranes, and not the outer membranes of the mitochondria, are the sites of these assemblies can be gleaned from the observation that heart mitochondria, which have more cristae per mitochondrion than liver mitochondria, also have a higher content of the respiratory pigments and a

(See Chapter 17.)

higher oxidative rate. These cristae are, like other membrane structures, lipoproteins; hence the respiratory assemblies are thought to be actually protein (enzyme)-lipid complexes. It is visualized that the lipid acts like a specific cementing substance, binding one particular carrier rigidly to its next neighbor. Thus, electron transport is thought to occur not by molecular collision between the components of the chain, during which electrons are transferred, but by means of an actual flow of electrons, possibly through the protein matrix of the carrier, from one rigidly positioned carrier to the next. These miniature mitochondria, lipoprotein in nature, have been called ETP, or electron-transport particles. Adjacent to these particles, but still visualized as being part of the membranes, are most of the dehydrogenases of the Krebs cycle.

Recently, attempts have been made to fractionate further this miniature multienzyme complex. Smaller particles have been obtained that contain some of the components and lack others. In this way, working with this jumble of particles and getting to know some of their properties, we hope to put together a meaningful chemical picture out of what is now simply a jigsaw puzzle. Some of these mitochondrial membrane fragments are themselves a complete assembly of the electron-transport and phosphorylation chains. Similarly, the phosphorylation-sequence system can be dismantled into fragments, as for example, those that contain only the dinitrophenol-activated adenine triphosphatase.

Lately, it has been possible to subfractionate mitochondria and to isolate both the inner, cristae membranes and the outer membranes. As expected, it seems that the electron-transport chain and some of the dehydrogenases are localized on the inner membrane; others of the dehydrogenases are in the soluble matrix. Some of the enzymes of the outer membrane resemble enzymes found on the membranes of the endoplasmic reticulum (Chapter 4), whereas others are distinctive to mitochondria. As yet, we do not know the integrated function of this outer membrane, but it is probable that it acts as a discriminating septum between the inner workings of the mitochondrion and the remainder of the cell.

A wonderful example of the functional architecture of the cell is illustrated by the mitochondrial cristae. Figure 13-1 shows the structural complexity of these membranes at which oxidative phosphorylation occurs. For some years now biochemists have wondered about the functional meaning of this organization, particularly the inner membrane spheres clearly seen in Fig. 13-1C. At first, it was thought that these

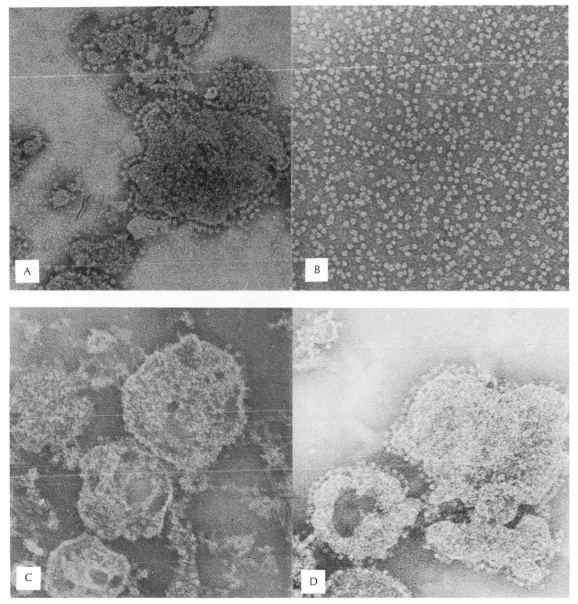

Fig. 13-5. Electron micrographs of mitochondrial subfractions (× 120,000), A. Inner mitochondrial membranes lined with spherical particles. B. Inner membrane particles; isolated. C. Inner membranes devoid of spheres; isolated. D. Reconstitution of spheres attached to membranes. (Courtesy of E. Racker, Cornell University.)

particles might contain the tight complex of electron-transfer pigments outlined in Fig. 13-3. However, recent work by Racker has shown that these spheres are probably the site of the coupling between oxidation and phosphorylation.

Racker has been able to take a broken mitochondrial preparation such as seen in Fig. 13-5A and further break it apart by various means. By further separations, five fractions were obtained, two of which are seen in Figs. 13-5B and 13-5C. The original preparations are capable of electron transport and of coupled phosphorylation, but none of the separated subfractions were capable of these activities. One of the subfractions (Fig. 13-5B) is a preparation of inner membrane spheres, and its only activity is an ATPase. Another preparation is that of membranes devoid of spheres (Fig. 13-5C), with the oxidation of succinate as its only activity, uncoupled to phosphorylation. When the spheres and the membranes are mixed in a certain way, we get the picture seen in Fig. 13-5D, a reconstitution of the spheres attached to the membranes; when the other, soluble subfractions are added, a return of coupled phosphorylation is found, together with the ability to oxidize NADH. In other words, a reconstitution of architecture has been accompanied by a reconstitution of function. Thus, it appears that the spheres pictured in Fig. 13-5B are the sites of the coupling of phosphorylation to oxidation and that their attachment to the membranes convert the spheres from the hydrolytic activity of an ATPase to the synthetic activity of the synthesis of ATP coupled to the production of high-energy intermediates by the electron-transport system localized on the membranes. Thus, the attachment of the particles onto the membranes drastically alters their enzymatic behavior, and it could well be that the regulation of coupled phosphorylation and of the synthesis of ATP is precisely at this organizational level of particle attachment to the membrane. There are also indications that the other energy-transducer system in the cell (the plant cell), the chloroplast, has similar particles on the inner side of the chloroplast membrane. While later illustrations show organization at the enzymatic level, Fig. 13-5 shows this organization extended to a higher level, of macromolecules coming together to form integrated enzyme systems, the first rudiment of a mitochondrion. One can now appreciate the structure of the mitochondrion; we can liken it to an efficient, structurally organized machine for the rapid transduction of oxidative to utilizable chemical energy.

Physiological Behavior of Mitochondria

The mitochondrial membranes are similar to others in being semipermeable, that is, some substrates, notably citrate, have difficulty in transversing the membranes whereas ions like K^+ are seemingly taken up against a concentration gradient. Again, being bounded by such a structure, the mitochondria behave like osmometers in that they take in water and swell when placed in hypotonic solutions. Under certain conditions of swelling they lose some of their low-molecular weight soluble components, like the adenine nucleotides and the coenzymes NAD and NADP, and hence lose their oxidative ability. Lehninger has found that certain kinds of swelling—that caused by the addition of phosphate ions, for example—are reversible; a very good reversing agent is ATP. Mitochondria swollen under these conditions will be contracted by the additions of ATP, extruding water. Indeed, on the basis of very fine measurements, it has been concluded that during the passage of electrons that goes on when substrate is oxidized and hence when phosphorylation takes place, there are distinct changes in the volume—presumably the shape—of the mitochondrion. It is intriguing that the mitochondria might contain a contractile protein that responds to ATP addition, for (See Chapter 16.) this is somewhat the same situation found in the contraction of muscle fibrils. What this alternate swelling and contraction of mitochondria has to do with the functioning of this organelle is not certain, but the main function of the mitochondria is not just the manufacture of ATP but its secretion to other parts of the cell. Thus, it is believed that these changes in the permeability properties of the mitochondrial membrane, as indicated by the volume changes undergone by the mitochondria, have something to do with the need for the secretion of ATP. In addition to the phosphate acceptor effect noted above, any intra- or extramitochondrial influence on the permeability properties of the mitochondrial membrane could be a regulatory device for mitochondrial function.

Another intriguing result of this type of study is that mitochondria can undergo swelling and contraction cycles. In Fig. 13-6 we see one of these cycles, as shown by electron micrographs of mitochondria fixed and set in structure immediately at certain stages of the cycle, and as shown by turbidity (light-scattering) measurements of the mitochondria (inserts on the Fig.). Both of these determinations are illustrative of conformational changes taking place. In Fig. 13-6A we note the appearance of isolated mitochondria just after a substrate (succinate) has been added; they are contracted

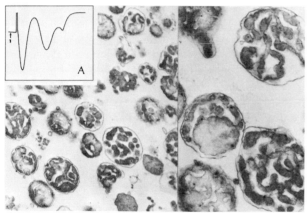

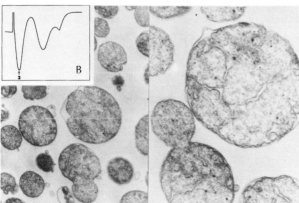

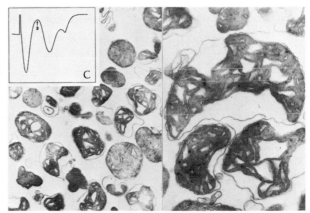

Fig. 13-6. Changes in conformation of mitochondria. Electron micrographs show contracted (A, C) and swollen (B) states of isolated rat liver mitochondria, whereas inserts show light-scattering measurements of the mitochondrial suspension. A description of the changes is given in the text (low mag. × 22,000; high mag. × 56,000). (Courtesy of L. Packer, University of California, Berkeley.)

in that the cristae membranes are drawn toward each other, away from the outer membrane, producing a dense matrix in the intracristae space. The insert in this figure indicates the level of light scattering at the moment the mitochondria were fixed at point "1." Fig. 13-6B shows what has happened to the mitochondria just after the addition of sodium phosphate; the downward deflection in the light-scattering curve indicates swelling, and when mitochondria were isolated at point "2," they do indeed appear quite swollen as compared to Fig. 13-6A. However, soon after they begin to contract again, as shown by the upward deflection in the insert, and if mitochondria are fixed for electron microscopy at point "3," most of them are again in the contracted configurations (Fig. 13-6C). A possible explanation for these changes in conformation and for the cyclic modality of these changes (note that the inserts show these changes taking place, although in an attenuated form, for several minutes afterwards) is that using substrate energy, mitochondria take up a permeant anion as phosphate, and since water comes in at the same time, they swell. At some point the maximum efficiency of the membrane to transport these ions has been reached, and the accumulated ions leak out more readily than they can come in; the mitochondria, losing water also, then contract. This cycle continues for as long as the mitochondria have, and can use, oxidizable substrates for energy.

Recently, much thought has been given to the idea that these conformational changes that occur in mitochondria are not merely passive responses to osmotic changes in mitochondria, but are instead intimately connected to the energy transformations that occur during substrate oxidation. It is as if we might picture these morphological changes as being counterparts of energy flow through the postulated $X{\sim}Y$ and $Y{\sim}P$ intermediates. Some experimenters picture the process as being one in which one form of chemical energy, the electron-transport system, is converted to another form, ATP, through the intermediacy of conformational changes in the proteins of the cristae membranes; in other words, that $X{\sim}Y$ and $Y{\sim}P$ are not identifiable chemical compounds such as ATP but are instead descriptions of the energy stored when a protein, for example, is made to contract from an extended form. Thus, Fig. 13-6A might be a morphological description of an intermediate high-energy state, the contracted form attained just after substrate has been added and electron flow has begun. If these ideas are proved correct, they would be some verification of the Mitchell hypothesis stated above. But much more difficult work will have to be done before that occurs; all that we can say at present is that this area is a fruitful one for current and future research in the field of coupled phosphorylation.

A perplexing problem concerning the permeability of mitochondria has been the observation that they seem to be impermeable to NADH. Since this coenzyme is produced extensively through glycolysis (Fig. 12-6) in the extramitochondrial cytoplasm and since its oxidation by mitochondria could engender phosphorylation energy, biochemists have pondered the ways and means of its becoming available to the mitochondrial enzymes. This problem seems somewhat solved now, and its solution relies on the observations that some enzymes catalyzing the same reactions are found in the mitochondrial and in the extramitochondrial cytoplasm. One of the enzymes (Fig. 13-7)

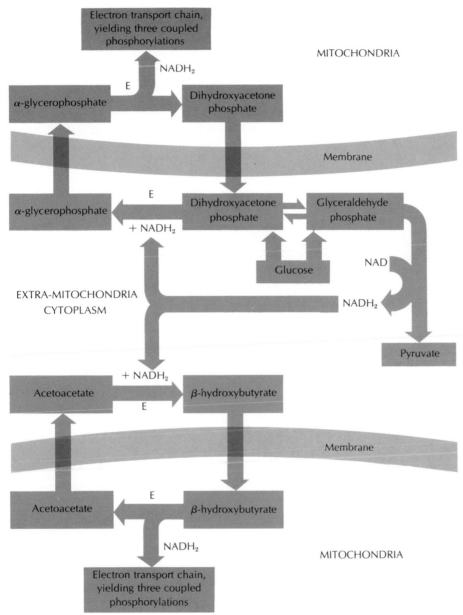

Fig. 13-7. Postulated mechanism of transport of reducing equivalents across mitochondrial membranes. (After Sacktor, Bücher, Boxer, and Devlin.)

is α-glycerophosphate dehydrogenase; in the cytoplasm the enzyme reduces dihydroxyacetone phosphate to α-glycerophosphate, using $NADH_2$; the α-glycerophosphate, which now contains the hydrogens and electrons from the extramitochondrial $NADH_2$, freely penetrates the mitochondria where a similar dehydrogenase oxidizes it back again to dihydroxyacetone phosphate, producing $NADH_2$ in the mitochondria;

meanwhile, the electrons originating in the $NADH_2$ are utilized for energy production by passage along the electron-transport chain. Similarly, the existence of malic dehydrogenase and glutamic-aspartic transaminase in and outside the mitochondria has given rise to a similar scheme (Fig. 13-7). Other examples can be given, one involving the existence of two malic dehydrogenases—one mitochondrial and one extramitochondrial. Some or all of these cycles are perhaps operating in the cell, but, overall, it would appear that in the matter of regulation of metabolic pathways via the compartmentation of enzymes and substrates, we biochemists are just at the beginning of our understanding. Even within the mitochondrion, some enzymes are undoubtedly localized in the inner cristae membranes whereas others are situated in the outer membranes. We should remember that we are always dealing in the cell with regulated enzymatic reactions that occur many times a second and that must insure a free flow of chemical molecules along pathways from one part of the cell to another.

Efficiency of Phosphorylation as an Energy Transducer

Let us now look back at what actually happens to the energy contained within a glucose molecule in the cell. If this glucose is completely burned in a test tube, it will give about 690,000 cal of energy per mole, all of it as heat. In the cell, however, some of this energy is not lost as heat but is retained in the form of ATP. The first reaction that glucose undergoes in the cell is the series of reactions called the Embden-Meyerhof scheme of anaerobic glycolysis (see Fig. 12-6). The first step in this process is the phosphorylation of glucose by ATP via the enzyme hexokinase to form glucose-6-phosphate. The latter compound is then enzymatically converted to fructose-6-phosphate, which then reacts with another mole of ATP to form hexose diphosphate or fructose-1,6-diphosphate. This compound then goes through a series of reactions and eventually ends up as two molecules of 3-phosphoglyceraldehyde. This compound in turn is oxidized by NAD to give 3-phosphoglyceric acid and NADH. This reaction is actually a coupled phosphorylation step, for it occurs in the presence of ADP and inorganic phosphate, and ATP is thereby formed. All of the above enzymatic steps are performed by soluble enzymes not found in mitochondria. Two moles of ATP are needed to start off the glycolysis of glucose, and during the oxidative reaction of glycolysis, these two moles of ATP are regained. Two additional moles of ATP are formed in the phosphoenolpyruvate to pyruvate reaction.

The next steps all occur in the mitochondria. The $NADH_2$ that is formed as a result of the oxidation of 3-phosphoglyceraldehyde is oxidized via the electron-transport chain to form three more ATP molecules. The end result of the glycolytic reactions is the production of pyruvic acid, in that two moles of pyruvic acid are formed from each mole of glucose. This pyruvic acid enters the mitochondria and is oxidized by NAD, giving overall 3 moles of ATP. The 2-carbon compound that is formed, acetyl coenzyme A, condenses to form citrate as explained in the previous chapter. The next energy-yielding step is the oxidation of isocitrate to oxalosuccinate by NAD forming

$NADH_2$, whose oxidation then gives three more ATP molecules. The next step is a substrate-level phosphorylation, the oxidation of α-ketoglutarate to succinyl coenzyme A, and then to succinate, to give one more ATP, via GTP. The $NADH_2$ that is formed as a result of this oxidation is oxidized along the electron-transport chain to give 3 ATP molecules. Adding up all these phosphorylations gives 19 moles of ATP formed from the oxidation of pyruvate to CO_2 and water; but, since there are two pyruvate molecules formed from 1 mole of glucose, this 1 mole of glucose, on glycolysis and oxidation, gives 38 ATP molecules.

ATP is a so-called "high-energy" compound. By this we mean that if an ordinary phosphate ester link is broken, as that in glucose-6-phosphate, about 3000 cal of energy per mole are obtained, but if the ATP is hydrolyzed, the energy released is about 10,000 cal per mole of the terminal phosphate group in ATP. Multiplying all the ATP's formed from the metabolism of glucose gives $38 \times 10,000 = 380,000$ cal. This is about 55 percent of the total energy in glucose; in other words, the process is 55 percent efficient — a high rate. Probably about 90 percent of the energy liberated during the breakdown of foodstuff takes place during the process of electron transport. Thus, we can see that most of the energy conversion — the conversion of that energy inherent in a chemical structure of the substrate to the energy inherent in the configuration of ATP — takes place in mitochondria.

The membrane is a decidedly functional structure. The mitochondria, and particularly their membranes, are a fine example of nature's conjoining of design and function to produce a structure that provides the greatest efficiency. No wonder mitochondria have been called the "powerhouse of the cell."

CHLOROPLASTS

Life would not be possible without the light energy of the sun; living forms have evolved because a process of capturing and utilizing this energy has been perfected these past millions of years. This process is called photosynthesis; a simple definition would be that it is a series of events in which light energy is converted to chemical energy that can be used for the energy requirements of the cell. These biosynthetic processes eventually build up an accumulation of organic matter, the magnitude of which can be visualized by the estimate that the annual yield of organic matter rendered by photosynthesis is about 10^{10} tons, a truly astonishing figure.

As we know it today, photosynthesis is quite a complicated process, involving the interaction of many compounds, and it is still far from being understood. From the standpoint of evolution it is difficult for us to visualize how this process could have originated, what the first event was that had to occur in the "primeval slime." But somehow a compound must have evolved that was capable of capturing light quanta, thereby being raised to an "excited" state in which it was capable of passing on the energy of this excited state to other molecules; then it became nonexcited and capable of once again being raised by the capture of another quantum of light. We do not know what the evolutionary forebears were, but among the present-day excitable molecules only the chlorophylls are involved in photosynthesis. These are compounds that absorb in the blue and red regions of the spectrum and therefore appear green.

This absorption of light by chlorophyll is the first process in photosynthesis. Along with a description of this will be given a summary of the three other general processes involved: the migration of the light energy captured by chlorophyll to an "active center" where the chemical events of photosynthesis occur; the chemical events themselves, which include electron-transfer reactions among various molecules, and along the way, a change in the form of some of the energy into phosphate-bond energy; and finally, the fixation or reduction of carbon dioxide leading to the synthesis of carbohydrate, the end product of photosynthesis.

There are five classes of chlorophyll in nature, differing very little in molecular structure. Chlorophyll *a* is present in higher green plants, algae, and certain protozoa; chlorophyll *b* also occurs in higher green plants and algae; chlorophyll *c* is found in brown algae, diatoms, and dinoflagellates; and chlorophyll *d* is found in red algae; in addition there is a bacteriochlorophyll found in photosynthetic bacteria. Chlorophyll *a* (Fig. 13-8), the one that has been most studied, is made up of four pyrrole rings bound together to each other; these form a porphyrin, with a long phytyl side chain on one of the rings and a Mg atom complexed to the rings. The common property of all these chlorophylls is their change in electronic structure when light of a certain wavelength strikes them. This change can be measured with a spectroscope, for they become deformed, so to speak; they become "excited." A rough analogy can be made to a coiled spring. Energy is put into the spring to coil it in the first place; its structure is deformed because of this input of energy, for as soon as we release the pressure on the spring, it will come back to its resting, noncoiled, nonexcited state. In addition, we

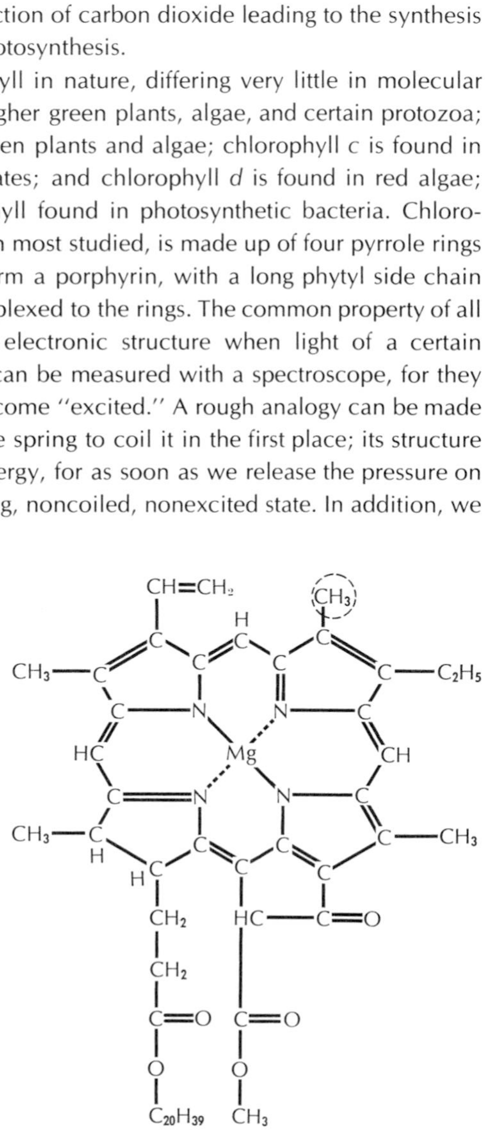

Fig. 13-8. Structure of chlorophyll *a*. Chlorophyll *b* differs in having a —CHO group in place of the —CH₃ group enclosed by the dotted circle.

can couple the coiled spring to other systems that can retrieve the energy of the spring, systems that can be used to do work such as turning the gears in a watch. Thus, as with the biological photochemical events, the light energy that has activated the chlorophyll is being used by the cell via a series of intermediary processes to furnish the necessary energy for certain synthetic processes.

Photosynthesis — The Light Reactions

The energetic aspects of the photochemical events can best be described as a "pumping up" of electrons from a low-potential region (that of the oxygen electrode) to a high-potential region (that of the hydrogen electrode), a difference of 1.2 v as illustrated in Fig. 13-9. Specifically, the photosynthetic process is essentially a two-quantum booster system, in which an electron is given two successive shoves to lift it to the required potential. Now, it has been known for years that one of the results of overall photosynthetic activity is the breakup by oxidation of water to "nascent" oxygen and hydrogen atoms. In fact, it is believed that the present-day content of oxygen in the atmosphere is the result of these photosynthetic events having occurred during the past millennia. However, another product of photosynthesis is the production of hydrogen in the form of reducing potential, in this case NADPH. Thus, while it is certain that any

Fig. 13-9. Current concept of electron flow during photosynthesis.

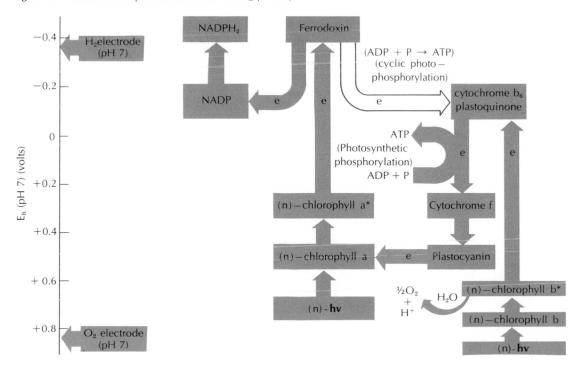

photochemical event that produces a reductant must at the same time produce an oxidant, it was hard for plant physiologists to formalize these events into a general chemical picture. This dilemma has been somewhat resolved by the discovery that two photochemical events must occur to describe the complete electron-flow picture of photosynthesis. For example, it has been found that light of two different wavelengths will have an additive effect upon chlorophyll activation when shown on leaf extracts, and this can best be explained by postulating the existence of two different light-absorbing pigments, each with a peak absorption at its own characteristic wavelength. Thus, we now know that probably another chlorophyll, *b*, is involved in the initial photochemical event. In addition, other components found in plants have been strongly implicated in the flow of electrons during the initial photosynthetic process, such as a quinone peculiar to plants, plastoquinone, two cytochromes peculiar to plants, b_6 and *f,* and a nonheme iron compound, ferredoxin. Although we have some idea as to how these compounds are involved in electron flow, there are other compounds present in plants that are also probably implicated, either in the initial photochemical activation or the subsequent electron-transfer reactions, such as modified chlorophyll molecules, carotenoids, and phycobilins, of which we have very little knowledge. The sum of all these compounds, known and unknown, arranged in fixed positions relative to each other has been called a photosynthetic unit. This photosynthetic unit can be thought of as a sort of molecular lens, funnelling excitation energy from a large cross-section of the bulk of the chlorophylls into traps or sinks where the electron-transport processes occur. The tentative present-day thought (though parts of the scheme are still speculative) concerning the passage of electrons through this unit, following ideas of Robert Hill, might briefly be described as follows (Fig. 13-9).

Chlorophyll *a* absorbs light of the far-red region ($\sim$ 700 mμ), and in doing so it becomes activated to a higher energy state; in this state it can reduce ferredoxin. This latter compound is re-oxidized by a passage of electrons to NADP, forming $NADPH_2$. This segment of the chain has been described as the photochemical reduction of NADP. The $NADPH_2$ so formed can be used in a variety of biochemical reactions that require reducing potential, but in the case of plant cells, photoreduced NADP is used as a reductant primarily in the synthesis of carbohydrate, as will be described below. One quantum of light is probably required to move one electron from activated chlorophyll *a*, and, therefore, two quanta are required in the production of $NADPH_2$. Now, the oxidized chlorophyll *a* must be brought back to that state where it once again can absorb light; it must be reduced, and this is accomplished by its reaction with a system containing reduced cytochrome *f* and probably plastocyanin. These latter compounds, having become oxidized in this process, are once again reduced by another system containing reduced plastoquinone and possibly reduced cytochrome b_6. The reduction of the oxidized states of the latter two compounds is mediated by the second photochemical event, the excitation by red light ($\sim$ 650 mμ) of chlorophyll *b* to emit an electron. The light energy absorbed by chlorophyll *b* is used at still some unknown trapping center to effect by unknown mechanisms the transfer of electrons from water to the oxidized plastoquinone and/or cytochrome b_6. When electrons are removed from two water molecules in this fashion, the ultimate product released is O_2. Thus, in the overall process we can state that the two photochemical events store energy, whereas the reactions between the cytochromes release energy.

The splitting of water by activated chlorophyll is called the Hill reaction, after its discoverer, Robert Hill. Essentially, it is the evolution of oxygen by illuminated chloroplasts in solution. It is measured by having light shine on chlorophyll in a leaf extract in the presence of a suitable oxidant, usually a colored dye whose reduction can be measured as a change in the specific absorption of the compound. Thus, in equation form the Hill reaction of photosynthesis can be visualized as a series of reactions: (1) two compounds, X and YH_2, in the presence of light are reacted to form a photoreductant XH_2 and a photooxidant Y as products; (2) Y reacts with H_2O to form $1/2\ O_2$ and to regenerate YH_2; (3) when an added electron acceptor, usually a dye, is added, this dye (A) oxidizes XH_2 to form AH_2 and to regenerate X; thus, in the presence of a dye and of continuous light, the dye is being continually reduced, and oxygen is being continually evolved. It is easy to fit the scheme presented here with that given in Fig. 13-9, for if the whole photosynthetic apparatus is intact, one can add NADP as the final Hill oxidant, $NADPH_2$ being produced, with the H extracted from H_2O finally ending up in $NADPH_2$.

In the overall photosynthetic scheme, for every O_2 liberated or for every CO_2 reduced, probably eight light quanta are needed—four red and four far-red—and as mentioned above, two molecules of $NADPH_2$ are formed in the process. This efficiency, close to 25 percent, of the conversion of light energy to chemical energy in the form of the reducing potential of $NADPH_2$ is quite remarkable, many orders of magnitude higher than nonbiological, photochemical energy-storage processes. In addition, during this process of electron transport, a synthesis of ATP occurs, called photosynthetic phosphorylation; it occurs apparently at the site indicated in Fig. 13-9 where there is a drop in energy levels between the two cytochrome systems. This drop in energy levels is salvaged in the form of ATP; thus, no extra light energy is needed for the synthesis of ATP. These two cofactors, ATP and $NADPH_2$, are the only ones known at present to be generated by the light reactions of photosynthesis and that are required for the reduction of CO_2. Exactly how electron-flow energy is coupled to ATP synthesis is not known, but the mechanism will probably be similar to the one being worked out in the case of mitochondria-coupled phosphorylation. There is a difference, however, between photosynthetic phosphorylation and mitochondrial oxidative phosphorylation, in that in chloroplasts we can obtain what has been called cyclic photosynthetic phosphorylation. If a suitable exogenous electron acceptor such as phenazine methosulfate is added, the electrons can be cycled around and around the electron-transport system at the expense of light energy as indicated in Fig. 13-9, with a continuous generation of ATP. It is likely that this process occurs in the cell; indeed, in some plant cells under certain conditions 50–100 percent of the ATP generated comes via cyclic electron transport, and in photosynthetic bacteria it may be the only good way the cell has of making ATP.

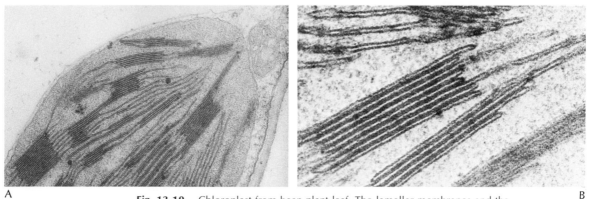

A

B

Fig. 13-10. Chloroplast from bean plant leaf. The lamellar membranes and the membrane stacks are clearly visible. A. Part of a chloroplast; mitochondrion is also visible at top (× 75,000). B. (× 300,000). (Courtesy of T. E. Weier, University of California, Davis.)

Chloroplast Structure

The photosynthetic electron-transport system of plant cells is contained in a membrane structure, the chloroplast (Fig. 13-10A), just as the electron-transport system of animal cells is contained in the membranes of the mitochondria. Even in the photosynthetic bacteria, as *Chromidium* or *Rhodospirillum,* the system is found in the membranes of a vesicular reticulum spread throughout the cell. In the blue-green algae these vesicles have become much enlarged and flattened, and the membranes of two adjacent vesicles have, at some places along their lengths, become fused together. In the green algae, these fused membranes have become encircled by another membrane, which has effectively separated them from the rest of the cell; a new intracellular organelle that is about 4 μ long, the chloroplast, has been formed. In the higher plants, the internal membranes or lamellae of the chloroplasts have become more complicated, with differentiations appearing among the internal membranes. Thus, extensive fusion of membranes has occurred, so much so that microscopic examination of the chloroplasts reveals that many of these fused lamellae are piled one on top of the other, resembling layered discs with long solitary membranes connecting these discs (Figs. 13-10 A and B). Viewed from above, these discs would appear to be large balls that are about 0.3 μ in diameter and connected by strands; these flattened balls have been called grana. (A diagram of current ideas concerning this structure is given in Fig. 13-11). It is here in the grana that the chlorophyll is found and that the photochemical events take place. From a comparative viewpoint the whole chloroplast of the green algae can be considered to be analogous to the individual grana found in the chloroplasts of higher plants because these latter contain many grana connected by membranes. Recently, even finer substructures of the grana membranes have been described, the so-called "quantasomes," (Fig. 13-12) but at present the exact function of such particles in the cell is still in doubt.

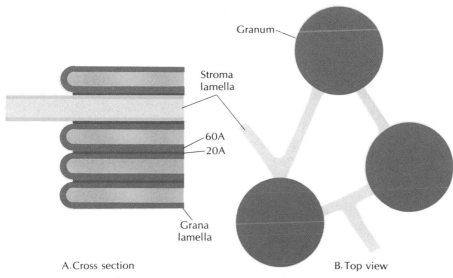

Granum

Stroma
lamella

60A
20A

Grana
lamella

A. Cross section   B. Top view

Fig. 13-11.   Diagram of lamellar structure in chloroplasts of higher plants.

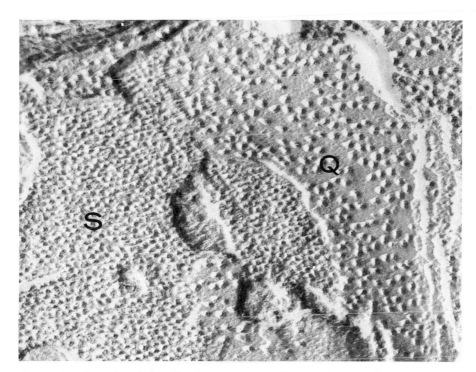

Fig. 13-12.   Choroplast of plant showing quantasomes. Freeze-etch preparation showing inner surface of chloroplast lamellae, showing large particles (quantasomes), and showing small particles (S) (× 150,000). (Courtesy of D. Branton, University of California, Berkeley.)

The stacked appearance of the grana membranes (Fig. 13-10) gives us some clue to the nature of the photochemical events. We do know that four chlorophyll molecules have to be excited for each water molecule split in a photosynthetic unit that contains around 2500 chlorophyll molecules. This means that light quanta absorbed anywhere within the photosynthetic unit containing these 2500 chlorophylls can cooperate in oxygen evolution and electron flow. Within this unit are the trapping sinks mentioned above, perhaps 40 of them, and in these centers are found the "active" chlorophyll molecules, in contrast to the bulk of chlorophyll molecules. Energy transfer could occur more easily if the "active" chlorophylls in the chloroplast were in a state approaching the semisolid state. Thus, the meaning of the membrane structure of the grana could be that the chlorophyll molecules are associated with lipids, proteins, and the other catalysts of Fig. 13-9 into some sort of semicrystalline lattice and that this lattice is fixed into the planar surfaces of the grana disc membranes. The whole picture of grana lamellar structure can be viewed as one in which light energy excites the electronic structure of the chlorophyll molecules embedded in the matrix of the membranes to so change the configuration of the molecules that a separation of charge may occur, positive from negative. This separation of charge is facilitated by the lattice structure of the whole photosynthetic unit, enabling the electrons to "wander" through the entire structure until they are trapped in an "active center" where the various chemical events can occur. This tightened structure seems to be very efficient for the conversion of light energy to electron-flow energy, which in turn is converted to chemical-reducing potential (NADPH) and chemical-synthesizing potential (ATP). Much more work has to be done before we can have any definite knowledge concerning the chemical architecture of the grana structure.

The chloroplasts of higher plants and of green algae are even more complicated than has been described above. The grana and connecting membranes sit in a matrix within the chloroplast. Embedded in this matrix are innumerable small particles, of about the size and density of ribosomes (see Fig. 13-10A); indeed ribosomes have been isolated from chloroplast preparations. Also in the matrix are globules which are probably lipid droplets, as well as starch granules, and in some cases other bodies that have been called pyrenoid bodies and eye-spots whose function is not known. A word can be said about the development of the grana. If young leaf tissue is examined, the colorless chloroplasts look quite different; instead of grana, all they contain are isolated vesicles. When this tissue is illuminated, some remarkable changes take place, all culminating not only in a proliferation of membranes, but in the typical appearance of the architecture of the grana. Thus, light seems to be necessary both for the synthesis of chlorophyll (which cannot occur in the dark in the chloroplasts of higher plants) and for the synthesis of the chlorophyll-containing membranes. These newly formed membranes then gather themselves together to form the determined structure typical of the grana. The mechanism of these events is entirely unknown.

Photosynthesis — The Dark Reactions

The electron-transport events described above are only part of the process of photosynthesis, for we have been describing only the light reactions of photosynthesis;

another part is what has been called the dark reactions, not because they can only occur in the dark, but because they do not require light for the chemical reactions to take place. Thus, in the dark reactions, the hydrogen atoms generated during the light reactions are used to reduce carbon dioxide eventually to carbohydrate, giving us the equation of photosynthesis: 6 CO_2 plus 12 H_2O plus light energy yields $C_6H_{12}O_6$ plus 6 O_2 plus 6 H_2O. In addition, 672,000 cal of energy are formed and stored in the form of one molecule of carbohydrate. In some bacteria, the sulfur-bacteria, H_2S can replace H_2O, with sulfur being formed instead of oxygen, but the photosynthetic events are quite similar. In general, from a metabolic standpoint we can say that all photosyntheses involve an anaerobic oxidative metabolism, using light as energy, which can be linked to a reductive assimilation of CO_2, using the hydrogen generated in the former step as a reducing agent.

The enzymatic mechanisms whereby carbohydrate is formed from CO_2 are shown diagrammatically in Fig. 13-13. This scheme is the result mostly of the work of M. Calvin and his collaborators, who, by using radioactive CO_2 and very short-time exposures in the light, discovered and identified the initial steps in the reduction of CO_2. Later on, other workers such as Horecker and Racker began to identify and isolate the various enzymes involved so that at present we believe that the scheme as shown is essentially correct.

In its initial steps, the carbon-reduction cycle is the coupling of CO_2 to a 5-carbon compound, ribulose-diphosphate, to form an ephemeral 6-carbon compound, which immediately breaks down to a triose phosphate, phosphoglyceric acid. Indeed, if radioactive CO_2 is used, it was found that the first compound to be labeled (in the first few seconds) was phosphoglyceric acid. The enzyme that does this coupling is called carboxydismutase; it is a very large enzyme complex, catalyzing a series of reactions not all of which have as yet been worked out. Overall, six of the CO_2 molecules end up in the chloroplast as a molecule of hexose. For this process to take place, ribulose-diphosphate must be continually generated, so that the most important function of the cycle in its complicated enzyme reactions must be this continuous regeneration to insure a continuous fixation of CO_2.

Let us cursorily travel through the scheme along its pathways to examine how this is accomplished. Six moles of CO_2 combine with 1 mole of ribulose-diphosphate to give 12 moles of a 3-carbon compound, phosphoglyceric acid. This latter compound is reduced (by NADPH) to give 12 moles of glyceraldehyde-3-phosphate. Of the 12 moles formed, it is thought that 2 moles combine with two 2-carbon fragments of fructose-6-phosphate to give 2 moles of a 5-carbon compound, xylulose-5-phosphate; that 2 moles combine with two 2-carbon fragments of sedoheptulose-7-phosphate to also give xylulose-5-phosphate; that 5 moles are dismutated to give dihydroxyacetone-phosphate; and that the 3 remaining moles combine with 3 moles of the dihydroxy-acetone phosphate to form 3 moles of fructose-6-phosphate, via the intermediary of fructose-diphosphate. Of these 3 moles, two 2-carbon fragments are split off, as mentioned above, to form xylulose-5-phosphate; two 4-carbon fragments are split to give a 4-carbon sugar, erythrose-5-phosphate; and only the one remaining mole of fructose-6-phosphate goes to form hexose. Thus, six CO_2 will give one 6-carbon hexose. The

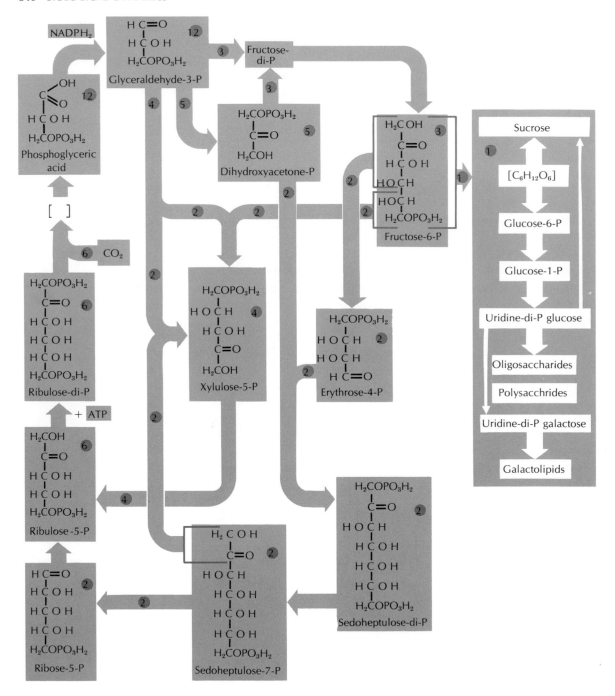

Fig. 13-13. Carbon-reduction cycle in chloroplasts (figures in circles indicate number of molecules participating in one turn of the cycle).

other parts of the cycle are a continuation of the effort to regenerate ribulose-diphosphate. The combination of 2 moles of erythrose-4-phosphate with 2 moles of dihydroxyacetone phosphate leads to 2 moles of a 7-carbon compound, sedoheptulose-diphosphate, which is then dephosphorylated to sedoheptulose-7-phosphate. This latter compound is then split into a 2-carbon and a 5-carbon fragment. The 5-carbon fragment goes to ribose-5-phosphate and then to ribulose-5-phosphate, which is then phosphorylated by ATP to ribulose-diphosphate. The 2-carbon fragment combines with glyceraldehyde-3-phosphate to form xylulose-5-phosphate, which goes then to ribulose-5-phosphate directly, and then on to ribulose-diphosphate. Thus, in the scheme there are presented three pathways for the regeneration of ribulose-diphosphate; in this way, it would seem, the cell insures that this compound is being continuously formed and that no bottleneck is present in the fixation of CO_2.

From fructose-6-phosphate are formed, through the steps given in Fig. 13-13, sucrose, the oligosaccharides and polysaccharides such as starch, and the galactolipids, compounds found abundantly in the chloroplast membranes. For each complete turn of the cycle, to form 1 mole of a 6-carbon sugar, 6 moles of CO_2 are required, in addition to 12 moles of NADP and 12 moles of ATP. Half of the latter are required to phosphorylate ribulose-5-phosphate to ribulose-diphosphate; and it is obvious that for every CO_2 being fixed, 2 hydrogens are required (in the form of 2 moles of NADPH) in order to obtain the general formula $C_6H_{12}O_6$.

Now, in addition to the early formation of the compounds indicated in Fig. 13-13, other compounds are detected in rapid radioactive-labeling experiments in the light. These include acids such as malic, fumaric, and phosphoenolpyruvic and amino acids such as alanine, serine, and aspartic acid. The early formation of compounds such as these would indicate that some of the carbon atoms of the 6 moles of CO_2 are not directly converted to hexose or do not regenerate ribulose-diphosphate but are sent through some bypasses for the syntheses of these compounds. For example, we do know of enzymatic mechanisms that can form phosphoenolpyruvic acid from phosphoglyceric acid; once the former compound is formed, it is possible that alanine is formed from it by amination and that malic acid is formed by carboxylation. Since these steps all require reductions, it is probable that part of the NADPH generated during the light is used in the formation of these carboxylic and amino acids. Also of interest is the very distinct possibility that acetyl CoA is formed from one of the compounds in the carbon-reduction cycle, and once this compound is synthesized, we are on the pathways toward the syntheses of fatty acids, steroids, and carotenoids. In addition, another possibility is that one of the byproducts of the carbon-reduction cycle is the formation of the porphyrins, via a condensation of glycine with succinic acid, both of which could arise from the cycle.

What is the sum of all the pictures we have presented regarding the chloroplast and photosynthesis? We have seen that the membrane system in the chloroplast is responsible for the light reaction of photosynthesis, including the production of NADPH and ATP. In the stroma of the plastids, in which the membranes are embedded, are located the enzymes of the carbon-reduction cycle, plus possibly all the enzymes involved in the assembling of the porphyrins, the fats, the carotenoids, and the poly-saccharides. In addition, we can observe that when plants are kept in the light, starch accumulates within the chloroplast, and when they are transferred to the dark, this accumulated starch breaks down to sucrose, which is exported to the rest of the cell.

How much energy is processed by plant cell chloroplasts relative to plant cell mitochondria? It is difficult to estimate this accurately, but some idea may be had from the finding that the photosynthetic rates of green cells are, at a maximum, some 10–15 times their maximum respiratory rates. We thus think that in the light the plant cell derives most of its energy via the light reactions of photosynthesis; we know that in general there is more chloroplast-membrane area within the plant cell than there is mitochondrial membrane area; we can guess that even in the dark the plant cell obtains a good deal of its energy from the breakdown of starch in the chloroplast and subsequent glycolysis in the soluble cytoplasmic matrix. However, we should also consider that the amount of ATP produced by glycolysis is much less, per mole of carbon, than that produced by mitochondrial oxidative phosphorylation. We also surmise that in the plant cell, the chloroplast-produced ATP, like the mitochondrial-ATP, is secreted to the rest of the cell. Finally, we do find eucaryotic cells having mitochondria but no chloroplasts, but we never find cells having chloroplasts but no mitochondria. Nevertheless, we can say that in an evolutionary sense, chloroplast-less cells would never have evolved without the necessary existence of plant cells containing an apparatus capable of utilizing light energy and converting it to stored chemical energy.

MITOCHONDRIA AND CHLOROPLASTS AS ORGANELLES

The fact that both mitochondria and chloroplasts are intracellular bodies, each surrounded by a membrane and each with a definite structure, has led to interesting speculation, which in turn has prompted experiments, the results of which are very intriguing. The question is: are these bodies autonomous, can they exist outside the cell? The answer is probably no, for most probably they require for their function compounds that can only be generated by other parts of the cell. But, are they semi-autonomous, that is, can they, like viruses, reproduce and divide by themselves but only within the confines of the living cell? The answer, at present, is a possible yes. For example, it has been recently found that both chloroplasts and mitochondria contain DNA; furthermore, this DNA is different from nuclear DNA in having a different purine and pyrimidine base composition (see Chapter 8). Also, mitochondria have a DNA polymerase different from the nuclear enzyme. The differences in DNA base composition are reflected in the densities of the nuclear and mitochondrial DNA's, and thus they can be separated by means of density-gradient centrifugation (see Figs. 8-16 and 8-17) as shown in Fig. 13-14. This DNA, like phage DNA, is in the form of circles that are of uncommonly uniform 5 A circumference, (Fig. 13-15). The physiological meaning of this circular form is unknown at present. In mammalian cells mitochondrial DNA constitutes less than 1 percent of the amount of nuclear DNA, with a very low-molecular weight of about 10^6. However, we do not know at present the function of this chloroplast or mitochondrial DNA, or whether it can "code" for chloroplast or mitochondrial proteins as described for nuclear DNA in Chapters 8 and

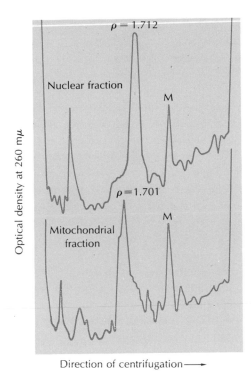

Direction of centrifugation ⟶

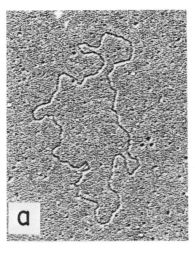

Fig. 13-14. Sedimentation diagram in a CsCl gradient showing difference in density (ρ) between Neurospora nuclear DNA (1.712) and mitochondrial DNA (1.701). (*M* denotes a bacteriophage DNA put in as a marker.) (Data from D. Luck and E. Reich, *Proceedings of the National Academy of Science 52*, 931, 1964.)

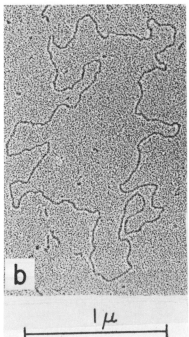

Fig. 13-15. Circular mitochondrial DNA. The figure shows a preparation of DNA from leucocyte mitochondria. A. Single length. B. Circular dimer form. (Courtesy of D. A. Clayton and J. Vinograd, California Institute of Technology.)

14. However, mitochondria can use their DNA to synthesize complementary RNA as described in Chapters 8 and 14. We also know that both the chloroplasts and the mitochondria have the machinery for protein synthesis, containing as they do distinctive ribosomes, distinctive transfer RNA's, and the enzymes necessary for "activating" amino acids. Moreover, isolated mitochondria can synthesize some, but not all, of their constituent proteins. For example, we know from direct biochemical evidence that they cannot synthesize cytochrome c, the synthesis of which takes place on ribosomes in the extramitochondrial cytoplasm, and from genetic evidence we infer that nuclear DNA, not mitochondrial DNA, contains the genomes for the other cytochromes. Indeed, it is thought that the only proteins the mitochondria are able to synthesize, via their own DNA, RNA, and ribosomes, are the "structural" proteins of the cristae membranes, proteins that are thought to act as self-assembly organizers for the lipids, the enzymes, and other proteins such as the cytochromes, which constitute a complete membrane structure.

It does appear that during cell division the mitochondria of the daughter cells are produced by division of the mitochondria of the mother cell and do not arise *de novo* from nonmitochondrial precursors in the cytoplasmic matrix. The evidence at present does therefore point to some sort of semiautonomous existence for chloroplasts and mitochondria, that is, the continuity of these bodies from one generation to another depends on something within their own structures, but the existence of the finished organelles depends on a collaboration between themselves and the rest of the cell.

Finally, speculation has arisen on the possible relationships between mitochondria and unicellular bacteria, on the one hand, and chloroplasts and unicellular blue-green algae on the other. The first two are about the same size, their structure is somewhat similar, and the inner cristae correspond in function to the bacterial membrane. The second two, again, have similarities in size, structure, and function. They all seem to have DNA molecules much smaller than nuclear DNA; their DNA is not organized in the form of chromosomes and probably is circular; and their ribosomes are all about the same size, distinctly smaller than the cytoplasmic matrix ribosomes of plant and animal cells. It is of some speculative interest that the DNA-containing organelles of a cell are all surrounded by what we may consider to be a double-membrane envelope. The nucleus has its own membrane, but in addition, this membrane is in close apposition to, and in the case of the nuclear "pores" in tight junction with, an outer membrane that is in reality in continuity with the "rough" endoplasmic reticulum membranes. The mitochondrion has an inner membrane with infolded cristae and an outer membrane that has some biochemical similarity to, again, the endoplasmic reticulum membranes. The chloroplast has an inner membrane that possibly is in continuity with the grana membranes while, again, it has an outer membrane enveloping the whole organelle. It is as if these bodies, all containing DNA in the matrix, are membrane-bounded insertions in the cell, all surrounded by a "real" cellular membrane. These considerations have led some to consider present-day mitochondria and chloroplasts as greatly modified descendents of an ancestral free-living bacterium or alga that at one time came together with the ancestral eucaryotic cell to form a symbiotic relationship.

SUGGESTED READING LIST

Offprints

Arnon, D. I., 1960. "The Role of Light in Photosynthesis." *Scientific American* offprints. San Francisco: W. H. Freeman & Co.

Bassham, J. A., 1962. "The Path of Carbon in Photosynthesis." *Scientific American* offprints. San Francisco: W. H. Freeman & Co.

Calvin, M., 1962. "The Path of Carbon in Photosynthesis." *Science*, pp. 879–889. Bobbs-Merrill Reprint Series. Indianapolis: Howard W. Sams & Co.

Green, D. E., 1954. "The Metabolism of Fats." *Scientific American* offprints. San Francisco: W. H. Freeman & Co.

Green, D. E., 1960. "The Synthesis of Fat." *Scientific American* offprints. San Francisco: W. H. Freeman & Co.

Hendricks, S. B., 1968. "How Light Interacts with Living Matter." *Scientific American* offprints. San Francisco: W. H. Freeman & Co.

Hogeboom, G. H., Schneider, W. C., and Palade, G. E., 1948. "Cytochemical Studies of Mammalian Tissues: Some Biochemical Properties of Mitochondria and Submicroscopic Particulate Material." *Jour. Biol. Chem.*, pp. 619–636. Bobbs-Merrill Reprint Series. Indianapolis: Howard W. Sams & Co.

Lehninger, A. L., 1960. "Energy Transformation in the Cell." *Scientific American* offprints. San Francisco: W. H. Freeman & Co.

Lehninger, A. L., 1960. "Oxidative Phosphorylation in Submitochondrial Systems." *Fed. Proc.*, pp. 952–962. Bobbs-Merrill Reprint Series. Indianapolis: H. W. Sams.

Lehninger, A. L., 1961. "How Cells Transform Energy." *Scientific American* offprints. San Francisco: W. H. Freeman & Co.

Lipmann, F., 1941. "Metabolic Generation and Utilization of Phosphate Bond Energy." *Advances in Enzymology*, F. Nord, ed., Interscience Publishers, Inc., pp. 99–162. Bobbs-Merrill Reprint Series. Indianapolis: Howard W. Sams & Co.

Palade, G. E., 1953. "An Electron Microscope Study of the Mitochondrial Structure." *Journal of Histochemistry and Cytochemistry*, pp. 188–211. Bobbs-Merrill Reprint Series. Indianapolis: Howard W. Sams & Co.

Rabinowitch, E. I., and Govindjee, 1965. "The Role of Chlorophyll in Photosynthesis." *Scientific American* offprints. San Francisco: W. H. Freeman & Co.

Racker, E., 1968. "The Membrane of the Mitochondrion." *Scientific American* offprints. San Francisco: W. H. Freeman & Co.

Siekevitz, P., 1957. "Powerhouse of the Cell." *Scientific American* offprints. San Francisco: W. H. Freeman & Co.

Stanier, R. Y., 1961. "Photosynthetic Mechanisms in Bacteria and Plants: Development of a Unitary Concept." *Bacteriological Reviews*, pp. 1–17. Bobbs-Merrill Reprint Series. Indianapolis: Howard W. Sams & Co.

Articles, Chapters and Reviews

Calvin, M., 1962. "Photosynthesis." *Science 135*, p. 879.

Chance, B., and Williams, G. R., 1956. "Respiratory Chain and Oxidative Phosphorylation." In: *Advances in Enzymology*, Vol. 17, p. 65. New York: Interscience.

de Duve, C., and Berthet, J., 1954. "Use of Differential Centrifugation in the Study of Tissue Enzymes." *International Review of Cytology,* Vol. 3, p. 225.

Ernster, L., and Lindberg, O., 1958. "Animal Mitochondria." *Annual Review of Physiology*, Vol. 20, p. 13.

Gibbs, M., 1967. "Photosynthesis." *Ann. Rev. Biochem. 36,* 757–784. Palo Alto, California: Annual Reviews, Inc.

Granik, S., 1961. "The Chloroplast, Inheritance, Structure and Function." In: *The Cell,* J. Brachet, and A. E. Mirsky, eds. New York: Academic Press, Inc.

Granik, S., and Gibor, A., 1964. "The Plastids." *Science 145,* p. 820.

Green, D. E., 1959. "Electron Transport and Oxidative Phosphorylation." In: *Advances in Enzymology,* Vol. 21, p. 73. New York: Interscience Publishers, Inc.

Green, D. E., and Fleischer, S., 1960. "Mitochondrial System of Enzymes." In: *Metabolic Pathways,* Vol. 1, p. 41. D. M. Greenberg, ed. New York: Academic Press, Inc.

Green, D. E., and Hatefi, Y., 1961. "Mitochondrion and Biochemical Machines." *Science 133,* p. 13.

Hogeboom, G. H., and Schneider, W. C., 1955. "The Cytoplasm." In: *Nucleic Acids,* Vol. 2, p. 199. E. Chargaff, and J. N. Davidson, eds. New York: Academic Press.

Howatson, A. F., and Ham, A. W., 1957. "Fine Structure of Cells." *Canadian Journal of Biochemistry and Physiology,* Vol. 35, p. 549.

Novikoff, A. B., 1961. "Mitochondria (Chrondriosomes)." In: *The Cell,* Vol. 2, p. 299. J. Brachet, and A. E. Mirsky, eds. New York: Academic Press, Inc.

Pullman, M. E., and Schatz, G., 1967. "Mitochondrial Oxidations and Energy Coupling." *Ann. Rev. Biochem. 36,* pp. 539–610. Palo Alto, California: Annual Reviews, Inc.

Schneider, W. C., 1959. "Mitochondrial Metabolism." In: *Advances in Enzymology,* Vol. 21, p. 1. New York: Interscience Publishers, Inc.

Slater, E. C., 1959. "Constitution of Respiratory Chain in Animal Tissues." In: *Advances in Enzymology,* Vol. 20, p. 147. New York: Interscience Publishers, Inc.

Wadkins, C. L., Cooper, C., Devlin, T. M., and Gamble, J. M., 1958. "Oxidative Phosphorylation." *Science 128,* p. 450.

Books

Boyer, P. D., 1967. *Biological Oxidation,* T. P. Singer, ed. New York: John Wiley.

Calvin, M. D., and Bassham, J. A., 1962. *Photosynthesis of Carbon Compounds.* New York: W. A. Benjamin, Inc.

Clayton, R. K., 1965. *Molecular Physics in Photosynthesis.* New York: Blaisdell.

Kamen, M. D., 1963. *Primary Process in Photosynthesis.* New York: Academic Press, Inc.

Lehninger, A. L., 1964. *The Mitochondrion.* New York: W. A. Benjamin, Inc.

Lehninger, A. L., 1965. *Bioenergetics.* New York: W. A. Benjamin, Inc.

Racker, E., 1965. *Mechanisms in Bioenergetics.* New York: Academic Press, Inc.

Ray, P., 1963. *The Living Plant.* New York: Holt, Rinehart and Winston, Inc.

The Nucleus and the Replication and Transmission of Information

So far we have talked about events that take place in the cytoplasm; we now turn our attention to the nucleus. All animal and plant cells contain well-defined nuclei with a circumscribed membrane. The nuclei can be isolated fairly easily from cells, still retaining most of their structural features (Fig. 14-1). Some cells have two nuclei, and some have many. Bacteria, however, contain a "nuclear" region, or nucleoid, unbounded by any membrane. Unlike the cytoplasm, which seems to have many functions, all the activities of the nucleus are geared to the timeless preservation and the exact reproduction of information. It is now a well-established fact that the nucleus is the depository of the Mendelian factors, or genes; that these genes are located in chromosomes; that the primary chemical substance of the genes is deoxyribonucleic acid (DNA) (Chapter 8); and that it is this very large molecule, with a molecular weight numbering into millions, that carries the information enabling the cell to express its individuality. How this is done we do not know exactly, but we are on the road to knowledge. This chapter will describe what might be called the initial travel along the highway.

From what we already know about the structure of DNA, we can surmise that it must have two properties, two functions: it must contain the information to code for the development of the cell, and it must be reproducible in exact replicas for the transmission of this code to future generations of cells. These two functions of the DNA reside in the arrangements of nucleotides: in one of its roles it acts as a "template" for the exact duplication of its nucleotide arrangements; in its other role it serves as a template for "messenger" RNA production, which in turn directs the synthesis of specific cell proteins. We will take up each of these functions separately.

There is a particular stain, Feulgen's stain, that reacts specifically with polymerized deoxyribonucleotides, that is, DNA. By means of this stain it has been shown countless

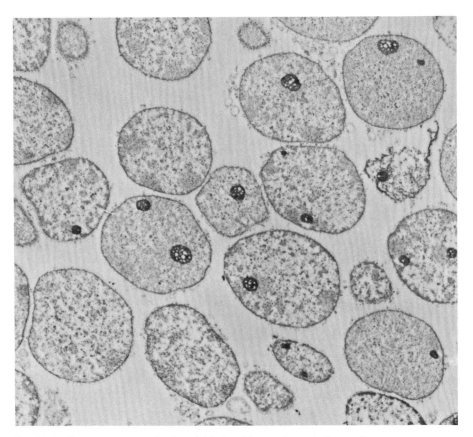

Fig. 14-1. Electron micrograph of nuclei isolated from guinea pig liver cells (× 5000). The nuclei still retain their envelope, most of their contents, and their nucleoli (dark bodies inside the nucleus). See Chapter 4 for details of nuclei *in situ*. (Photograph courtesy of G. E. Palade, Rockefeller University.)

times that DNA resides solely in the nucleus. However, we now know that a small part, probably less that 1 percent, of the cell's DNA resides outside the nucleus, in mitochondria, and, in cells that contain them, in chloroplasts. Actually, the nuclear DNA is complexed with a basic protein, a histone, forming the naturally occurring nucleohistone.

The amount of DNA is constant; in all the cells of a tissue from an animal of a certain species, the amount of DNA per chromosome set is the same. More precisely, the amount of DNA per nucleus is the same for all somatic cells of the particular animal, and this amount is twice that found in the germ cells of that particular species. We can go further and say that the amount of DNA per particular chromosome is a fixed amount and is characteristic of that species. This concentration of DNA remains the same no matter under what metabolic or nutritional strain the cells or the animal may be put. Another constant of the DNA of the nucleus is its metabolic stability. If radioactive inorganic phosphate is injected into an animal, the radioactive label will be found in high amounts in all sorts of phosphorus-containing compounds, but there will be no radioactivity in the phosphates of the DNA. It now appears that once the DNA of the chromosomes is synthesized, it does not break down. It is a stable molecule, and this is precisely what one would expect of a molecule that carries the information about the properties of the cell. This is not to say that the DNA is inert—in fact, we think it is not; it is to say rather, that once DNA is made, its molecular pattern is set, and it does not undergo any further metabolic transformations. Being part of the chromosomes, the DNA does participate in cellular events (in fact, we can "see" it do so) when the chromosomes replicate and divide during cell division.

CELL DIVISION

This brings up the matter of cell division; why should a cell divide at all? We do not know for certain, but we can make good guesses. It may be that the growth of a cell, its increase in mass, has a certain upper limit and that by dividing, a cell can obtain more surface area per mass of cell substance. In order to survive the cell must have a constant interchange with its environment, be its environment the immediate, minute one of a droplet of water or the intermediate one of the circulating blood. There probably is a limiting ratio of cell surface to cell mass, different for various kinds of cells, and when this limit is reached, the cell divides. Another reason for division might be that there occurs a kind of "aging" of the cytoplasm; cell division would thus allow a cell nucleus, or nuclear material, to gather around it fresh cytoplasm and still not exceed the limiting ratio of surface to mass. Overall, of course, the main reason for cell division during growth is to make new but similar cells until the adult size of the organ is reached. What causes a cell of a young tissue to divide and what causes a cell of an adult tissue virtually to stop dividing are both enigmas.

Mitosis

We do know that in most cases the visible result of cell division is chromosomal substance replication. In the nucleus of a nondividing, "resting" cell—the so-called "interphase" nucleus (see Fig. 4-30)—all that appears is a dense, definite but unbounded area, which is the nucleolus; other less dense areas, not well-defined, lie in the nucleoplasm, that area of the nucleus outside of the nucleolus. However, although we cannot "see" structures in the nucleoplasm, we do know that some molecular organization is present. During cell division, mentioned in Chapters 3 and 4, this organization comes into prominence. Centrosome regions appear, one at each end of the dividing cell, which seem to act as focal points for the spindle fibers. The spindle fibers are part of an organization, the mitotic apparatus, that can be separated intact from dividing cells. In the nucleoplasm exist the proteins that later, during cell division, become condensed and so structured that they form the visible spindle fibers along which the chromosomes move to each end of the dividing cell. Clearly, there is much more work to be done in determining the molecular suprastructure of the nucleus and how this structure changes dramatically during various stages of the nuclear cycle.

The division of a somatic cell into its two daughter cells is described as a process called mitosis (see Figs. 4-30 and 4-31.). Essentially what happens in mitosis is that the DNA that had been precisely duplicated previously is now exactly distributed to the two new cells. Mitosis begins with the condensation of DNA, protein, and possibly RNA into visible chromatin threads, which now become the entwined, coiling chromosomes that can be seen for the first time. After division, all that can be seen of the chromatin material in the interphase nucleus are dense bodies (heterochromatin) which, however, do stain with Feulgen's stain, indicating the continued presence of the DNA-containing threads. Thus, we can say that if a cell is partitioned, it should be done in such a manner that the daughter cells have exactly the same characteristics as the mother cell. The significance of mitosis is that it provides a precise mechanism for this to be accomplished. In morphological terms, this means an exact duplication of the chromosomal number, size, and architecture, for the chromosomes carry the Mendelian genes. In biochemical terms, this must mean, as we shall see, an exact duplication of the chemical structure of the DNA.

DNA Duplication

The initial event in cell division thus seems to be the doubling of the DNA content of the nucleus. How this is initiated we have no idea, but it occurs sometime before the visible condensation of the chromatin material into the chromosomes. We are well on the way to finding how the DNA is duplicated, for biochemists led by Kornberg have isolated enzymes from the cell that will actually synthesize DNA (Fig. 14-2 A).

$$\begin{matrix} n \text{ dTPPP} \\ n \text{ dGPPP} \\ n \text{ dAPPP} \\ n \text{ dCPPP} \end{matrix} + \text{DNA} \xrightarrow{\text{DNA polymerase}} [\text{—dTP—dGP—dAP—dCP—}]_n + 4(n)\text{PP}$$

A

$$\begin{matrix} n \text{ UPPP} \\ n \text{ GPPP} \\ n \text{ APPP} \\ n \text{ CPPP} \end{matrix} + \text{DNA} \xrightarrow{\text{RNA polymerase}} [\text{—UP—GP—AP—CP—}]_n + \text{DNA} + 4(n)\text{PP}$$

B

Fig. 14-2. A. Equation for synthesis of DNA. (After A. Kornberg.) B. Equation for synthesis of DNA-patterned RNA. (After S. Weiss and J. Hurwitz.)

Now, even before the enzyme called DNA polymerase was discovered, biochemists had made some guesses as to its probable mode of action, guesses based on the properties of the DNA. It was inferred, from the Watson-Crick formulation (Chapter 8), that one chain of the double helix should serve as a template for the aligning of the bases of the presumptive other chain by specific hydrogen-bond base-pairing as depicted in Figs. 8-6 and 8-7. The enzyme was visualized to hook up the nucleotides in a phosphodiester bond; the energy for the synthetic reaction was thought to come from the energy inherent in the high-energy phosphate compounds, specifically in the deoxynucleotide triphosphates. Both of these expectations were met so that now we know that DNA polymerase acts as shown in Fig. 14-3.

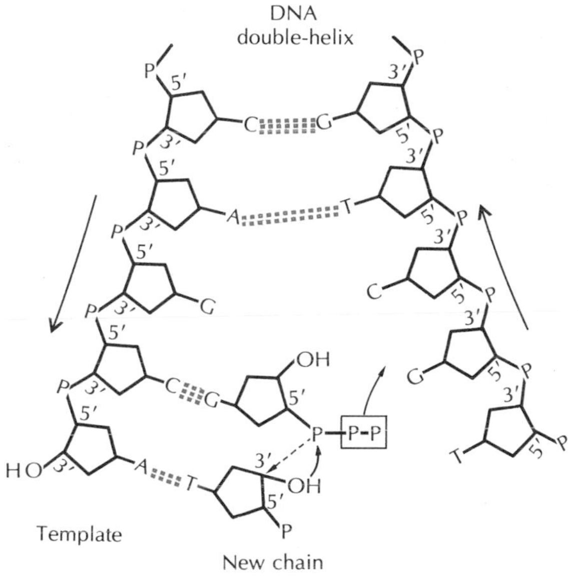

Fig. 14-3. Mechanism of enzymatic replication of DNA. (From A. Kornberg.) Two DNA strands are shown: the one on the left runs from 5'-phosphate to 3'-OH, and the other runs antiparallel from 3'-phosphate to 5'-phosphate (direction of arrows). The strands are unwinding, and the left one is being "copied," starting from the bottom. Thus, the newly synthesized chain starts with a 5'-phosphate, in this case thymidine deoxyribose phosphate. The thymine is base-paired (two hydrogen bonds) to adenine, and the guanine is base-paired by three hydrogen bonds to cytosine. The phosphodiester bond is made between thymidine phosphate and guanosine triphosphate by an attack by the 3'-OH group of the deoxyribose on the bond between the inner phosphate and the outer two phosphates of the adjacent nucleotide, with the phosphates being split off as inorganic pyrophosphate.

Let us first state our current knowledge concerning the enzymatic replication of DNA and then give some of the evidence that led to these generalizations. The enzyme acts by adding on monomeric units to a growing chain, these units being the triphosphates of the deoxyribonucleotides; inorganic pyrophosphate is split off at the same time that a phosphodiester bond is formed between the 3'-OH group of the deoxyribose in the DNA chain and the 5'-OH group of the deoxyribose of the incoming nucleotide (Fig. 14-3); the chain is elongated from the 5'-OH end of the deoxyribose as depicted in Fig. 14-3. There is an absolute requirement for a template that directs which nucleotides are placed next to each other along the opposite, presumptive DNA chain, strictly according to the Watson-Crick formulation of DNA structure (that is, of base-pairing of adenine with thymine and of guanine with cytosine); however, there is growing evidence that the DNA polymerase is involved, along with the template, in a replication of DNA which is error-free.

Because the first attempts at showing DNA synthesis *in vitro* were with crude extracts, the only way to show a synthesis of DNA was by using radioactive nucleotides; for while no net synthesis occurred (indeed, because of the presence of nucleases in the preparations there was a net breakdown of the added DNA), it could be shown that radioactive nucleotides were incorporated into a large, acid-insoluble DNA fraction. Using such a system, it was quickly found that DNA had to be present in the reaction mixture and that the three other deoxyribonucleotides, in addition to the radioactive one, had to be present in order for synthesis to occur. The substrates had to be the triphosphates, and only the deoxyribonucleotides, not the ribonucleotides, were active in the system. Using an enzyme preparation from *E. coli*, Kornberg showed that the DNA could be of animal, plant, bacterial, or even of viral origin. Using a more purified enzyme preparation it was later found that the amount of DNA synthesized was more that twenty times that added as a primer or template. This latter finding was important, for up to this time it was not known whether the DNA that had to be added was acting as a primer, in much the same way as does glycogen in the glycogen synthetase reaction (that is, by adding on nucleotides to loose ends) or whether it acted as a true template in the Watson-Crick sense.

Because of the successful net synthesis of DNA *in vitro*, we now know that the added DNA acts as a true template, and we know it because of the following reasons. The synthesized DNA has many of the physical properties of size, sedimentation rate, viscosity, as well as optical properties of double-stranded DNA; hence, it is assumed that each of the single-stranded DNA chains in the double-stranded DNA that was added as a primer really acted as a template for an exact duplication of itself, with one strand of the synthesized DNA and one strand of the template DNA forming a DNA that had all the characteristics of a double-stranded helix. Thus, bonding of nucleotides to the end of the primer DNA could be ruled out. Another piece of evidence for the template hypothesis was the finding that certain chemical analogues could be substituted for the natural nucleotides in the enzymatic reaction; thus, deoxyuridine triphosphate could substitute for deoxythymidine triphosphate but not for the other nucleotides, and 5-methyldeoxycytosine triphosphate could substitute for deoxycytosine triphosphate. Both of these indicate that as long as the chemical analogue could make a base-pair with the natural nucleotide in the template DNA, it could serve in the enzymatic reaction.

More to the point, two questions were asked and answered affirmatively; does the enzymatically produced DNA have the equivalence of adenine to thymine and of guanine to cytosine that characterizes the template DNA, and does the composition of the template DNA, the sequence of bases therein, determine the composition of the product DNA? Table 14-1 indicates that the answer to the second question is yes; it also shows that each of the two strands of the primer DNA acts as a template for the synthesis of a single-stranded DNA, for this is the only way that this base equivalence can be realized. When the primer is a copolymer of adenine and thymine (dAT copolymer), even the DNA synthesized in the presence of all four nucleotides contains only deoxythymine and deoxyadenine in its structure. Also, no trace of deoxyguanine and deoxycytosine can be found because these cannot be base-paired in any way to the dAT copolymer; hence, they are not taken up in the enzymatic reaction although they are in the reaction mixture in large amounts and the enzyme can react with them.

Table 14-1. COMPARISON OF BASE COMPOSITION OF *IN VITRO* SYNTHESIZED DNA WITH THAT OF THE PRIMER DNA*

SOURCE OF PRIMER DNA	ADENINE PLUS THYMINE / GUANINE PLUS CYTOSINE	
	OF PRIMER DNA	OF SYNTHESIZED DNA
Micrococcus lysodeikticus	0.39	0.41
Mycobacterium phlei	0.49	0.48
Aerobacter aerogenes	0.82	0.80
Escherichia coli	0.97	1.01
Bacteriophage	1.06	1.00
Bacillus Subtilis	1.29	1.26
Calf thymus	1.25	1.32
Hemophilus influenzae	1.64	1.62
Bacteriophages T2, T4, T6	1.84	1.76
dAT copolymer	⟩40	⟩250

(See Table 8-1.)

* (Data after A. Kornberg, *Enzymatic Synthesis of DNA,* John Wiley & Sons, 1961.) The data of the table is explained in the text. In all cases there occurred a large net synthesis of DNA, using the primers noted on the left; thus, the base composition of the DNA isolated at the end of the reaction was largely that of the DNA synthesized in the reaction.

Another very elegant way to determine whether the *in vitro* enzymatic reaction obeys the Watson-Crick formulation, both in base-pairing and in polarity of strands of the helix, has been by the "nearest neighbor" technique worked out by Kornberg. By polarity we mean the nature of the aligning of the two strands of the DNA helix; if the two strands are parallel, they both start from the 5'-phosphate end and end up in the 3'-OH end, but if they are antiparallel, the one strand goes from 5'-phosphate to 3'-OH, and the other from the 3'-OH to the 5'-phosphate end; in both cases the hydrogen-bonded base-pairing would still be apparent. Figure 14-3 shows the two antiparallel strands in DNA, the left one going from the 5'-phosphate to the 3'-OH while the right strand goes in the opposite direction. We now know that the two strands are in this sense opposite to each other, antiparallel, and we know this because of the results of the following experiment.

The deoxyribonucleotide triphosphates are all chemically synthesized with a radioactive phosphate label on the innermost phosphate. Each of these labeled triphosphates is incubated together with the three other triphosphates (but unlabeled) in separate experiments. After the DNA has been synthesized in the four experiments (each with a different labeled nucleotide), the synthesized DNA is isolated, purified, and is enzymatically hydrolyzed to its constituent nucleotides. The enzyme used is one that cleaves the DNA at the 5'-phosphate bond, leaving the 3'-deoxyribonucleotide phosphates as shown in Fig. 14-4. Thus, it is clear that the radioactive phosphate that was attached to the 5'-end of the substrate nucleotides has now been transferred to the 3'-end of the nucleotides isolated after the enzymatic hydrolysis of the DNA, that is, the radioactive nucleotides isolated in each of the above experiments must have been adjacent to the particular labeled substrate of that particular experiment (it must have been its "nearest neighbor"). In Fig. 14-3, for example, if the innermost phosphate of the deoxyguanosine triphosphate were labeled, after hydrolysis the radioactive phosphate would be recovered in the deoxythymidine phosphate.

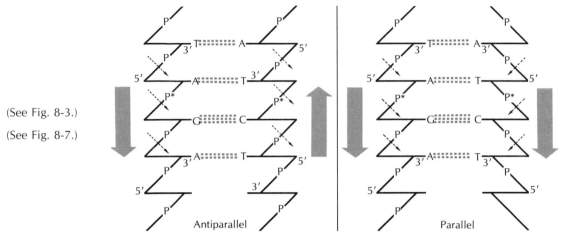

(See Fig. 8-3.)

(See Fig. 8-7.)

Fig. 14-4. Illustration of detection by "nearest-neighbor" analysis of the polarity of the two strands of DNA in the double-helix. The experiment is explained in the text. The starred *P*'s represent radioactive phosphates. The dotted lines with arrows show the points of cleavage by the enzyme used to hydrolyze the DNA, giving 3'-deoxyribonucleotide monophosphates. The full arrows show the direction of each of the strands, from 3'- to 5'-phosphate ends.

Now to the experiment. In Fig. 14-4 we give an example of two DNA helices, one parallel and one antiparallel. Suppose in the *in vitro* experiments we added labeled deoxyguanosine triphosphate in one case, labeled deoxythymidine triphosphate in another, and labeled deoxycytidine triphosphate in the third. All the nucleotides resulting from hydrolysis were isolated, but let us narrow them down to the 3'-deoxyadenosine nucleotide in the first case (nearest neighbor in Fig. 14-4), labeled 3'-deoxycytidine nucleotide in the second, and labeled 3'-deoxythymidine nucleotide in the third case. If we examine Fig. 14-4, we can notice that if the strands were antiparallel, the amount of labeled deoxyadenine nucleotide should be equal to the amount of the labeled deoxycytidine nucleotides and not equal to the amount of the labeled deoxythymidine nucleotide; if the strands were parallel, the dA and dT should be equal in radioactivity. It was actually found that the first two were equal. Indeed, when all possible combinations (sixteen in all) were examined as shown in Table 14-2, it was found that the equivalences in radioactivity were found only in those cases that could be predicted by the theorem that the two strands are antiparallel. This type of experiment very strongly indicates that the DNA polymerase was using the DNA template to align the sequences of the nucleotides in the synthesized DNA exactly as they were in the template DNA, as if the base-pairing formulations of the Watson-Crick hypothesis were obeyed.

Finally, the theorem was proven by the exceptional case in which a single-stranded DNA was used as a primer, in this case a DNA from a bacteriophage. This DNA, being a single strand, has no equivalence of base composition between a deoxyadenosine and deoxythymidine nor between a deoxycytidine and deoxyguanosine. If only a limited amount of synthesis was allowed to occur in the presence of the primer and in the presence of radioactive nucleotides, it was to be expected that the newly synthesized chain should have radioactive deoxyadenosine equal in amount to the deoxythymidine of the primer and radioactive deoxyguanosine equal to that of the deoxycytosine of the primer; and indeed, this was the case. In this instance, a newly synthesized chain equal in length to that of the DNA template seemed to have a base composition that obeyed the base-pairing hypothesis.

The evidence is impressive that the characteristics of the nuclear DNA polymerase, which have been unearthed *in vitro,* mirror quite faithfully that of the *in vivo* enzyme. Perhaps most amazing is that these enzymic properties, derived from biochemical research with the enzyme, are a faithful counterpart to the biophysical properties of the DNA postulated by Watson and Crick; this coming together of results from two different fields to provide a unified picture of one of the most important problems in biology is one of the highlights of modern biology.

Table 14-2. "NEAREST NEIGHBOR" FREQUENCIES OBTAINED AFTER
IN VITRO INCUBATION*

LABELED TRIPHOSPHATE	ISOLATED 3'-DEOXYRIBONUCLEOTIDE			
	T	A	C	G
dATP	.012	.024	.063	.065
dTTP	.026	.031	.045	.060
dGTP	.063	.045	.139	.090
dCTP	.061	.064	.090	.122
Sum	0.162	0.164	0.337	0.337

* (Data after A. Kornberg, *Enzymatic Synthesis of DNA,* John Wiley and Sons, 1961.) The experiment is explained in the text. The values are the radioactivities found in each isolated nucleotide in the four experiments in which a different labeled triphosphate was added to each. The dotted lines indicate the two values which should be equal to each other if the base-pairing of the Watson-Crick model were obeyed and if the two strands of the DNA were of opposite polarity. The identities between what should be equal and what the values actually were shows the proof of the antiparallel nature of the base-paired DNA strands. Notice that T = A and C = G in the sums.

But let us not think that all the problems are solved with regard to DNA duplication. For example, earlier work led to the finding that many of the synthesized DNA chains were branched; that they had free 5'-phosphate ends; and that they behaved not quite the same in a physical sense as did natural double-stranded DNA. Natural bacteriophage DNA, when isolated, is in the form of closed double-stranded circles. When this DNA is used as a template in the DNA polymerase reaction, it was found that indeed such circles were formed, but in all cases these circles seemed to be incomplete or showed a branching out of some of the chains. However, recently a solution to this particular problem was found by Kornberg and his associates, in the form of using an additional enzyme in the reaction, an enzyme that can reattach loose DNA ends. In the presence of this enzyme, of purified DNA polymerase, and of purified phage DNA, these workers have recently shown a net synthesis of a biologically active DNA molecule, a DNA composed of perfect double-stranded circles. Because this DNA was of phage origin, its biological activity could be tested; when this was done, it was found that indeed there had occurred a net synthesis *in vitro* of many copies of the active phage template.

But other problems abound. One has to do with the meaning of circular DNA in a biological sense and the mechanism of its replication in a biochemical sense. Based on elegant autoradiographic work by Cairns, it is most likely true that the enzyme involved with the double-stranded circular DNA of *E. coli,* for example, duplicates both strands of the DNA starting from a single point on the circle. That means that the two strands of the double helix must unwind at this point (and the mechanism of this unwinding is itself entirely unknown), and the two strands must be synthesized from this point. This is theoretically feasible, but the isolated and purified DNA polymerase does not seem to have the properties required. This enzyme seems to account for the replication of only the chain ending in 3'-OH (Fig. 14-3), for if this enzyme also could replicate the 5'-phosphate chain, in which the triphosphate is put in place in the opposite chain, the mechanism of its action would have to be different from what it actually is. In Fig. 14-3 we see that the 3'-group of the nucleotide already in the chain is the site of formation of the phosphodiester bond; whereas, to produce a chain elongation of the 5'-phosphate chain, the 3'-OH of the incoming triphosphate should be the site of the enzymatic attack. There are two, and possibly three, solutions to this problem: there is another enzyme which replicates the other chain, which starts with the 5'-deoxyribonucleotide triphosphates, whose mechanism of action is dissimilar to the unearthed DNA polymerase, and which has not been found as yet (compare Fig. 14-5 with Fig. 14-3); or else there is another enzyme which synthesizes DNA in the same way as does the DNA polymerase which has been found but which uses the 3'-deoxyribonucleotide triphosphates (Fig. 14-5). The latter possibility is most unlikely, for no one has found 3'-triphosphates in the natural world. At present, we have no firm evidence that the first possibility exists either. Thus, we are left with a baffling mystery, whose solution will probably depend on the solution to the problem of the unwinding of the DNA double helix and on the meaning of the circular form of DNA for its replication. Although this circular form has only been found in the case of rather small DNA molecules such as mitochondrial and bacterial DNA's, it is possible that the DNA in the chromosomes is also in the form of closed circles, but because of its huge size, it is easily clipped during isolation and circles have not been detected as yet. The third possibility rests upon the discovery of an enzyme which seemingly ties together loose ends of DNA polynucleotide chains by forming 3',5'-phosphodiester bonds between these ends, an enzyme, called appropriately enough, "ligase," which was used by Kornberg to synthesize net amounts of biologically active phage DNA. This third possibility is also depicted in Fig. 14-5; the DNA polymerase copies one chain, making a

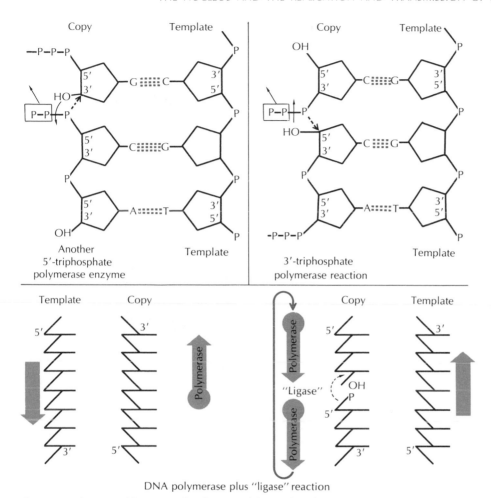

Fig. 14-5. Three possible means of replication of the 5'-phosphate DNA chain. (After A. Kornberg.) The last diagram indicates the presently accepted scheme: the synthesis of short polynucleotide chains by DNA polymerase, starting from the 5'-phosphate ends of the copy, and then the condensation of these chains by the "ligase" enzyme. Further details are in the text.

chain going from the 5'-end (bottom to top in Figs. 14-3 and 14-5), and this same enzyme begins to also copy the other chain, starting at about the same point and also starting to make a copy with a free 5'-phosphate end. However, while the polymerase copying the left-hand chains in Figs. 14-3 and 14-5 continues the whole length of the DNA, that copying the right-hand chain (from top to bottom in the same figures) only goes a short way and then stops. As the DNA double helix unwinds, the enzyme copying this chain goes back up again to start to make another short copy with a free 5'-phosphate end. These short segments with free 5'-phosphate and 3'-OH ends are then covalently linked together by the "ligase." In this way it might be possible that as the

DNA unwinds, both chains are copied together by the same type of enzyme but in slightly different ways. The solution to this problem is, as you may guess, eagerly awaited.

Finally, it is thought that the DNA in the chromatin material of the nucleus is in the double-stranded configuration and that DNA duplication in the nucleus takes place in much the same way as can be brought about in the test tube. However, DNA replication is not a requisite to cell division, for in certain cases the duplication takes place without cell division occurring. But the reverse is not true; cell division does not take place without a previous duplication of the DNA content.

Chromosome Duplication

The next general event in cell division is the prophase stage; this stage marks the convenient way of recognizing cells that are undergoing division. Here for the first time can be seen the condensation of the chromatin into coiling, spiraling threads, the chromosomes. It is generally agreed that the chromosome, that which is seen in the light microscope, is itself at least a duplex structure in the sense that it contains two, or perhaps more, equivalent strands called chromatids. These chromatids are the structures that will be separated from each other at a later stage. Is one of these chromatids the new DNA that was synthesized upon the old DNA as template? It would appear that in some cases each of these chromatids is itself duplex, being made up of "half-chromatids" and that each of these half-chromatids is replicated so that in time a chromatid coil contains a new DNA strand and an old DNA strand.

The following experiment, first performed by Taylor (Fig. 14-6), permits us to observe what might be happening in the cell. The rapidly dividing cells were given radioactive, tritium-labeled thymidine and then fixed for microscopic examination. In cells taking up the thymidine—those producing a new generation of chromosomes preparatory to division—all the chromosomes were found to be labeled, and radioactivity was equally distributed between the two chromatids of each chromosome. The cells were then placed in a medium of nonradioactive thymidine, and a second generation of chromosome duplication was allowed to take place. Taylor found by autoradiography that one chromosome was labeled and one was not. The simplest way of explaining this, particularly in the light of present-day biochemical knowledge of DNA duplication, is illustrated by the diagram in Fig. 14-6. A chromosome consists of two parts or strands, each of which acts as a template for the production of another part. In the radioactive medium, each of the chromosomes, after splitting in two, acts as a template for the production of a radioactive partner. Therefore, all the new chromosomes are labeled. However, when these labeled chromosomes duplicate in a nonradioactive medium, the chromosomes will split to give a radioactive and a nonradioactive partner. Each of these then acts as a template for new chromatid formation, yielding chromosomes half of which are labeled and half are not.

This seems clear enough, but when we reach down into smaller dimensions we run into difficulty. We have no idea as yet of the molecular organization of the chro-

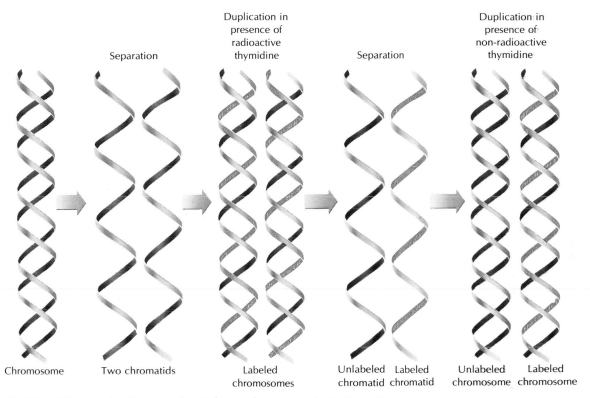

| Separation | Duplication in presence of radioactive thymidine | Separation | Duplication in presence of non-radioactive thymidine |

Chromosome   Two chromatids   Labeled chromosomes   Unlabeled chromatid Labeled chromatid   Unlabeled chromosome Labeled chromosome

Fig. 14-6. Diagram of radioautographic studies on chromosome duplication. (After H. Taylor.) The diagram is explained in the text.

matid, of how many strands of DNA it contains, and of how this DNA is arranged within the chromatid. It appears that, somehow, the double-helix configuration of individual DNA molecules is reflected in the coiling of the thousands of times larger chromosomes. Thus, whereas at the biochemical level we can recognize that it is individual DNA molecules that are being duplicated, at the morphological level we are not sure what the corresponding duplicating structure might be—whether it is a whole chromosome, a chromatid, a half-chromatid, or an even finer subdivision of the visible complex.

(See Fig. 8-21.)

This type of experiment can be done in another way, as shown by Meselson and Stahl, using cells of the bacteria *Escherichia coli;* the complications of chromosomal structure are avoided, for bacteria have no chromosomes. *E. coli* cells can be grown in a medium containing "heavy" nitrogen (N^{15}); thus, all the DNA of the multiplying cells becomes labeled with N^{15}. If the growing culture is then switched to a normal N^{14}-medium, the DNA now produced will contain N^{14} in its bases. The DNA of all these cells can then be extracted, and by centrifuging it for a long period of time in a density-gradient system containing cesium chloride, the N^{15}-DNA can be separated from the N^{14}-DNA. During the centrifugation the heavy cesium chloride molecules begin to be sedimented through the aqueous solution, setting up a density gradient.

The heavier N^{15}-DNA molecules will be separated from the lighter N^{15}-DNA molecules in just such a system. If this separation is done at different times during the experiment outlined above, the following results are obtained. At first, only one DNA band is visible, that of N^{15}-DNA. After one doubling of cell number, in the presence of N^{14}-thymidine, again only one band is found, but in a new position, intermediate between the positions that N^{15}-DNA and N^{14}-DNA would have occupied. After a second doubling, two bands of DNA are found, one in the same position as the previous intermediate band and the other where N^{14}-DNA would have appeared. The intermediate band can then be isolated and separated in essentially the same manner into two bands, one made up of N^{15}-DNA and the other of N^{14}-DNA. The explanation of all these results seems obvious. The intermediate band came about because the N^{15}-DNA synthesized in the N^{15}-medium then served as a template for a newly synthesized N^{14}-DNA molecule. In the second division, both the N^{15}-DNA and the N^{14}-DNA chains served as templates for the synthesis of new DNA molecules. The N^{15}-DNA became linked to an N^{14}-DNA to form another intermediate band, while the N^{14}-DNA served as a template for the laying down of another N^{14}-DNA to form a new N^{14}-DNA band. Thus, it would appear, in bacteria at least, and probably in all cells, that each DNA chain can serve as a template or a primer for the synthesis of another DNA chain.

On the biochemical level we seem to have unearthed the mechanism of the exact duplication of an existing DNA molecule to form a new DNA molecule destined for the new daughter cell. On the morphological level, the mechanism is somewhat obscure, but the end result is the same: formation of a new chromosome that is the same as the one in the mother cell. It is thought that the inexactness of the duplication results in what can be observed grossly as a mutation; the new DNA chain is not exactly the same as the old one. But what exactly does this duplication of a structure within the nucleus portend with regard to the metabolism of the cytoplasm? We have said that both the nuclear and cytoplasmic characteristics of a cell are determined by the genetic material of the nucleus, the chromosomes. In other words, the synthesis of specific cellular protein, enzymes, and cellular structures is determined by the information in the genes. The question is how are the instructions in the chromosomes transmitted to the recipient protein-synthesizing mechanism in the cytoplasm? The nucleus itself has, of course, some of the same reaction patterns as the cytoplasm. It can apparently make its own energy in the form of ATP, for it seems to have many of the glycolytic enzymes described in Chapter 12. It can synthesize various small molecules, and it has various enzymes to perform tasks whose relevance to the general metabolism of the cell is unclear. The nucleus has, then, its own complement of enzymes, in some cases the same types as found in the cytoplasm, in other cases different. Nevertheless, these enzymes must be present in the nucleus of the daughter cell. The DNA of the chromosomes undoubtedly has the specification for the synthesis of these enzymes. But most of the enzymatic material and much of the cellular synthetic machinery is in the cytoplasm, and although mitochondria and chloroplasts both contain DNA (Chapter 13), it is still believed that nuclear DNA contains some of the information for the duplication of some of the mitochondrial and chloroplast proteins.

ROLE OF RNA

The means by which transfer of information occurs between DNA in the nucleus and the synthesizing machinery in the nucleus and cytoplasm is tied up with the metabolism of the other nucleic acid, ribonucleic acid (RNA). In the cells of mammalian tissues about 10 percent of the RNA of the cell is in the nucleus. Of this, about 20 percent is in the dense nucleolar material (the nucleolus contains no DNA although it does have some DNA material concentrated at its borders). The rest of the RNA is in the nucleoplasm, some of it being intimately connected with the chromatin material, and the rest possibly being in the form of small particles, ribosomes or their precursors. It has been known for some years that nuclear RNA, taken as a whole, has a high turnover. What this means can be illustrated by the following types of experiments. Radioactive inorganic phosphate can be injected into an animal, and the liver then removed and fractionated into its various morphological entities such as nucleus, mitochondria, microsomes, and cell sap. The RNA is then extracted from all these fractions and tested for radioactivity. Invariably, the total nuclear RNA has been found to be the most radioactive when measured at intervals soon after the injection. In fact, in most instances it is the RNA in the nucleolus that is the most radioactive as indicated by autoradiographic experiments. In some cases, at later times after the injections, it has been found that the radioactivity of the total nuclear RNA has declined while that of the total cytoplasmic RNA has increased. The same sort of experiment has been done with the technique of autoradiography, using such favorable specimens as the large oöcyte cells; it is easy to determine in these cells whether the silver grains lie over the nucleus or the cytoplasmic areas. In these cases it was again found that the nuclear RNA became radioactive more quickly than the cytoplasmic RNA.

The results of both these kinds of experiments led to the hypothesis that RNA is synthesized in the nucleus and that these completed RNA molecules, called "messenger" RNA, then move out into the cytoplasm. There is no doubt that something like this occurs, but there still remains the question as to whether all the RNA of the cell is synthesized in the nucleus or whether there are two independent sites of cellular RNA synthesis, the other being in the cytoplasm. Certain experiments have raised the latter possibility. For example, using favorable types of cells where such a microoperation can be performed, it has been possible to remove the nucleus from the cell and still have a somewhat viable cell, at least for a short period of time. In most cases it has been found that there is no net increase in cytoplasmic RNA after the removal of the nucleus while in a few cases there does seem to be cytoplasmic RNA synthesis going on in the absence of the nucleus.

However, we have mentioned (Chapter 8) that there are different types of RNA molecules with differing functions, and the question whether the nucleus synthesizes

(See Chapter 8.)

(See Chapter 15.)

(See Chapter 15.)

(See Chapter 15.)

all these types in a finished state will not be answered in full until finer techniques have been developed. But at present, it appears that not only is "messenger" RNA made in the nucleus, but so are ribosomal and transfer RNA. Indeed, there is strong evidence that the nucleolus is the site of synthesis of ribosomal RNA, based on various kinds of experiments, including one with a certain frog mutant whose embryo can develop only to an early tadpole stage, presumably because the cell nuclei lack any nucleoli. The cells of this mutant were found to be unable to synthesize ribosomal RNA. In normal cells, indeed, it appears that a large RNA molecule is made in the nucleolus and that stepwise this breaks down to form the RNA of each of the ribosomal subunits; these subunits, probably in the form of ribonucleoproteins, then go out into the cytoplasm.

(See Chapters 8 and 15.)

"Messenger" or "Informational" RNA

As will be discussed later, it is known that RNA is intimately connected with protein synthesis; this seems to be the sole role of the RNA molecules in the cell. Most of this synthesis of proteins takes place in the cytoplasm. From other lines of work involving viruses containing RNA and not DNA it is clear that the genetic material in these viruses is RNA. In these cases the RNA of these particular viruses performs the same function as does the DNA in most other organisms and cells, that is, it contains the information for the synthesis of a new viral protein. Thus, not only is RNA similar in structure to DNA, but in some instances it can perform the same function. From these lines of evidence it has been postulated that the intermediate between the information in the DNA in the nucleus and the synthesis of protein in the cytoplasm is the RNA that is synthesized in the nucleus. There really has been no definite proof for this until recently. As first shown by Astrachan and Volkin, when bacteriophage containing DNA as genetic material infect bacteria, the infected cells synthesize a new kind of RNA molecule. This RNA seems to have the characteristics of the phage DNA in the sense that the base ratios of the newly synthesized RNA mirror the base ratios of the DNA. In other words, there seems to be information in the DNA for the synthesis of DNA-like RNA molecules.

It was also discovered, by Weiss and Hurwitz, that extracts can be made from bacteria, or from mammalian cell nuclei, that contain an enzyme that will make RNA. More to the point, the synthesis of this RNA was dependent on the presence of DNA (Fig. 14-2B). The enzymatic reaction utilizes the nucleoside triphosphates, ATP, CTP, GTP, and UTP; inorganic pyrophosphate is split off each of these, concomitant with the formation of the phosphodiester bond linking the resultant nucleoside monophosphates to form the RNA chain. The enzyme, called RNA polymerase, is specific for the ribonucleoside triphosphates; the deoxyribonucleoside triphosphates are not suitable. Moreover, as in the case of DNA synthesis, the nature of the RNA product formed is dependent on the DNA that is added. It was found that the base ratios of the RNA synthesized (the relative proportions of the nucleotides to each other) were the same as the base ratios of the added DNA as in the case of the phage-infected bacteria. When a different DNA, having different base ratios of its nucleotides, was added, the resultant synthesized RNA contained the same base ratio as the added DNA. The enzyme was therefore making an RNA molecule that was the image of the DNA molecule. It can be envisaged that where there is a deoxyguanylic acid residue in the DNA, there will be cytidylic acid in the RNA and where there is thymidylic acid in one, there will be adenylic acid in the other. Indeed all the types of experiments performed with the DNA polymerase and outlined above were also done with the RNA polymerase. All the criteria mentioned then as being used to prove the characteristics of the reaction of the DNA polymerase also held up for the RNA polymerase reaction. All the findings would neatly fit into a scheme that envisages the transfer of information from the DNA to the newly synthesized RNA and, thence, from the RNA to the mechanism that synthesized protein both in the nucleus and cytoplasm, but mostly in the latter. From histochemical and autoradiographic evidence it is clear that DNA duplication and RNA synthesis in the nucleus take place at different times during the nuclear division cycle. At one time the DNA acts as a template for the enzymatic synthesis of a duplicate DNA molecule; at another time it acts as a template for the enzymatic synthesis of an RNA molecule similar to it. The means of the regulation of this twofold functioning of DNA is not known, but it is thought to be bound up with the macromolecular structure of the DNA and possibly with the presence of the histone that is in combination with it.

Does the DNA-mediated RNA polymerase function in the cell as it does in the test tube? In the last few years many laboratories, notably those of Roberts, Spiegelman, and Gros, have been examining the *in vivo* behavior of the RNA that is probably synthesized by this enzyme in the cell. It has been called "messenger" or "informational" RNA, a name coined to denote the idea that it could fulfill the role of conveying a message from the nuclear DNA to the cytoplasmic protein-synthesizing ribosomes (See Chapter 15.) as to the specificity of the proteins these ribosomes will manufacture. When a bacterial culture is given a single shot of an RNA precursor, labeled with P^{32} or C^{14}, for a short period of time, one can easily notice the existence of an RNA fraction with a high turnover. The system favored for the separation of the macromolecules is a density-gradient system, using either sucrose or the cesium chloride mentioned earlier. When sucrose is used, ribosomal RNA separates in such a density-gradient system in the form of ribonucleoprotein particles, mostly in the form of polyribosomes with single (See Chapter 15.) ribosomes sedimenting to a lighter region designated as 70 S particles (of bacterial origin) or 80 S particles (of eucaryotic origin), as shown in Fig. 14-7. However, the radioactivity does not follow congruently the absorbance at 260 mμ, but sediments over a broad base in the polysome region. This radioactive RNA is not transfer RNA, which is very small and would appear at the top of the gradient, nor is it ribosomal RNA, for no radioactivity appears in the 70 S or 80 S region where single ribosomal particles (monomers) would lie. The interpretation first preferred, and still held by many workers in the field, was that the quickly labeled RNA was all messenger RNA; after it was synthesized, it moved onto the ribosomes, forming them into polysomes (See Chapter 15.) and there, attached, acted as a specifying template for protein synthesis. Instead

Fig. 14-7. Appearance of "messenger" RNA. Explanation of the figure is given in the text.

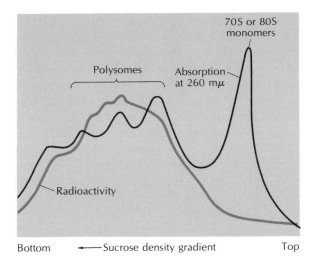

of the normal cell, phage-infected cells can be used, as with Astrachan and Volkin; the result is the same sort of picture as is illustrated in Fig. 14-7. Furthermore, as mentioned above, this newly synthesized RNA did seem to have the same base ratios as did phage DNA; it varied markedly in this regard from the bulk of the ribosomal RNA. In other words, this newly synthesized RNA fulfilled one of the requisites to qualify it as messenger. However, the thought among many workers today — but perhaps not tomorrow — is that this newly synthesized RNA is partly of the messenger type and partly a precursor of ribosomal RNA and that the difference between these two types of RNA is not simply that one contains a code message from the DNA and (See Chapter 8.) that the other acts as a structural component of the ribosomes. What the relationship between the two is we do not know; perhaps, what we are witnessing is a counterpart, but on an unbelievably miniature scale, of some sort of punch-card indexing machine with these two kinds of RNA macromolecules acting as the tapes.

However, at the same time investigators are firmly bound to the notion that such a messenger exists. One of the other properties of this postulated messenger — outside of its quick formation in the cell in certain circumstances, of its having the same base composition as the DNA, and of there existing enzymes that can make an RNA molecule based on a DNA template — is that this messenger RNA has arrangements of bases along its length that constitute a chain complementary to that of the DNA (Fig. 14-8). If this is so, it would be expected that hydrogen bonds would be formed between these two complementary chains in the same way as the two chains interact in DNA synthesis to form the tightly coupled double helix. That this occurs has been shown recently by Spiegelman and Hall. They used a technique devised by Doty and Marmur, that of making one polynucleotide chain become hydrogen bonded to another by heating a solution containing these macromolecules to about 40°C and then slowly cooling the solution (Chapter 8). Doty and Marmur found that during this cool-

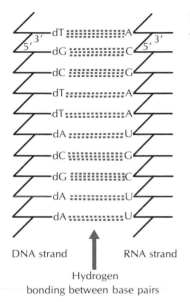

Fig. 14-8.   Complementary bonding between segments of DNA chain and segments of RNA chain.

DNA strand   RNA strand

Hydrogen
bonding between base pairs

ing, the single strands of DNA began to attach to each other, forming the double helix; this happens, however, only if there exist large stretches along these chains where the bases of one are complementary to the bases of the other—the prime requisite for formation of hydrogen bonds between them. This experiment can likewise be performed with a mixture of DNA and RNA macromolecules. If *E. coli* bacteria are infected with phage in a medium containing tritium-labeled thymidine, a precursor of DNA, phage DNA labeled with tritium (H[3]) can be obtained. If P[32] is also included in the medium, highly labeled "messenger" RNA can be obtained. The H[3]-labeled DNA and the P[32]-labeled RNA can be isolated and then separated from each other by means of a cesium chloride gradient as shown in Fig. 14-9A. If, however, the solution containing the labeled DNA and RNA is given the Doty treatment and the DNA and RNA are centrifuged in the gradient, there is a different result, that shown in Fig. 14-9B. Some of the newly labeled RNA, denoted by its P[32] label, moves over into the H[3]-labeled DNA region. A complex has thus been formed, which behaves as a single sedimenting unit in the density-gradient system. This complex formation is specific; if other DNA's—from other bacteria, from *E. coli* itself, from other phage— are treated under the same experimental conditions with the P[32]-labeled RNA, the result is that shown in Fig. 14-9A, not that in Fig. 14-9B.

In the light of all that has been said about the structure of DNA and RNA, the only interpretation that can be made at present as to why a DNA-RNA hybrid is obtained in the above experiment is that along a large portion of the DNA and RNA chains there must be regions where the bases are arranged in such a sequence that the DNA bases are complementary to those of RNA. In other words, a real complementary copy of the RNA has been synthesized in the *E. coli* phage-infected cell, which mirrors a good

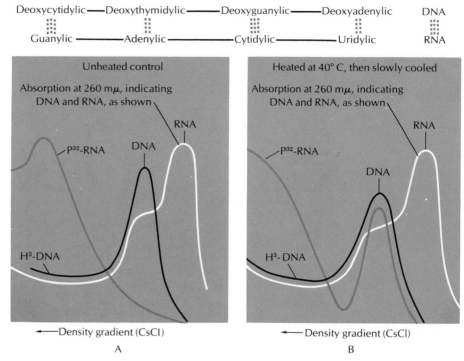

Deoxycytidylic——Deoxythymidylic——Deoxyguanylic——Deoxyadenylic   DNA

Guanylic————Adenylic————Cytidylic————Uridylic   RNA

Fig. 14-9. Formation of complementary helices between DNA and "messenger" RNA (P³²-RNA). Figure is explained in the text. (Data from B. Hall and S. Spiegelman.)

deal of the DNA. Finally, Spiegelman's laboratory has shown, by similar techniques, that a complex of DNA and RNA can be extracted from the phage-infected cell, this complex containing the labeled RNA and hence the presumed messenger RNA.

Recently, the technique of complexing regions of DNA with regions of RNA, called "hybridization," has been greatly improved by Spiegelman. Firstly, it is clear that the DNA has to be single-stranded, for only denatured DNA (single-stranded) and not native DNA (double-stranded) will work. Secondly, the DNA is "immobilized" by adsorbing it on nitrocellulose filters. The radioactive RNA whose hybridization property is to be tested is then poured through the filter; some of it will hold onto the filter. It cannot be washed off, and indeed this holding back of RNA by the filter is dependent on the DNA immobilized on the filter. Since the specific radioactivity of the RNA was known, the amount of radioactive RNA held onto the DNA-impregnated filter is a measure of the length of RNA bound to the length of DNA on the filter. If the length of the DNA is also known (from the molecular weight), a measure can be obtained of the length of DNA which has sequences complementary to those of the bonded RNA as indicated in Fig. 14-8. Control experiments can be performed to cut down the "noise level" of the experiment: how much RNA is bound to the filter in the absence of DNA. Since the RNA in the DNA-RNA complex is resistant to the action of

RNase, RNase treatment of the complex will get rid of nonspecifically adsorbed RNA.

This hybridization between DNA segments and complementary RNA segments is very specific. This specificity has been used, as mentioned above, to determine that transfer RNA has base sequences complementary to nuclear DNA; hence, tRNA is probably synthesized in the nucleus. This technique has also been widely used to test for homologies between two RNA molecules. For example, a radioactive RNA species known to be base-paired to regions of DNA is bonded to the immobilized DNA. In another tube, a mixture containing this radioactive RNA and a nonradioactive RNA species to be tested is passed through the DNA filter. If the unknown nonradioactive RNA has base sequences similar to that of the radioactive RNA, it will compete with the latter for the same segment of the DNA, and hence the amount of radioactive RNA bound to the DNA will have decreased. If the two RNA species have no base sequences in common, then no matter how much nonradioactive RNA is added, there will be no effect of it on the amount of radioactive RNA hybridized to the DNA. In this way it has been shown, for example, that purified tRNA's have no similar sequences, no homologies, with either of the two ribosomal RNA's. Thus, even though our chemistry is not yet refined enough to determine base sequences of such large molecules as ribosomal RNA's, we can still learn something about them by the use of the hybridization technique. Radioactive polynucleotides, containing known sequences of 10 to 20 nucleotides, can be chemically synthesized, and if mixed with ribosomal RNA's and passed through DNA filters, then by a measure of dilution of radioactivity we can determine whether ribosomal RNA's contain base sequences similar to those of the chemically synthesized polynucleotides.

(See Chapters 8 and 15.)

Altogether, this technique is being used as a measure of how much of the DNA is being used as a template for the synthesis of certain RNA species; ribosomal RNA's, for example, seem to be synthesized off a DNA segment that occupies just a few tenths of a percent of the whole DNA length. This technique can also be used as an indicator of where certain RNA molecules are being made. As an example we can ask the question whether the ribosomes that are found in mitochondria and chloroplasts have RNA in them that is synthesized off a nuclear or a mitochondrial or chloroplast DNA. We can partially solve this problem, not by trying to isolate the enzymes responsible for synthesizing RNA off a certain DNA template, but simply by obtaining radioactive RNA from these organelles and observing whether it forms a hybrid specifically with one of the DNA's mentioned above; it does so with the organelle DNA. Finally, the technique is being used in trying to find homologies between the DNA's obtained from different species and genera in order to construct what may be called a microevolutionary scale. This can be done by immobilizing some standard DNA onto the filter, hybridizing this with a radioactive DNA from another source, and then using the dilution technique mentioned above, to observe the effects of adding nonradioactive DNA's from various sources on the binding of the radioactive DNA. Already, using this idea, it has been found that probably all the cells of a multicellular organism have very similar DNA molecules. The uncanny specificity of this technique is overshadowed only by its beauty.

We suspect that only single-stranded DNA acts as a template for messenger RNA. Thus, it is clear that the double helix of the DNA must unwind before this RNA synthesis can occur. While *in vitro*, both DNA strands are copied by the RNA polymerase enzyme, *in vivo*, it is becoming clear that only one of the DNA strands is involved in RNA synthesis. This is the strand on which the enzyme begins to work at its 3'-end (Fig. 14-2); the RNA is synthesized beginning at its 5'-end. Firstly, on theoretical grounds based on our present knowledge of the amino acid code, it is difficult to assume that both an RNA chain and its complement (Chapter 8) serve as messenger for the same protein. This is because the two DNA chains differ in the order of the nucleotides; hence, the two RNA chains should differ in this order. We know, however, that there is only one order involved in protein synthesis. Secondly, it has been recently found possible to separate the two complementary chains of DNA (in this case it was a phage DNA), and it was shown by techniques similar to the one described in Figs. 14-7 and 14-8 that the newly synthesized RNA that resulted from the mediation of phage DNA was only capable of interacting with one of the DNA strands; hence, in this phage-infected cell it was clear that only one of the phage DNA strands was used in the synthesis of the resultant RNA.

(See Chapter 15.)

(See Chapter 15.)

It should be stressed that not all the DNA is used as a template for the synthesis of messenger RNA, for it is now clear that both kinds of ribosomal RNA (small and large) and transfer RNA can all be synthesized by an RNA polymerase using only a very small portion of the nucleotide sequences in the DNA chain. To summarize, it appears at present that all the cellular RNA, messenger, ribosomal, and transfer, is synthesized in the nucleus by an RNA polymerase working off one of the DNA strands of the double helix. However, it is not certain if these RNA molecules are finished or if they have to be completed in the cytoplasm. The purpose of the other DNA strand is not known. Also, the means whereby the cell controls this type of synthesis is not known. Again, where this messenger RNA synthesis takes place in the nucleus — whether in the nucleolus border where there is chromatin material or in the chromatin material of the rest of the nucleoplasm — is not certainly known, but it would appear that ribosomal RNA is synthesized off that DNA lying adjacent to the nucleolus whereas the bulk of the RNA (messenger) is synthesized off other parts of the chromatin. How this RNA acts in the synthesis of proteins will be the subject of the next chapter.

In summary, the main biochemical function of the nucleus is twofold. It acts as a storehouse in the cell for the genetic information, which, upon cell division, is duplicated exactly for the daughter cells. It also acts as a storehouse for the information that during the lifetime of the cell, informs the enzymatic mechanisms in the cytoplasm exactly which, and how many of, the protein enzymes the cell should synthesize. All the metabolism of the nucleus seems to be geared to these functions. Where they take place in the nucleus, we do not know. We are not sure of the biochemical "meaning" of the nucleolus, of the nucleolar-chromatin apparatus. All we can surmise is that, in tune with the rigid exactness of DNA duplication and RNA synthesis, there must be a structural framework that is just as exact and upon which the biochemical mechanisms must rest.

SUGGESTED READING LIST

Offprints

Beerman, W., and Clever, U., 1964. "Chromosome Puffs." *Scientific American* offprints. San Francisco: W. H. Freeman & Co.

Gibor, A., 1966. Acetabularia: A Useful Giant Cell." *Scientific American* offprints. San Francisco: W. H. Freeman & Co.

Hayashi, M., Spiegelman, S., Franklin, N. C., and Luria, S. E., 1963. "Separation of the RNA Message Transcribed in Response to a Specific Inducer." *PNAS*, 729–736. Bobbs-Merrill Reprint Series. Indianapolis: Howard W. Sams & Co.

Horowitz, N. H., 1956. "The Gene." *Scientific American* offprints. San Francisco: W. H. Freeman & Co.

Hurwitz, J., and Furth, J. J., 1962. "Messenger RNA." *Scientific American* offprints. San Francisco: W. H. Freeman & Co.

Kornberg, Arthur, 1960. "Biologic Synthesis of Deoxyribonucleic Acid." *Science*, pp. 1503–1508. Bobbs-Merrill Reprint Series. Indianapolis: Howard W. Sams & Co.

Kornberg, A., 1968. "The Synthesis of DNA." *Scientific American* offprints. San Francisco: W. H. Freeman & Co.

Taylor, J. H., 1958. "The Duplication of Chromosomes." *Scientific American* offprints. San Francisco: W. H. Freeman & Co.

Taylor, J. H., Woods, P. S., and Hughes, W. L., 1957. "The Organization and Duplication of Chromosomes as Revealed by Autoradiographic Studies using Tritium-labeled Thymidine." *PNAS*, pp. 122–128. Bobbs-Merrill Reprint Series. Indianapolis: Howard W. Sams & Co.

Articles, Chapters, and Reviews

Allfrey, V., 1960. "Isolation of Subcellular Components." In: *The Cell*, Vol. I, p. 193. J. Brachet, and A. E. Mirsky, eds. New York: Academic Press, Inc.

Briggs, R., and King, T. J., 1960. "Nucleocytoplasmic Interactions in Eggs and Embryos." In: *The Cell*, Vol. I, p. 537. J. Brachet, and A. E. Mirsky, eds. New York: Academic Press, Inc.

Crick, F. H. C., 1966. "The Genetic Code—Yesterday, Today and Tomorrow." In: *Cold Spring Harbor Symposia on Quantitative Biology, 31.* Cold Spring Harbor, L. I., New York.

Mazia, D., 1961. "Mitosis and the Physiology of Cell Division." In: *The Cell*, Vol. 3, p. 77. J. Brachet, and A. E. Mirsky, eds. New York: Academic Press, Inc.

Mirsky, A. E., and Osawa, S., 1961. "The Interphase Nucleus." In: *The Cell*, Vol. 2, p. 677. A. E. Mirsky, and J. Brachet, eds. New York: Academic Press, Inc.

Prescott, D. M., 1960. "Nuclear Function and Nuclear-cytoplasmic Interaction." *Annual Review of Physiology, 22,* p. 17.

Stern, H., 1962. "Function and Reproduction of Chromosomes." *Physiological Reviews, 42,* No. 2.

Swann, M. M., 1957. "Control of Cell Division." *Cancer Research 17*, p. 727.

Books

Cold Spring Harbor Symposia on Quantitative Biology, 31, 1966. *The Genetic Code.* Cold Spring Harbor, L. I., New York.

Ingram, W., 1965. *The Biosynthesis of Macromolecules.* New York: W. A. Benjamin Inc.

Kornberg, A., 1962. *Enzymatic Synthesis of DNA.* New York: John Wiley & Sons, Inc.

McElroy, W. D., and Glass, B. (eds.). 1957. *Chemical Basis of Heredity.* Baltimore: The Johns Hopkins Press.

Petermann, M., 1964. *Physical and Chemical Properties of Ribosomes.* New York: Elsevier Publishing Co.

The Ribosome and the Translation of Information

About 15 years ago biochemists began the use of radioactive amino acids in the study of the processes involved in protein synthesis. Up to this time there was thought to be very little protein synthesis and breakdown in the cells of the adult, nongrowing organism. It was assumed that once the cellular proteins were formed, they lasted the lifetime of the cell. There is, however, a constant excretion of ammonia and urea in the urine, and since these nitrogen-containing compounds could have come only from the catabolism of proteins, it was assumed that this steady excretion represented that small proportion of cells that were breaking down, their proteins being hydrolyzed, and the resultant amino acids deaminated to form urea and ammonia. This hypothesis of the relative metabolic stability of proteins was known as the "wear and tear" theory, that is, the only proteins being metabolized were those resulting from the breakup of dead and dying cells. It was thus a distinct surprise to find that if one injected radioactive amino acids into an animal and then isolated proteins from the various tissues, these proteins were found to be radioactive. The radioactive amino acids had become incorporated into the protein molecules of the cell, proving that there is synthesis of proteins in the cells of the adult organism.

This finding fitted in with the concept of the "dynamic state of body constituents," introduced by the biochemist Schoenheimer, to explain earlier findings on fat metabolism. According to this view, all the large compounds of the cell—not only proteins, but carbohydrates, fats, and nucleic acids—are constantly being broken down and re-synthesized in the cells of a nongrowing organism. The present view is somewhat in between these two extremes; namely, that certainly there is some protein synthesis going on in all cells, but that most of it is due to the synthesis of protein for new cell formation and the synthesis of proteins for export from the cell. For example, when

371

radioactive amino acids are injected into the animal and the proteins isolated from the various tissues, it is found that the tissues most active in protein synthesis are those tissues whose cells make protein for export purposes. Liver cells make most of the blood proteins; pancreas cells make most of the proteolytic enzymes destined for the gut; intestinal mucosal cells make digestive enzymes and some protein hormones; and some endrocine glands make protein hormones for secretion into the blood. All these tissues are those that were found to contain the most radioactive proteins. On the other hand, tissues like skin and muscle were found to contain very little radioactivity in their proteins. In fact it was very clear that the second largest class of proteins in the body, the muscle proteins, exhibit very little breakdown and synthesis, or, as it has been called, protein turnover. In most tissues, however, there was turnover, not for secretion nor for daughter cell protein, but simply a breakdown and resynthesis of a major part of the cell's own protein. What this represents is not certain at the present time.

Cytological Basis of Protein Synthesis

We can go further and delineate in the following way those cellular structures that are functioning in protein synthesis. The liver homogenate can be fractionated by differential centrifugation into the morphological components of the liver cell. Thus, the nuclear and whole cell fraction and the mitochondrial fraction are obtained. After the heavy nuclei and mitochondria have been spun out from the liver homogenate, the supernatant remaining from the mitochondrial centrifugation is spun at high speed; in this manner another pellet is obtained at the bottom of the tube. This pellet is called the "microsome" or small particle fraction. When this pellet is analyzed morphologically and chemically, it is found to contain most of the RNA and phospholipid of the cell and to be composed of fragments of the endoplasmic reticulum. Its importance to the economy of the cell was realized through the discovery that most protein synthesis of the cell took place in this fraction. After the animal is injected with radioactive amino acids, the liver is extirpated and fractionated into the nuclear and whole cell fraction, the mitochondrial fraction, the microsome fraction, and the supernatant fraction. The proteins from these different fractions are obtained, and their radioactivity is measured. Figure 15-1 shows, from a very early experiment, that the microsome fraction is the one with the highest specific radioactivity, that is, radioactive counts per minute per milligram of protein.

But what is this endoplasmic reticulum? For years cytologists had noticed in the cytoplasm of most cells a fine network or mesh, too diffuse and small to be made out clearly with the light microscope. When the electron microscope began to be used for biological specimens, it was quickly found that this network could be resolved into a number of membranes enclosing vesicular, or tubular, or flattened spaces in the cell and that these spaces were probably interconnected with each other. In some cells such as secretory or glandular cells there is an immense profusion of these membrane-limited spaces; in other cells such as muscle cells they are diminished in number; in still others such as bacteria they are nonexistent. Furthermore, in some cases the mem-

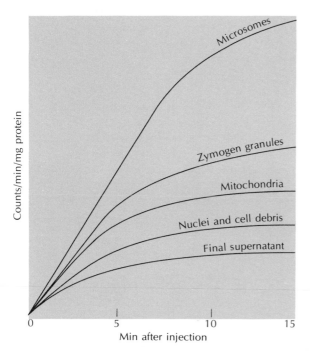

Fig. 15-1. *In vivo* incorporation of radioactive amino acids into proteins of various morphological fractions of pancreas tissue in guinea pig.

branes are lined with tiny particles whereas in other instances the membranes are bare. In general, we can describe the extramitochondrial part of the cytoplasm in terms of these structures as follows: (1) cells with membranes free of particles; (2) cells with membranes all of which have particles on their surfaces; (3) cells with membranes with and without particles; (4) cells with no membranes at all, just particles; and (5) in a few cases cells with a very few membranes and very few particles. In other words, the whole gamut of possible combinations is found.

Actually, "microsome" is a not very descriptive term, for it only describes what results from a method, that of spinning at high speed the supernatant from the mitochondrial fraction. Thus, if a cell has membranes with particles in its cytoplasm, then the microsome fraction is found to consist of the fragments of these membranes, still with particles on them. If the cell has only particles, then the microsome fraction consists of only particles; if only membranes, then the microsome fraction is a membranous fraction; and so on. For example, in pancreatic acinar cells the cytoplasm consists of membrane-bound vesicles and spaces, most of these membranes having particles on their surfaces but some being bare as shown in Fig. 15-2. The microsome fraction from pancreas is therefore composed of fragments of these membrane-bound vesicles, some with particles on their surfaces (Fig. 15-2). In the liver half these membranes have particles; thus, the liver microsome fraction is markedly heterogeneous morphologically, being made up of membranes, of materials enclosed by membranes, and of particles on these membranes. It is also heterogeneous chemically, for most, if

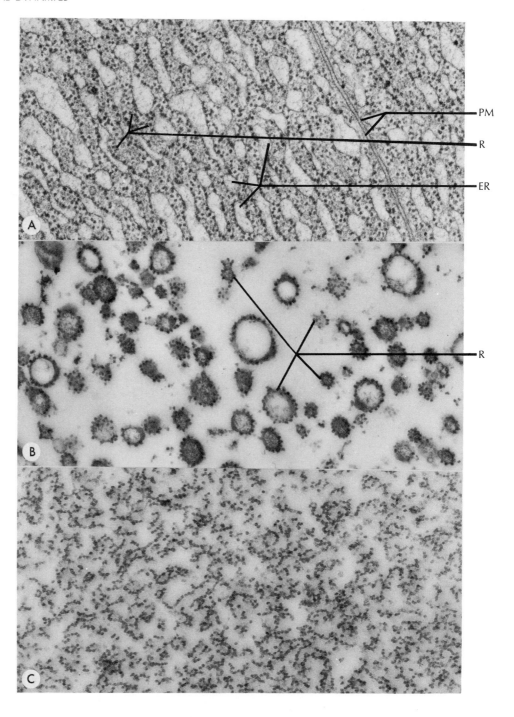

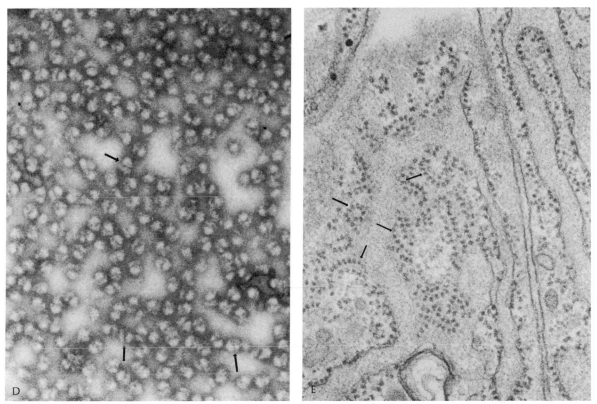

Fig. 15-2.   A. The endoplasmic reticulum in a pancreas acinar cell (×46,000). R-ribosomes; ER-endoplasmic reticulum; PM-plasma membrane. B. Isolated microsomes from a pancreatic acinar cell (×39,000). R-ribosomes. C. Ribosomes isolated from pancreatic acinar cell microsomes (×52,000). (A, B, and C courtesy of G. E. Palade, Rockefeller University.) D. Ribosomes isolated from *E. coli* and negatively stained by phosphotungstate. The dark groove (arrows) between the larger (50 *S*) and smaller (30 *S*) subribosomal particles can be clearly seen (×215,000). (Courtesy of H. Huxley, Cambridge University.) E. Endoplasmic reticulum in a plasma cell (×100,000) showing polysome clusters on membranes (arrows). (Courtesy of G. E. Palade, Rockefeller University.)

not all, of the phospholipid is found in the membranes whereas most of the RNA is found in the particles. It is heterogeneous biochemically because the membranes have various functions, whereas the particles are intimately involved in protein synthesis. These particles, composed of RNA and protein, have been named "ribosomes"; their significance for our story is that they seem to be the intracellular structures that are involved in the synthesis of cellular proteins.

It seems strange that 15 years ago no biochemist had a really good idea as to the function of RNA. The existence of these macromolecules had been known for a long time, and it was even known that RNA was present in both nucleus and cytoplasm but mostly in the latter. For many years it has been observed that certain cells, indeed only certain parts of these cells, are responsive to being stained with basophilic dyes; these cells are said to be basophilic because in them are concentrated large amounts of macromolecular acidic compounds that combine with basic dyes. We now know that it is the RNA in these cells that does the combining with these dyes. It was further observed that cells which were basophilic are exactly those which are known to be active in protein synthesis. This observation by Brachet was the first glimmer that RNA is involved in protein synthesis. Furthermore, it is the particles in the cell that are responsible for the basophilia, for it is the RNA of these ribosomes that combines with the basic dyes. Thus, those basophilic cells that are highly active in protein synthesis are just those cells that contain a large number of particles, some attached to membranes and some lying freely in the cytoplasm.

Protein Synthesis *in Vitro*

Next, it was found that this incorporation of radioactive amino acids into protein could take place in the test tube. A homogenate can be made of an active tissue such as liver, and if the appropriate factors are incubated with this homogenate, the added radioactive amino acids are found to be incorporated into liver proteins. There actually is a net breakdown of protein during this incubation, but the presence of the radioactive amino acids makes it possible to detect that there is a small amount of protein synthesis; it is so small that only by the very sensitive radioactivity method could it be detected in the midst of the much larger breakdown. The most necessary additions were found to be oxidizable substrate and cofactors for the oxidation. In other words, the conditions that were found necessary for the incorporation of radioactive amino acids into protein were just those required for oxidative phosphorylation for synthesis of ATP (see Table 15-1). This finding was important because for years it had been thought that the mechanism for protein synthesis and protein degradation was one and the same, that protein synthesis was just the reversal of protein degradation. The main reason why this was thought to be the case has to do with the need for the cell to synthesize not just any protein, but to make specific proteins. We also know of the existence of many proteolytic enzymes that are very specific in hydrolyzing peptide bonds only between certain amino acids. These proteolytic enzymes, like all enzymes, could theoretically catalyze a reversible reaction; thus, hypothetically, they could not only degrade specific proteins to amino acids but also synthesize them from amino acids. But we now know that protein synthesis is not the reversal of protein breakdown; that, whereas it does not take energy to hydrolyze a protein, it does take energy to build one up.

The homogenate that had been incubated with substrate, cofactors, and radioactive amino acid can, after various times of incubation, be fractionated into its different subcellular fractions. Again, the proteins of the microsome fraction are found to be the most radioactive of all the cell fractions (Fig. 15-3); since the picture

Table 15-1. ABILITY OF VARIOUS ISOLATED MOR-
PHOLOGICAL LIVER CELL FRACTIONS TO
OXIDIZE SUBSTRATE, FORM ATP, AND
INCORPORATE RADIOACTIVE AMINO
ACIDS INTO THEIR PROTEINS*

	OXYGEN CONSUMPTION	ATP FORMED	COUNTS/MIN/ MG PROTEIN
Homogenate	37.4	4.0	10.8
Mitochondria	8.2	4.2	1.3
Microsomes	0.4	0.0	1.1
Supernatant	0.8	0.0	0.4
Mitochondria plus microsomes	14.0	4.1	10.2
Mitochondria plus supernatant	9.7	4.1	1.5
Mitochondria plus microsomes plus supernatant	18.8	3.8	4.3
Mitochondria plus boiled microsomes	9.7	4.4	1.2

* (Data from Siekevitz.)

in this *in vitro* experiment is the same as in the *in vivo* experiment (Fig. 15-1), it seems
certain that experiments with the microsome fraction *in vitro* are parallels of what
happens inside the cell.

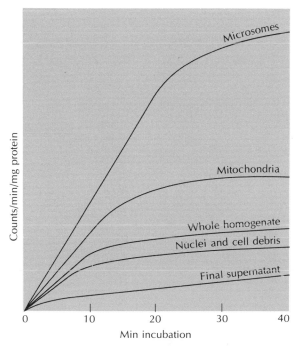

Fig. 15-3. *In vitro* incorporation of radioactive amino
acids into proteins of various morphological fractions of
rat liver homogenate. (Data after P. Siekevitz.)

Going further, incorporation of radioactive amino acids into protein can be obtained if there are mitochondria oxidizing substrate in the presence of the necessary phosphorylation factors and if there are microsomes containing the ribosomes, plus some other factors, in the supernatant. This is shown in the early work of Table 15-1 where it can be seen that both mitochondria, to supply the energy in the form of ATP, and microsomes, to supply the protein-synthesizing machinery, are necessary for the incorporation to take place; either alone will not work. Later, modifications were made upon this basic system. Instead of mitochondria, ADP, phosphoenolpyruvate, and the appropriate purified kinase enzyme can be added to make ATP. Instead of whole microsomes, a pure preparation of ribosomal particles obtained from the fragmenting of microsomes can be used. In some cases, as with bacteria or reticulocyte cells, a reasonably pure preparation of ribosomes can be obtained by homogenizing the cells and differentially centrifugating the suspension. Instead of obtaining amino acid radioactivity in a mixture of degraded proteins, we can isolate, as in the case of the reticulocyte cell, only one radioactive protein, hemoglobin, for these cells synthesize mostly this one oxygen-carrying heme protein. Thus, as can be seen in Table 15-2, from isolated ribosomes, an energy source like phosphoenolpyruvate and pyruvic kinase, amino acids, and various other cofactors such as those contained in the "pH 5" fraction, we can make protein—in some cases, a specific protein.

Table 15-2. REQUIREMENTS OF ISOLATED RIBOSOMES FOR THE INCORPORATION OF RADIO-ACTIVE AMINO ACID INTO PROTEIN*

SYSTEM	CPM/MG PROTEIN
Complete	66
Minus phosphoenolpyruvate and pyruvic kinase	7
Minus GTP	3
Minus ATP	3
Minus "pH 5" fraction	6

* (Data from Kirsch, Siekevitz, and Palade.) The complete system consists of liver ribosomes, ATP, GTP, $MgCl_2$, the "pH 5" fraction (which contains the activating enzyme, the transfer enzyme, probably other protein factors, and transfer RNA), an ATP-generating system (phosphoenolpyruvate, ADP, and pyruvic kinase), and radioactive leucine.

Nature of Ribosomes

Because of the newly found importance of the ribosomes, much work has been done with them lately. These particles are ubiquitous elements found in all cells that synthesize protein, from bacteria to cells in the organs of mammals (see Fig. 15-2). As already mentioned, they are sometimes bound to membranous elements in the cell and in other cases lie freely in the cytoplasm. Their disposition in the cell seems to make no difference in their function; bacterial ribosomes, seemingly existing freely in the cell, as do ribosomes of reticulocytes, make protein as well as do ribosomes from other cell types. There is some evidence now, though, that ribosomes attached to membranes in mammalian and plant cells are more active in protein synthesis than those not so attached; even in bacteria, the ribosomes associated in some way with the outer bacterial membrane are more active than those free in the cytoplasm. The size of the ribosomes varies somewhat, being ~150 A in bacteria and chloroplasts and 200 A in the cytoplasm of plant and animal cells.

Their constitution is about the same in all cases; they contain from 40 to 60 percent RNA, the rest being practically all protein. The proteins of the ribosomes seem to be basic in nature, of about 20,000 mol wt; there seem to be many different kinds of proteins making up the ribosomal structure. The RNA of the ribosomes has a very large molecular weight, estimated to be about 2,000,000 in various types of particles. As far as is known, further specifics about this RNA have been noted in Chapter 8. The involvement of this ribosomal RNA and indeed of ribosomal proteins in specific protein synthesis is at present not clear at all. They may act as somewhat specific "cementing" substances in the ribosome, or they may act, in conjunction with messenger RNA (see below), in a more specific metabolic role in the synthesis of proteins. The lifetime of ribosomal RNA in the cell is longer than that of "messenger" RNA or that of transfer RNA (see below); thus, like DNA, it seems to be rather metabolically stable, but still less stable than is the DNA. The forces that hold the RNA to the ribosomal protein are either electrostatic in nature, forming salt bonds between the phosphate groups of the RNA and the amino groups of the basic amino acids in the protein, or are bonds involving magnesium complexing between the same groups; they are most likely a combination of both. The RNA can be removed from the protein by agents that break just such bonds, as strong salt solutions, or magnesium-complexing reagents. The proteins are thought to be on the inside of the ribosome, almost entirely covered by the RNA.

The ribosomal structure and appearance are strongly dependent on the presence and amount of magnesium. In the absence of magnesium in the isolation medium, the larger particles fall apart into a family of smaller particles, but mostly two smaller particles, each of these having the same RNA-to-protein ratio as do the original particles. (See Chapter 8.) Bacterial ribosomes, and ribosomes from chloroplasts and mitochondria, probably exist in the cell, except for the qualification given below, as 70 S particles (S = Svedberg unit, Chapter 9), but they can be broken down mostly to 50 S and 30 S particles (see Fig. 15-2D and Table 15-3). Ribosomes from the cytoplasmic matrix of eucaryotic cells exist probably as 80 S particles, and these can be split to 60 S and 40 S particles (Table 15-3). Each of the smaller subunit particles has a distinctive RNA. Thus, the 50 S and 60 S particles have a 23 S and a 28 S RNA respectively and also a newly discovered 5 S RNA of unknown function; while the 30 S and the 40 S particles have (See Fig. 8-24.) associated with them respectively a 16 S and an 18 S RNA (see Chapter 8). Another difference between the 70 S and 80 S ribosomes is that protein synthesis conducted by the former is inhibited by the antibiotic, chloramphenicol, whereas that of the latter is inhibited by the antibiotic, cycloheximide; the reverse does not hold. The larger subunit contains about 30 proteins, all different from each other, while the smaller subunit contains about 20 different proteins; the functions of these are entirely unknown.

Table 15-3. DIMENSIONS OF RIBOSOMES, SUBUNITS, AND rRNA

SOURCE		RIBOSOME	LARGE SUBUNIT	SMALL SUBUNIT	RNA FROM LARGE SUBUNIT	RNA FROM SMALL SUBUNIT
Procaryotes, and probably chloro-plasts and mito-chondria	daltons	2.7×10^6	1.8×10^6	0.9×10^6	1.2×10^6	0.6×10^6
	S values	70	50	30	5 and 23	16
Eucaryotes (in cytoplasmic matrix)	daltons	$4.5–5.0 \times 10^6$	$3.0–3.6 \times 10^6$	$1.5–1.8 \times 10^6$	$1.5–1.8 \times 10^6$	$0.8–1.0 \times 10^6$
	S values	75–80	50–60	32–40	5 and 28	18

It is not known exactly how the smaller particles come together to make up the larger one, but again magnesium complexing is thought to be the chief factor involved. By themselves, the smaller particles are not active in protein synthesis; it is only the 70 *S* and the 80 *S* particles that are active. The reason why the cell finds it necessary to have two particles brought together to form the active protein-synthesizing unit is not known. However, we do know of some morphological and biochemical differences between the larger and smaller subunits of the active particles. For example, in those cells where the particles exist on the membrane, it seems to be the larger particle that is bound to the membrane, the smaller subunit being then bound to the larger. Further-more, if we allow cells to synthesize radioactive protein and then we isolate the ribo-somes, we find that not only is the newly synthesized radioactive protein bound to the ribosomes, but it seems to be bound specifically to the larger of the two subunits. Of the other kinds of RNA's already mentioned, messenger and transfer RNA's, more will be explained below concerning their biochemical involvement in protein syn-thesis, but their morphological involvement seems to be such that the messenger RNA is specifically held by the smaller subunit while the transfer RNA seems to be specifically held by the larger subunit. More specifically, a portion of the tRNA mole-cule is bound to the larger subunit, but a portion is bound to the mRNA on the smaller subunit, the latter binding being between the "codon" and "anticodon" sites on the mRNA and tRNA, respectively (see below). Indeed, there seem to be two transfer RNA's bound to the ribosome, one at one "slot," carrying the lengthening polypeptide chain, and one at another, adjacent "slot," carrying the amino acid next in line. Figure 15-4 shows all these structural features of the ribosomes as we know them at present and the mechanical features of the structure as we visualize them at present. The biochemical meaning of this relationship will become clearer later on, but the morphological meaning may be that this structure provides a rigid framework of exact dimensions upon which a complicated process can take place.

There has been much recent work leading to the idea that the 70 *S* or 80 *S* particles are active when they are attached together in rows or in clusters. These strings, called polysomes, can be seen in the cell, mostly attached to membranes (Fig. 15-2A); they have even been isolated from many cell types, separated from the 70 *S* or 80 *S* particles

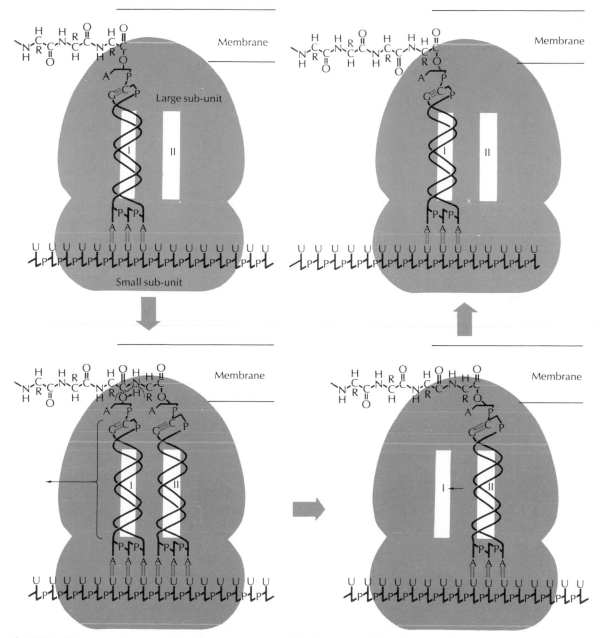

Fig. 15-4. Diagrammatic presentation of current concept of the functioning of the ribosome. The diagram indicates the attachment of the small subunit to the large subunit, with the latter being attached to the membrane. The messenger RNA, in this case polyuridylic acid, is attached to the small subunit. The growing polypeptide chain, in this case polyphenylalanine, is attached by its terminal phenylalanyl tRNA to site I on the large subunit; the anticodon of the tRNA (AAA) is hydrogen-bonded to the codon (UUU) of the messenger. The incoming next amino acid, in the form of phenylalanyl tRNA becomes attached at site II. A peptide bond is enzymatically formed between the phenylalanines, and the free tRNA is released from this site, leaving the polyphenylalanyl tRNA, lengthened by one more phenylalanine, attached by means of its tRNA to site II. This entire complex moves over to site I, leaving site II free and ready for the initiation of another cycle of bonding and polymerization. Fuller details are in the text. (Modified from F. Lipmann.)

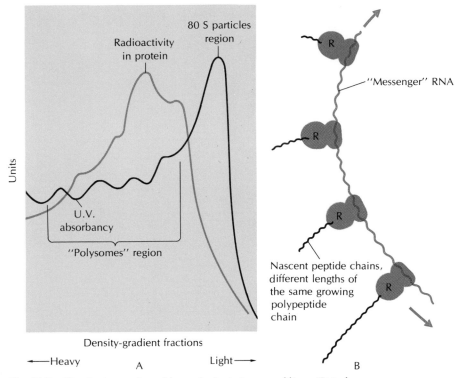

Fig. 15-5. The "polysome" — evidence for it. A. In case of liver. (Data from P. Siekevitz.) B. The concept itself. (Based on ideas of J. Watson, W. Gilbert, A. Rich, and H. Noll.)

by density-gradient centrifugation (Fig. 15-5A). If ribosomes are made to engage in radioactive protein synthesis, either *in vitro* or *in vivo,* and then separated by density-gradient centrifugation, it can be seen from Fig. 15-5A that most of the radioactive protein is not associated with the single 80 *S* particles, but with heavier structures, which because of their size have sedimented faster during the centrifugation. These latter structures are clusters, or even chains, of ribosomes (Fig. 15-2E). What holds them together is not completely certain; it could be the newly synthesized proteins themselves; it could be some bonds in the ribosomal protein or RNA; but the presently accepted theory is that it is the messenger RNA that binds the ribosomes, as depicted in Fig. 15-5B, the number of ribosomes depending on the length of the messenger RNA. Thus, the reason that preparations consisting mostly of single ribosomes are not very active in protein synthesis is probably that when they are isolated, they contain very little bound messenger RNA. It has been postulated that a single specific messenger RNA contains the "code" (see below) for a single specific protein. A number of ribosomes bind onto the messenger RNA through some attachment sites on the smaller subunit, and each of these ribosomes, in conjunction with the messenger and with the transfer RNA's, begins to synthesize that messenger-specific protein. It is even thought that there is a relative movement between messenger RNA and ribosomes, either the ribosomes moving along the messenger (though this is not likely in view of the fact that in many cells the ribosomes seem to be anchored to membranes), or else the messenger moving, like a coded tape, among the ribosomes. As this movement goes on, the ribosomes attach a specific amino acid, in turn and in order, to an existing polypeptide fragment. Thus, according to this theory, each messenger "serves" many ribosomes, and each ribosome synthesizes one protein chain. As the ribosome comes to the end of the messenger RNA, it "falls off," and the newly synthesized protein somehow becomes discharged from the particle. In the case of those cells that make secretory proteins, the proteins are released across the membranes of the endoplasmic reticulum upon which sit the ribosomes. As can be noted, this extremely mechanical model has many hiatuses, particularly concerning the biochemical mechanisms involved. Although it may be wrong partly or completely, it is a good working model, and as such, is receiving much experimental attention. It should be noted that were this model approximately correct, it would be a good example of an intermediate stage in the chemical architecture of the cell.

Mechanism of Protein Synthesis

What are the factors necessary for protein synthesis to occur? To answer this we must go into some detail of what is known about the mechanism of protein synthesis. A few years ago Lipmann presented the idea that, in preparation for their role in protein synthesis, the different amino acids must be brought into a "high-energy" state. It was further hypothesized that this could be done only through the mediation of ATP. Looking for just such a reaction, M. Hoagland discovered the amino acid activating enzymes. ATP reacts with an individual amino acid, in the presence of this enzyme, to split off inorganic pyrophosphate and to form an amino acyl adenylate compound (Fig. 15-6). This compound remains tightly bound to the enzyme. There are specific enzymes catalyzing the reaction for each individual amino acid. The high-energy character of the amino acyl adenylate compound is indicated by the observation that when the enzyme is incubated with the amino acid, ATP, and the radioactive inorganic pyrophosphate, there is radioactivity in the ATP that can be isolated. Thus, the reaction is easily reversible, showing that the energy content of the acyl bond in the amino acyl adenylate is approximately the same as that of the penultimate pyrophosphate bond in the ATP. The amino acid has become "activated."

The next step in protein synthesis was actually hypothesized by Crick before it was discovered. The reasoning was that if the ribosomes are involved in protein synthesis by containing informational or template RNA, there should be a carrier of the activated amino acid to the ribosomes, this carrier operating in such a way as to be able to "recognize" the RNA of the ribosomes. The RNA of the ribosomes was at that time postulated to have the information to make a specific protein — and what other

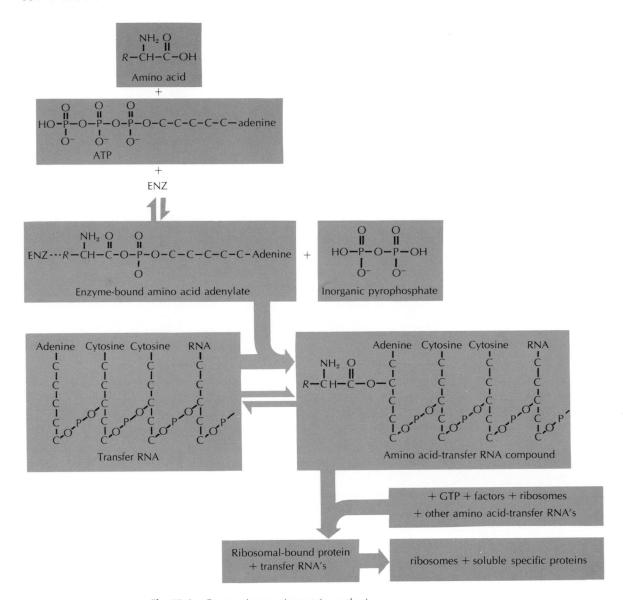

Fig. 15-6. Enzymatic steps in protein synthesis.

molecule except another molecule of RNA was better suited for this role? The next step (Fig. 15-6) is the transfer of the amino acid from its acyl binding onto the adenylate moiety at the end of a certain type of RNA molecule, the so-called soluble, or transfer, RNA. This step is catalyzed by the same specific amino acid activating enzyme, and in the process, AMP is liberated from the amino acyl adenylate. There are specific RNA molecules for each of the amino acids, just as there are specific activating enzymes. Thus we can say that for each amino acid there is a specific activating enzyme and a specific transfer RNA molecule. Each of these transfer RNA molecules is much smaller than the RNA of the ribosomes, but it still is a relatively large molecule, being about 23,000 molecular weight and having from 70 to 80 nucleotides in its chain. The last three nucleotides (—cytidylic—cytidylic—adenylic) are the same in each of the specific amino acid transfer RNA's. In the transfer from amino acyl adenylate, the amino acid hooks up with the ribose moiety of the terminal adenylic acid, to form an ester bond. Again the reaction seems to be reversible; hence we are left to suppose that this ester bond is of a high-energy nature. This is the only known instance in which an ester bond is in equilibrium with a pyrophosphate bond of ATP; the reason why this is the case is not known at present. If, however, all the three terminal nucleotides of the different transfer RNA's are the same, then to account for the specificity of the RNA we must assume that in the rest of the RNA chain there are certain arrangements of the nucleotides that make for the amino acid specificity. And indeed, experiments that show exactly this point have recently been performed by Lipmann's laboratory.

(See Chapter 8.)

The experiment bears a telling. First, the transfer RNA for alanine was isolated with radioactive alanine still attached to it; it can be designated as (alanyl tRNA^{ala}). At the same time, the transfer RNA specific for cysteine was isolated still with radioactive cysteine attached to it (cysteinyl tRNA^{cys}). The sulfhydryl group of the cysteine was then chemically removed, converting it to alanine, with no apparent effect on the structure of the transfer RNA. The result was an alanine residue attached to the specific transfer RNA for cysteine, that is, alanyl tRNA^{cys}. This alanyl tRNA^{cys} was then added to an *in vitro* amino acid incorporating system. As a cofactor was added a polynucleotide, a polyuridylicguanylic acid, which causes the incorporation of cysteine and several other amino acids, but not of alanine, into protein. It was found that, in the presence of this cofactor, the alanine of alanyl tRNA^{cys} was incorporated into protein, but the alanine of alanyl tRNA^{ala} was not. This result indicates very strongly that the amino acid specificity for incorporation into protein resides not in the amino acid attached to the RNA but in some specific portion of each individual transfer RNA molecule. The transfer RNA thus acts not only as a carrier, but as a specific "key and lock" carrier. The nature of the key undoubtedly rests in the sequence of the nucleotides making up the "anticodon" site (see Table 15-7) on the tRNA. It is these sites that bind specifically to "codon" sites on the messenger RNA.

(See Fig. 8-25.)

(See Fig. 8-26.)

(See Fig. 8-27.)

The next step or steps in protein synthesis (Fig. 15-6) are by far the most baffling. We can set up *in vitro* test tube experiments in which the radioactive amino acid is linked to its specific transfer RNA; if ribosomes and certain factors are added, the result is radioactive protein bound to the surface of the ribosomes. We need, as cofactors, another high-energy compound, GTP, and two or three soluble proteins. One of the latter is probably the peptide-bond forming enzyme; what the other two are and what GTP does is a mystery. In addition, there seem to exist distinct steps in initiating the synthesis of a protein, in terminating its synthesis, and releasing it from the ribosome-complex; what these steps might entail will be described below.

From the work of Dintzis and Schweet, we know that in the reticulocyte ribosome system that synthesizes hemoglobin—and probably in the syntheses of all proteins—the protein chain of the hemoglobin molecule is synthesized in a linear fashion, starting from the free amino-group end of the molecule. The reaction probably proceeds as follows:

$$
\begin{array}{l}
\qquad\qquad\quad \overset{R^1}{\underset{|}{\;}}\quad \overset{O}{\underset{\|}{\;}} \qquad\qquad\qquad\qquad\qquad \overset{R^2}{\underset{|}{\;}}\quad \overset{O}{\underset{\|}{\;}} \\
1.\ NH_2\!-\!CH\!-\!C\!-\!O\!-\!\text{transfer RNA}^1 + NH_2\!-\!CH\!-\!C\!-\!\text{transfer RNA}^2 \rightarrow
\end{array}
$$

$$
\overset{R^1}{\underset{|}{\;}}\quad \overset{O}{\underset{\|}{\;}}\quad \overset{R^2}{\underset{|}{\;}}\quad \overset{O}{\underset{\|}{\;}}
$$
$$
NH_2\!-\!CH\!-\!C\!-\!NH\!-\!CH\!-\!C\!-\!\text{transfer RNA}^2 + \text{transfer RNA}^1
$$

$$
\qquad\quad \overset{R^1}{\;}\quad \overset{O}{\;}\quad \overset{R^2}{\;}\quad \overset{O}{\;}
$$
$$
2.\ NH_2\!-\!CH\!-\!C\!-\!NH\!-\!CH\!-\!C\!-\!\text{transfer RNA}^2 +
$$

$$
\overset{R^3}{\;}\quad \overset{O}{\;}
$$
$$
NH_2\!-\!CH\!-\!C\!-\!\text{transfer RNA}^3 \rightarrow
$$

$$
\overset{R^1}{\;}\quad \overset{O}{\;}\quad \overset{R^2}{\;}\quad \overset{O}{\;}\quad \overset{R^3}{\;}\quad \overset{O}{\;}
$$
$$
NH_2\!-\!CH\!-\!C\!-\!NH\!-\!CH\!-\!C\!-\!NH\!-\!CH\!-\!C\!-\!\text{transfer RNA}^3 + \text{transfer RNA}^2
$$

3. repeated *n* times

It is interesting that these results on the linearity of protein synthesis agree with the results of work on the genetic code where the arrangement of genes is linear along the chromosome and where the messenger RNA seems to be synthesized linearly from a DNA template. We would like to know more about this reaction, including questions as to the fate of the transfer RNA molecules, as to the forces holding the first amino acid onto the template, and as to why the amino group is the reactive group in peptide-bond formation since the amino acid is "activated" at its carboxyl end. The latter feature of the reaction mechanism is similar to what occurs in fatty acid activation in the synthesis of long-chain fatty acids, for here too the carboxyl end is activated, but the reaction occurs at the other, methylene end of the molecule.

Specific Protein Synthesis

The relative stability of ribosomal RNA, while comforting from a genetic point of view, pointed up a new paradox. If DNA is the genetic code in the nucleus and the ribosomes are the protein-synthesizing machinery in the cytoplasm, then it should be the ribosomal RNA that has the genetic information, derived from DNA, which translates blueprint into concreteness. The finding of enzymes that synthesize RNA in the presence of DNA and whose RNA has the same base ratios as does the DNA serves to strengthen this argument (see Chapter 14). In bacteria the synthesis of certain new proteins—in this case, new enzymes—can be induced by the addition of their substrates to the culture medium; the details of this induced enzyme synthesis are given in *Microbial Life* (Holt, Rinehart and Winston Modern Biology Series). The synthesis of new protein starts immediately upon addition of substrate and ceases very quickly after the bacterial cells are washed free of the substrate; since the new protein is an enzyme it can be measured very conveniently. Also, when bacteria are infected by bacterial viruses or bacteriophages, the infected cells begin very quickly to make new enzymes necessary for the new synthesis of specific viral constituents. In the former case, the addition of substrate has somehow caused the cell to make new protein; in the latter case the DNA of the phage is the predetermining agent. However, in neither case does there seem to be any turnover of the bacterial ribosomal RNA; it is stable. Yet, if we have postulated it is the ribosomal RNA that has the necessary information to make specific protein, how is it that when the cell is induced to make new protein, there still is no perceptible change in the ribosomal RNA?

Faced with this problem, several investigators, led by Jacob and Monod, have come up with the idea of a messenger RNA (see Chapter 14). According to the hypothesis, there exists an RNA molecule that is synthesized in the nucleus, or nuclear region in the case of bacteria, perhaps by enzymes using the triphosphates and DNA; it can be said that the RNA is "transcribed" from a DNA template. This "messenger" goes to the cytoplasm, becomes bound to the ribosomes, and directs the ribosomes already in the cell to make specific protein. According to this view, the ribosomes are nonspecific protein-synthesizing machines. In the second case cited above, it is the DNA of the infecting phage particles that was conceived to be the directing force in the synthesis of the messenger. It has been found that when a bacterial cell becomes infected with a DNA-containing phage, the synthesis of RNA of the bacteria ceases, but a small percentage of the host-virus system RNA becomes rapidly synthesized. Furthermore, if the base ratios of this rapid-turnover RNA are determined in a certain way, it turns out to have the same composition as does the DNA of the infecting virus. This type of RNA is probably a composite of RNA molecules of various sizes, and all of these molecules seemingly end up becoming bound to the preexisting ribosomes of the bacteria. Therefore, as already mentioned in Chapter 14, there seems to be some experimental evidence for the existence of an RNA molecule that does have the properties that have been proposed for messenger RNA.

However, as in all rapidly expanding subfields of science, there is also some evidence against this notion, or at least against the rather simplified idea propounded above. For example, the ribosomes from the reticulocyte cells—cells that are the precursors of red blood cells and that have no DNA—do have the capacity to synthesize hemoglobin; thus, here it is not necessary for DNA to make messenger RNA that acts immediately as a template for the synthesis of a specific protein. This discrepancy can be hypothetically overcome by the assertion that bacterial cells, doubling their number every 20 to 30 min and being very responsive to changes in the environment, are in need of an unstable messenger. Cells of the multicellular organism, on the other hand, have a much longer doubling time and have no need to be responsive to a changing environment since their environment is a rather stable blood supply; thus, they really have no need for an unstable messenger, and this messenger remains attached to the ribosomes as long as the cell lives. Some investigators, notably Roberts, have contended that what really happens when the bacteria are infected by virus is not the synthesis of new messenger molecules that later become bound to preexisting ribosomes, but the synthesis of a very small number of new ribosomes, with the so-called messenger really being the RNA of these new ribosomes. Obviously, at present the picture is not too clear. Undoubtedly, there is a fraction of cellular RNA that has a high turnover under certain conditions, and seemingly this fraction has the base ratios of the relevant DNA; a fraction like this has recently been uncovered in thymus cell nuclei and in the nuclei of regenerating liver cells. Its relationship to ribosomal RNA is unclear. However, some investigators look upon it as being, in part, a precursor for ribosomal RNA; most look upon ribosomal RNA not as being a genetically coded carrier, but only as having a structural role in being able to bind the messenger RNA to it.

Furthermore, there is a dictum in biology, at present incapable of proof, that the DNA in all the somatic cells of the multicellular organism is the same. This may be the case; for example, many plants and some animals can regenerate an entire organism from buds cut off the original adult organism or even from single cells from some adult plant tissue. If so, and if the genetic information resides only in the base sequence of the DNA, then the genetic information is the same in all somatic cells. We do know, however, that the cells of one organ, of one tissue, are different from those of another; for one thing, they have different enzymes, and thus different information is transmitted in each case to tell a cell that, for example, it is a liver cell and not a heart muscle cell. If the genetic information in the DNA is the same, there must be an intermediate in the chain between nuclear DNA and cytoplasmic synthesizing machinery that is not messenger RNA and that is not merely a replication of the DNA. This hypothetical intermediate could be either a differential subtraction or differential masking of the DNA from one cell to another. It has been postulated that the histone that is in a complex with the DNA in the nucleus acts as a masking agent in that only the "naked" DNA acts as a template for RNA synthesis. Complications like these cast doubt on the hypothesis that there is a simple linear relationship between the genetic information in the DNA and the recipient of this information in the cytoplasmic ribosomes. It should be mentioned that the nucleus appears to have ribosomal-like particles in it also, and from the limited amount of work done with these particles it appears that these too are capable of synthesizing protein.

Nucleic Acid "Code"

We have been speaking of genetic information in the DNA, this information being precisely that necessary for the synthesis of specific proteins. We can say that in the DNA is a series of nucleotides that signifies that a certain amino acid be put in a certain place in a protein during the synthesis of that protein; in other words, the arrangements of nucleotides in the DNA might be a code for the arrangement of amino acids in a protein. Until 1962, biologists in general had very little hope of even attempting to "break" this "code." However, physicists like Gamow and Crick began to draw up mathematical postulates as to what kinds of theoretically possible codes would relate the four different nucleotides in the DNA, or in the RNA, to the 20 different amino acids in proteins. At present the most accepted code is that worked out by Crick and Brenner in which a "word" of three "letters," or nucleotides, arranged in a certain sequence signifies a certain amino acid. There is now a large experimental and theoretical basis for just such a code. This code is believed to be nonoverlapping, that is, the code reads linearly along the DNA, or along the "messenger" RNA, and hence any one nucleotide along the nucleic acid chain can be part of only one "three-letter word."

It is now thought that this code has been broken in the following way. It has been possible to isolate from bacteria an enzyme that synthesizes RNA from added nucleotide diphosphates. The enzyme is nonspecific in that it can synthesize any kind of poly- nucleotide molecule, the kind being dependent on the kind of nucleotides that are added to the incubation mixture. For example, if uridine diphosphates are added, the enzyme will make a polyuridylic acid, that is, uridylic acid residues linked to each other by $3', 5'$-phosphodiester bonds. What the function of this enzyme is in the bacteria and what the function of the RNA it makes is in the economy of the cell are not known. Nevertheless, the products of the enzymatic reactions have proved very useful, for now we can make many kinds of RNA-like molecules of specified composition, practically at will, by merely varying the nucleotide diphosphates that are added as precursors to the incubation medium containing this useful enzyme. Thus, in addition to polyuridylic acid, other polynucleotides containing adenylic, guanylic, and cytidylic acids can be made in various and known proportions of nucleotides.

One of these polymers, polyuridylic acid, was first used by Nirenberg in an *in vitro* radioactive amino acid incorporation system containing ribosomes from *E. coli* and the appropriate cofactors for amino acid incorporation to take place. When various radioactive amino acids were used, it was found that only in the case of phenylalanine was there a large increase in the incorporation of the amino acid into protein in the presence of polyuridylic acid. Thus, this polyuridylic acid acted like a messenger RNA. Similarly, when others of these synthetically made polyribonucleotides were added to the system, other amino acids were found to be specifically incorporated into protein. Here, for the first time, was a tool with which nucleotides could be experimentally related to amino acids. The results of using different polymers containing various pro- portions of the four different nucleotides and using all the 20 amino acids as radio- active precursors in individual experiments can be set up in a table; from it can be ascertained that certain combinations of nucleotides seem to be the necessary ingredi- ents for the incorporation of particular radioactive amino acids into the proteins of *E. coli* ribosomes. If three nucleotides in a row "mean" or signify a certain amino acid, then we can say that three uridylic acids in a row "mean" phenylalanine. Table 15-4, which is a compilation of early results from the laboratories of Ochoa and Nirenberg, shows that certain combinations of nucleotides signify the various amino acids. As can be seen from Table 15-4, some amino acids have more than one code word or "codon." Indeed, there is some evidence that there is not just one, but, in some cases, two or even several transfer RNA's for one amino acid, and that each of the transfer RNA's for the same amino acid has a different coding feature. That is to say, each of these transfer RNA's must have the same specific recognition site for the amino acid in question, but they have different coding sites (for attachment to messenger RNA)

Table 15-4. RELATIONSHIP OF NUCLEOTIDE COMBINATIONS TO INCORPORATION OF AMINO ACIDS INTO PROTEINS*

NUCLEOTIDES IN POSTULATED TRIPLETS	INCORPORATED AMINO ACIDS
3 uridylic acids	Phenylalanine
2 uridylic acids, 1 adenylic	Isoleucine, leucine, tyrosine
2 uridylic acids, 1 cytidylic	Leucine, serine
2 uridylic acids, 1 guanylic	Valine, cysteine, leucine
1 uridylic, 2 adenylic	Asparagine, lysine
1 uridylic, 2 cytidylic	Proline
1 uridylic, 2 guanylic	Tryptophan, glycine
1 uridylic, 1 adenylic, 1 cytidylic	Threonine, histidine, asparagine
1 uridylic, 1 adenylic, 1 guanylic	Methionine, glutamic acid, aspartic acid
1 uridylic, 1 cytidylic, 1 guanylic	Glutamine, alanine, arginine

* (Data from Nirenberg and Ochoa.)

for the particular amino acid (Table 15-7). The present assumption, based also on other lines of evidence, is that the whole code may be redundant, that is, a single amino acid may be specified by more than one set of nucleotides. This redundancy can also be seen in the data of Table 15-5, which show the results of later work of Nirenberg, with information on the actual order of nucleotides within each three-nucleotide code word. As can be seen from a comparison of Tables 15-4 and 15-5, some of the earlier work concerning the code words is somewhat wrong, but the important point is that the order of the nucleotide letters is beginning to be known. The redundancy has the feature that while the first two nucleotide letters of the code word are the same in the redundant words, the third letter can vary and, apparently in some cases, can be any one of the four nucleotides. The significance of the redundancy for biological function is not known, but it could be that the redundancy reduces the chance of error in the translation from nucleic acid to protein; or it may be that a certain code word is used for the same amino acid in one kind of protein and another code word in another type of protein; or it could be that the same amino acid, but in different positions along the polypeptide chain, is coded for by different code words, that is, the neighbors of the amino acid in question have some directing influence on the specificity of the code

Table 15-5. ORDER OF NUCLEOTIDES IN "CODE WORDS"*

ACU ⎫		AUG ⎫		GCU ⎫	
ACC ⎬ Threonine		GUG ⎭ Methionine		GCC ⎬ Alanine	
ACA ⎪				GCA ⎪	
ACG ⎭		GAU ⎫ Aspartic acid		GCG ⎭	
		GAC ⎭			
CAU ⎫ Histidine				CGU ⎫	
CAC ⎭		GAA ⎫ Glutamic acid		CGC ⎬ Arginine	
		GAG ⎭		CGA ⎪	
AAU ⎫ Asparagine				CGG ⎭	
AAC ⎭					
				CAA ⎫ Glutamine	
				CAG ⎭	

* (Data from Nirenberg.)

words. Indeed, there is experimental evidence that the three arginine residues in the hemoglobin α-chain are coded for by three different arginine code words.

Independent evidence has verified some of these code words for certain amino acids. For example, mutants of tobacco mosaic virus can be produced by chemical means through deaminating with nitrous acid the RNA of the virus and thus changing the cytidine of the RNA into uridine, and the adenine to guanine. In the resultant protein coat of these mutant viruses it can be seen that some of the amino acids of the protein have been altered; some amino acids have been replaced by others in certain places along the protein chain, presumably as a result of the mutation caused by nitrous acid treatment of the RNA. Knowing what kind of changes in the RNA were produced by nitrous acid and knowing what amino acids were replaced by others, we can make a correlation between these two events. Table 15-6A gives such a correlation.

Armed with this code, investigators can make predictions as to what nucleotides in a combination coded for a certain amino acid might be changed into another combination coded for another amino acid. For example, different kinds of hemoglobin molecules can be isolated from humans with diseases characterized by specific changes in red blood cell hemoglobin proteins. It can be determined which amino acids in these proteins differ from their counterparts in normal hemoglobin molecules; thus, in some cases a glutamic acid in a certain position in normal hemoglobin has been changed to a glycine molecule in that same position. It was then predicted that these amino acid replacements could be due to changes in single nucleotides in the three-letter code word for certain amino acids, changing this into another code for another amino acid (see Table 15-6B). In other words, the most common amino acid replacements could be predicted simply on the basis of the probable frequency of changes of one nucleotide in a three-nucleotide code—changing, for example, uridylic-adenylic-guanylic acids to uridylic-cytidylic-guanylic acids.

Similarly, in the mold Neurospora, mutants have been obtained and isolated, either by spontaneous selection or by induction with chemical mutagens such as nitrous acid, in which mutant a certain enzyme, tryptophane synthetase, has been changed

Table 15-6. EXAMPLES OF AGREEMENT BETWEEN AMINO ACID REPLACEMENTS AND EXPERIMENTALLY DETERMINED NUCLEOTIDE "CODE WORDS"

A. IN TOBACCO MOSAIC VIRUS MUTANT PROTEINS*

Postulated Code:

aspartic acid → GAU,GAC	leucine CUU	proline CCU

Nitrous acid produced changes:

glycine	phenylalanine	leucine

Postulated Code:

glycine → GGU,GGC	UUU	CUU

(aspartic acid → glycine; leucine → phenylalanine; proline → leucine)

B. IN VARIOUS SPECIFIC HEMOGLOBIN MOLECULES FROM VARIOUS SOURCES

Amino acid replacements:

histidine → arginine	glutamic acid → glycine

Postulated changes in "code words":

CAC,CAU → CGC,CGU	GAA,GAG → GGA,GGG

Amino acid replacements:

glutamic acid → lysine	histidine → tyrosine

Postulated changes in "code words":

GAA,GAG → AAA,AAG	CAC,CAU → UAC,UAU

C. IN NEUROSPORA TRYTOPHANE SYNTHETASE MUTANTS†

	DETERMINED GENETICALLY	DETERMINED BIOCHEMICALLY
Glycine	GGA,CGG	GGU,GGC, GGA,GGG
Glutamic acid	GAA	GAA,GAG
Alanine	GAC,GCA	GCU,GGG, GCA,GCG
Valine	GAU	GUU,GUC, GUA,GUG
Arginine	CGA	CGU,CGC, CGA,CGG
Isoleucine	CUA	AUU,AUC
Serine	CGU	UCA,UCG, AGU,AGC
Threonine	CCA	ACU,ACC, ACA,ACG

* The mutants were obtained by treating the virus with nitrous acid, which changes the RNA base cytosine to uracil, and the base adenine to guanine.
† In spontaneous and chemically-induced mutants, eight different amino acids were found at the same site in one of the protein chains of the enzyme.

from the wild type counterpart. Very good experimental work by Yanofsky has pin-pointed many of these mutant effects at certain positions along the polypeptide chains of this enzyme. For example, at a certain point along one of the chains there have occurred many substitutions of one amino acid for another. Again, by guessing wisely about the probable effects of the chemical mutagens and by logical analyses of the genetic results, similar to the case of the hemoglobins just mentioned, a nucleotide code can be and has been worked out. Table 15-6C shows a correlation between the two types of independent experiments, the amino acid incorporation experiments and the biochemical genetics experiments; for the eight amino acids in question, the agreement is quite marked.

Another different kind of correlation between codons and amino acids can be observed when one looks closely at the three-dimensional structures of the tRNA's. Table 15-7 shows the codons for five amino acids. In Chapter 8, we have seen the possible conformations of the tRNA's that are the carriers for these amino acids; they can be all folded into a cloverleaf pattern. When this is done, certain nucleotide triplets

(See Fig. 8-26.)

Table 15-7. CORRESPONDENCE BETWEEN OBSERVED AND EXPECTED "ANTI-CODONS" OF FIVE tRNA MOLECULES*

AMINO ACID	CODONS	ANTICODONS IN tRNA's	
		EXPECTED	OBSERVED
Alanine	GCU	CGA	
	GCC-----	CGG-----	CGI
	GCA	CGU	
	GCG	CGC	
Tyrosine	UAU	AUA	
	UAC-----	AUG-----	AUG
Serine	AGU	UCA	
	AGC	UCG	
	UCU	AGA	
	UCC-----	AGG-----	AGI
	UCA	AGU	
	UCG	AGC	
Phenylalanine	UUU	AAA	
	UUC-----	AAG-----	AAG
Valine	GUU	CAA	
	GUC-----	CAG-----	CAI
	GUA	CAU	
	GUG	CAC	

*The five amino acids are the ones whose codons are known from the genetic code. The expected anticodons are obtained from these (see text). The observed "anticodons" are obtained from the three-dimensional structures of the tRNA's when these are set into a pattern so that the anticodon site is the same for all the tRNA's (see Chap. 8).

occupy one of the exposed sites. These triplets could be the "anticodons" on the tRNA for the "codons" in the mRNA; thus, the anticodon for GCU is CGA. Table 15-7 indicates the close fit (inosine can replace G) between the expected "anticodons" (known from the genetic code) and the observed anticodon in the tRNA site. That the fit is so good seems a confirmation of the structures of the tRNA's.

Thus, a code initially postulated that was based on the work with bacteria has been verified, using quite different laboratory manipulations, by work on viruses, on molds, and on mammalian hemoglobins. We can say that a universal code is being unveiled in which singular sequences in the DNA molecule can code, probably through an intermediary RNA molecule, for the placement of specific amino acids into particular positions in a protein.

Now, laboratory experiments performed with many varieties of synthetic polynucleotides, similar to those mentioned above, have produced three ancillary and intriguing footnotes to our knowledge of the code involvement in protein synthesis. Firstly, it is now known that in translating the nucleotide sequence in messenger RNA into amino acid sequence in protein, the ribosome complex begins the translation at the 5'-end of the messenger RNA. This is precisely the end of the messenger RNA that is initially synthesized by the RNA polymerase transcribing the DNA template (Chapters 8 and 14); the vectorial linearity in the transcription of messenger RNA is followed by the translation of it into protein synthesis.

Secondly, one of the questions that was quite bothersome in the past now seems on its way to solution. The question is how does the ribosome complex "know," in the process of translation of a messenger RNA that may contain the codons of many specific proteins, precisely where one "message" begins and the previous "message" ends—where, in other words, to initiate the synthesis of a new protein. For, it is thought that one length of messenger RNA contains the coding, not for only one, but for many different protein molecules; such a linkage group, going back to the genetically related linkage group in the DNA, has been called a "cistron." It now appears, at least in the case of bacteria, that specific codons, AUG and GUG in the messenger RNA, signify the initiation of a new protein chain. It does this by coding for a specific transfer RNA, one of the two methionine transfer RNA's; one is coded for by a AUG internal in the chain, the other by initial AUG or GUG. (Table 15-5). Moreover, the AUG or GUG-coded one has the property of having the methionine attached to its RNA being capable of becoming formylated on its amino group, to give an N-formylmethionine-transfer RNA, while the other methionine-transfer RNA cannot be formylated. The formylation of the amino group probably blocks this group from participating in peptide bond formation; therefore, initial AUG or GUG in messenger RNA is a signal for initiation of a new protein chain. Thus, in bacteria, N-formylmethionine becomes the initial amino acid at the amino-group terminal end of the peptide chain. Furthermore, since all bacterial proteins do not begin with methionine, much less N-formylmethionine, it appears that enzymes, peptidases, must exist, which chop off not only the formyl group, but in some cases also the methionine and possibly the adjacent amino acids. Finally, it has recently been found that an "initiator complex" exists, which consists of the smaller ribosomal subunit, to which is attached specific mRNA, specific initiating aminoacyl tRNA, and protein factors (role unknown). When this complex combines with the larger subunit, in presence of GTP and other protein factors, protein synthesis ensues. This does sound complicated, perhaps unnecessarily so, but it must be realized that the correct initiation of protein chains is very important, and that a precise mechanism must exist so that overlapping amino acid sequences do not occur in adjacent proteins during translation.

The third point, that there should exist a mechanism for the correct termination of translation into protein, necessarily follows the previous one. And while the evidence is not as good as that above, it does appear that there exist distinct codons for chain termination. Thus, from genetic work it is probable that UAG and UAA and possibly UGA, are in fact "nonsense" codons, that is, they do not code for any amino acid, but instead their presence at certain places in the messenger RNA is somehow a "signal" for the ending of that particular protein code, for the termination of that particular translated protein chain. It is thought that at the end of the synthesis of a protein, the ribosomes fall off the mRNA, and are split into their two subunits. The free smaller subunit then forms an "initiator complex," as mentioned above, combines with a free larger subunit, and the synthesis of a new protein begins. We thus have a glimmer; we think we are right; but how actually these protein chain-initiating and -terminating mechanisms work is far from being understood.

Finally, Table 15-8 gives the present-day thoughts on the code. The data in the table were obtained by many workers doing many kinds of experiments, some of which are explained above. In general, there is a close agreement between all the results. Thus, we think that a real picture has been obtained of the translation of information in the cell. Even though some corrections will undoubtedly have to be made in the future, the data in Table 15-8 represent one of the great accomplishments of biology, the code by which a cell can reproduce its proteins into perfectly, or almost perfectly, duplicated copies.

Workers in this particular field have reached the experimental point where they can take the ribosomes from E. coli, add the necessary cofactors for radioactive amino acid incorporation into proteins, add a certain kind of RNA, and obtain the synthesis of a specific protein. For example, if a certain phage RNA is added, a certain phage protein appears to be synthesized. It is not yet known what the E. coli ribosomes do, specifically, or what the role of ribosomal RNA is in these syntheses. Possibly, it simply acts as a structural feature of the ribosomes upon which the large protein molecules are to be built; or it may modify the RNA messenger in such a way that the ribosomes from liver, for example, make—in conjunction with the universal messenger from the universal DNA in all the cells of the organism—not universal protein, but specifically liver cell proteins.

Synthesis of Structure-Specific Proteins

So far our discussion has centered on the amino acid sequence in proteins as determining protein structure; actually, a protein is a three-dimensional structure with

Table 15-8. THE GENETIC CODE AS IT IS KNOWN AT
PRESENT WITH THE BEST ALLOCATIONS
OF THE 64 CODONS*

1ST ↓	2ND →	U	C	A	G	↓ 3RD
U		PHE	SER	TYR	CYS	U
		PHE	SER	TYR	CYS	C
		LEU	SER	Terminator††	Terminator††	A
		LEU	SER	Terminator††	TRP	G
C		LEU	PRO	HIS	ARG	U
		LEU	PRO	HIS	ARG	C
		LEU	PRO	GLUN	ARG	A
		LEU	PRO	GLUN	ARG	G
A		ILEU	THR	ASPN	SER	U
		ILEU	THR	ASPN	SER	C
		ILEU	THR	LYS	ARG	A
		MET, Initiator†	THR	LYS	ARG	G
G		VAL	ALA	ASP	GLY	U
		VAL	ALA	ASP	GLY	C
		VAL	ALA	GLU	GLY	A
		Initiator†	ALA	GLU	GLY	G

*(Table modified from F. C. Crick.) Notice that many amino acids are coded for by more than one trinucleotide codon (redundancy); it is usually the third nucleotide that is changed.
† In bacteria, the initiator amino acid seems to be formyl met.
†† These "nonsense" codons do not code for any amino acid.

specific configuration. This configuration is very important because some of the properties of proteins as enzymes are due not only to the amino acid sequence, but also to the way the protein is folded, folded to accommodate the substrate of the enzymatic action. It is presently thought that once a specific amino acid sequence of a protein has been synthesized, it will naturally fold into a particular configuration and that this depends in some proteins, rather more or less specifically, on the distribution and conjunction of cysteine residues in the protein to provide disulfide bonds. However, all proteins do not have intramolecular disulfide bridges, and it is currently thought that the most important forces holding proteins in a three-dimensional configuration are the hydrophobic bonds between the hydrophobic amino acids. These amino acid side-chains, by excluding water, are forced together into the center of the molecule, thus imposing a structure onto the protein. It is thought that once a specific protein is synthesized, it will then naturally fold into its correct shape. Moreover, not only must the cell synthesize a specific protein having particular properties, but also this protein has a predestined localization within the cell. Cytochrome oxidase, for example, is a mitochondrial protein. How this enzyme is finally placed in its correct position in the cell is completely unknown. Thus, although we are well on our way in deciphering the way the genetic material is translated into protein enzymes in the cell, we are completely in the dark as to how these proteins come together to form the beautiful specific organizational patterns of the variety of different cell types. This will probably prove to be a very difficult problem to solve, but once solved, the solution will open up to us the real meaning of a living cell: the nature of the organization that can duplicate itself; the nature of a factory that has utilized simple chemicals to become larger in a specific way and to give rise to other duplicate factories; and the nature of a machine, finally, that knows itself.

Summary

Now let us recapitulate our present ideas on specific protein synthesis as detailed in this and the preceding chapter. We think that in the nucleus one of the strands of DNA acts as a template for the synthesis of possibly all the RNA in the cell. A part of this RNA is messenger, a part is ribosomal, and a part is transfer RNA. There are indications that ribosomal RNA is synthesized in conjunction with the nucleolar-associated DNA. All these RNA species then move into the cytoplasm, a part possibly staying behind to act in nuclear protein synthesis. The informational RNA either forms part of the ribosomes, or else it becomes attached to an already formed ribosome. Investigators cannot at present distinguish between these two alternatives, nor is it known whether ribosomal RNA picks up its protein in the nucleus or whether the completed ribosome forms in the cytoplasm. However, the ribosome, with its component or attached messenger RNA, has become a template for the synthesis of specific proteins. Meanwhile, the amino acids have become activated and subsequently have become bound to their own specific transfer RNA. The example given in Fig. 15-4 is phenylalanine; since the code word for phenylalanine is known to be three uridylic acids (UUU), it is assumed that the transfer RNA has a region that "says" phenylalanine and that this region is a three-adenylic acid (AAA) region complementary to the UUU region on the messenger RNA. The amino acid is thus now aligned in its correct position according to the code for the specific protein that is contained in the messenger RNA. Thus set, the peptide bond-forming enzyme hooks up this amino acid with its adjacent amino acids. It appears that more than one enzyme is involved in this final step; what GTP does is still an enigma.

At this time, some unknown release mechanism operates to peel the protein off its template. What happens to the messenger RNA and to the transfer RNA is not known for certain. In bacteria it appears that the lifetime of messenger RNA is very short, only lasting for the synthesis of a few proteins; in mammalian cells it appears to be much longer. We are equally uncertain concerning transfer RNA, although here it does appear that the three-end nucleotides (—CCA) have to be renewed each time this RNA acts as a carrier for the amino acid. The final three-dimensional configuration of the synthesized protein is probably determined by the amino acid sequence; we can say that the primary structure of the protein, its amino acid sequence, is directly responsible for its secondary structure.

Although it is clear that there are many hiatuses in this scheme, we believe that what we have here is at least a glimmer of the truth. To complete the picture, we have the tools of genetics to supplement our biochemical knowledge. Of course, even if we solve the mysteries of how the cell can unerringly continue to synthesize its very own specific proteins, we will still be left with the very important problem of how cell structures are formed (see Chapters 4, 10, and 18). These problems include the formation of subcellular structures, the formation of cell structure that is distinctive for a given tissue or organ, and the factors determining the form and size of individual organs and even of the whole organism.

SUGGESTED READING LIST

Offprints

Allfrey, V. G., and Mirsky, A. E., 1961. "How Cells Make Molecules." *Scientific American* offprints. San Francisco: W. H. Freeman & Co.

Brenner, S., Jacob, F., and Meselson, M., 1961. "An Unstable Intermediate Carrying Information from Genes to Ribosomes for Protein Synthesis." *Nature,* pp. 576–581. Bobbs-Merrill Reprint Series. Indianapolis: Howard W. Sams & Co.

Crick, F. H. C., Barnett, L., Brenner, S., and Watts-Tobin, R. J., 1961. "General Nature of the Genetic Code for Proteins." *Nature,* pp. 1227–1232. Bobbs-Merrill Reprint Series. Indianapolis: Howard W. Sams & Co.

Crick, F. H. C., 1962. "The Genetic Code." *Scientific American* offprints. San Francisco: W. H. Freeman & Co.

Crick, F. H. C., 1963. "On the Genetic Code." *Science,* pp. 461–464. Bobbs-Merrill Reprint Series. Indianapolis: Howard W. Sams & Co.

Crick, F. H. C., 1966. "The Genetic Code: III." *Scientific American* offprints. San Francisco: W. H. Freeman & Co.

Dintzis, H. M., 1961. "Assembly of the Peptide Chains of Hemoglobin." *PNAS,* pp. 247–261. Bobbs-Merrill Reprint Series. Indianapolis: Howard W. Sams & Co.

Fraenkel-Conrat, H., 1964. "The Genetic Code of a Virus." *Scientific American* offprints. San Francisco: W. H. Freeman & Co.

Hoagland, M. B., 1959. "Nucleic Acids and Proteins." *Scientific American* offprints. San Francisco: W. H. Freeman & Co.

Hurwitz, J., and Furth, J. J., 1962. "Messenger RNA." *Scientific American* offprints. San Francisco: W. H. Freeman & Co.

Ingram, V. M., 1958. "How Do Genes Act?" *Scientific American* offprints. San Francisco: W. H. Freeman & Co.

Littlefield, J. W., Keller, E. B., Gross, J., and Zamecnik, P. C., 1955. "Studies on Cytoplasmic Ribonucleoprotein Particles from the Liver of the Rat." *Jour. Biol. Chem.,* pp. 111–123. Bobbs-Merrill Reprint Series. Indianapolis: Howard W. Sams & Co.

Nirenberg, M. W., Matthaei, J. H., Jones, O. W., Martin, R. G., and Barondes, S. H., 1963. "Approximation of Genetic Code via Cell-free Protein Synthesis Directed by Template RNA." *Fed. Proc.,* pp. 55–61. Bobbs-Merrill Reprint Series. Indianapolis: Howard W. Sams & Co.

Rich, A., 1963. "Polyribosomes." *Scientific American* offprints. San Francisco: W. H. Freeman & Co.

Warner, J. R., Knopf, P. M., and Rich, A., 1963. "A Multiple Ribosomal Structure in Protein Synthesis." *PNAS*, pp. 122–129. Bobbs-Merrill Reprint Series. Indianapolis: Howard W. Sams & Co.

Yanofsky, C., 1967. "Gene Structure and Protein Structure." *Scientific American* offprints. San Francisco: W. H. Freeman & Co.

Zamecnik, P. C., 1958. "The Microsome." *Scientific American* offprints. San Francisco: W. H. Freeman & Co.

Articles, Chapters and Reviews

Campbell, P. N., 1960. "Synthesis of Proteins by Cytoplasmic Components of Animal Cells." *Biological Reviews,* Vol. 35, p. 413.

Gros, F., 1960. "Biosynthesis of Proteins in Intact Bacterial Cells." In: *Nucleic Acids,* Vol. 3, p. 409. E. Chargaff, and J. N. Davidson, eds. New York: Academic Press, Inc.

Hoagland, M. B., 1960. "Relationships of Nucleic Acid and Protein Synthesis as Revealed by Studies in Cell-free Systems." In: *Nucleic Acids,* Vol. 3, p. 349. E. Chargaff, and J. N. Davidson, eds. New York: Academic Press, Inc.

Hultin, T., 1961. "On the Function of the Endoplasmic Reticulum." *Biochemical Pharmacology, 5,* p. 359

McQuillen, K., 1962. "Ribosomes and Synthesis of Proteins." *Progress in Biophysics and Biochemistry, 12,* p. 69.

Palade, G. E., 1955. "A Small Particulate Component of the Cytoplasm." *Journal of Biophysical and Biochemical Cytology, 1,* p. 59.

Porter, K. R., 1961. "The Ground Substance: Observations from Electron Microscopy." In: *The cell,* Vol. 2, p. 621. J. Brachet, and A. E. Mirsky, eds. New York: Academic Press, Inc.

Siekevitz, Philip, 1959. "The Cytological Basis of Protein Synthesis." *Experimental Cell Research,* Supplement 7, p. 90.

Books

Brachet, J., 1960. *The Biological Role of Ribonucleic Acids.* New York: Elsevier Publishing Co.

Cold Spring Harbor Symposia on Quantitative Biology, 28. Synthesis and Structure of Macromolecules. Cold Spring Harbor, L.I., New York.

Cold Spring Harbor Symposia on Quantitative Biology, 31. The Genetic Code. Cold Spring Harbor, L.I., New York.

Chantrenne, H., 1961. *Biosynthesis of Proteins.* New York: Pergamon Press, Inc.

Ingram, V., 1965. *The Biosynthesis of Macromolecules.* New York: W. A. Benjamin, Inc.

Peterman, M., 1964. *Physical and Chemical Properties of Ribosomes.* New York: Elsevier Publishing Co.

Roberts, R. R. (ed.), 1958. *Microsomal Particles and Protein Synthesis.* New York: Pergamon Press, Inc.

Chapter 16

The Cytoplasmic Matrix and the Conversion of Chemical Energy into Work

(See Chapters 13 and 17.)

Mechanical work is a universal property of living things. Such diverse phenomena as the injection of DNA into a bacterium by the phage T2, the wriggling of the bacterial flagella of *Salmonella*, the vibrations of the blue-green alga *Oscillatoria*, the beat of the cilia of *Paramecium*, the flow of the cytoplasm of the slime mold *Physarum*, the cyclosis of the cytoplasm of the water plant *Elodea*, the movement of chromosomes during cell division, the constriction of the animal cell during cell duplication, and the contraction of smooth or striated muscle are but a few examples of the great variety of mechanochemical transductions occurring in nature. Indeed, it is entirely possible that phenomena such as active transport, bioelectric phenomena, and oxidative phosphorylation may well include some mechanochemical transitions in their molecular machinery.

Although mechanochemical phenomena are numerous and varied in nature, it is likely that the basic molecular machinery is fundamentally the same, and it should therefore be our aim to understand in molecular terms the precise mechanism whereby chemical energy can be converted into work.

As we shall see, a great deal of progress has been made in our understanding of the contraction of striated muscle. We shall therefore concentrate our attention on this system, realizing full well that in spite of its staggering efficiency it is just one of the many mechanochemical systems found in nature.

Biophysics of the Striated Muscle Cell

In Chapters 12 and 13 we discussed how the muscle cell oxidizes foodstuffs and utilizes the energy released to synthesize ATP. The striated muscle cell is a highly evolved, extremely effective piece of machinery for the conversion of some of the energy of

ATP into mechanical work. Thus, for instance, the flight muscle that acts on the wing of the bee has a continuous power output equivalent to that of a piston aircraft engine — it hydrolyzes about one-half of its weight of ATP per minute (2400 kcal/kg/hr). It can do that at a speed equivalent to ten times its length per second, reaching its maximum power output in milliseconds. A muscle 1 cm in cross section can exert a tension equivalent to 3 kg. We have here not only a very efficient piece of transducing machinery, but also, as we shall see, a delicately regulated mechanism in which the molecular events must be interconnected with the greatest of precision.

Striated muscle, the voluntary muscle that moves the bones of animals, is composed of numerous elongated cells called muscle fibers. Motor nerves are attached at various points on the muscle fiber and transmit to it the electric impulse that initiates the contraction. Under the microscope, especially the phase-contrast microscope, muscle fibers show a series of highly regular and distinct bands and zones (Fig. 16-1), and these can be resolved with even greater clarity under the electron microscope.

Muscle has some unique physical properties, which were studied by A. V. Hill and others in the earlier part of this century. They found the following.

1. Maximum tension is exerted by muscle when it is held at constant length. Although in the formal sense work is force times distance, muscle maintaining tension at constant length does "internal work" and gives off heat.

2. If stimulated muscle is allowed to shorten, the tension it can exert is less than that exerted at constant length. The higher the rate of shortening the lower the tension it can exert.

3. Muscle that is allowed to shorten liberates more heat than muscle held at constant length. The difference, called the shortening heat, is proportional not to the rate of shortening but to the distance of shortening.

Since a muscle lifting a heavy weight does more work than when it is lifting a light weight and since for a certain distance of contraction the shortening heat remains constant, the total energy (heat plus work) that a muscle must expend will vary with the weight it lifts. Thus, the machinery determining the energy release in muscle is controlled not only by the distance of contraction, but also by the tension the muscle experiences during contraction. This is a beautiful example of machinery that is so regulated as to adjust energy expenditure to the work that must be performed.

These interesting physical parameters, which are responsible for the remarkable efficiency of muscle, have in recent years been rendered less mysterious by a most ingenious theory of muscle contraction proposed by Hugh Huxley and Jean Hanson. Their "sliding" theory of muscular contraction can best be understood if we first discuss the mechanochemical proteins of the muscle fiber, then describe the fine structure of the muscle cell, and then try to unify the information obtained from these two levels into a coherent description of muscle contraction. As will be seen, there are still many gaps in our knowledge of the mechanism of muscle contraction, especially at the highest resolution, molecular level. So much progress, however, has been made in recent years that one can begin to recognize the direction in which our future efforts must be exerted.

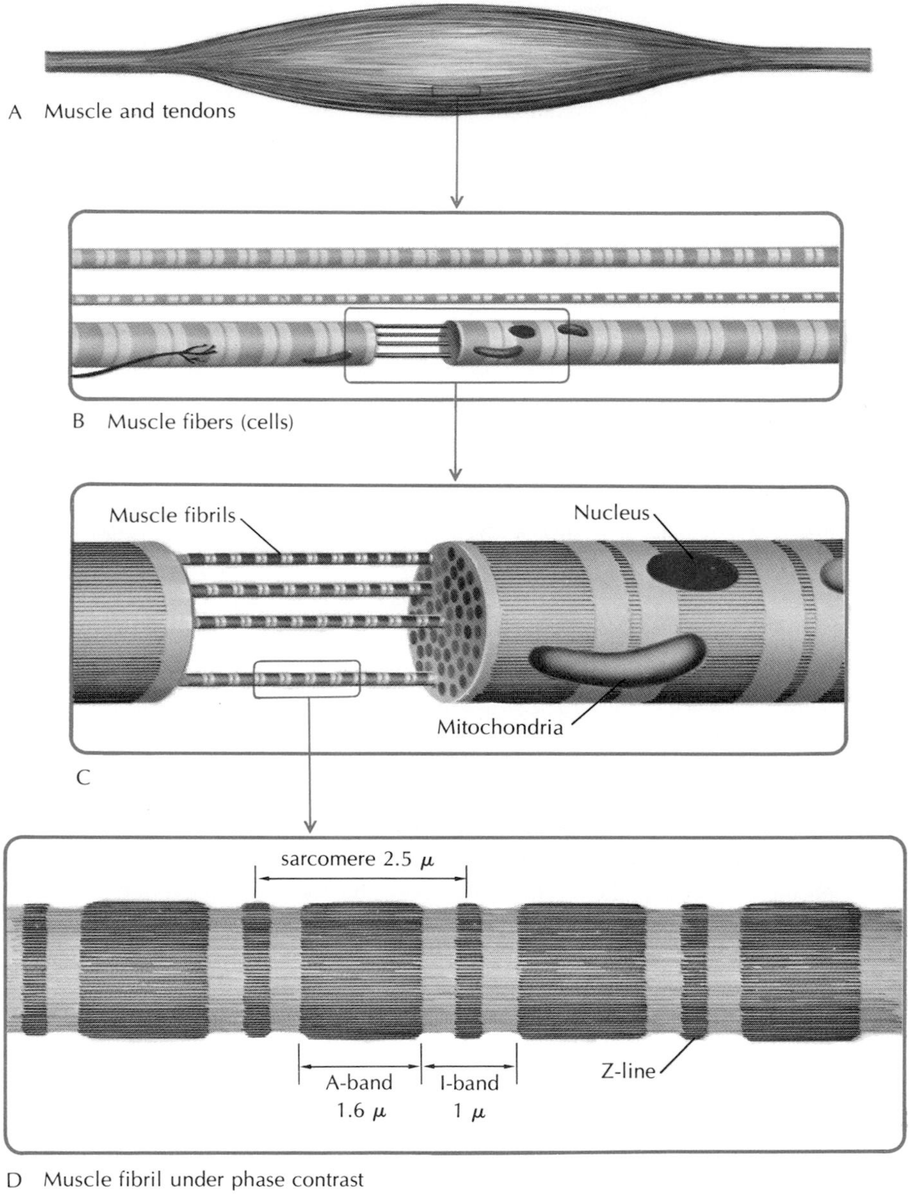

A   Muscle and tendons

B   Muscle fibers (cells)

C

Muscle fibrils

Nucleus

Mitochondria

sarcomere 2.5 μ

A-band
1.6 μ

I-band
1 μ

Z-line

D   Muscle fibril under phase contrast

Fig. 16-1.   The organization of striated muscle. The muscle (A) is an organ composed of numerous elongated cells or fibers (B). The fiber in turn is composed of numerous contractile elements or fibrils (C), which under the phase-contrast microscope (D) can be seen to have a striated structure of repeating units (sarcomeres) composed of two types of bands—the *A*-band and the *I*-band. The latter is divided by a structure called the *Z*-line. In the electron microscope the band structure can be seen in far greater detail.

The Mechanochemical Proteins of Muscle

Since the muscle cell is such a highly efficient chemical-mechanical work transducer, it seems reasonable to assume that a considerable proportion of its proteins are concerned with this process. Indeed, as early as 1864 the great physiologist Kuhne suggested a method of extracting muscle utilizing 10 percent NaCl, which yielded a concentrated protein extract. Unfortunately Kühne considered muscle contraction to be analogous to blood clotting, and this analogy, which also dominated the thinking of other muscle physiologists, seriously interfered with further progress in the study of muscle proteins. Danilevskii (1881–1888) made important contributions by distinguishing among various classes of proteins that could be extracted from muscle using water, dilute alkali and acid, and 6–12 percent NH_4Cl solutions. He showed great foresight by extracting muscle fibrils with various solutions and noting the disappearance of the A bands under the microscope. At the beginning of this century a number of workers (von Fürth, Halliburton, H. H. Weber, and Edsall) did much to increase our understanding of the protein component of muscle that was extracted most conveniently at high ionic strength and alkaline pH. This material, which von Furth called "myosin," was readily extracted at high ionic strength (0.6–1.2) and pH 7–8.5. It precipitated at ionic strength 0.05 and could thus be purified from other globulins, which do not precipitate as readily at low ionic strength. Physical studies such as ultracentrifugation (Weber) soon showed that "myosin" was not a homogeneous substance, and in 1942 Straub discovered actin and thus showed that "myosin" is really composed of two proteins, which he renamed actin and myosin. These two proteins have very different physical properties but interact with each other very strongly. In 1948 Bailey isolated from muscle another major protein component, which he called tropomyosin. Although we still do not know what role tropomyosin plays in muscle contraction, its ubiquitous and abundant (15 percent) presence suggests that it must be of importance in muscle contraction. Actin, myosin, and tropomyosin make up 85–90 percent of the total protein of the myofibril. Recently, Ebashi has identified some minor components (actinin, troponin) that appear to have a regulatory function in the contraction-relaxation process.

The discovery of myosin and actin, though important, was soon overshadowed by the excitement generated by a number of discoveries relating to the interaction of these proteins to each other and to the high-energy compound ATP.

In 1939 Englehardt and Lyubimova made the dramatic discovery that "myosin" (really actomyosin) has ATPase activity. After the discovery of actin, Banga purified myosin and showed that this enzymatic activity resided in the myosin component of actomyosin. Because the ATPase activity of purified myosin is 100 times slower than that of other ATPase enzymes, a number of workers considered the possibility that myosin and the ATPase activity could be separated from each other. These efforts proved unsuccessful, and we can now write the following equation:

$$\text{ATP} \xrightarrow{\text{myosin}} \text{ADP} + \text{Pi}$$

where Pi is used as a symbol for inorganic phosphate. The importance of the discovery of Englehardt and Lyubimova cannot be overemphasized. Although it may yet turn out that the myosin-catalyzed hydrolysis of ATP does not as such have physiological significance in muscle contraction, the discovery of the ATPase activity of myosin focused the attention of muscle physiologists on the molecular properties of myosin and actin and thus helped transform muscle physiology into a molecular science.

Another important discovery regarding the interaction of these proteins with ATP was made by Albert Szent-Györgyi and his co-workers. They showed that mixing actin with myosin in concentrated salt solutions produced a great increase in viscosity, which suggests that these molecules interact to form large polymers. Furthermore, the addition of ATP causes the viscosity to drop back to the original level, that is, that of the actin and myosin solution (Fig. 16-2). These results can be summarized in the following way:

$$\text{actin} + \text{myosin} \longrightarrow \text{actomyosin}$$
$$\text{actomyosin} + \text{ATP} \longrightarrow \text{actin} + \text{myosin}$$

Of course, since myosin hydrolyzes ATP, the drop in viscosity is not permanent but slowly reverses as the ATP disappears from the preparation.

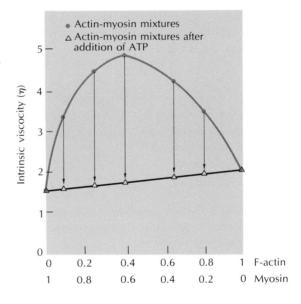

Fig. 16-2. The effect of ATP on the viscosity of actomyosin. (From Kerekjarto, 1952.) The graph is a summary of experiments in which ATP was added to different mixtures of *F*-actin and myosin (actomyosin) after which the decrease in viscosity was measured. The results show that pure actin and pure myosin do not decrease in viscosity upon addition of ATP, but mixtures containing varying amounts of both proteins do. The fact that the values for actomyosin after the addition of ATP lie on a straight line connecting the pure actin and pure myosin values is consistent with the interpretation that ATP acts by dissociating the actomyosin complex into actin and myosin. Other physical measurements have confirmed this interpretation.

These viscosity studies showed that ATP has a fundamental effect on the physical relationship between actin and myosin, although what the dissociation of actin and myosin by ATP precisely means in relation to muscle contraction is by no means clear even today.

Since, as we shall see later, myosin and actin are not found in solution in the muscle cell, it would seem more relevant to study the myosin-actin-ATP interaction in the precipitated or solid state. H. H. Weber (1934) had shown that if a solution of actomyosin is extruded from a capillary tube into a solution of low ionic strength, it forms a threadlike precipitate. A number of workers then studied the effect of ATP on these threads, but it was not until Albert Szent-Györgyi (1941) used actomyosin preparations richer in actin and precipitated in 0.05M KCl and 10^{-4}M Mg^{2+} that something truly exciting happened—the threads on addition of ATP contracted! To be sure, these early threads contracted in all directions rather than just shortened, but more recent threads, prepared from higher actomyosin concentrations and extruded at sufficiently high rates so as to orient the actomyosin molecules turned out to be much better models of the muscle fibril since they truly shortened and were capable of performing work!

The analogy between the actomyosin thread and the muscle fiber was strengthened even further by the preparation of glycerinated fibers (Szent-Györgyi, 1949). When a muscle fiber is placed into glycerol at 0°C, the membrane breaks down, and the soluble proteins are slowly extracted. What remains behind is strictly the contractile machinery of muscle composed almost entirely of actin, myosin, and tropomyosin. Glycerinated muscle fibers were found to have properties analogous to the synthetic actomyosin threads on the one hand and to muscle fibers on the other, and this provided further support for the molecular approach to this problem: i.e., that muscle contraction can be studied by purifying the protein components and examining their interactions.

Having briefly surveyed the history of the molecular approach to muscle contraction, let us now turn to a more systematic examination of the mechanochemical molecules of muscle and their interaction.

Actin can be extracted from muscle with solutions of high ionic strength and slightly alkaline pH, but since it is extracted more slowly than myosin, one finds that short-term extractions are rich in myosin whereas long-term extractions contain considerable amounts of actin. Straub developed a highly imaginative procedure for preparing actin. He extracted the muscle fibrils first to remove some of the myosin; then after treating with alkaline solution, he dried the residue with acetone, thereby denaturing most of the proteins including some myosin that remained in the residue. The actin was then extracted from the powder with water. This form of actin, which we shall call G-actin, is a relatively small globular molecule (68,000 mol wt).

The interesting thing about G-actin is that in the presence of ATP, KCl, and Mg^{2+} it polymerizes into long threads of F-actin with the conversion of ATP to ADP. The ADP is strongly bound to the F-actin, and it turns out that the number of ADP molecules produced is equal to the number of G-actin units in the F-actin polymer. Thus we can write that

$$n \text{ G-actin} + n \text{ ATP} \xrightarrow{\;Mg^{2+}\;} n \text{ F-actin-ADP} + n \text{ Pi}$$

The ADP can be released from F-actin by reacting it with myosin. Since a phosphate bond is split during the polymerization, one cannot escape the conclusion that actin has a catalytic function. Actin differs from a normal enzyme in that each G-actin unit splits only one phosphate bond. However, Asakura and Oosawa have demonstrated

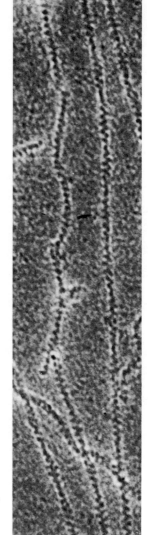

Fig. 16-3. The structure of *F*-actin. (Courtesy of Hanson and Lowy.) A. Electron micrograph of *F*-actin negatively stained with uranyl acetate. If one interprets this picture to be produced by two filaments of globular units twisted around each other, then one can count the number of subunits per turn of the helix as well as identify the places where the two strands cross over one another. B. Model of *F*-actin filaments according to Hanson and Lowy. Their present view is that the helix is "nonintegral" with 13–14 subunits per turn (pitch 2x 360–370A).

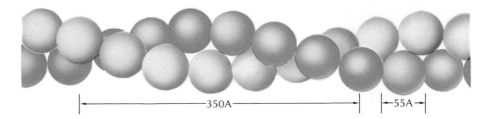

|←————————— 350A —————————→|←— 55A —→|

that if *F*-actin in the presence of ATP is subjected to sonic vibration, the ATP continues to be split to ADP and Pi at a rate that, within limits, is proportional to the period of vibration! This unusual observation clearly shows that actin has enzymatic properties even though under normal conditions it behaves as a "half enzyme": only one ATP molecule is split per molecule of *G*-actin.

Examination of *F*-actin under the microscope shows a rigid filament of helical structure (Fig. 16-3). X-ray analysis of *F*-actin prepared *in vitro* and also as it is found in filaments isolated from the muscle cell suggests that actin is a polymer of two strands wrapped around each other. The precise structure of this two-stranded helix is still under debate. Figure 16-3 shows an electron micrograph by Hanson and Lowy as well as a model giving a plausible interpretation of what can be seen in the micrograph.

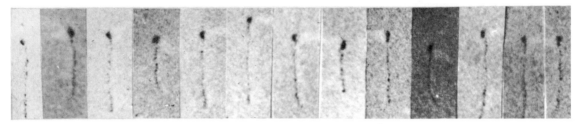

Fig. 16-4. Electron micrograph of a number of myosin molecules showing bulbous heads and long tails. More recent pictures show that the head is split forming two bulbous structures of equal dimensions. (Courtesy of H. E. Huxley.)

Myosin can be purified by short-term extraction in solutions of high ionic strength at pH 6.8. It can be purified further by precipitation at low ionic strength, and in recent years ultracentrifugation has been used to remove the much larger actomyosin complexes contaminating the myosin preparation. The molecular weight of myosin is still debated, with values ranging from 470,000 to 600,000 and even higher, but the majority of measurements cluster around 500,000. The most recent work by Dreizen and his co-workers suggests that the molecule is composed of two identical large chains of about 215,000 mol wt and two small chains of 25,000, giving an overall molecular weight of about 500,000.

The myosin molecule is highly elongated, being approximately 1550 A long. In the electron microscope it can be seen to consist of a bulbous head 150–200 A long and 40–50 A wide and a long tail 15–20 A wide (Fig. 16-4). Andrew Szent-Györgyi discovered that when myosin is treated with the proteolytic enzyme trypsin for a short time, the molecules split into two particles, which he called light meromyosin (126,000 mol wt) and heavy meromyosin (324,000 mol wt). Light meromyosin retains the low solubility properties of the myosin molecule, being insoluble at low ionic strength and aggregating with ease to form long fibers (Fig. 16-5). It would therefore appear that the light meromyosin represents the thin "tail region" of the myosin molecule. Heavy meromyosin includes the more bulbous head region of the myosin molecule; it is water

Fig. 16-5. Aggregation of light meromyosin to form filaments. Notice that the surface of these filaments is smooth and lacks the projections observed with filments aggregated from myosin. Light meromyosin retains the ability to aggregate and form filaments at low salt concentrations. (Courtesy of H. E. Huxley.)

soluble and retains the ATPase properties of myosin as well as the affinity for actin. Figure 16-6 shows a model that summarizes the properties of myosin.

Purified myosin does not appear to change in molecular structure upon addition of ATP since careful studies using a variety of methods such as viscosity, light scattering, and optical rotation suggest that ATP does not affect the conformation of the myosin molecule. For ATP to have a physical effect, both actin and myosin must be present in some form of association. Thus, we must conclude that it is the actomyosin complex which constitutes the primary mechanochemical transducing system.

Actomyosin is formed when a solution of F-actin is mixed with a solution of myosin. As we have seen, the viscosity of such a solution rises dramatically as a result of the association of myosin with the F-actin filaments to form very large fibers of variable length. Actomyosin acts on ATP to form ADP and Pi, but this enzymatic activity is so vastly different from that of the myosin ATPase that one hesitates to consider them as analogous reactions. Thus, the actomyosin enzyme activity requires Mg^{2+} and is inhibited by EDTA and high ATP concentrations; myosin ATPase, however, is inhibited by Mg²⁺, activated by EDTA, and not inhibited by high ATP concentrations. The pH maxima of the two activities are also different.

The mechanism of the action of ATP on actomyosin remains shrouded in mystery. Both actin and myosin play a catalytic role in their action on ATP. ATP promotes the polymerization of G-actin—so much we know. We also know that ATP brings about a far-reaching physical change in actomyosin, and this conversion of ATP to ADP and Pi has an entirely different property from the hydrolysis of ATP by myosin. But what in fact happens to ATP, actin, and myosin when they are locked to each other in intimate contact is still unclear to us. However, the above facts are precisely what we shall have to explain in the future if we are to develop a molecular insight into muscle contraction.

Tropomyosin represents some 15 percent of the total structural protein of the muscle fibril. Since we do not as yet know its function, we shall avoid discussing it in detail and simply state that it is an elongated molecule with a molecular weight of 54,000 and appears to be associated with actin in the muscle cell.

One question that has puzzled muscle researchers a long time is how muscle relaxes after contraction. Living muscle, when not stimulated, is soft and stretchable; we say that it is in a *relaxed state*. Some time after the death of an animal, however, it becomes brittle and inextensible, and then we say that it is in *rigor*. We now know that rigor is due to the disappearance of ATP from muscle. Interestingly, it turns out that

Fig. 16-6. Structure of myosin molecule. The molecule appears to be composed of two large chains and two small ones. It would appear that the large chains form the long "tail" of the molecule and have a very high helical content. The "head" of the molecule is composed in part of the remainder of the large chains and in part of two small chains. The molecular weight of the chains can be measured by disaggregating the molecule in 5M guanidine hydrochloride and using the ultracentrifuge.

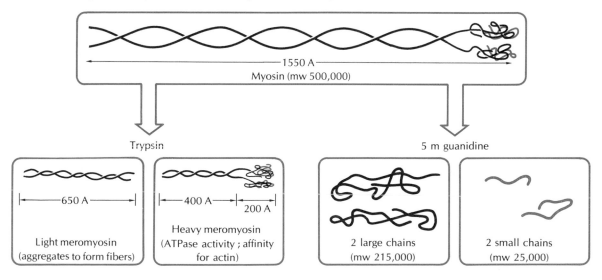

actomyosin requires ATP to maintain its relaxed state. Whether ATP brings about contraction or maintains the muscle in a relaxed state depends on whether Ca^{2+} is present. These insights into the control of muscle contraction were obtained by studying glycerinated fibers. Thus, if one takes glycerinated fibers and washes out the glycerol, one obtains brittle threads that are in a state of rigor. If one removes all traces of Ca^{2+} with EGTA, [a Ca^{2+}-chelating reagent the full name of which is ethylenedioxybis-(ethylenamino)tetraacetic acid] and adds ATP and Mg^{2+}, one obtains soft extensible threads that are in a relaxed state. If one now adds Ca^{2+}, one obtains contraction. This contraction can be reversed by removing the Ca^{2+} and adding ATP. When the ATPase activity of the myosin in the fibers breaks down the ATP, the fibrils will become brittle again, having returned to the state of rigor.

$$\text{Rigor} \xrightarrow[\text{ATP}]{Mg^{2+}} \text{Relaxed} \xrightarrow[]{\overset{\displaystyle \text{Removal of } Ca^{2+} \text{ but}}{\underset{Ca^{2+}}{\text{ATP and } Mg^{2+} \text{ present}}}} \text{Contracted} + \text{ADP} + \text{Pi}$$

The above behavior of the glycerinated fibers differs from that of threads prepared with highly purified actin and myosin. Ebashi has demonstrated that the difference seems to be due to yet another muscle protein he has called troponin, which renders the contracting system sensitive to calcium. According to Ebashi, it is only in the presence of troponin that calcium triggers the contraction of glycerinated fibers. What precisely happens here is a mystery compounding a pre-existing mystery, and it is unnecessary to speculate about it prematurely.

Fine Structure of the Muscle Fibril

Although we still do not understand in precise molecular terms how the chemical energy of ATP is converted into work by the muscle proteins, a great deal has been learned in recent years about the contraction of striated muscle by studying the fine structure of the muscle cell with a variety of optical techniques. Much of what follows has been worked out by H. E. Huxley and Jean Hanson, although a number of other workers such as A. F. Huxley, S. Page, J. Lowy, and Keith Porter have also made important contributions.

Figure 16-7 is an electron micrograph of a striated muscle fiber. Note that one can discern three major systems, each playing an important role in muscle contraction.

Fig. 16-7. (Right) The fine structure of muscle. Three different electron microscopic views show the following. A. Muscle fibrils composed of muscle filaments. The fine structure of the bands is clearly visible and will be described in Fig. 16-8 and 16-9. (Courtesy of H. E. Huxley.) B. Muscle fibrils of cardiac muscle in relation to mitochondria. These organelles are the source of ATP and provide a continuous energy input to power the work output of the muscle cell. (Courtesy of D. Fawcett.) C. Muscle fibrils in relation to the membrane system (sarcoplasmic reticulum). Note both the longitudinal and the transverse elements of the sarcoplasmic reticulum, and the precise matching of their periodicity in relation to that of the muscle fibrils. (Courtesy of D. Fawcett.) D. Schematic representation of the structural elements of striated muscle. (Courtesy of Fawcett and Bloom.)

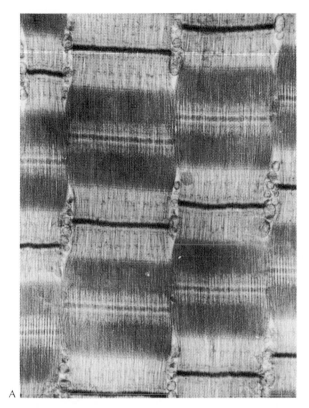

A

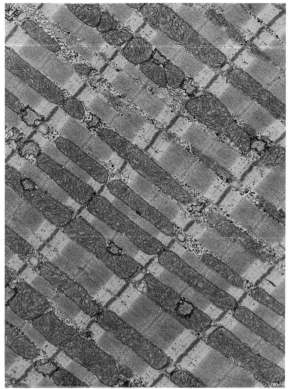

B

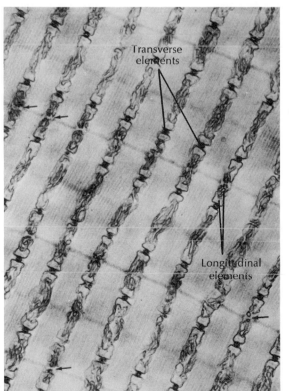

Transverse elements

Longitudinal elements

C

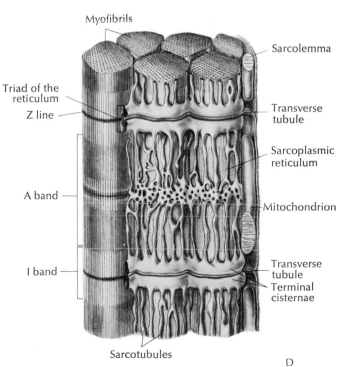

Myofibrils

Sarcolemma

Triad of the reticulum

Z line

Transverse tubule

Sarcoplasmic reticulum

A band

Mitochondrion

I band

Transverse tubule

Terminal cisternae

Sarcotubules

D

1. A system of elongated elements called myofibrils, parallel to the muscle fiber, can be discerned. As we shall see, the myofibrils are the contractile machinery of muscle.

2. Between the myofibrils are located numerous mitochondria that supply the myofibrils with ATP. The more active the muscle is, the more numerous the mitochondria are, another beautiful example of the regulatory system that governs the relations between cellular organelles.

3. Also lying between the myofibrils one finds two sets of vesicles that do not appear to connect with each other (Fig. 16-7 C and D). These are (1) the thin transverse tubules, which appear to be invaginations of the plasma membrane, and (2) the sarcoplasmic reticulum with its interconnected transverse and longitudinal vesicles.

The interconnection of the transverse tubules with the plasma membrane and the lack of connection between the transverse tubules and the sarcoplasmic reticulum was demonstrated by H. E. Huxley, who immersed some living muscle fibrils in a solution of ferritin, an iron-containing protein that is readily seen in the electron microscope as a small dense granule. After a short period of immersion he sectioned the muscle fibril and found the ferritin throughout the muscle in the interior of the transverse tubule but not in the sarcoplasmic reticulum. This result suggests that the transverse tubule is continuous with the surface membrane of the muscle fibril but does not connect with the sarcoplasmic reticulum.

The muscle cell or fiber is the smallest unit of muscle that can give a normal physiological response when activated by a nerve cell. A. V. Hill pointed out that the time interval between excitation by the nerve cell and the response of the muscle is so short that there is not enough time for a substance to diffuse from the surface of a fibril into its interior.

The connection of the transverse tubules to the plasma membrane as well as the physical proximity of the transverse tubules to the sarcoplasmic reticulum can explain the rapidity of the response of the muscle fibril to stimulation. It is now believed that the wave of excitation traveling down the nerve leads to the release of acetyl choline at the neuromuscular junction, which then generates another excitation wave that travels rapidly along the membrane system of the muscle cell (both up and down into the myofibril), presumably causing contraction by releasing Ca^{2+}. The release of Ca^{2+} by the sarcoplasmic reticulum during excitation is being actively studied at present. A great deal of evidence has accumulated to show that preparations of sarcoplasmic reticulum vesicles bind Ca^{2+} very tightly. These preparations can in fact be used to promote relaxation of glycerinated fibers in the presence of ATP.

The next question we must consider is what happens when Ca^{2+} is released from the sarcoplasmic reticulum in the presence of ATP produced by the mitochondria. In order to understand the contraction which follows, we must describe the myofibrils of the muscle cell in greater detail.

Figure 16-8 shows diagrams and micrographs of the bands, zones, and lines of the myofibril at two levels of resolution. It can be seen that these various bands, zones,

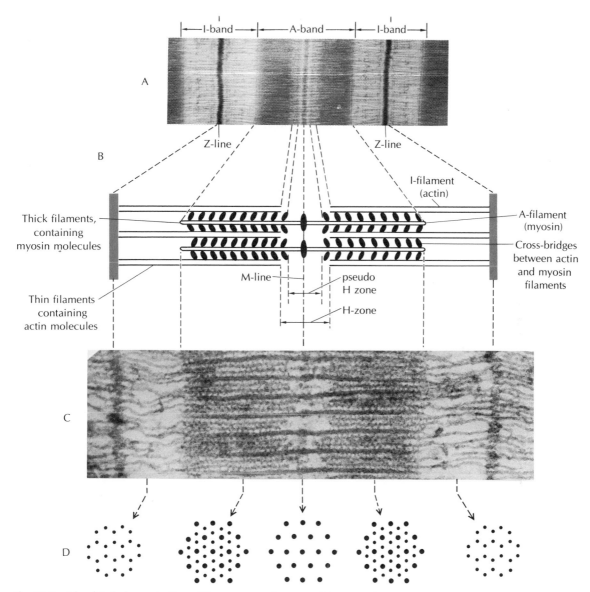

Fig. 16-8. The detailed organization of the sarcomere interpreted in terms of overlapping thick and thin filaments. A. A low-resolution electron micrograph showing the band structure of the sarcomere. The A-band, I-band, H-zone, pseudo H-zone, M-line, and Z-line can all be seen. B. Interpretation of banding in terms of structure of overlapping thin and thick filaments. C. High resolution electron micrograph showing details of filament structure. Notice that M-line can be seen to be due to interconnections between thick filaments; the pseudo H-zone, due to the absence of cross-bridges; the H-zone, due to the absence of thin filaments; the I-band, due to the absence of thick filaments; and the darkest part of the A-band, due to the presence of thick filaments, thin filaments, and crossbridges. The Z-line is due to interconnections between these filaments. D. Cross sections of the sarcomere at different levels showing the organization of the thick and thin filaments as well as the crossbridges. They show why when the fibril is cut parallel to the thick filaments, one can see two thin filaments between every thick filament. (Courtesy of H. E. Huxley.)

and lines under the light microscope can be interpreted in terms of two sets of filaments (thin and thick) overlapping partially with each other.

The sarcomere, the smallest repeat unit in the contractile process, lies between two Z-lines. Connected on either side of the Z-line are the thin I-filaments. Overlapping with two sets of thin filaments are the thick A-filaments, the center of which is demarcated by the M-line. The thick filaments carry projections or cross bridges on their surface that are absent near the center of the filaments, giving rise to the pseudo-H-zone. The H-zone is a region of no overlap of thick and thin filaments at the center of the sarcomere. Thus, the A-band is the region represented by the A-filaments (thick), and the I-band is the region where the I-filaments (thin) of two adjacent sarcomeres do not overlap with thick filaments. This description may at first seem involved, but if the student will study Fig. 16-8 patiently, these simple geometrical relationships will soon become clear.

So far we have described what can be seen in a longitudinal section of a sarcomere. The precise interdigitation of thin and thick filaments can be appreciated even better if we describe the arrangement of these filaments in cross section. Figure 16-8 also illustrates the arrangement of the filaments in cross section and shows that every thick filament is surrounded by six thin filaments. It also shows why one can see in longitudinal section two thin filaments between every pair of thick filaments.

Now that we have a clear picture of the geometry of the filaments in resting muscle, the next question we must ask is what happens to the interdigitating structure during contraction? Since the electron microscope cannot utilize living material, what one must do is fix muscle that has been made to contract and observe the effect on thin sections of the material. The answer appears quite simple and straightforward: the degree of overlap between the interdigitating thin and thick filaments increases upon contraction. (Fig. 16-9). What turned out to be far from simple was to account for this phenomenon quantitatively and to correlate quantitatively what can be seen under the light microscope with the picture under the electron microscope.

The sliding theory of contraction of striated muscle makes the following minimum predictions.

1. Under the light microscope (especially using interference or phase contrast optics), it should be possible to show that upon contraction the width of the A-band remains constant, the width of the I-band decreases, and the distance between the Z-line and the edge of the H-zone remains constant.
2. Under the electron microscope the length of the thick filaments and of the thin filaments remains constant during contraction.
3. Contraction brings about an increase in degree of overlap between thin and thick fibers.

In spite of considerable technical difficulties and consequent controversy, we can say with some assurance that these predictions have been validated by the painstaking work of Hansen and Huxley, although final proof will rest on the more precise X-ray diffraction studies now in progress.

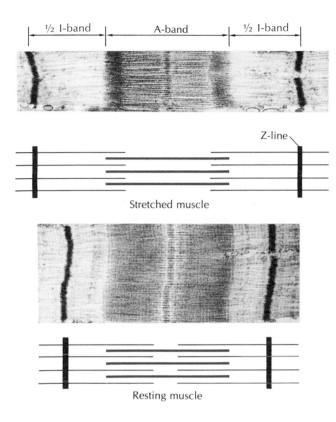

½ I-band	A-band	½ I-band

Z-line

Stretched muscle

Resting muscle

Fig. 16-9. The effect of contraction or stretching on the width of the *I*-band and the *H*-zone. The sliding theory demands that the sarcomere at various stages of concentration should show differences in the *I*-band and the *H*-zone, whereas the width of the *A*-band should remain constant. The above electron micrographs show that this is indeed the case, and the diagrams explain why this should be. (Courtesy of H. E. Huxley.)

To summarize, the sliding theory of striated muscle contraction states that during shortening of the muscle the overlap between the interdigitating thin and thick filaments increases without an overall change in length of these filaments.

The next question we must ask is of what are the thin and thick filaments composed? This question bridges the gap between the microscopic and biochemical work, and it is hoped that in this area important progress will be made in the near future.

Hanson and Huxley demonstrated very convincingly that after treatment of muscle with solutions that extract myosin, *A*-bands will disappear as observed in the phase-contrast microscope (an experiment which Danilevskii had anticipated in 1888), and thick filaments will vanish from electron micrographs. This led Hanson and Huxley to the conclusion that the thick filaments are composed mostly of myosin.

If one extracts the remaining material with 0.6M KI, which is known to remove actin, the thin filaments disappear; but quantitative estimation shows that in the thin filaments there is another major component, which we believe is tropomyosin. If one extracts for both tropomyosin and actin, the two proteins come off in constant proportions, suggesting that these two proteins in a constant ratio constitute the thin filaments. It has been suggested that tropomyosin fits into the two grooves of the two-stranded helix of actin, thus giving it additional rigidity.

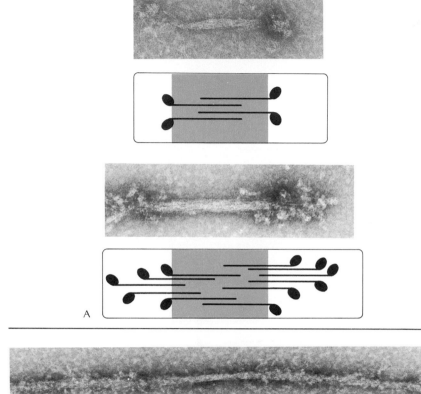

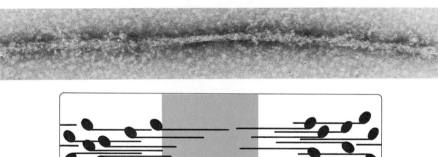

Fig. 16-10. Relationship between the structure of the myosin molecule and the morphology of the *A*-filaments. (Courtesy of H. E. Huxley.) A. *In vitro* aggregate of the myosin molecule. Note the smooth area at the center of the aggregate, which can be explained by assuming a mechanism of tail-to-tail aggregation as illustrated. B. The thick filament (myosin) has a smooth region (pseudo *H*-zone) at the center similar to the one seen in the aggregate above.

Let us look at the architecture of the thick filaments more closely. We have seen that the myosin molecule has a small bulbous head and a long tail and that *H*-mero-myosin (which contains the head) also has the ATPase and the actin-combining activity. Under the electron microscope we can see that the thick filaments have crossbridges that interact with the actin-containing thin filaments. It is therefore tempting to conclude that the crossbridges represent the bulbous heads of the myosin molecule. H. E. Huxley in a dramatic series of experiments was indeed able to demonstrate that this is so. He showed that when myosin is allowed to aggregate at low ionic strength, it forms elongated filaments of variable length but reasonably uniform diameter. Furthermore, just as is the case in the thick filaments, there is a smooth region (1500 A long) in the middle of the filament corresponding to the pseudo-H-zone, beyond which one finds the same projections or crossbridges. The presence and length of this smooth region can be explained by assuming that the myosin molecules begin their aggregation by arranging themselves tail to tail with their heads pointing in opposite directions, with further aggregation just increasing the length of the filament carrying the crossbridges (Fig. 16-10). On the other hand, if L-meromyosin is polymerized, then entirely smooth filaments are obtained (Fig. 16-5), a fact which should not surprise us since by now the reader should be convinced that the crossbridges are indeed the bulbous heads of the myosin molecule.

Now, one consequence of the sliding theory is that there should be an opposite polarity in the two halves of the thick filament since the thin filaments slide over it in opposite directions. It is also possible that the opposite polarity could reside in the two sets of thin filaments of a sarcomere, or it might be that the opposite polarity is found in both the thick and the thin filaments.

The Huxley aggregation experiments described above indicate the geometric basis for opposite polarity in the thick filaments, but what about the thin filaments? To answer this question, H. E. Huxley again performed an ingenious experiment. He mixed actin filaments with H-meromyosin and showed with the electron microscope that the latter attaches itself to actin to form "arrowheads" (Fig. 16-11), which are a good index of the polarity of the actin filament. One can now ask whether the thin filaments (composed of actin and tropomyosin) on opposite sides of the Z-line have opposite polarity. When a test was made using thin filaments extracted from muscle (but still attached to their Z-lines) and H-meromyosin, the arrowheads pointed in different directions on different sides of the Z-line. Therefore, it became clear that the thin filaments on either side of the Z-line are of opposite polarity (Fig. 16-12). Thus Huxley was able to show that both types of filaments have the requisite polarity to account for the sliding theory.

One puzzling feature of Huxley's aggregation experiments is the nature of the in vivo mechanism for generating filaments of constant length. Since the mechanism of assembly of the thick and thin filaments appears to be relatively uncomplicated, the study of the in vivo assembly of the myofibrils may be a convenient way of attacking the problem of assembly of cell structures.

Since the measurements of the thick and thin filaments suggest that no change of filament length occurs during contraction, the sliding model requires that conformational changes occur in the crossbridges in order to bring about movement. Recent X-ray diffraction studies of living muscle by H. E. Huxley indicate that this is indeed what happens. X-ray diffraction studies of resting muscle show that the crossbridges

Fig. 16-11. Formation of "arrowheads" demonstrating polarity of thin filaments (actin). A. Actin filament showing two-coiled structure of globular units. B. Electron micrograph of a thin filament forming an aggregate with *H*-meromyosin. The meromyosin seems to attach at an angle, showing that the actin has an inherent polarity. (Courtesy of H. E. Huxley.)

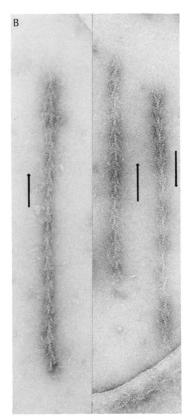

Fig. 16-12. Opposite polarity of thin filaments on opposite sides of the Z-line. When thin filaments are still attached to the Z-line, it can be shown that they have opposite polarity because arrowheads point away from the Z-line on either side of it. This opposite polarity is consistent with the sliding filament model since during contraction the thick filaments are thought to converge towards the Z-line in adjacent sarcomeres. (Courtesy of H. E. Huxley.)

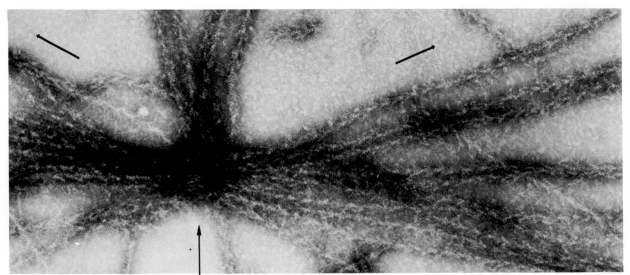

Z-line attachments of thin filaments

are arranged in an approximately helical pattern (which differs from the helical pattern that can be observed in the thin filaments). Detailed studies of the crossbridges suggest that they are organized in a helical array, there being six crossbridges per turn of helix, and each helix being about 400 A long. Similar studies of contracted muscle seem to show considerable changes in the configuration of the crossbridges without any significant changes in the length of the filaments. The nature of the changes in the crossbridges and how they bring about the sliding motion must still be determined.

Let us now return to our earlier description of the biophysical parameters of muscle contraction. Can we explain these in terms of the sliding theory?

We shall assume that the directional motion of the filaments is due to the cycle of reaction between the crossbridges and the actin filaments involving (1) attachment, (2) conformational change producing a directional force, and (3) detachment. Thus, the more crossbridges attached at any one time, the greater the force. The greatest number of attachments occurs therefore at rest. If the muscle is allowed to shorten, then the number of attachments that can take place at any one time will decrease with increasing rate of shortening, and the force that is generated will be correspondingly lower. This is then how one might explain why the force exerted by contracting muscle decreases with increasing rate of shortening.

According to the above model, it would seem reasonable that energy is liberated only when a cycle of attachment-detachment occurs. Thus, only when a crossbridge can contribute to tension is work being done, and thus there is a necessary connection between the work done by the muscle and the energy released.

Finally, the sliding mechanism can explain very nicely the shape of the length-tension curve observed in muscle. The student should study Fig. 16-13 carefully to see if he can explain the precise features of this curve in terms of the sliding theory. It can be seen that the amount of rest tension the muscle can exert depends on the degree to which it is stretched. Maximum tension is obtained when there is maximum overlap between thin and thick filaments. If muscle is stretched beyond this length, then the tension exerted decreases with increase in distance of stretch. This can be explained by the decreasing amount of overlap between the thick and thin filaments. Beyond a certain degree of stretch the tension comes down to zero, and this can be demonstrated to represent the degree of stretch when the filaments cease to overlap. At the other extreme, when muscle is allowed to shorten considerably, then tension decreases, and this can be shown to be due to the overlap of thin filaments coming from opposite directions. Tension comes down to zero when thick filaments collide with the Z-line and begin to crumple.

We have come a long way in obtaining an understanding of the events bringing about muscle contraction. Let us summarize what we have learned by describing what we think does happen in a cycle of contraction.

The muscle is at rest (relaxed state); ATP is present in the myofibril, but Ca^{2+} is absent because it is actively bound by the sarcoplasmic reticulum. A nerve impulse causes a wave of excitation to travel through the sarcoplasmic reticulum, Ca^{2+} is released, and the myosin crossbridges begin to utilize ATP in their cycle of attachment,

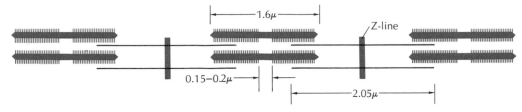

A. Precise dimensions of relevant parameters.

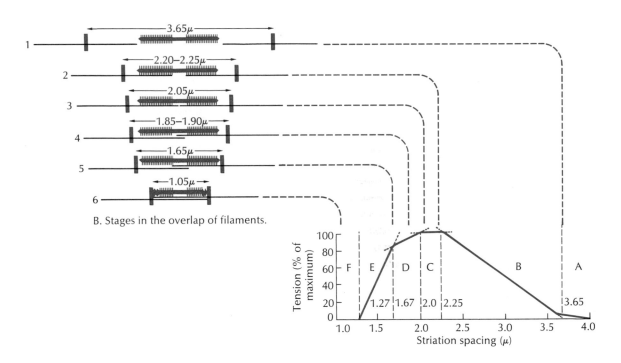

B. Stages in the overlap of filaments.

C. Length-tension diagram in relation to filament overlap.

Fig. 16-13. Length-tension diagram of striated rabbit muscle in relation to degree of filament overlap. (Redrawn from Gordon, Huxley, and Julian 1966.) A. Precise dimensions of relevant filament parameters. B. Six critical stages in overlap of filaments which can explain the shape of the length-tension diagram: (1) no overlap between thin and thick filaments; (2) maximum overlap between thin filaments and region of thick filaments with crossbridges; (3) maximum overlap between thin and thick filaments; (4) overlap between thin filaments equivalent to width of bridgeless region of thick filaments; (5) degree of contraction causing thick filaments to collide with Z-lines; and (6) degree of overlap causing crumpling of tips of thick filaments. C. Length-tension diagram showing how the shape of curve is related to degree of overlap between thin filaments and region of thick filaments carrying crossbridges.

conformational change, and detachment, thus bringing about a sliding motion. When excitation stops, Ca^{2+} is pumped out of the sarcoplasm by the sarcoplasmic reticulum, and the crossbridges stop going through their cycle of attachment but remain in the unattached position. The muscle can now relax in the presence of ATP and assume a rest length, which can be brought about by a very low force. If the supply of ATP disappears because of anoxia, poisoning of respiration, or death, the muscle will assume a state of rigor.

Much of what we have discussed regarding the mechanism of muscle contraction is still under active debate. It may be that a future edition of this book will have to take back a great deal of what was said here, yet we feel that the risk is worth taking.

The Metabolic Regulation of Contraction

As we have mentioned before, ATP is the primary energy source of a number of transductions occurring in the cell. ATP also appears to exert a regulatory role on a number of metabolic processes. This fact must mean that the exact level of concentration of ATP is of critical importance. The question therefore arises as to how the muscle cell can make available to the contractile machinery large amounts of energy from ATP, given the fact that the nature of muscle tissue is such that huge amounts of energy are frequently released in sudden bursts of activity.

By conducting experiments with excised muscle that no longer is connected to an external supply of blood, it can be shown that rapid and repeated contraction of the muscle does not at first bring about a reduction of ATP concentration. However, there is another high-energy phosphate compound, phosphocreatine, which under the above conditions does decrease. Such experiments have taught us that the muscle cell can utilize phosphocreatine to convert ADP to ATP with the aid of the enzyme phosphocreatine kinase. Since ATP is necessary for the synthesis of phosphocreatine, we must conclude that phosphocreatine is a high-energy storage compound that ensures the rapid availability of large amounts of energy to the muscle.

$$
\begin{array}{ccc}
\underset{\text{phosphocreatine}}{\overset{\displaystyle \begin{array}{c} OH \\ | \\ H{-}N{-}P{=}O \\ | \\ OH \\ | \\ C{=}NH \\ | \\ CH_3NCH_2COOH \end{array}}{}} + ADP
& \underset{\text{kinase}}{\overset{\text{phosphocreatine}}{\rightleftharpoons}}
& \underset{\text{creatine}}{\overset{\displaystyle \begin{array}{c} NH_2 \\ | \\ C{=}NH \ + \ ATP \\ | \\ CH_3NCH_2COOH \end{array}}{}}
\end{array}
$$

We conclude that phosphocreatine (or phosphoarginine in invertebrates) is the first, most readily available reservoir of energy of the muscle cell. The ready availability of energy from phosphocreatine, however, is only of value to the exercising muscle if it can be replaced by the production of ATP by the metabolism at rates which ultimately must be equivalent to the rate of ATP utilization. At moderate rates of exercise, the aerobic metabolism of muscle is indeed able to use the glycogen stored in muscle to synthesize sufficient ATP for the contractile machinery. But during rapid exercise the oxygen supply to the muscle is not sufficient to maintain the requisite rate of aerobic metabolism.

This phenomenon was discovered long ago with muscle fibers that were stimulated in the absence of oxygen. It turned out that muscle fibers are capable of contraction under anaerobic conditions and that they accumulate large amounts of lactic acid. When the stimulation was continued for a while, the state known as "muscle fatigue" ensued. If this fatigued muscle were then exposed to oxygen, the lactic acid disappeared, and the muscle regained its ability to contract. If the muscle were to contract at a moderate rate in the presence of air, no lactic acid would ever accumulate. When muscle was stimulated in the presence of iodoacetate—a poison that inhibits glycolysis, or the accumulation of lactic acid—it still contracted, but at the same time it was noticed that phosphocreatine broke down to creatine and inorganic phosphate. When this breakdown was complete, no further muscular contractions took place. Under all these conditions of anaerobic contractions and relaxations, the concentration of ATP remained unchanged.

The release of lactic acid from muscle under anaerobic conditions turned out to be an important effect that biochemists such as Carl and Gertrude Cori, and Meyerhof and Embden utilized to study the anaerobic metabolism of the cell. In fact, the study of glycolysis in muscle and fermentation in yeast represented two major areas of biochemical research in the 1930s, and the discovery of the near congruence of these two phylogenetically distant pathways represented one of the early triumphs of cell biology.

(See Chapter 12.)

The importance of the phenomenon of glycolysis in muscle contraction is that during violent exercise of a given muscle, glycolysis is capable of producing ATP locally but at the expense of generating large amounts of lactic acid, which, in the intact organism, is removed by the blood supply and oxidized by other tissues. Under violent and generalized exercise, the aerobic metabolism of the whole organism may lag behind the energy release of the muscle, and lactic acid may accumulate throughout the organism. Under these conditions, the organism adapts by raising its rate of breathing (panting). The student is reminded that a detailed discussion of glycolytic metabolism and its basic energetics can be found in Chapter 12.

Thus, in summary, we can conclude (Fig. 16-14) that (1) muscle uses ATP to convert chemical energy into work; (2) ATP is maintained at a constant level in the muscle cell by the presence of phosphocreatine, the energy of which can be utilized to synthesize ATP; and (3) at moderate rates of exercise, the muscle can make up the energy utilization through its aerobic metabolism, but when the oxygen supply to the muscle is insufficient to balance the utilization of ATP, anaerobic metabolism at very high rates is capable of supplying sufficient ATP to maintain muscle work. After muscular exercise, the lactic acid that has accumulated is aerobically converted back into glycogen.

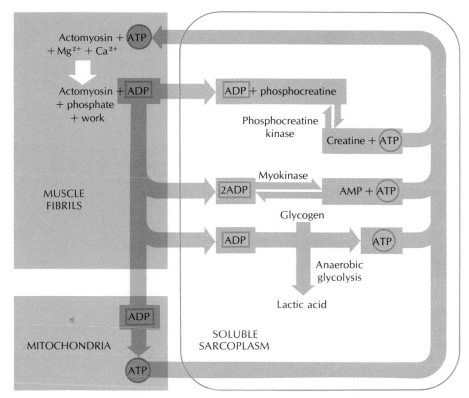

Fig. 16-14. Energy flow in muscle. The muscle is seen here as a mechanochemical transducer using energy from ATP to produce work. The ATP is generated locally by four separate systems: (1) by the mitochondria, (2) from phosphocreatine, (3) from ADP and myokinase, and (4) from the anaerobic glycolysis of glycogen to lactic acid. (Redrawn from Siekevitz, 1959.)

The contraction of muscle is a most exciting biological phenomenon. Consider insect flight muscle capable of sudden and rapid release of large bursts of energy, or heart muscle, continually in motion for the entire life span of the organism, or the catch muscle of the clam, able to maintain great tensions for long periods of time with considerable energetic economy. Modern muscle research is a truly integrated field of inquiry, which utilizes the entire armory of modern experimental approaches and seeks to unify the whole spectrum of phenomena — mechanical, electrical, and chemical — into one piece of functioning machinery. Although much has been learned in the last 20 years, we have yet to understand in molecular terms how the energy derived from ATP is converted into mechanical work by living systems.

Finally, the muscle cell is an excellent example of the synergistic relationship between two intracellular organelles, in this case the mitochondria and the fibrils. First, it is well known that muscle tissue is full of mitochondria, as seen for example in

Fig. 16-7B. Second, the relationship between the two is that mitochondria provide the energy of aerobic metabolism, and the fibrils use this energy. Third, both organelles are very responsive, and responsive in the same manner, to the cellular levels of that system they share in common, namely, the ADP-ATP interchange.

This entire system is probably a finely meshed one. Experiments on isolated mitochondria indicate that these mitochondria last longer, in a metabolic sense, when they are performing work than when they are not. If mitochondria are isolated and simply let stand at room temperature, they soon lose their ability to oxidize most substrates. If, however, they are made to oxidize substrate while they are left standing, they can continue to do so for a long period of time. Muscle tissue seems to be of the same nature: if the nerve connections to muscle are severed, making the muscle unable to function, in a relatively short time the muscle tissue starts to degenerate. We also know that muscle tissue can be made to "develop" by being made to work. What this means biochemically is that when muscle contracts and breaks down ATP to ADP, the ADP enters the mitochondria, and there acts like a phosphate acceptor, as explained in Chap. 13. If substrate is available, the release of ADP from the fibrils to the mitochondria acts as a stimulant to the mitochondria; substrate is oxidized and ATP is produced, and once again it can be recycled through the muscle fibril system. In other words, the mitochondria seem to be able to respond to ADP, to make ATP, as long as there is a need for ATP. If the muscle contracts continuously, a continuous supply of ADP will be fed into the mitochondria, which will respond by supplying ATP. If there is no need for contraction to continue, if there is no need for ATP, no ADP is formed, and hence no ATP will be generated. Because energy is precious, nature seems to have constructed a system in which no more biological energy is produced than can be used. It should be pointed out, however, that this idea of the symbiotic relationship between muscle fibrils and muscle mitochondria is based on the results of experiments with isolated mitochondria and may have little to do with cellular reality. Nevertheless, it does appear to many biochemists to have some relevance to what happens *in situ* in the muscle cell.

Other Mechanochemical Transducers of the Cytoplasmic Matrix

As we have pointed out at the beginning of this chapter, the contraction of striated muscle is just one of many examples of mechanochemical transduction found in nature. In recent years cell biologists have begun to study a large variety of phenomena involving the conversion of chemical energy into work, utilizing some of the techniques and insights gained from the study of muscle. Since it is very likely that the same basic molecular phenomena are operative in all mechanochemical systems, it is reasonable to expect that a broader study of these phenomena might in the long run help to clear up some of our present difficulties in understanding the basic molecular events involved.

One of the earliest observations in this broader approach to mechanochemical phenomena was the extraction in 1952 of an actomyosinlike protein from the plasmodium of the slime mold *Physarum polycephalum*. This organism has the appearance

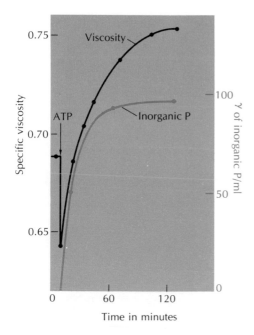

Fig. 16-15. Evidence for the presence of an actomyosin-like system in the protein extract of a primitive organism (the slime mold *Physarum polycephalum*). Notice that upon addition of ATP one obtains a very rapid drop in viscosity, followed by a slow rise. ATP hydrolysis, as indicated by the appearance of inorganic phosphate, which parallels the rise in viscosity. (From fig. 2, A. G. Loewy, *J. Cell Physiol.*, **40**:132.)

of a gigantic yellow amoeba — it can weigh several grams — and is capable of very rapid protoplasmic streaming, rates as high as 1 mm per second having been observed. It was shown that the actomyosinlike extract from this organism exhibited a very rapid drop in viscosity upon addition of ATP, followed by a slower rise in viscosity (Fig. 16-15). It was also possible to show that the rise in viscosity paralleled a release of inorganic phosphate from the ATP, suggesting that the ATPase activity of the extract accounted for the viscosity increase. Nakajima purified the protein responsible for the ATP-induced viscosity change and showed that the very same protein also carried the ATPase activity. He also found that properties such as effects of divalent cations on ATPase activity and the inhibitory effects of sulfhydryl poisons greatly resemble the properties of striated muscle actomyosin. Was this protein extracted from the slime mold one protein, or was it, as is the case in muscle, a complex of an actinlike and myosinlike protein?

Oosawa and his co-workers have recently shown that the latter is the case by purifying from the slime mold a protein that has almost identical properties to those of muscle actin. The myxomycete actin occurs in the G-form (57,000 mol wt) in distilled water and polymerizes to an F-form upon the addition of KCl and ATP. The F-form appears under the electron microscope as a two-stranded helix closely resembling the structure of F-actin (Fig. 16-16). Just as is the case with muscle actin, the plasmodial actin converts 1 mole of ATP to ADP for every mole of G-actin.

$$n \text{ G-actin} + n \text{ ATP} \xrightarrow{\text{0.3 M KCl}} n \text{ F-actin-ADP} + n \text{ Pi}$$

Fig. 16-16. *F*-actin from the slime mold *Physarum polycephalum*. A high-resolution electron micrograph negatively stained with phosphotungstate showing a two-stranded helical structure. The white bars indicate the positions where the two strands cross over, this distance corresponding to about 120–350A. (Courtesy of Hatano and Oosawa, 1967.)

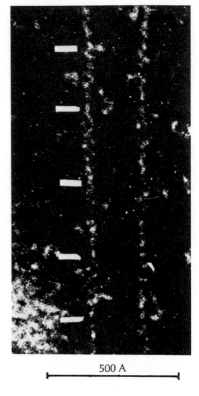

500 A

But interestingly, when *G*-actin polymerizes in the presence of Mg^{2+}, the actin becomes a genuine ATPase, continuing the breakdown of ATP at a steady rate (Fig. 16-17).

$$ATP \xrightarrow[\text{2mM MgCl}_2]{\text{actin} \quad \text{0.1 M KCl}} ADP + P_i$$

Fig. 16-17. The ATPase properties of *F*-actin. A. The broken line shows the amount of inorganic phosphate which is equimolar with *G*-actin. Thus, in the absence of Mg^{2+}, slime mold *G*-actin polymerizes to form *F*-actin while hydrolyzing an equimolar quantity of ATP to ADP. B. In the presence of Mg^{2+}, however, slime mold actin becomes an ATPase, hydrolyzing ATP to ADP at a steady but lower rate than the initial rate of hydrolysis occurring during actin polymerization. Oosawa and his colleagues were able to show independently that actin polymerization does indeed go to completion in this first 15-min interval of rapid ATP hydrolysis. (Courtesy of Oosawa, 1967.)

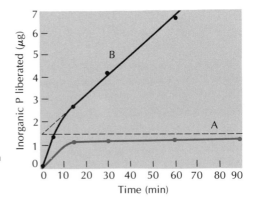

Thus, plasmodial actin is able to do, under normal ionic conditions, what muscle actin can do only when subjected to ultrasonication. This most interesting observation reenforces the view that actin is indeed a catalytic protein and illustrates the value of studying basic biochemical phenomena over a broad range of different living systems.

In addition to drawing attention to this interesting phenomenon, Oosawa and his colleagues have demonstrated the similarity of myxomycete and muscle actin by showing that (1) the amino acid composition of the two proteins is strikingly similar, (2) the two forms of actin copolymerize to form one F-actin polymer with intermediate viscosity, and (3) myxomycete actin interacts with muscle myosin to form an actomyosin complex retaining the property of decreasing in viscosity on addition of ATP (Fig. 16-18).

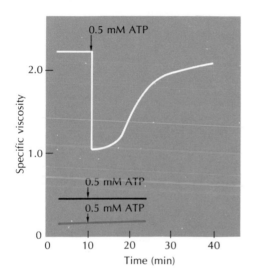

Fig. 16-18. The effect of ATP on the viscosity of actomyosin in which the actin was derived from the slime mold and the myosin from striated rabbit muscle. Solid color—actin purified from the slime mold *Physarum polycephalum*. Black—myosin prepared from striated rabbit muscle. White—mixture of plasmodial actin and rabbit myosin. Notice the large increase in viscosity obtained upon mixing the two proteins, as well as the viscosity-lowering effect of ATP. (Courtesy of Oosawa, 1966.)

Very recently Hatano has succeeded in purifying slime mold actomyosin, which under the electron microscope appears as long filaments 100 A wide. In the presence of ATP and Mg^{2+} myosin appears to go into solution leaving behind 75 A-wide filaments that seem to correspond to actin filaments, which have been prepared by polymerizing purified actin (Fig. 16-16). Hatano and his colleagues were able to purify slime mold myosin from slime mold actomyosin dissociated by ATP. Slime mold myosin differs from muscle myosin in that it is soluble at low ionic strength. However, its sedimentation constant and ATPase activity—especially in relation to Mg^{2+} and Ca^{2+}—are the same as that of muscle myosin.

To what extent then is the presence of actinlike and myosinlike mechanochemical proteins a general phenomenon of nature? Recently Poglazov and his co-workers utilized the effect of ATP on the viscosity of protein extracts as a criterion for the presence of actomyosin and investigated a number of cell types and cell organelles.

They found the characteristic rapid viscosity drop followed by a slower viscosity rise in such diverse living systems as the green alga *Nitella flexilis* (Fig. 16-19), and in extracts of animal tissues such as liver, brain, kidney, as well as of organelles such as mitochondria.

Fig. 16-19. The effect of ATP on the viscosity of an actomyosinlike material from the plant *Nitella flexilis*. This plant exhibits very rapid "cyclosis" (circular protoplasmic streaming), and it is therefore not surprising that it should contain measurable amounts of actomyosinlike mechanochemical proteins. Notice that the effect of viscosity is reversible and can be repeated more than once. (Courtesy of Poglazov, 1967.)

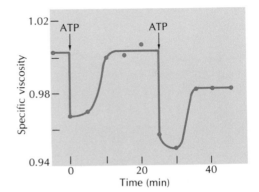

In the case of the bacteriophage T2, from the tail or "caudal sheath" (Fig. 16-20A) one can prepare a protein that in the presence of Ca^{2+} or Mg^{2+} aggregates into long spiral threads.(Fig. 16-20B). This material does not decrease in viscosity on addition of ATP, but Poglazov showed that when muscle myosin is added, a high viscosity aggregate is formed that upon addition of ATP behaves in a typical actomyosinlike manner. The presence of an actinlike protein in the tail sheath of T2 bacteriophage is not surprising since it has been shown that this structure is capable of contracting upon attachment to the bacterial cell wall, a phenomenon which presumably initiates the injection of the virus DNA into the host cell (Fig. 16-20C). Here then we seem to have the actin component of the system while the myosin component has as yet not been isolated.

In the case of cilia the reverse seems to be true. Gibbons has isolated and purified a myosinlike protein that has ATPase activity. The protein comes in two sizes that appear to be two states of aggregation of the same molecule, the smaller one having a molecular weight of 600,000 while the larger one is an elongated rod of 5.4×10^6 molecular weight. The extraction of the protein from cilia causes the disappearance of the two arms stretching from one of the tubular elements in the outer ring of fibers. Gibbons found that by incubating the denuded cilia with the protein extract, it is possible to reassemble the arms in their proper position.

It is not yet known, even in gross mechanical terms, how the movement of cilia is brought about. It is conceivable that the ciliary microtubules are composed of actin and that some sliding motion of the myosin arms with respect to the tubules brings about a distortion or bending of the cilium.

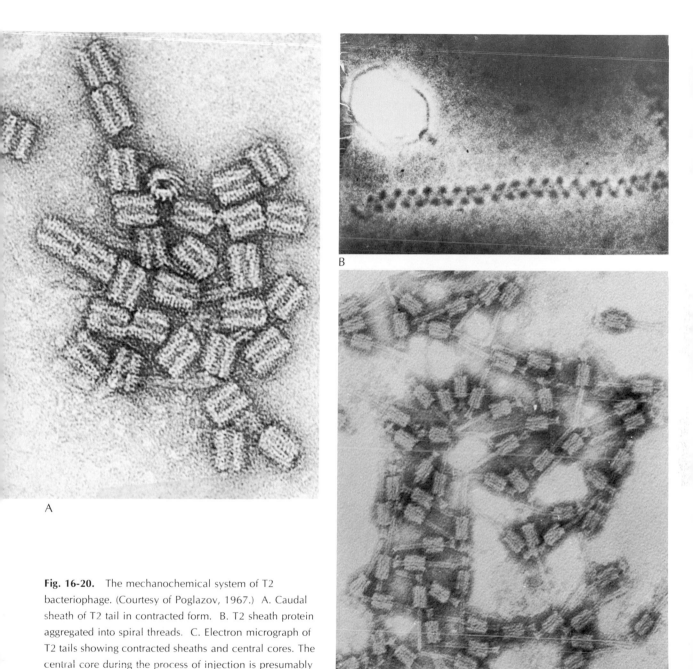

Fig. 16-20. The mechanochemical system of T2 bacteriophage. (Courtesy of Poglazov, 1967.) A. Caudal sheath of T2 tail in contracted form. B. T2 sheath protein aggregated into spiral threads. C. Electron micrograph of T2 tails showing contracted sheaths and central cores. The central core during the process of injection is presumably pushed through the wall and membrane of the bacterium.

It is so far premature to outline a coherent description of the variety of phenomena of cell motility. Nevertheless, it is possible to make a few general statements, which at least indicate the direction in which the search should be pointed.

The phenomena of motility show great variability in outward form. There is amoeboid streaming, which looks as if the movement of the fluid is the result of some active changes in pressure, occurring presumably in the gelated cytoplasm surrounding the stream. On the other hand, we have streaming in plant cells, in which the motile force appears to be exerted at the interface between the moving and stationary phases. But we also have autonomous movement of organelles such as chloroplasts, mitochondria, and perhaps of membranes in pinocytosis. And then, of course, we have the chromosomes at anaphase sliding along the spindle in opposite directions. Although these phenomena appear different in their outward morphology, it would seem reasonable to assume that they are all powered at a molecular level by sliding motions between actinlike and myosinlike proteins.

In striated muscle, at physiological ionic strength and pH, both these proteins occur in polymerized form as fibers. In other cells such as smooth muscle, amoeboid cells, and actively streaming plant cells, the myosinlike proteins probably occur either in solution or in association with the polymerized actin. Certainly, it seems a universal fact that all mechanochemical systems contain either filaments or microtubules. Furthermore, it would appear that in the cases in which the presence of microtubules is correlated with motion, they seem to occur in association with projections or sidearms. In the case of nuclear division, for instance, the material often observed in association with the centromere of the chromosome may be the myosinlike protein interacting with the actinlike proteins of the spindle microtubules.

Thus, we might enunciate a general hypothesis, which would assert that mechanochemical systems are composed of two proteins—actin, a fibrous protein forming either filaments or tubules, and myosin, present in various degrees of polymerization and in close association with actin. Motion, then, would be brought about by sliding of these two proteins with respect to each other, or at least by conformational changes induced in the actomyosin complex by ATP. Such a hypothesis should cause us to search for these types of proteins in a variety of living systems, for such a "broad-gauge" approach might well bring about a better understanding of the molecular details of the mechanism whereby chemical energy is transformed into mechanical work.

We have discussed mechanochemical systems in some detail because they represent an excellent illustration of an important and complex cellular phenomenon that is becoming increasingly susceptible to analysis at a molecular level. Although the exact molecular details of these mechanochemical conversions still elude our understanding, we are beginning to see the outlines of a general mechanism coming into increasingly sharper focus.

SUGGESTED READING LIST

Offprints

Hayashi, T., 1961. "How Cells Move." *Scientific American* offprints. San Francisco: W. H. Freeman & Co.

Huxley, H. E., 1956. "Muscular Contraction." *Endeavour,* pp. 177–188. Bobbs-Merrill Reprint Series. Indianapolis: Howard W. Sams & Co.

Huxley, H. E., 1958. "The Contraction of Muscle." *Scientific American* offprints. San Francsico: W. H. Freeman & Co.

Huxley, H. E., 1965. "The Mechanism of Muscular Contraction." *Scientific American* offprints. San Francisco: W. H. Freeman & Co.

Huxley, A. F., and Taylor, R. E., 1958. "Local Activation of Striated Muscle Fibres." *Jour. Phys.,* pp. 426–441. Bobbs-Merrill Reprint Series. Indianapolis: Howard W. Sams & Co.

Porter, K. R., and Franzini-Armstrong, C., 1965. "The Sarcoplasmic Reticulum." *Scientific American* offprints. San Francisco: W. H. Freeman & Co.

Smith, D. S., 1965. "The Flight Muscles of Insects." *Scientific American* offprints. San Francisco: W. H. Freeman & Co.

Articles, Chapters, and Reviews

Gibbons, I. R., 1968. "The Biochemistry of Motility." *Annual Review of Biochemistry, 37,* p. 521. Palo Alto, California: Annual Reviews, Inc.

Huxley, H. E., 1960. "Muscle Cells." In: *The cell,* Vol. 4, p. 365. J. Brachet, and A. E. Mirsky, eds. New York: Academic Press, Inc.

Huxley, H. E., 1966. "The Fine Structure of Striated Muscle and Its Functional Significance." In: *The Harvey Lecture Series, 61.* New York: Academic Press, Inc.

Seifter, S., and Gallop, M., 1966. "The Structure Proteins." In: *The Proteins,* Vol. 4. H. Neurath, ed. New York: Academic Press, Inc.

Books

Allen, R. D., and Kamiya, N., 1964. *Primitive Motile Systems in Cell Biology.* New York: Academic Press, Inc.

Finean, J. B., 1962. *Chemical Ultrastructure in Living Tissues.* Springfield, Ill.: Charles C Thomas, Publisher.

Gergely, J., 1964. *Biochemistry of Muscular Contraction.* Boston: Little, Brown & Co., Inc.

Poglazov, B. F., 1966. *Structure and Function of Contractile Proteins.* New York: Academic Press, Inc.

Szent-Györgyi, A., 1951. *Chemistry of Muscular Contraction,* 2nd ed. New York: Academic Press, Inc.

Chapter 17   The Membrane System
and The Exchange of Materials

The cell is a locus of chemical structure and function in which a continuity of properties is maintained in the midst of a drastically different and ever-changing environment. One important mechanism by which the cell achieves this constancy is the regulation of the movement of materials into the cell or out of it. Even within the cell, materials are not uniformly distributed, and here also we encounter a precise regulation of the interchange of materials. To achieve this regulatory control the cell utilizes a delicate membrane 75 A in width, which can engage in two distinct activities: (1) it can distinguish among different molecular species, slowing down the diffusion of some substances through it while allowing others to pass almost unimpeded; and (2) it can, with the help of the necessary energy sources, bring about the transport of material against diffusion gradients either inward (accumulation) or outward (excretion and secretion).

How Cell Permeability Is Studied

When in the early 1950s the electron microscope first revealed the precise silhouette of the plasma membrane, cell biologists were not at all surprised by this discovery because the existence and approximate dimensions of the plasma membrane had been deduced by generations of workers since Overton's classical studies during the last decade of the 19th century. Indeed even before Overton, Pfeffer had shown that cells behave like little osmometers, shrinking and swelling in relation to the concentration of the solution in which they are placed. Figure 17-1 records the basic observations regarding the osmotic behavior of cells when these are placed in solutions of nonpene-

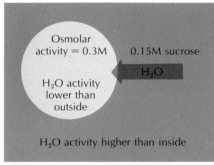

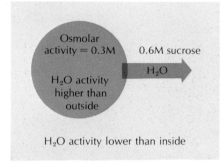

A. Water enters the cell   B. Water leaves the cell

Fig. 17-1. The cell as an osmometer. In the presence of a nonpenetrating solute like sucrose, water diffuses from a region of high thermodynamic activity to a region of low thermodynamic activity. The osmolar activity of the cell is the sum total of all the osmolar activities of impermeable ions and molecules capable of diffusing freely inside the cell. This includes mostly small molecules since the molar concentration of macromolecules is very low. A. When the cell is placed in a sucrose concentration that is lower than the osmolar activity of the cell interior (hypotonic), water will diffuse into the cell. B. The opposite happens when the cell is placed into a solution of sucrose of higher concentration than the cell's osmolar activity (hypertonic).

trating solute molecules. The changes in cell volume shown in Fig. 17-1 are due to the fact that solute molecules lower the thermodynamic activity of the water (which readily permeates the cell membrane), causing it to diffuse across the membrane from a region of high to a region of low thermodynamic activity. After equilibrium has been reached, we say that the osmolar activity inside the cell is the same as outside. Therefore, one can measure the osmolar activity of a cell under normal growth conditions by determining the sucrose concentration in which no change in cell volume occurs (Fig. 17-2).

Using techniques of measurement such as described in Fig. 17-2, Pfeffer was able to show that the osmolar activity of 0.3 M sucrose is equivalent to that of 0.15 M NaCl, which in turn is equivalent to the osmolar activity of 0.1 M $CaCl_2$. This fact puzzled Arrhenius and eventually contributed to the formulation of his theory of dissociation of electrolytes: one instance in which a biological observation led to the formulation of an important physical theory.

Although most cell physiologists interpreted the volume responses of cells to various concentrations of external solutions as being osmotic in nature, a number of workers insisted that the swelling and shrinking of colloidal gels could not be entirely excluded as explanations. Figure 17-3 illustrates one of the most convincing demonstrations of the selective permeability of the cell membrane. By using cells (such as those of the beet root or red cabbage) that contain *anthocyanins,* which are natural pH indicators, or by allowing the penetration of an indicator (such as *neutral red*), it is possible to estimate the approximate intracellular pH and show that under certain

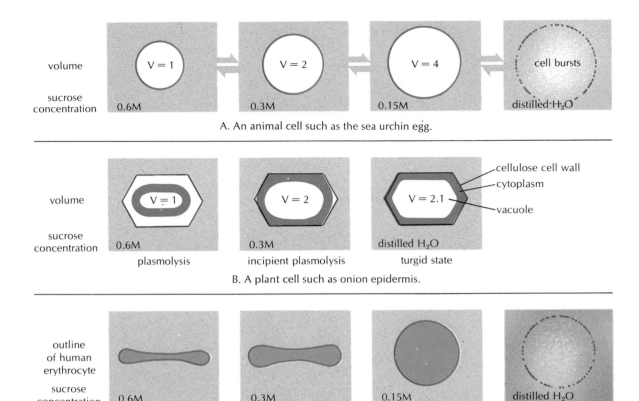

A. An animal cell such as the sea urchin egg.

B. A plant cell such as onion epidermis.

C. A human erythrocyte (red blood cell).

Fig. 17-2. Osmotic behavior of plant and animal cells. The osmolar activity of the cell under normal growth conditions can be measured by determining the concentration of the sucrose solution that causes no volume change in a cell placed in this solution. A. In the presence of a nonpenetrating solute like sucrose, a deformable animal cell acts like an osmometer and will swell and shrink in approximately inverse proportion to the molar concentration of sucrose. This change in volume occurs because sucrose lowers the thermodynamic activity of water, which will move across the membrane from regions of high to regions of low thermodynamic activity. B. In the case of plant cells placed in distilled water (or water of low solute concentrations such as soil water), the diffusion pressure exerted by the water trying to enter the cell is counteracted by physical pressure (turgor pressure) of the cytoplasm against the rigid cellulose wall. This phenomenon gives nonwoody plants the rigidity they require to stand erect. The osmolar concentration of the cell is slightly less than the osmolar concentration of the surrounding solution at incipient plasmolysis. C. The normal human erythrocyte is a biconcave disk. When placed into a solution of lower osmotic activity, it expands until it reaches a spherical shape (maximum volume per unit area). After this time, any further volume increase causes the membrane to become leaky, thus allowing the hemoglobin molecules to diffuse out. At this point, the turbid red blood cell suspension suddenly becomes a clear hemoglobin solution, a very useful "end point" for permeability studies. The red cell "ghosts" are almost pure membrane and have been used extensively for studies on the chemical and physiological properties of membranes.

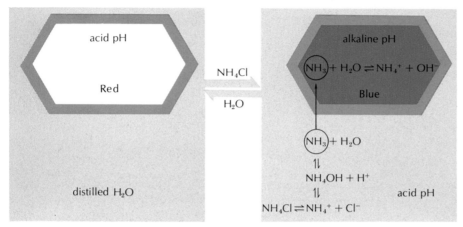

Fig. 17-3. Demonstration of the selective permeability of the cell membrane. If a thin slice of tissue of a beet root or of red cabbage leaf is placed in distilled water, the anthocyanin pigments inside the cells will remain red, the color they assume at slightly acid pH. We can therefore conclude that the pH inside the cells is slightly acid. If the tissue is now transferred to an ammonium chloride solution, the tissue will turn blue; and microscopic observation will reveal that the entire cytoplasm and vacuole is uniformly blue, showing that the whole interior of the cell is now alkaline despite the fact that ammonium chloride has an acid pH. The explanation for this striking phenomenon is that the cell is surrounded by a thin plasma membrane that is selectively permeable to some molecular or ionic species and not to others. In this instance NH_3 and H_2O can diffuse through the membrane much more rapidly than any of the other molecules or ion species. The NH_3 will therefore bring about an alkaline reaction inside the cell while the NH_4^+ ion produces an acidic reaction outside the cell — a clear-cut case of selective permeability.

conditions it differs from the pH outside the cell. The experiment described in Fig. 17-3 shows that of the different ions and molecules surrounding a cell, some will penetrate the cell while others won't or do so very much more slowly. The most persuasive explanation of this observation is that the cell is surrounded by a very thin membrane, invisible under the light microscope, that is selectively permeable. In other words, it can distinguish among molecules and allow some to penetrate easily while hindering the penetration of others either partially or almost totally.

This experiment, by demonstrating the selective permeability of the cell membrane, dramatizes the important observation made by Overton and others — that substances differ in their ability to enter the cell (permeability). At the most rapidly penetrating end of the scale we have water and dissolved gas molecules such as oxygen and nitrogen; at the other end of the scale we have ions such as Cl^- or K^+ which penetrate only one ten-thousandth as fast.

The rate of penetration of various substances into the cell can be measured in a variety of ways. The classical approach was to assume that the rate of penetration of water was very much greater than that of the substance studied so that it was possible

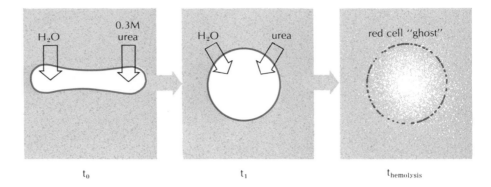

| t_0 | t_1 | $t_{hemolysis}$ |

Fig. 17-4. The measurement of the human erythrocyte permeability to urea. Zero time, when the erythrocytes are placed into an 0.3 M urea solution, is t_0. This concentration is equivalent to the osmolar activity of the material inside the cell, and therefore there is no initial movement of water across the cell membrane. However, since the cell membrane is slowly permeable to urea, it will gradually begin to enter the cell because at first there is no urea inside the cell and so the activity of urea outside the cell is higher. But as urea diffuses into the cell, the osmotic activity of the cell interior becomes greater than that of the exterior, and water diffuses in to equalize the situation. Since water is very much more permeable than urea, one can assume that the rate of swelling of the erythrocyte is proportional to the urea permeability.

to use the rate of cell volume increase as a measure of the rate of penetration of the substance (Fig. 17-4). In more recent times the use of compounds labeled with radioactive isotopes has greatly facilitated permeability measurements.

The rate of penetration of a substance into the cell is expressed as the permeability constant (K), which has the dimensions of velocity, that is cm/sec. If one uses large volumes of medium surrounding the cells being studied so that the concentration of the penetrating substance outside the cell does not decrease appreciably, and if by the use of tracer methods one can perform short-term experiments during which there is no appreciable change in volume or surface area, it is possible to express the permeability constant in very simple terms:

$$K = \frac{V}{At} \ln \frac{C_{Out} - C_{In}}{C_{Out} - C'_{In}}$$

where K = permeability constant;

V = volume of the cell;

A = area of the cell membrane in cm²;

t = time in seconds

ln = natural logarithm;

C_{Out} = concentration of the penetrating
substance outside the cell;

C_{In} = concentration of the penetrating substance
inside the cell at 0 time; and

C'_{In} = concentration of the penetrating substance
inside the cell after t seconds.

Before the use of the electron microscope, the membrane area A was estimated by measuring the surface area of the cell. This is a satisfactory procedure for some cells such as the erythrocyte, but in the case of cells such as those of the intestinal mucosa, which show under the electron microscope an extensive folding producing tremendous enlargement of the cell membrane, it is necessary to make special estimates of the size of the plasma membrane by using the electron microscope.

Passive Diffusion through the Membrane

Using a variety of plant cells, Overton studied the rate of penetration of numerous substances. He noticed that, in general, the rate of penetration of a substance was related to its lipid solubility, which he measured by determining the partition coefficient of that substance between olive oil and water; that is, after shaking the substance in an olive oil-water mixture, he determined the concentration of the substance in the two phases.

$$\text{Partition coefficient} = \frac{\text{Concentration in oil}}{\text{Concentration in water}}$$

Overton concluded from his observed lipid solubility-cell permeability correlation that the cell membrane was constructed of a thin film of lipid. Although this generalization has been subjected to continued modification in the intervening years, the basic conclusions of Overton's theory, namely that the membrane is thin and that lipid is a major component of it, have held up remarkably well.

Studies performed by Overton's successors confirmed his results, but they also discovered that Overton's lipid solubility rule broke down with very small molecules such as water, methanol, formamide, and so forth (Fig. 17-5). These small molecules penetrate the cell much more rapidly than could be explained on the basis of their solubility in lipid. From such results cell physiologists concluded that Overton's lipid membrane was interrupted by small aqueous pores that permitted the rapid penetration of small polar substances such as water or methanol.

If water penetrates the cell through small pores, one might ask what proportion of the membrane is taken up by such pores. Measurements of water permeability fall somewhere in the 10^{-4} cm/sec range, which turns out to be one hundred-thousandth of the rate at which water diffuses through a water layer 75 A in thickness. Thus, we can conclude that if special pores for water penetration exist in the cell membrane, they represent a very small percentage of the area of the cell surface. In fact, recent experiments by Solomon and his co-workers suggest that these pores are 7–8.5 A in diameter, and they occupy only 0.06 percent of the erythrocyte surface area. This gives us an idea of how very highly isolated a cell is from its environment. Thus, urea, which is considered a readily permeable substance, penetrates 100,000 times slower than water; some ions penetrate cells 100,000 times more slowly than urea, and yet, as we shall see, these rates of penetration serve definite physiological functions in the life of the cell.

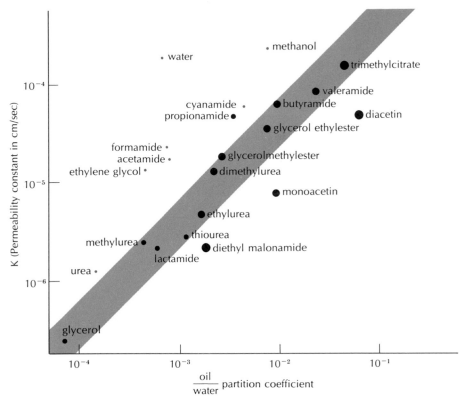

Fig. 17-5. The effect of lipid solubility and molecular size on the rate of penetration of different substances into the alga *Chara*. The size of the points is roughly proportional to the size of the molecule. Note that on the whole, small molecules (colored points) lie above the line, and large molecules below it, showing that the membrane behaves like a solvent for nonpolar molecules and like a molecular sieve for polar molecules. (After Collander, 1947.)

We shall discuss the rate of penetration of ions at greater length later, but suffice it to say here that ions are among the most impermeable of substances and that cells usually differentiate between anions and cations, most cells being more permeable to cations although the erythrocyte is a well-known exception to this rule—anions penetrate almost a million times faster than cations. The difference in passive diffusion between anions and cations can be explained by assuming that pores carry a charge. If, for instance, the charge on the pore is negative, this will reduce its effective diameter towards anions but will present no obstacle to the diffusion of cations.

We have considered so far only the so-called "passive" diffusion phenomena in which the direction and the rate of diffusion are related to the concentration gradient between the cell exterior and the cell interior. Consideration of passive diffusion phenomena has led us to conclude that the cell is surrounded by a thin membrane that has the dual properties of being a solvent for nonpolar molecules and a molecular sieve for polar molecules.

Catalyzed Diffusion through the Membrane

Let us consider the rate of penetration of glycerol into a plant cell such as given in Fig. 17-5. Glycerol, being a highly polar substance that forms hydrogen bonds with water, has a very low K of 2×10^{-7} cm/sec. If we now measure the K of glycerol in the human red cell, we find it to be 2×10^{-5} cm/sec, or about one hundred times greater; yet, in the beef erythrocyte it is even lower than in the plant cell. Since the human erythrocyte is generally not more permeable than the beef erythrocyte, we must conclude that some special mechanism is involved in the penetration of glycerol into the human erythrocyte. In such a case, when the substance is penetrating with unusual rapidity, it can be shown that this type of permeability has properties that are similar to enzyme catalysis. That is, it shows great pH dependence, competitive inhibition by structurally similar compounds, noncompetitive inhibition by trace concentrations of certain compounds, and finally a concentration dependence that is similar to an enzyme reaction.

Let us take these up in turn. Figure 17-6 shows the pH dependence of the rate of penetration of glycerol into the human erythrocyte. It shows that between pH 6.5 and 6.0 there is a hundredfold decrease in permeability to glycerol. Below pH 6 one obtains a value for K that is comparable to glycerol K values in other cells such as beef erythrocytes or plant cells. This latter rate of penetration we therefore define as the uncatalyzed

Fig. 17-6. The effect of pH and of Cu^{2+} on the catalyzed penetration of glycerol into the human erythrocyte. Note that the pH of blood plasma is normally 7.2–7.3. Lowering the pH to 6.0 decreases the glycerol permeability one hundredfold. The Cu^{2+} ion in trace quantities has the same effect. (From a class experiment at Haverford College.)

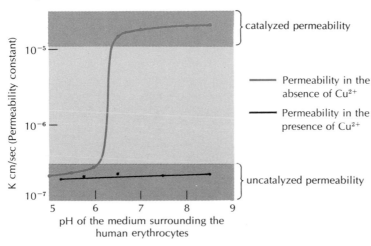

or unfacilitated value, whereas the abnormally high value of 2×10^{-5} cm/sec we say is catalyzed or facilitated. The pH value at which the transition occurs is close to the pK of the imidazole side chain of histidine. The Cu^{2+} ion complexes with the imidazole group very well, and it is interesting to note that very small amounts of Cu^{2+} eliminate the catalyzed permeability of glycerol into the human red cell. In fact, Jacobs showed that as few as two ions of Cu^{2+} per human red cell are sufficient to eliminate the catalyzed permeability to glycerol, and it is therefore entirely possible that a human erythrocyte membrane has only one catalytic site for glycerol penetration.

The competitive inhibition of glycerol penetration into human erythrocytes by structural analogs of glycerol is another property that is similar to enzyme action. Thus, the presence of ethylene glycol markedly reduces the catalyzed penetration of glycerol, yet ethylene glycol has no effect on the catalyzed penetration of sugars such as glucose and galactose into the red cell.

Finally, studies of the effect of concentration on the rate of penetration of glycerol into the human red cell show that it greatly resembles Michaelis-Menten kinetics, suggesting that there are very few sites in the cell surface with which the glycerol interacts, the rate of penetration being controlled by one or more rate-limiting steps in this interaction.

(See Chapter 18.)

Another well-documented instance of catalyzed permeability is found in bacteria. As we shall learn, the bacterium *E. coli* when grown on a galactose-containing medium will induce the formation of the enzyme β-galactosidase, which is necessary for the hydrolysis of the sugar. Certain mutants of *E. coli* are unable to be induced to form β-galactosidase in this manner. Nevertheless, these mutants are permeable to galactose as can be shown by digesting the bacterial cell wall away with lysozyme, after which the bacteria round up into spheres (spheroplasts) and behave like perfect little osmometers. Spheroplasts from these mutants, when placed into a solution of galactose, will expand and lyse because galactose penetrates into the cell and is then followed by water. However, other mutants upon exposure to galactose become induced to produce the enzyme β-galactosidase as in the wild-type cell; yet, they are incapable of utilizing galactose at any appreciable rate. That the β-galactosidase enzyme is present can be shown by breaking up the cells and testing for enzyme activity. However, spheroplasts from these mutants will not lyse in galactose, and thus one concludes that these mutants are not permeable to galactose. Furthermore, it was possible to show that like the enzyme β-galactosidase, the permeability to galactose must also be induced. Wild-type cells not grown on galactose do not have galactose permeability, and it takes a short period of exposure to galactose to induce the permeability.

These ingenious studies performed by G. Cohen, Rickenberg, and Monod show that the cell membrane contains "catalysts" that they named permeases, which play a specific role in mediating the penetration of certain compounds. Like many enzymes inside bacteria, the synthesis of these permeases is induced by the presence of the penetrating substance in the medium.

The phenomenon of catalyzed permeability adds a new parameter to the classical picture of membrane structure: the presence in membranes of specific catalytic entities, no doubt made of protein, which like enzymes interact specifically with given penetrants. Like enzymes they do not, as such, change the equilibrium of a reaction but merely increase the rate at which equilibrium is reached. By this we mean that the catalysis of permeability does not by itself accumulate or extrude molecules against concentration gradients. What it does is allow concentration equilibria of certain penetrants to be reached more rapidly than would be possible by ordinary passive diffusion through the membrane.

Active Transport

Active transport is transport that requires energy to effect it. It can be shown in numerous instances that cells are able to modify diffusion equilibria, just as they are able to modify chemical equilibria, by coupling energy-yielding reactions to these processes. Thus, if we argue by analogy, we can expect that active transport utilizes specific interactions between penetrants and specific proteins in the membrane, but in addition to that, active transport "couples" to the catalytic transport process a source of energy that makes it possible to move the penetrants against diffusion gradients. Let us consider some of the better-documented examples of active transport.

Table 17-1 shows the concentrations of various ions in the human erythrocyte and in the blood plasma surrounding it. It shows that the human erythrocyte is much higher in K^+ and much lower in Na^+ than the plasma surrounding it.

The dramatic difference in concentration of Na^+ and K^+ inside and outside the human erythrocytes can be eradicated, reversibly, by a number of agents.

1. Upon cooling to 2°C, erythrocytes will release K^+ and pick up Na^+ until ionic equilibrium is established. This process can be reversed if the red cells are restored to 37°C.
2. Treatment with certain metabolic poisons such as cyanide or iodoacetate also brings about a release of K^+ and an absorption of Na^+.
3. When erythrocytes are stored for a while in plasma at 37°C, a time comes when K^+ begins to leak out and Na^+ begins to enter the cells. Upon addition of glucose, the erythrocyte will again resume the active extrusion of Na^+ and accumulation of K^+.

Table 17-1. IONIC CONCENTRATIONS IN THE ERYTHROCYTE AND THE SURROUNDING BLOOD PLASMA

| | CONCENTRATIONS IN MILLIEQUIVALENTS PER LITER | | | |
	K^+	Na^+	Cl^-	Ca^{2+}
Erythrocyte	150	26	74	70.1
Blood plasma	5	144	111	3.2

These and other experiments show that the erythrocyte accumulates K^+ and extrudes Na^+ by an active process requiring metabolic energy. Similar observations have been made on numerous other systems such as *E. coli,* yeast, and plant cells (*Nitella*), and other animal cells such as muscle and nerve, though variations in some of the details have been noted from one system to another.

Experiments on red cells by Harris, starting in the early 1950s, have suggested that the accumulation of K^+ and the extrusion of Na^+ are linked. Perhaps the most convincing experiments were those by Glynn and Post in which they studied the accumulation and extrusion process of red cells that were returned to 37° after they had lost their K^+ and then picked up Na^+ during incubation at 2°C. It was possible to show that by reducing the external K^+ concentration one also reduced the Na^+ efflux. The exact stoichiometry of the exchange was worked out by these workers and found to be three Na^+ extruded for every two K^+ accumulated.

Sen and Post in a series of ingenious experiments demonstrated that the extrusion of Na^+ and the accumulation of K^+ were linked to a special Mg^{2+}-activated ATPase, one ATP molecule being hydrolyzed to ADP for every two K^+ and three Na^+ being transported (Fig. 17-7). They were able to demonstrate this by using ouabain, a poison of K^+ and Na^+ transport, on red cells that had been artificially made Na^+-rich by cooling and Mg^{2+}-ATP-rich by introducing Mg^{2+}-ATP into the cell with a special technique of reversible hemolysis. As a result of subsequent experiments, there is now little doubt that this Mg^{2+}-activated "$(Na^+ + K^+)$-ATPase" is indeed involved in active transport. As experiments with washed red cell membranes show, it is located on the membrane. Furthermore, it has definite vectorial properties (Fig. 17-7), being sensitive to ATP on the inside but not on the outside. Studies of the kinetics of activation also show that the Na^+ and K^+ sites are different; the Na^+ site is competitively inhibited by K^+, and the K^+ site competitively inhibited by Na^+. Finally, the membrane ATPase can be stimulated by a mixture of Na^+ plus K^+, and this stimulated ATPase activity is inhibited

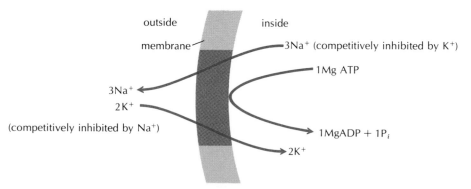

Fig. 17-7. Stoichiometry and localization of active transport of Na^+ and K^+ across the membrane of the human erythrocyte. (After Sen and Post, 1964.) The hydrolysis of one ATP phosphate bond apparently provides energy for the linked transport of two K^+ in and three Na^+ ions out. The tinted region in the membrane symbolizes the Mg^{2+} activated $(Na^+ + K^+)$-ATPase enzyme. This enzyme activity is poisoned by ouabain, which also inhibits active transport of Na^+ and K^+.

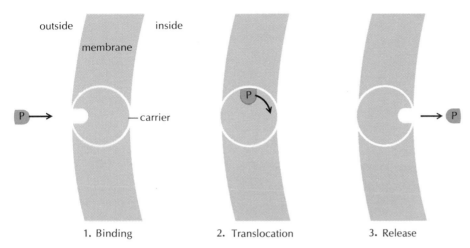

outside   inside

membrane

P ⟶   — carrier

1. Binding   2. Translocation   3. Release

Fig. 17-8. Generalized scheme for transport of a penetrant (P) across a membrane by a carrier system. The scheme involves three steps: (1) binding of penetrant to carrier, (2) translocation of penetrant across membrane, and (3) release of penetrant on the other side of the membrane. This scheme in its simplest form does not involve an energy source, and this explains catalyzed transport. For active transport, which involves the accumulation or excretion of penetrants against concentration gradients, an energy source supplied by the metabolism of the cell must be coupled to it. Translocation need not be an energy-requiring step. Thermal motion may be sufficient to bring about the two states of the carrier, and direction of transport would depend purely on concentration difference on either side of the membrane.

by the transport inhibitor, ouabain. This $(Na^+ + K^+)$-stimulated ATPase activity, which was originally discovered by Skow in crab nerve, has now been shown to occur in all mammalian tissues, especially brain and kidney. Extremely active pumping organs such as the electric organ of the electric eel and the salt gland of the herring gull are especially rich sources of this enzyme system.

The vectorial properties of the $(Na^+ + K^+)$-ATPase system we have just described can best be explained by assuming that its action in the membrane is that of a carrier. In its simplest form a carrier consists of (1) a molecule X (probably a protein or lipoprotein) with a specific combining site for the penetrant; (2) a physical translocation mechanism involving the carrier, which moves the penetrant from one side of the membrane to the other; and (3) a specific change in the carrier bringing about the release of the penetrant. One can imagine in the carrier molecule a conformational change that alters the active site so that instead of a high association constant, it now has a low association constant for the penetrant.

In the case of Na^+ and K^+ transport we must assume a more complex carrier system than the one depicted in Fig. 17-8. First, the carrier must translocate Na^+ specifically out and K^+ specifically in; that is, it has two vectors working in opposite directions. Second, the carrier is coupled to an energy-releasing mechanism to allow it to move

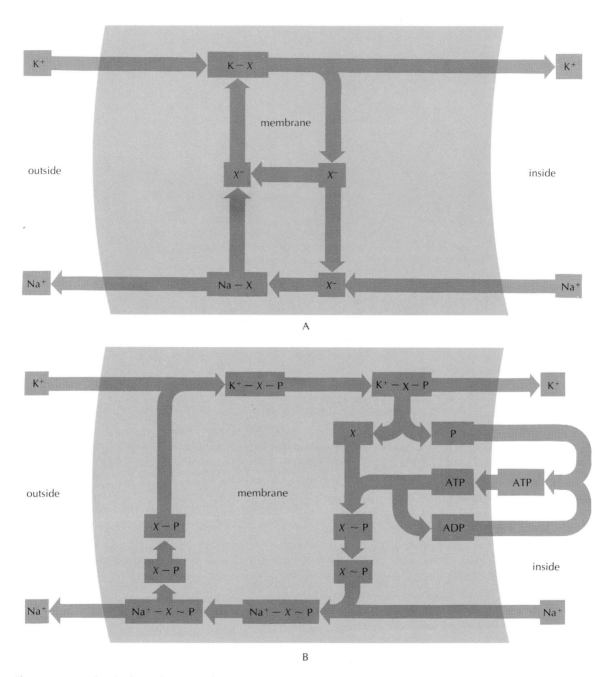

Fig. 17-9. Postulated scheme for Na$^+$ and K$^+$ transport across membranes. Details are explained in text.

these ions against concentration gradients. The mechanism shown in Fig. 17-9 is based on the ideas of Post, Judah, and Shaw. K^+ combines with a carrier, the phosphorylated protein, to form K^+-X-P, which then moves across the membrane and unloads the K^+ to the inside of the cell; at the same time P is split off the enzyme, X. X must then be phosphorylated by ATP to form the active carrier protein, $X\sim P$. This then combines with the Na^+ on the inner side of the membrane, forming Na^+-$X\sim P$, which moves across the membrane, releasing Na^+ to the outside. At the same time the energy state of the protein is lowered, giving a low-energy form of the phosphoprotein, X-P, which goes through the cycle once again. ATP is necessary to form the high-energy form of the carrier; the breakdown of the ATP in forming this carrier is what is observed as an ATPase. The ATP can be generated by some energy-yielding reaction within the membrane itself, but more likely by some ATP-generating systems in the cytoplasm as depicted in the Figs. 12-4 and 13-3.

Although the details of such a scheme are far from proven, evidence has been accumulating in recent years for the presence of a number of phosphorylated intermediates that appear to be involved in the $Na^+ + K^+$ transport system. This has been a fertile area of research, and rapid developments can be expected in the near future.

We have discussed the active transport of Na^+ and K^+ but have so far failed to mention the passive diffusion of these ions through the cell membrane. That there is some passive diffusion can be shown in various ways such as poisoning the active transport with ouabain, slowing it down by cooling to 2°C, or using isotopes to measure the rate of exchange in the reverse direction (inward for Na^+ or outward for K^+) when the pumping mechanism has produced a steady-state concentration of ions. Although the values obtained vary somewhat depending on the methods used, it seems clear that Na^+ ($K\approx 5 \times 10^{-10}$ cm/sec) is slower than K^+ ($K\approx 5 \times 10^{-9}$ cm/sec). This difference might be explained by the fact that Na^+ is more hydrated than K^+ and is in fact a larger ion (diameter of $Na^+ = 5.1$ A and of $K^+ = 4.0$ A). Indeed, we find that Li^+ (the most hydrated of the alkali metal ions) is slower than Na^+, and Rb^+ (the least hydrated) is faster than K^+.

This very low permeability to cations has an important physiological function for the erythrocyte: it saves the cell a great deal of energy that the cell would otherwise have to spend to hold the ions it has accumulated or to keep out the ones it has extruded. A physical analogy would be that of a reservoir into which water has been pumped. If the reservoir is very leaky, it takes much more energy to maintain a certain level of water in it than if the reservoir is relatively tight. Thus, it can be calculated that if K^+ had a permeability constant like that of urea (8×10^{-5} cm/sec), the work necessary to maintain the K^+ inside the cell would be 1.8×10^6 cal/kg/hr; whereas, the work actually necessary to maintain K^+ with its own permeability constant of 1×10^{-9} cm/sec is only 13 cal/kg/hr. From the O_2 consumption of the human erythrocyte, assuming 100 percent efficiency, one can calculate that the maximum amount of work the cell can do is equivalent to 40 cal/kg/hr. We can therefore conclude that the passive permeability to ions is just low enough to make it energetically feasible for the human erythrocyte to engage in the active transport of ions.

We have already emphasized that the facts regarding ion transport in the erythrocyte are generally the same in a wide variety of cell systems from microorganisms to plant cells and in other animal cells such as muscle and nerve. We shall have occasion to discuss nerve in some detail later on, but in the meantime we should refer to a number of other important examples of active transport that have been studied in some detail.

Because all membranes contain phospholipids as an intimate part of their structure and because these phospholipids, though structural components, are metabolically active (their phosphate groups, as measured with radioactive phosphate, are in constant turnover), attempts have been made to link these active phospholipids to ion transport. One such scheme is shown in Fig. 17-10, taken from the work of Hokin and Hokin and based on their experiments with the exotic salt-secreting gland of the albatross. Briefly, the idea is that the four underlined intermediates (phosphatidic acid, CDP-diglyceride, phosphatidyl inositol, and diglyceride) exist in the membrane and are in a constant flux depending on whether or not the membrane has been stimulated to act as an active membrane in the transport of ions. Based on experiments with radioactive compounds such as inorganic phosphate and ATP, based on work with such naturally secretory substances as acetylcholine, based on the existence of enzymes which catalyze all the reactions depicted in the scheme and which we think are on the membrane, and based on the effects of such transport inhibitors as ouabain, the Hokins have postulated the existence of an active membrane, most likely a lipoprotein, which has, at one moment, phosphatidyl inositol in its "resting" state, and, at another moment phosphatidic acid in its "active" state. These states are the possible sites for Na^+ transport and would be inoperative when the lipoprotein at the site is in the phosphatidyl inositol form, operative when the site is converted (by means of the mechanisms shown in Fig. 17-10) to the phosphatidic acid form. A comparison of Fig. 17-10 with Fig. 17-9 shows many points of similarity: the need for a lipoprotein or phosphoprotein, the need for energy in the form of ATP to phosphorylate the protein, and the turnover of phosphate during the reaction. We feel that both schemes have a modicum of truth in them; in addition, both schemes could accommodate, with modifications, the acknowledged relevance of the $(Na^+ + K^+)$-activated ATPase, mentioned above, to Na^+ transport.

So far we have discussed osmotic properties of the plasma membrane, the membrane surrounding the cell. However, it was known even before the advent of electron microscopy that membrane systems occur inside the cell—for example, the system surrounding the interphase nucleus and the one surrounding plant vacuoles. In fact, as early as 1885, DeVries was able to isolate plant vacuoles with membranes around them by microsurgical techniques and to demonstrate that they had osmotic properties. In recent years the most thoroughly studied membrane-limited organelles are the mitochondria. It has been found that the electron-transport assemblies of the mitochondria are parts of structures that are osmotically active, can act in selective permeability, and can accumulate ions. The latter process is essentially a vectorial one for one ion, that is, only one ion is actively transported, the other electrically balancing ion coming in via the electrochemical potential gradient so established. If Na^+ is

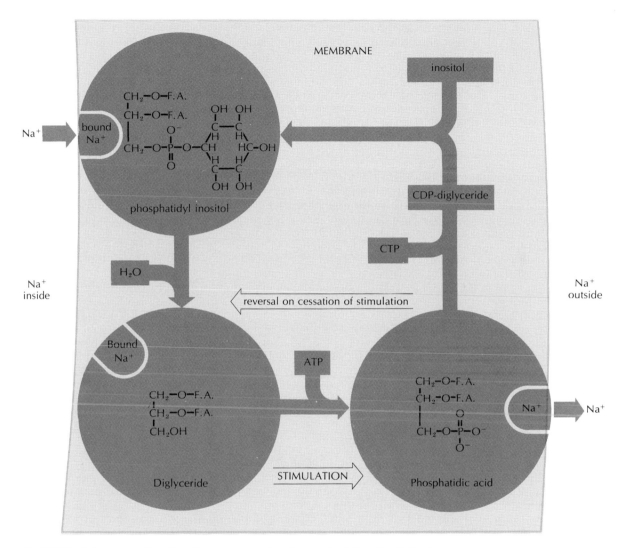

Fig. 17-10. Scheme postulated for the events occurring during stimulation of salt gland to secrete salt.

actively transported, Cl^- will diffuse in to balance the electric charges on that side of the membrane. There is thus initially established a separation of negative from positive charges. It has been known for a long time that in the oxidation-reduction couples of the electron-transport system there is also a separation of charges. The hydrogen ions and the electrons travel via different pathways, the former via the aqueous soluble phase and the latter by somehow moving from carrier to carrier directly. Recently, Davies and Ogston, Conway, Robertson, and particularly, Mitchell have postulated that the two processes, electron transport and ion transport, are really the same mechanism but viewed from different experimental vantage points.

This hypothesis is that the energy used for the transport of certain ions lies not in the formation of ATP directly, though this certainly happens concomitantly, but in the energy gained from electrons going from a reduced carrier to an oxidized carrier, from a higher to a lower potential. Figure 17-11 illustrates this hypothesis in relation to "acid" secretion by various cell types. The mitochondrial electron-transport process itself supplies the energy. Also, the vectorial quantity is somewhat increased because electron transport is an isotropic process, H^+ going one way and the electrons going the other. Since the correct positioning of the electron carriers is very important for the occurrence of this process, once again—assuming that this scheme does indeed reflect reality—the necessity for supramolecular structure in the mitochondrial electron-transport chain is pointed up. This is not to say that all ion transport takes place via the mitochondrial cristae; the intriguing fact is that the endoplasmic reticulum, or microsomal, membranes also contain a portion of an electron-transport chain, specific-

Fig. 17-11. Postulated scheme of shift of "acid" across membrane. The "secretion" of H^+ is coupled to electron transport. (Modified after P. Mitchel, *Journal of General Microbiology.*) Details are explained in the text.

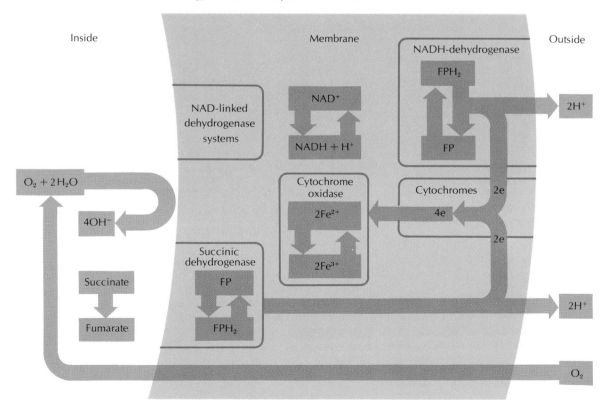

ally an enzyme or enzymes that can transfer electrons from $NADH_2$ to a distinctive microsomal cytochrome, cytochrome b_5. That this electron-transport fragment can also act as an anion-transport machine is a possibility; as will be mentioned below, there is cytological evidence to suggest that these endoplasmic reticulum membranes are in continuity with the plasma membrane of the cell; hence, its components might have access to the environment outside the cell.

In contrast to the difficulties that cells encounter in accumulating these cations — a process requiring energy — anions like Cl^- seemingly get across membranes without metabolic assistance. An exception is inorganic phosphate, for energy is required to move phosphate across the red blood cell membranes, into muscle cells, into marine eggs, into yeast cells and kidney tubule cells. All the experiments with inorganic phosphate point to the conclusion that this molecule gets across as the anion PO_4^{3-}. It is thought that in some cases it moves across as part of the ATP molecule, that the energy requirement lies in the making of ATP, and that this takes place at the cell membrane. If there is an adenosine triphosphatase at the cell membrane, splitting off phosphate, then the adenylate moiety can be considered as a carrier of phosphate. The evidence does point to the conclusion that there is an adenosine triphosphatase at the membrane of the yeast cell as well as at the bacterial and red blood cell surfaces. In some cases the transport of K^+ inward has been linked with the transport of phosphate inward. Thus, the common carrier in this scheme, an "ATP-X," is a carrier both for K^+ and for phosphate. The energy need in this case would be for the resynthesis of ATP.

In Chapter 13 the observation was made that the isolated mitochondrion can couple its ATP formation to the accumulation of ions like K^+. Isolated kidney mitochondria also have an active mechanism for concentrating SO_4^{2-} against a concentration gradient; this process depends on oxidative phosphorylation. The interesting point is that the accumulation of SO_4^{2-} by kidney mitochondria is very similar in many respects to the uptake of the same ion by kidney slices. One wonders, then, if it is not the mitochondrial membranes rather than the plasma membrane that are responsible for the active transport of SO_4^{2-}, and perhaps of other ions, in these and possibly other types of cells. In this respect, it is noteworthy that in kidney tubule cells the mitochondria are located very close to the plasma membranes and follow every convolution of the membranes.

Ions are not the only substances that are accumulated or excreted by energy-requiring processes. Sugars are also actively transported, and this is not surprising, for they are important metabolites that permeate very slowly by passive diffusion through the cell membrane — slowly, because they are large, very water-soluble substances. In some cells such as red blood cells, ascites tumor cells, muscle cells, or liver cells, glucose seems to penetrate by a catalyzed diffusion process but is not accumulated against a concentration gradient. In the epithelial cells of the small intestine and of the proximal convoluted kidney tubule, glucose and galactose are actively accumulated, and a mobile carrier has been proposed (Fig. 17-12). The evidence for this is as follows.

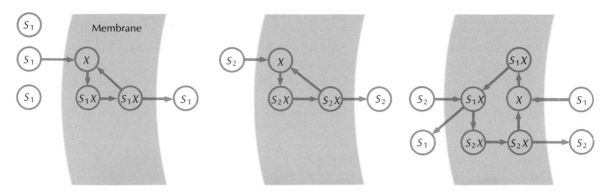

Fig. 17-12. Sugar transport across membranes. Details are explained in text.

First, it can be shown that in spite of their larger size, galactose and glucose are transported more rapidly than pentose sugars (Table 17-2). Second, measurements of glucose concentration in the intestine and in the epithelial cells of the small intestine show that it is concentrated by these cells severalfold. Third, metabolic poisons inhibit the process of accumulation.

The so-called kinetics of penetration fits in well with the way the process should behave if a mobile carrier were involved. For example, we can visualize two sugars, S_1 and S_2 (Fig. 17-12), combining with the mobile carrier, X, and being ferried across the membrane as shown in the figure. If one of the sugars has already been equilibrated with the carrier, forming S_1X, and another sugar, S_2, is then added, the first sugar will come out into the medium—against its own concentration gradient. This result can be explained by assuming that S_2 competes successfully with S_1 for X and combines with X at the same time S_1 is released from X. We know something about this carrier from the properties of the sugars that do get across: for example, their configuration (D or L) and the way they are folded into a three-dimensional structure. The need now is to find some compound or compounds that have the structural configurations to make them fit the structures of the specific sugars and also to be soluble in the lipophilic membrane structure. This mobile carrier somehow requires energy in order to work, but exactly where in the scheme the energy requirement comes, we do not know.

However, recent evidence has thrown some light onto this problem—the exciting finding that transport of sugar into cells, and probably of amino acids also, is dependent on the presence of sodium ions. Figure 17-13 is a schematic view of a possible mechanism. A mobile carrier in the membrane, moving from one side of the membrane to the other as mentioned earlier, has binding sites for both the sugar and Na^+; the sugar and Na^+ on the inside of the cell are in equilibrium with the sugar and Na^+ on the outside via the mobility of the carrier. What bestows an asymmetric character to the system, what moves the sugar from the outside toward the inside, is the operation of an outwardly directed, energy-dependent Na^+ pump, which constantly moves Na^+ from inside the cell out concomitant with the movement of K^+ from outside to the

Table 17-2. RELATIVE RATES OF ABSORPTION OF SUGARS FROM SMALL INTESTINE*

SUGARS	RELATIVE RATES
Hexoses	
Galactose	110
Glucose	100
Fructose	43
Mannose	19
Pentoses	
Xylose	15
Arabinose	9

* (From Cori, 1926.)

inside as in Fig. 17-9. The picture is undoubtedly more complex than that shown, for there seems to be no stoichiometry between the number of sugar molecules and the number of sodium ions moved by the carrier; but, in the face of the energy-driven Na^+ pump, in the presence of Na^+ on the outside, sugar will move inward as though it is coupled to Na^+ transport.

Now we have shown several schemes, each of which tries to explain some feature of either ion or sugar transport. At this moment, it appears that in general a part of all the schemes is correct, though some may be limited in scope, as the one in Fig. 17-11 is limited to mitochondrial transport. How to fit all the observations and ideas into one coherent diagram (though to be sure we do not know that only one scheme can draw together sugar, ion, and amino acid transports, the involvement of ATPase, and the involvement of phospholipids) is not an easy task; but all in the field feel, perhaps intuitively, that the general ideas concerning these topics are correct and that the true road to a rational explanation of the transport of ions and molecules has been opened.

Bulk Transport

In most cells, excluding bacteria and some fungi, there exist intracellular membranes. In some cases these membranes form an extensive intracellular network as can be seen

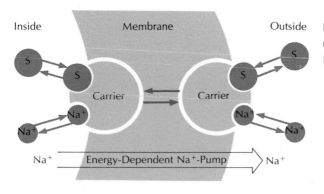

Fig. 17-13. Na^+-dependent sugar transport. (Diagram modified after R. K. Crane, Rutgers University, New Brunswick, N. J.) Details explained in text.

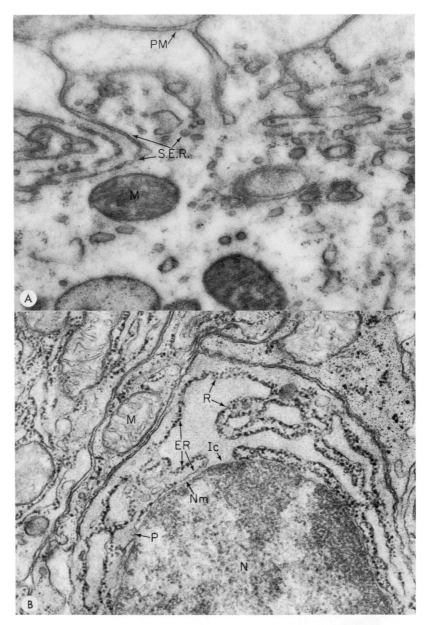

Fig. 17-14. A. Macrophage of spleen, showing indentations of plasma membrane in continuity with intracellular membranes (× 37,500). B. Intestinal smooth muscle cell showing continuity of the outer nuclear membrane with the endoplasmic reticulum membranes (× 52,500). PM, plasma membrane; M, mitochondria; N, nucleus; Ic, connection between cytoplasmic and perinuclear space; P, nuclear pore; Nm, inner nuclear membrane; ER, profiles of the endoplasmic reticulum; R, ribosomes; S.E.R., profiles and longitudinal sections of endoplasmic reticulum membranes which are in continuity with plasma membrane. (Courtesy of G. Palade, Rockefeller Institute.)

from electron micrographs (see Chapter 3), whereas in other cases the membranes are rather sparse. In some cells and in some cases where the sectioning of the material has been favorable the continuity between these intracellular membranes and the plasma membrane surrounding the cell can be clearly noted (Fig. 17-14A). In other sections it can be seen that these membranes are also in continuity with one of the membranes surrounding the nucleus (Fig. 17-14B). By making serial sections of fixed tissues and then examining the electron micrographs of these sections, investigators have discovered several key points regarding the internal structure of the cell. One is that the membrane system, in those cells where it is extensive, divides a cell into two compartments: one inside the membranes and bounded by them, and one outside. There is quite a lot of continuity among the spaces bounded by these membranes. In some cases these spaces are in the form of long, fingerlike tubules; in other cases they are in the form of large globules; while in still other instances they seem to be great, flattened vesicles called cisternae. No matter what their form, however, all these spaces seem to be in continuity with each other. This system of interconnected spaces has been given the name of endoplasmic reticulum (see Chapter 15). The mitochondria and various cellular inclusions such as the secretory bodies in some cells lie outside these spaces. Inside them can sometimes be seen dense bodies of unknown function. But the most interesting point is that the nucleus, with its one bounding membrane, seems to sit in a greatly enlarged space inside these membranes. A qualification must be pointed out, though, that the membranous elements of the endoplasmic reticulum do not entirely enclose the nucleus; numerous openings or pores are left on the nuclear surface. However, it appears that these pores may be plugged, and thus we must assume that they are not openings through which materials might pass between nucleus and cytoplasm but represent the incompletion of the encirclement of the nucleus by the endoplasmic membranes.

As illustrated in the stylized, composite diagrams of Figs. 4-2 and 4-3, the cell is thought to be not a self-enclosed entity shut off from the outside by a membrane, but a structure with channels that communicate to the outside (perhaps only intermittently) and that run deep enough so that they can, and sometimes do, reach into the nuclear region, to the perinuclear space surrounding the nucleus. Thus, the amount of membrane accessible to the environment is greatly augmented, and much more of the cell is in intimate contact with the environment, blocked only by the active curtain of a single membrane. Do materials from the outside enter in this way? In some cells, a process that has been called pinocytosis takes place. The cell membrane forms an infolding, which gets larger and finally pinches off, enclosing within it a bit of the formerly outside environment. This vesicle moves inward, there breaks up, and presumably releases its contents to the cell interior. The process of pinocytosis can be seen beautifully in the amoeba, in which it apparently happens on a large scale; this is probably how the amoeba takes in much of its food. Amoebae, when grown in a medium containing glucose, take in this sugar inside the pinocytotic vesicles; by using radioactive glucose, as did Holter, we can determine by autoradiography what is happening. This process seems to be induced by the presence of some protein in the medium, for without it no pinocytosis takes place. Once inside the vesicles in the cell, the radioactive glucose appears in the general cytoplasmic milieu; presumably the membranes lining the vesicles somehow break up and release the glucose to the cell.

Not all compounds in the outside medium are taken in, however, so it must be that even here there is some selectivity. This is thought to come about because there are specific sites on the cell membrane for binding some compounds and not others, and when the membrane invaginates, these bound extracellular compounds come in with it. In the beautiful example of the very thin endothelial cells that line blood capillaries (Fig. 17-15), it seems that materials are transferred from the blood into the interstitial space by literally being ferried across the cell in small vesicles; these form at one border of the cell, travel across the narrow cell, fuse with the membrane at the other side, and thus discharge the contents outside the cell—but on the other side. Another example is the manufacture and secretion of the digestive enzymes chymotrypsinogen, trypsinogen, amylase, procarboxypeptidase, and ribonuclease by the

Fig. 17-15. Pinocytosis. Endothelial cell of blood capillary in diaphragm. L = lumen of blood capillary; F = injected ferritin particles; EC = endothelial cell lining capillary. Arrows point to uptake of ferritin particles into pinocytotic vesicles and presumed transport of ferritin particles to other side of capillary cell (arrows) (× 130,000). (Courtesy of G. Palade and R. Bruns, Rockefeller University.)

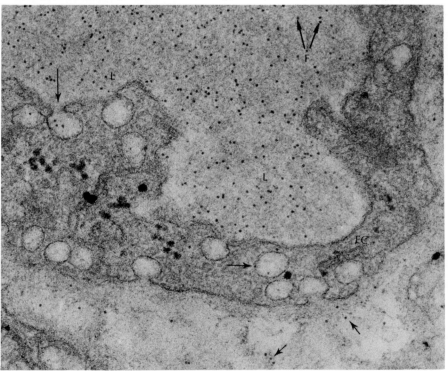

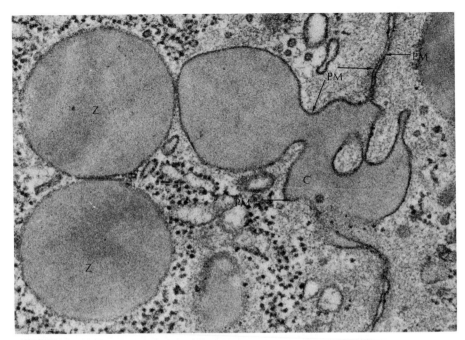

Fig. 17-16. Zymogen granules and discharge of contents into lumen (× 87,000). Z = zymogen granules; C = contents of zymogen granules *in lumen*; PM = plasma membranes. (Courtesy of G. Palade, Rockefeller University.)

acinar cells of the pancreas. These enzymes are synthesized on the ribosomes attached to the membranes of the endoplasmic reticulum. They are then somehow released from these particles and make their way through or across the membranes of the reticulum into the vesicular spaces. In these channels they travel toward the Golgi region in the basal part of the cell. Here they are seemingly packaged into large, dense bodies by being wrapped around by the membranes of the Golgi region. The latter smooth membranes are in continuity with the membranes that have ribosomes bound onto them, and hence there is a continuous channel from one part of the cell to another. These wrapped-around packets are now the mature zymogen granules peculiar to the pancreatic acinar cells. When the pancreas is stimulated to secrete, the zymogen granules move to the apical part of the cell, which borders one of the pancreatic ducts. The zymogen granule membrane then fuses with the cell membrane; a fissure is made in the latter, the enzymatic contents of the granules spill out into the glandular lumina, and by means of the pancreatic ducts are then finally discharged into the duodenum (Fig. 17-16).

All of these cases indicate that the intracellular membranes are not static, that the cell structure itself, even in the tightly restricted cells of a multicellular organism, is in a dynamic state. Indeed, observations of tissue culture cells show very clearly the constant movements of materials—even bodies like mitochondria—within these cells.

In some of the polar cells, which take in materials at one end or discharge materials at one side, there is even thought to be membrane flow, in one direction, from one side of the cell to another. The cellular structure is not a rigid body, but a constantly changing flux. Thus, in addition to the transport processes that utilize carrier systems to ferry materials into cells, there is yet another mechanism, that of membrane flow and vesiculation. Presumably, the latter process requires energy for the purpose of contracting and expanding membranes. Indeed, it may be that many cells use both these means of bringing materials into themselves.

The student will have realized from the preceding discussion that our knowledge concerning the mechanics of the transport of materials is far from extensive. The problem is the same as that involved in elucidating muscle contraction—the conversion of chemical energy into mechanical work. Until the complex morphological system can be broken down into its component parts, the questions concerning movement of materials will remain unanswered.

The Action Potential

Thus far we have considered the role of membranes in bringing about and maintaining a difference in concentration of diffusible substances between the cell and its surroundings or between different parts of the cell. We will now discuss an entirely different role played by the cell membrane—one which nevertheless depends on the membrane's permeability properties and its ability to move ions by active energy-requiring processes. We refer to the action potential, which finds its most specialized expression in nerve cells though it can be found in attenuated form in most cells, even in the unicellular green alga *Nitella*.

We shall restrict our discussion to a brief consideration of the resting potential and of the action potential and its propagation in a single cell system such as the giant axon of the squid. We shall not mention the synapse, the neuromuscular junction, nerve receptors, and other similar topics that are more properly the domain of integrative or organismic physiology. Fortunately, it is possible to refer the student to a truly inspiring little book by Katz, *Nerve, Muscle and Synapse,* which enlarges upon both the material treated here and that which we chose to omit.

Let us begin by introducing our experimental material. The phenomena we wish to describe can all be studied using a piece of the giant axon of the squid. This ideal experimental system is a piece of the cylindrical portion of a nerve cell. The central cell body and the synaptic contacts to other nerves found at one end of the nerve cell, as well as the motor nerve endings at the other end of the cell (see Fig. 1-2B), have been removed. This apparently mutilated piece of a cell can stay alive for several hours when placed in seawater and can conduct hundreds of thousands of nerve impulses. Its giant dimensions (up to 1 mm in diameter) make it easy to insert electrodes or remove material (axoplasm) from the inside without irreversible injury.

If one measures the concentration of ions inside the squid axon and in the body fluid surrounding it, one finds a distribution of concentrations (Table 17-3) that, though not numerically identical, is reminiscent of the situation observed in erythrocytes.

Table 17-3. IONIC CONCENTRATIONS IN THE SQUID
AXON AND THE SURROUNDING BODY
FLUID

	K+	Na+	Cl−	CONCENTRATIONS IN MILLIEQUIVALENTS PER LITER ORGANIC ANIONS (FOR EXAMPLE, ASPARTATE, ISOTHIONATE)
Squid axon	400	50	40–100	345
Body fluid	10	460	540	---

Numerous experiments have established for the squid axon (as indeed they have for other nerves and for muscle cells as well) that the distribution of ions shown in Table 17-3 can be explained in terms of (1) the presence of a thin semipermeable membrane surrounding the cell and (2) a pumping mechanism utilizing metabolic energy, which extrudes Na^+ and accumulates K^+. These conclusions that we have already reached in the case of the erythrocyte could be tested very conveniently with the squid axon, which because of its size lends itself especially well to direct chemical and electrical manipulation. The following few examples will serve to illustrate this fact.

By placing a small drop of radioactive K^+ on a squid axon, Hodgkin and Keynes were able to show that within 2 hours, 10 percent of the internal K^+ was labeled in a small region of the nerve directly opposite the drop, on the other side of the membrane. By measuring the rate at which the radioactive K^+ then spread inside the nerve, they were able to demonstrate that the diffusion of K^+ within the nerve cell was comparable with the diffusion of K^+ in the surrounding seawater. Similarly, inside and outside the cell the ionic mobility of K^+ was the same. They could measure the ionic mobility directly by inserting electrodes into the nerve cell and observing the longitudinal movement of the radioactive K^+. They were also able to show that the rate of exchange of radioactive K^+ inside with nonradioactive K^+ outside was very slow. Hodgkin and Keynes therefore concluded that K^+ moves very slowly across the membrane of the nerve cell but is able to move very freely inside the cell. The importance of the nerve cell membrane was also demonstrated dramatically by Hodgkin and his co-workers when they showed that it was possible to insert pipettes longitudinally into the axon, withdraw the axoplasm, and replace it with solutions of proper ionic concentrations without having an immediate effect on the axon's ability to propagate an impulse.

Recent studies of the electrical properties of nerve and muscle cells have confirmed the earlier work of Hodgkin and Keynes (Table 17-4). The data in Table 17-4 show that the membrane exerts a considerable resistance to the flow of ions. From the capacitance measurements it is possible to calculate (assuming a dielectric constant of 6, a reasonable figure for a lipid) that the thickness of the membrane is 50 A—again a good confirmation of the existence of a thin membrane and of its controlling influence on the movement of ions.

Table 17-4. ELECTRICAL PROPERTIES OF SOME NERVE
AND MUSCLE CELLS*

	CELL DIAMETER μ	MEMBRANE RESISTANCE ohm-cm²	MEMBRANE CAPACITANCE μfarad/cm²	RESISTIVITY OF CELL INTERIOR ohm-cm	RESISTIVITY OF MEDIUM ohm-cm
Squid axon	500	700	1	30	22
Lobster nerve	75	2000	1	60	22
Frog muscle	75	4000	2.5	200	87

* (After Katz, 1966.)

The above electrical measurements could also be confirmed by direct measurements of permeability using isotopic tracers (Table 17-5). The data in Table 17-5 show that in the nerve, just as in the erythrocyte, K^+ permeability is much greater than Na^+ permeability.

Table 17-5. PERMEABILITY VALUES FOR K^+ AND Na^+
IN THE SQUID AXON*

	PERMEABILITY cm/sec
K^+_{out}	6.2×10^{-7}
K^+_{in}	5.8×10^{-7}
Na^+_{in}	7.9×10^{-9}

* (After Katz, 1966.)

That metabolic energy is necessary for the transfer of Na^+ and K^+ was demonstrated by Hodgkin and Keynes, who showed that the extrusion of Na^+ was switched off reversibly by metabolic poisons such as dinitrophenol, azide, or cyanide. Furthermore, they showed that Na^+ extrusion would be resumed in the presence of metabolic poisons if ATP were injected into the squid axon. In addition, they pointed out that the poisons also prevented the accumulation but not the loss of K^+, and that Na^+ extrusion virtually ceases when K^+ is removed from the external medium. It is possible that a small amount of K^+ accumulation is not coupled to Na^+ extrusion. Thus, although most workers agree that some coupling between Na^+ extrusion and K^+ accumulation occurs in the nerve cell, the precise stoichiometry has yet to be worked out in detail.

So far we have demonstrated that the nerve cell is very similar to the erythrocyte in its ion-transport behavior. There is, however, a consequence of ion transport that we have not yet discussed—namely, the separation of positive and negative charges that occurs across the plasma membrane.

If one takes a pair of electrodes, carefully inserts one of them into the squid axon, and places the other one on the outer surface of the membrane, it is possible to show

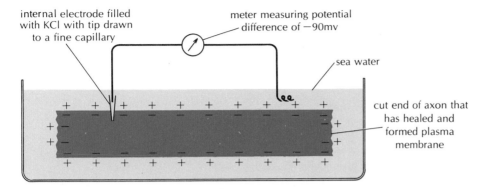

internal electrode filled with KCl with tip drawn to a fine capillary

meter measuring potential difference of −90mv

sea water

cut end of axon that has healed and formed plasma membrane

Fig. 17-17. The measurement of the resting potential of the squid axon. Modern devices can measure the potential difference without drawing any current, thus not disturbing the ionic steady state being measured. The convention is to express the potential difference as a negative quantity, the inside being negative with respect to the outside.

that there is a potential difference of 60–90 millivolts (mv) between the inside and the outside of the axon membrane, the outside being positive with respect to the inside (Fig. 17-17). This potential difference is called the resting potential.

Although the resting potential has been studied most extensively in nerve cells, we know that it is a widespread phenomenon among cells of widely differing origin. In nerve, however, it is especially important because it plays a role in impulse transmission, which after all is the main biological function of nerve tissue.

What is the explanation for the resting potential? It turns out that this important phenomenon has a decidedly unglamorous explanation: the resting potential appears to be simply a consequence of differential leakage of ions. Let us look again at the steady-state concentrations of ions inside and outside the squid axon in Table 17-3. The nerve membrane as we have learned differs markedly in its permeability to these ions. K^+ and Cl^- permeate most easily, Na^+ very much less easily, while the organic anions inside the cell are virtually unable to leave the cell. At the above steady-state concentrations we have a diffusion potential for K^+ outwards and Na^+ inwards, but since K^+ leaks out more readily than Na^+ leaks in, we have a net effect of positive charges attempting to escape the cell—leaving the inside of the cell negative with respect to the outside. Since the organic anions of the cell cannot leak out, a limit is placed on the escape of K^+, for at a certain point the electrical potential generated by the separation of charges will balance the diffusion potential of K^+ outward.

This simple qualitative argument works out very well indeed when expressed in quantitative terms. The relationship between the potential difference across a membrane and ionic activities is expressed by the Nernst equation. To illustrate how this equation works we shall take the simplified case that follows.

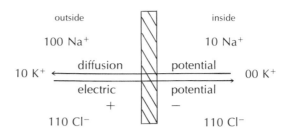

In this case we assume that the membrane is permeable only to K^+. At equilibrium

$$E = \frac{RT}{nF} \ln \frac{[K^+_{in}]}{[K^+_{out}]} = 58 \text{ mv}$$

where E = potential across the membrane in volts,
R = gas constant (8.312 joules/deg/mole),
T = absolute temperature,
F = the Faraday (96,500 coulombs/gram equivalent);
$\ln$ = the natural logarithm ($2.3 \times \log_{10}$),
$[K^+]$ = thermodynamic activity (but at physiological concentrations it is nearly equal to molar concentration),
n = charge per particle (here = 1).

The Nernst equation was derived from thermodynamic principles and simply states that in a situation such as the one above, an electrochemical equilibrium is set up between the electrical work necessary to transfer ions in one direction and the osmotic work needed to move ions in the other direction. It is important to recognize that the transfer of only a few ions is sufficient to set up an electric potential that can balance the diffusion potential. As we shall see later, this fact enables the nerve to "fire" repeatedly without greatly affecting the concentrations of ions on either side of the membrane.

The case we have considered above is of course an oversimplification of the real situation found in the nerve cell. Figure 17-18 is a more accurate representation of our present understanding of the nerve cell. It takes the form of a circuit diagram in which there are four different elements.

The first element is a membrane capacitance (C_m) due to a large and invariable area of thin membrane, which is composed of lipid and acts as an insulator. This lipid membrane has three different ionic leakage pathways, and ions travel at different rates in each of the three. In other words, each pathway allows leakage at a certain rate (the conductance value g, expressed in $\text{ohms}^{-1} \text{ cm}^{-2}$). The retardation exerted by these pathways is called the resistance $\frac{1}{g}$ in ohm-cm^2, which is the reciprocal of the conductance. Some of these resistance values, as we shall see later, are subject to physiological variation.

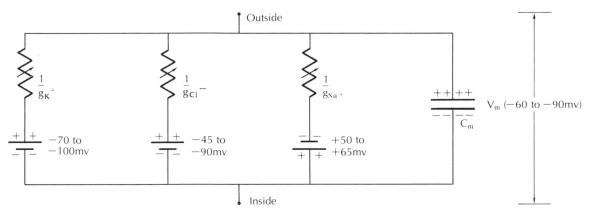

Fig. 17-18. Circuit diagram symbolizing the electrical properties of the nerve membrane in relation to the ions on both sides of it. (Modified from Hodgkin, 1958 and from Katz, 1966.) The nerve membrane is pictured as a leaky condenser that accumulates a resting potential of 60 to 90 mv. The leakage occurs in three separate pathways, each of which contributes to the overall potential difference across the membrane. In the case of the cations Na^+ and K^+ the potential difference opposes the diffusion potential; with Cl^-, however, the electrical potential and the diffusion potential have the same direction.

The pathways, symbolized in Fig. 17-18 as variable resistances, comprise the other elements of the circuit diagram. The second element is a leakage pathway of K^+_{out} $(\frac{1}{g_{K^+}})$ opposed by a potential difference (E_{K^+}). The third is a leakage pathway of Cl^-_{in} $(\frac{1}{g_{Cl^-}})$ assisted by a potential difference (E_{Cl^-}). The fourth is a leakage pathway of Na^+_{in} $(\frac{1}{g_{Na^+}})$ opposed by a potential difference (E_{Na^+}).

The resting potential then is the resultant of three leakage pathways and the electromotive forces they generate. Generally, the Na^+ leakage pathway can be ignored because, as we have seen, it is much smaller than that of K^+. And indeed it can be shown experimentally that removing Na^+ from the outside medium has very little immediate effect on the level of the resting potential. However, increasing the concentration of K^+ in the surrounding medium does markedly lower the magnitude of the resting potential. Thus, we find that the above formulation not only accounts quantitatively for the resting potential, but also permits us to predict the effects of experimentally varying the ionic concentrations.

In summary, the resting potential is a steady-state situation brought about by the active extrusion of Na^+ and accumulation of K^+ by the nerve cell. This activity is opposed by the relatively rapid leakage of K^+ and Cl^- and the relatively slow leakage of Na^+ (Fig. 17-19). Thus, in this somewhat indirect manner its own metabolism is able to convert the nerve cell into a tiny battery, capable of generating an electro-

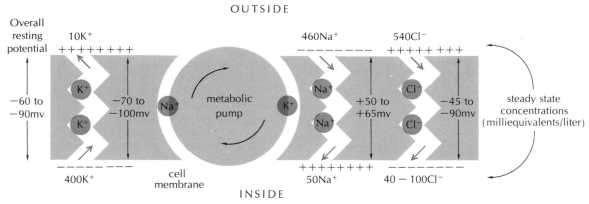

Fig. 17-19. The origin of the resting potential. This is an attempt to schematize the processes supporting the steady situation we call the resting potential. At the center we have the metabolic pump causing the K^+ influx and the Na^+ efflux. The steady-state levels of the Na^+, K^+, and Cl^- are indicated opposite the leakage pathways for each ion as well as the direction and magnitude of the potential difference each pathway is generating.

motive force, which as we shall see is utilized to initiate and propagate an electrical stimulus or action potential.

The nerve fiber is not a copper wire, which is capable of carrying a large current against a very small resistance. The fiber interior is an aqueous solution of ions, which has a considerable resistance to the passage of electricity, so that a current passing down the length of the fiber becomes greatly attenuated after even a few millimeters of travel. Yet we know that signals can travel along nerve fibers for several hundred centimeters without becoming attenuated. How then do nerves, which are such poor conductors of electric currents, manage to be such excellent transmitters of electric signals? The history of attempts to answer this question is an interesting one, and the student is referred to the excellent *Scientific American* articles listed in the "Suggested Readings" list. We shall restrict ourselves here to describing the present theory regarding the action potential and its propagation and to citing some of the evidence that supports it.

Let us begin by describing the electrical changes observed when a nerve is stimulated. Figure 17-20 illustrates the arrangement used for stimulating a nerve electrically and measuring the electrical consequences thereof. (It should be emphasized that many types of stimuli, including chemical and physical ones, will elicit a response from the nerve, but electrical stimulation is experimentally most advantageous, being physiologically more "natural" and physically easier to control and measure.)

If one takes a nerve preparation such as that shown in Fig. 17-20, one will observe of course the resting potential of 60–90 mv as soon as the recording electrode has penetrated the membrane. One can now stimulate by passing a brief current (2 milliseconds, 10^{-7} amp) through the membrane with the stimulating electrode. If the

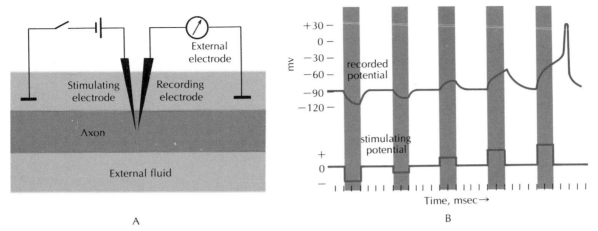

Fig. 17-20. Effect of direction and strength of stimulation on the potential difference across the membrane of a single nerve axon. (Modified from Katz, 1966.) A. Assembly showing stimulating and recording electrodes close to each other so that subthreshold perturbations can be recorded. B. Demonstration that the stimulating potential must be in the right direction and of sufficient intensity to produce a self-sustaining "spike" or action potential.

direction of the stimulating current (flow of negative charges) is from the outside to the inside of the membrane, then the membrane becomes locally even more polarized, creating a temporary local "disturbance" that can be recorded with the recording electrode (Fig. 17-20). If, however, the stimulating current is from the inside to the outside, the membrane will be depolarized; if the stimulus becomes sufficiently large to depolarize the membrane by some 50 mv (that is, reduce locally the resting potential from −90 mv to −40 mv), then a threshold is reached. Each time this occurs the nerve cell responds, independent of the stimulus current, with a stylized change in potential difference—which at its peak even reverses the polarity of the membrane (Fig. 17-21B). Furthermore, this response, which we call the action potential, does not diminish but instead travels from the stimulating electrode down the entire length of the nerve fiber without suffering any attenuation. Another unique characteristic of this system is that no matter how much one increases the intensity and duration of the stimulus beyond the minimum value, one does not affect the height, shape, or velocity of propagation of the action potential (Fig. 17-21C). This effect is one of the classical examples of a threshold response (all-or-none), a characteristic phenomenon frequently encountered in living systems.

Another interesting property of nerve is that numerous action potentials can be generated in close succession, but when the interval between them becomes less than a few milliseconds, one can observe a refractory period during which the nerve will not respond to the stimulation. The following is a resumé of our current explanation for the above phenomena.

1. The polarized state of the membrane (resting potential) determines its permeability properties to Na^+ and K^+.

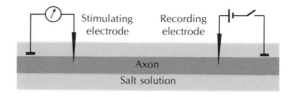

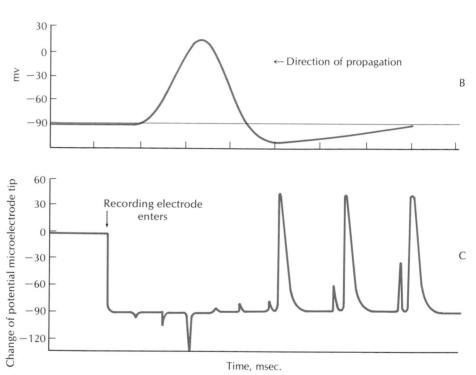

Fig. 17-21. The shape of the action potential and the all-or-none response. (Modified from Katz, 1966.) A. Assembly showing stimulating and recording electrodes sufficiently far apart to record the shape of a self-propagating action potential. B. Shape and dimensions of action potential. C. Demonstration that the nature of the action potential does not depend on strength of stimulus, displayed here as sharp spike.

2. When the membrane is even partially depolarized by electrical stimulation, the permeability to Na^+ rises, and the movement of Na^+ into the axon at the site of stimulation reinforces the depolarization.

3. At subthreshold stimulation the K^+ efflux and Cl^- entry balance the entry of Na^+ and quickly restore the potential difference across the membrane.

4. When the stimulus exceeds the threshold value so that the membrane becomes sufficiently depolarized after withdrawal of the stimulating pulse of

current, the rate of Na$^+$ entry is high enough to exceed the K$^+$ efflux and Cl$^-$ entry. At this point there begins a self-sustaining process, which continues until the high permeability to Na$^+$ is suddenly shut off. This transient increase in permeability to Na$^+$, called the "Na$^+$ gate," lasts only a few milliseconds. While the Na$^+$ gate is open, Na$^+$ rushes through the "open" portion of the membrane at a rate high enough to reverse the local polarity of the membrane, making the inside positive with respect to the outside (Fig. 17-21B).

5. While the Na$^+$ gate opens and shuts, a slower increase in K$^+$ permeability develops and continues to rise even after the Na$^+$ gate has shut. This K$^+$ permeability increase restores the original potential difference of the membrane even though in the meantime some Na$^+$ has leaked in and some K$^+$ has leaked out.

6. During the short interval when the Na$^+$ gate is already shut but the K$^+$ gate is still open, the potential difference temporarily reaches a value that exceeds the resting potential (Fig. 17-21B). This is the refractory period during which that particular portion of the membrane cannot be stimulated to open the Na$^+$ gate.

7. While the Na$^+$ gate is open at a particular point of the membrane, Na$^+$ from the adjoining region (which had not yet been stimulated) diffuses toward the region of the Na$^+$ gate, thus depleting the adjoining region of positive charge. This depolarization in the adjoining region brings about an increase of Na$^+$ permeability that eventually becomes self-sustaining, and so another Na$^+$ gate has been opened in the adjoining region. This process repeats itself over and over again in successive series of adjoining regions; thus, we have the propagation of a wave of reversal of polarization we call the action potential.

Let us now summarize this explanation of the action potential and its propagation. The permeability of the membrane to Na$^+$ is briefly increased by a reduction of the membrane potential, which ultimately brings about a brief localized reversal of polarity. This effect is propagated by the depolarization it induces in the neighboring region. The resting potential is restored by a temporary increase in K$^+$ permeability after which the nerve cell again becomes susceptible to stimulation.

Let us review some of the important lines of evidence that support the above "ionic theory" of the action potential and its propagation.

As far back as 1902, Overton had implicated Na$^+$ in the process of stimulation by showing that frog muscle loses its excitability when Na$^+$ is removed from the outside medium. Overton even suggested that excitation might be due to Na$^+$ and K$^+$ exchange. This interesting finding was confirmed in numerous experiments with nerve, particularly by Hodgkin and Katz, who showed that both the height of the action potential and the rate at which the peak is reached depend on the Na$^+$ concentration in the surrounding region. Thus, the magnitude of the resting potential depends mostly on the amount of K$^+$ inside, and the magnitude of the action potential depends on the amount of Na$^+$ outside. Using the postulates of the ionic theory, it is possible to calculate that the minimum amount of Na$^+$ that must enter the nerve to account for the depolarization of a certain area of membrane should be 10^{-12} moles/cm². Keynes and others have used tracers to make very precise determinations of this amount and have found, in fact, that the squid axon takes up 3 to 4×10^{-12} moles Na⁺/cm² per impulse and loses about the same amount of K⁺. Thus, there is sufficient transfer of ions to account for the magnitude of the electric current observed. It should also be noted that one impulse produces only a very minute change in ion concentration: in the squid axon only one-millionth of the K⁺ is lost. These lost ions are replaced by only a few seconds of metabolic pumping. Thus, an axon in which the metabolism has been stopped by poisons is capable of generating a few hundred thousand impulses before depleting its store of K⁺ and becoming loaded with Na⁺.

Curtis and Cole (1936–1949) did a number of remarkable studies of the electrical properties of the squid axon and demonstrated that during excitation there was a large and transitory decrease in resistance of the membrane but no change in capacitance. The resistance of the membrane fell from 1000 to about 20 ohm-cm² while the capacitance remained at 1 mf/cm². This meant that a major change occured not in the vast lipid area of the membrane, but only in the small portion of the membrane involved in the leakage of ions. In fact, one can calculate that less than 1 percent of the area of the membrane is involved in ion transport.

Fig. 17-22. The voltage-clamp method for studying the ionic conductance of the nerve membrane at a controlled potential difference across the membrane. (From Katz, 1966.) A. Apparatus shows disposition of two sets of electrodes and their connection to feedback amplifiers. B. Circuit diagram. C. Typical measurement obtained with voltage clamp method showing temporary inward current followed by outward current eventually reaching steady value.

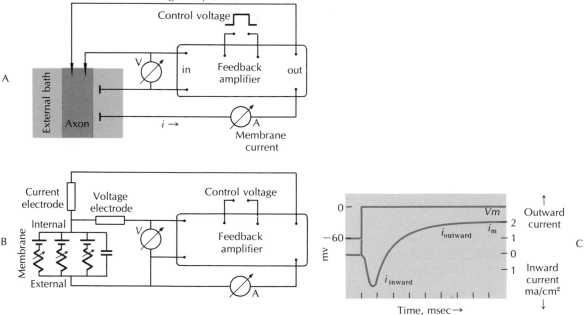

The most convincing evidence for the "ionic theory" of the action potential was provided by the voltage clamp technique developed by Cole and Marmont and used to great advantage by Hodgkin, Huxley, and Katz. This technique can prevent the "explosive" electrical events of the action potential by maintaining (that is, "clamping") a desired potential difference on the membrane with a feedback amplifier connected to two electrodes on each side of the membrane (Fig. 17-22). By combining this technique with the use of tracer methods and by replacing the outside Na^+ with the monovalent cation choline chloride, it was possible to disentangle the changes in conductance due to Na^+ influx and K^+ efflux. If the voltage clamp were used to reduce the potential so far that the nerve would ordinarily fire, it could be shown that the conductance due to Na^+ influx was of short duration while the conductance due to K^+ efflux was maintained as long as the potential difference was "clamped" on the nerve (Fig. 17-23).

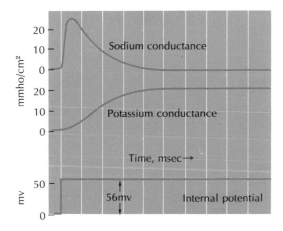

Fig. 17-23. Time course of Na^+ and K^+ conductance computed from voltage-clamp experiments in which external sodium was varied or replaced by choline chloride. (From Hodgkin, 1958 as shown in Katz, 1966.) This and other similar experiments made it possible to separate the conductance changes into Na^+ and K^+ conductances. Notice that the Na^+ conductance is short-lived, while the K^+ conductance reaches and maintains a steady value as long as the potential has been reduced, in this case to 56 mv.

By measuring the changes in ionic conductance brought about by Na^+ influx and K^+ efflux over a wide variety of membrane potentials maintained by the voltage clamp, it was possible to measure the amplitude and time course of Na^+ and K^+ currents over the entire range of physiological values (Fig. 17-24). The curves thus obtained enabled Hodgkin and Huxley to calculate a theoretical action potential, the properties of which duplicated remarkably well the experimentally observed phenomena (Fig. 17-25). These included the subthreshold events, the development of the all-or-none response, the charge reversal, the refractory period, the quantities of ion exchange, and even the velocity of propagation.

Subsequent work using more sophisticated voltage clamp apparatus and high-speed computing techniques has served to confirm and refine the ionic theory of the action potential.

The use of special microelectrodes has extended the method to a variety of nerve and muscle cells and also to other cells in the plant and animal kingdoms. Minor

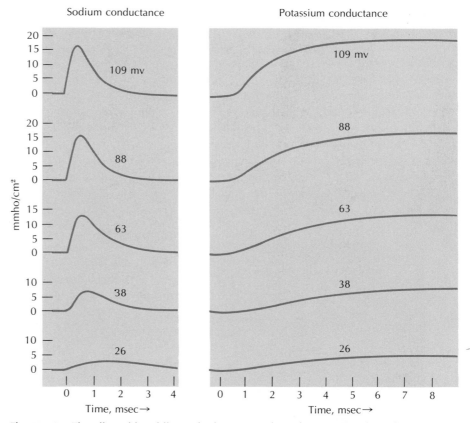

Fig. 17-24. The effect of five different displacements of membrane potentials on the Na⁺ and K⁺ conductances. (From Hodgkin, 1958 as shown in Katz, 1966.) The numbers indicate the depolarization step for each curve in mv.

Fig. 17-25. Theoretical reconstruction of change in conductance due to Na⁺ gate (g_{Na^+}) and K⁺ gate (g_{K^+}) as well as of resulting action potential. (From Hodgkin and Huxley, 1952 as shown in Katz, 1966.) A striking degree of agreement was obtained between observed and calculated properties of action potential, including even the velocity of propagation (calculated velocity = 18.8 m/sec; observed velocity = 21.2 m/sec).

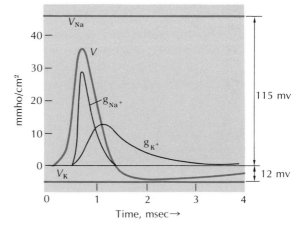

variations such as widely differing time constants have been observed; major variations such as the use of different ions (Ca^{2+} in the barnacle, and even Cl^- in *Nitella*) may be observed. However, it is likely that the basic principles of ionic pumping, selective leakage, and specific transitory changes in ionic conductance of the membrane are involved in all bioelectric phenomena.

Two major areas have to date eluded our detailed understanding, and in these areas we know little more than the early physiologists. One of these is the molecular mechanism of the active transport of ions; the other, the molecular mechanism of the highly precise, transitory changes in membrane conductance of various ions. What, for instance, determines the short-lived opening of the Na^+ gate and the more sluggish opening of the K^+ gate? In order to provide answers to these questions we will have to know much more about the molecular structure and properties of the membrane. And so we return to the plasma membrane for a final look at what we know of its structures and properties.

Theories of Membrane Structure

In the earlier portion of this book it has been our practice to proceed from a relatively rigorous treatment of structure to a discussion of biological function that relied heavily on our knowledge of structure. We did not follow this procedure in our discussion of the membrane because we do not as yet know enough about the structure of the cell membrane to illuminate our understanding of its function. In fact, the tendency has been so far the other way around. By studying the function of the membrane, we have through indirect means been able to make inferences regarding membrane structure. Thus, from permeability and electrical studies we have come to the conclusion that the membrane is a thin, lipid-containing layer surrounding the cell. A large proportion of the lipid component controls the movement of penetrants via their lipid-solubility properties. A small portion of the membrane also appears to consist of very small pores that permit polar molecules and ions of small dimensions to leak through. But in addition to these two types of unfacilitated ports of entry, the membrane also appears to have highly specific (catalytic) carrier systems that greatly amplify the selectivity of the membrane by allowing certain substances to penetrate at highly enhanced rates. Furthermore, as is the case with the action potential, the membrane is capable of varying its selectivity in time by opening and closing special "gates" with great precision. Finally, the membrane can couple some of these catalytic processes to metabolic energy sources and transport substances against chemical potential gradients. Thus, the membrane can act at a molecular level with carrier systems too small to be seen with the electron microscope or on a grosser level by folding of relatively large portions of membrane.

From the above properties we can infer that the membrane must be composed not only of lipid, but also of a large number of different proteins. These proteins endow the membrane with the wide range of specific functions that we know only the proteins can perform. Thus, at the very least, there must exist carrier proteins with highly specific catalytic functions, proteins that act as energy transducers, and possibly mechanochemical and structural proteins that endow the membrane with a variety of physical properties we shall discuss below. Let us therefore review briefly some physical studies on the membrane which will complement the insights we have gained so far.

While Pfeffer, Overton, and their successors studied the plasma membrane through its permeability and electrical characteristics, a number of other biologists were investigating the surfaces of cells by more direct methods. It should be pointed out, however, that until the electron microscope revealed the presence of a finite membrane surrounding all cells, we had scant evidence that the structure which accounted for the permeability characteristics of cells and the structure at the surface of the cell, which was found to have special physical properties, were one and the same. Since they turned out indeed to be the same, we shall use more recent insight to examine a number of fascinating observations of the cell surface that have been made over the years. We must, however, make one qualifying statement regarding the identity of the plasma membrane and the physical cell surface. The latter may include slightly more than the structure which regulates the exchange of materials. It may include a coating of variable thickness and chemical composition (mucopolysaccharides, and so forth) on the outside, and on the inside it may have a thin layer of gelated cytoplasm which, according to some physical experiments, endows the "liquid" membrane with some structural rigidity and elasticity.

One of the most dramatic experiments showing that the cell surface layer determines the permeability properties of the cell was performed by De Vries (1885). Using microsurgical methods, he was able to remove the outer membrane and the cytoplasm from plant cells, leaving behind an intact plant vacuole surrounded only by a vacuolar membrane. De Vries showed that this simple system retained its osmotic characteristics and behaved like an osmometer, swelling and shrinking in different sucrose solutions. De Vries' beautiful microsurgical techniques were developed further by a number of experimenters such as Janet Plowe, Scarth, Seifriz, and Chambers, who demonstrated that the cell surface was covered by a layer with very special physical properties. For instance, they showed that the interior of the cell could disperse in the aqueous medium surrounding the cell when the membrane was violently disrupted. This suggested at first that the membrane was like an elastic skin, keeping the liquid interior inside the cell. But subsequent work showed that most membranes were capable of rapid enlargement during osmotic swelling as well as of rapid healing when disrupted more gently. These and other experiments led Scarth to the notion that the membrane is really a liquid with the special property of spreading on the somewhat gelated cytoplasm below it and yet maintaining its integrity by not being miscible in the surrounding aqueous medium.

One of the earliest direct demonstrations that the plasma membrane contained considerable amounts of lipid was carried out by Gorter and Grendel (1925) who prepared red cell ghosts, the membraneous remains of erythrocytes, and extracted the lipid from them. By spreading the lipid extract on a "Langmuir trough," which is an apparatus capable of measuring the area of monomolecular films of lipid spread on water, they were able to show that there was just enough lipid in the red cell ghost to account for a bimolecular layer surrounding the cell. Gorter and Grendel, therefore, concluded that cells were covered with a bimolecular layer of lipid.

If the surface of the cell is an oily layer, it would seem useful to demonstrate this directly with surface-tension measurements. Harvey, Davson, Danielli, and others performed numerous ingenious experiments in which they (1) applied oil droplets to surfaces of cells and measured the wetting angle (from which one could calculate the interfacial tension between the oil and the cell surface), (2) measured the surface tension of cells directly with the du Noüy tensiometer by determining the attractive force between the surface of a sea urchin egg and a little platinum wire ring touching it, and (3) measured the tendency of cells that are being distorted by a centrifugal field to round up under the influence of its surface tension. To their surprise, they found that the surface tension of cells was considerably lower (1–2 dynes/cm) than that of oil layers (7–15 dynes/cm). Since it was known that the addition of protein to oil (for example, egg white to mackerel oil) lowered the surface tension to the low values observed on the surface of cells, Davson and Danielli concluded that protein was an integral part of the membrane. And thus a theory was born regarding the structure of the membrane which has dominated our thinking ever since (Fig. 17-26). This theory holds that the plasma membrane is a three-layered (protein-lipid-protein) sandwich. The lipid layer at the center is the bimolecular layer of Gorter and Grendel, the hydrophobic portions of the lipid molecules pointing toward each other and the hydrophilic portions binding to the protein layers on both sides.

When electron microscopic studies finally provided direct visual evidence for the existence and the properties of the plasma membrane, it appeared that the Davson and Danielli model had been completely confirmed. In 1954, Davson and Danielli published a final version of their model which included small aqueous (polar) pores, and Robertson (1959) elevated the model to a "unit membrane" theory which stated that all membranes of the cell were constructed of this protein-lipid-protein sandwich. And indeed, the evidence provided by the electron microscopy and X-ray diffraction of that period seemed to support these conclusions very well. Using $KMnO_4$ as a fixative, one could observe a layer approximately 75 A thick. The two electron-dense lines, 20 A wide, presumably of protein, were separated by a lighter layer, 35 A thick, presumably of lipid (Fig. 17-26). Much work has been invested in recent years to demonstrate that the two external electron-dense regions are protein and the internal light region is composed of lipid. In spite of sophisticated work with model systems such as artificial lipid membranes, however, the precise chemical nature and organization of the three layers observed under the electron microscope has still not been unambiguously identified. In fairness to the Davson and Danielli sandwich model, it should be pointed out that Branton and others, using the recently perfected freeze-etching technique that does not use any chemical fixative, have shown that the membrane is indeed a three-layered structure. The layer at the center seems to be the weakest, being frequently split by the fracturing process used to expose the interior

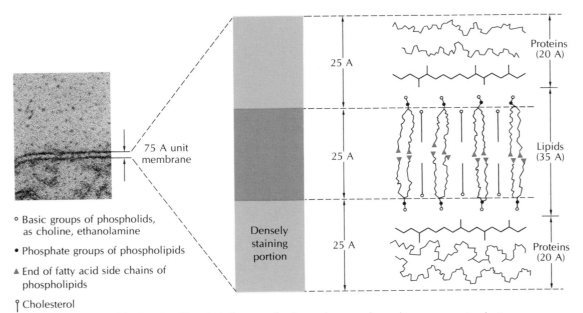

° Basic groups of phospholids,
  as choline, ethanolamine

• Phosphate groups of phospholipids

▲ End of fatty acid side chains of
  phospholipids

ʃ Cholesterol

Fig. 17-26. Chemical diagram of unit membrane and membrane as seen in electron microscope. (After Davson, Danielli, Robertson; micrograph courtesy of W. Stoeckenius, Rockefeller University.)

of the cell (Fig. 17-27). This is what one would expect if the middle layer of the sandwich is made of a bimolecular leaflet of lipid because such a layer would be held together only by the relatively weak van der Waals forces between the nonpolar lipid side chains.

In recent years, however, workers in the field of membrane structure began to have some second thoughts regarding the biological relevance of the Davson and Danielli model and the general applicability of the unit membrane theory. This has led to a variety of different, though as yet imprecise, models of membrane structure (Fig. 17-28). Let us summarize the evidence in favor of a more complex formulation of membrane structure.

1. *The chemistry of membranes varies considerably.* It must be obvious by now that membranes can be, and have been, isolated. We can strip off the plasma membrane of mammalian cells and we can get the outer membrane of the bacteria in the form of spheroplasts. We can isolate endoplasmic reticulum membranes in the form of microsomes, as detailed in Chapter 15; nuclei can be isolated and their membranes obtained. Mitochondria and chloroplasts have been isolated: in the case of the former the outer, limiting membrane of the mitochondria has been separated from the inner, cristae membranes with their knobs (see Chapter 13); whereas in the case of the latter the chlorophyll-containing lamellae have been obtained. Only a few membrane systems have been studied so far from the point of view of their protein-lipid content, but the

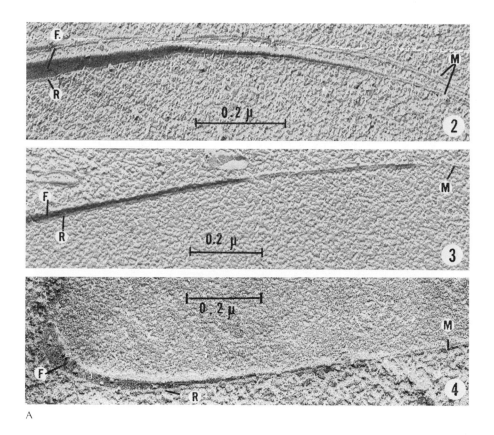

A

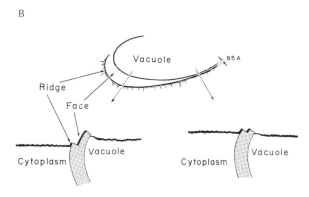

B

Fig. 17-27. Evidence for the existence of a sandwich (three-layered) structure in the vacuole membrane of the onion root tip using freeze-etching technique. (Courtesy of Branton.) The freeze-etching method does not use any chemical fixatives; it involves the rapid freezing of a specimen and subsequent "fracturing" with a microtome knife. The newly exposed surface is "etched" by briefly sublimating water from the frozen surface, after which the surface is shadowed and replicated. Electron microscope pictures are taken of the replica. One interesting result of this technique was the demonstration that the membrane seems indeed to be composed of three layers, the middle layer being weak with respect to the fracturing process. A. Electron micrograph of the replica showing "M" the three-layered membrane at right being continuous with the fractured membrane at left showing "R" the ridge of one outer layer and "F" the face of a middle layer. B. Explanatory diagram of the appearance of the ridge, face, and leaflet structure of a membrane.

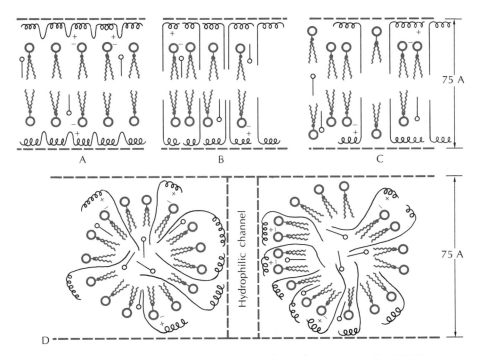

Fig. 17-28. Various other membrane models. (Symbols are the same as in Fig. 17.26.) (A) is a redrawing of the "conventional" membrane model in Fig. 17-26. (B) is one in which the hydrophobic side chains of the proteins penetrate into the interior of the membrane, along with the fatty acid side chains, cholesterol and neutral fats (not shown), resulting in an interior which excludes water. (The helical part of the proteins, including regions of polar amino acids, remain in an aqueous environment on the outside of the membrane). (C) is a modification of (B) in which the polar groups of the lipids also face the outer aqueous environment surrounding the membrane along with polar groups in the proteins, and still the fatty acids, cholesterol, neutral fats, and the hydrophobic side chains of the proteins face into the interior. (D) is a modification of (C) in which the associated proteins and lipids form globular clusters of lipoproteins within the membrane. These globular clusters still preserve the hydrophilic and hydrophobic characteristics of (C) but are so arranged that hydrophilic regions appear in the interior, between these clusters, allowing for the possibility of hydrophilic channels being present for the passage of H_2O and small hydrophilic molecules. The diagrams also illustrate some of the electrostatic binding, between positively charged groups in proteins (basic amino acids) and negatively charged groups in lipids (phosphates), which is thought to play a role in membrane structure.

data already suggest that membranes vary greatly in their protein-lipid ratio (Table 17-6). It is hard to reconcile such great variability in the protein-lipid ratio with a unit membrane theory requiring a uniform structure.

Since isolated organelles or structures containing membranes have vastly different enzymic compositions from each other, their membranes must have different enzymes as constituents. Two examples should suffice to illustrate this point. The inner mito-

Table 17-6. RATIO OF PROTEIN TO LIPID IN
MEMBRANES OF DIFFERENT ORIGINS

MEMBRANES	RATIO OF PROTEIN TO LIPID	(CALCULATED FROM PERCENT DRY WEIGHT)
Myelin	0.25	
Chloroplast lamellae	1.0	
Mycoplasma laidlawii	1.4	
Erythrocyte	1.5	
Mitochondrial inner membranes	3.0	
Micrococcus lysodeikticus	3.1	

chondrial membrane has an electron-transport chain (containing some dehydrogenases and all the cytochromes) that is greatly different from two other electron-transport chains found in endoplasmic reticulum membranes. The latter membranes in the liver cell have an enzyme, glucose-6-phosphatase, not found in any other membrane in the cell. However, it is possible, and probable since we know of some instances, that the same enzyme activity, and hence possibly the same enzyme, is found in two different membrane structures in the cell. Now, since we cannot purify many of these membrane-localized enzymes, we cannot determine how much of the protein content of any particular membrane is made up of enzymatically active proteins; that is, we do not know if all the proteins that constitute the structure of a membrane are enzymes in one form or another.

The precise chemical composition of the lipid component has also been shown to differ extensively (Table 17-7).

Table 17-7. LIPID COMPOSITION OF A VARIETY OF MEMBRANES*

LIPID	PERCENTAGE IN			
	MYELIN	CHLOROPLASTS	ERYTHROCYTE	MITOCHONDRIA
Phospholipids	32	10	55	95
Cholesterol	25	0	25	5
Sphingolipids	31	0	18	0
Glycolipids	0	41	0	0
Others	12	50	2	0

* (From Branton and Park, 1968.)

Phospholipids are the most common component of membranes whereas neutral lipids (triglycerides and cholesterol esters) are usually found in very low quantities. The plasma membrane is an exception—in mammalian cells neutral lipids can be as high as 25 percent.

The lamellae of the chloroplasts have a good deal of galactolipids (lipids in which the sugar galactose replaces the phosphorylated base); even among the phospholipids, while all membranes have phosphatidyl-choline (lecithin) as the predominant species, the relative amount can vary from the 60 percent found in endoplasmic reticulum membranes to the 40 percent found in mitochondrial membranes. Finally, some membranes might contain a lipid found nowhere else as is the case with the cardio-lipin found in the inner mitochondrial membranes. The difference in lipid composition of mitochondria and chloroplasts is indeed striking and is, no doubt, related to their particular biological function.

2. *Myelin* (a favorite object of membrane study) *differs greatly from other membranes.* The myelin sheath around certain nerve fibers appears to be formed from an extensive outgrowth of the plasma membrane, and its multilayered structure is very convenient for X-ray analysis. Robertson and others have performed very precise studies in which they compared X-ray and electron microscopic descriptions of myelin. Although this comparison did, on the whole, show good agreement between the sandwich seen under the electron microscope and the spacings obtained by X-ray diffraction, this work failed to provide precise data regarding the internal organization of the sandwich. More importantly, these studies indicated that the width of a single membrane in the myelin sheath was significantly less (60 A) than the width of the plasma membrane (75 A) of the same cell. All of this is not surprising since myelin has probably no other biological function than to act as an insulator of the nerve cell.

3. *Even the KMnO₄-fixed membranes observed under the electron microscope vary in thickness.* It is generally agreed the $KMnO_4$ fixation is likely to produce the least artifacts. Sjöstrand (1963) compared the thickness of membranes surrounding and inside the same cell preparations and found them to vary from 50–60 A in mitochondrial membranes to 90–100 A in plasma membranes and zymogen-granule membranes.

4. *Membranes differ in their appearance.* We now know a good deal concerning the morphological diversity of biological membranes. Grossly, some are sheets like the plasma membrane and membranes of the endoplasmic reticulum, while others such as the mitochondria cristae and the Golgi membranes are much more convoluted. Some, such as the mitochondrial cristae membranes, have particles adhering to them (Fig. 13-1C); while others, like the outer mitochondrial membrane, looked at under the same conditions of fixation, have none.

5. *Studies of artificial lipid membranes by electron microscopy and X-ray diffraction have yielded, so far, ambiguous results, especially since protein is usually lacking in these artificial membranes.*

6. *The sandwich model may explain the insulating properties of large portions of the membrane, but membranes have a variety of other biological functions.* Among these functions we can list:

(See Chapter 15.)

involvement in protein biosynthesis by membrane-bound polysomes;
facilitated transport of a variety of small molecules;
active transport by transduction of chemical energy into osmotic work;
precise opening and shutting of ionic gates;
pinocytosis and phagocytosis; and

mechanochemical transductions which may explain the independent movement of some organelles, such as chloroplasts and even sliding movement of
cells in development and wound healing of animals.

We now know enough about the biological function of proteins such as enzymes,
antibodies, and mechanochemical proteins to recognize that their specific properties
depend upon the three-dimensional folding of their polypeptide chains. It is difficult
to conceive that the complex biological properties of membranes can be carried out by
a uniformly thin (20 A) layer of protein, which is thought by some to consist of one or
two layers of polypeptide chains in the pleated sheet configuration.

7. *In fact, recent studies have revealed increasingly the presence of globular
subunits in a variety of membranes.* We have discussed in Chapter 13 the relatively
complex subunit structure of the inner membranes of mitochondria. Much evidence
has recently accumulated that the inner membranes of chloroplasts (lamellae) also
contain subunits (Fig. 17-29 A). A most convincing demonstration of the presence of
these subunits using the freeze-etching method is provided in Fig. 17-29 B, which com-
pares the smooth faces of the chemically inert myelin membranes with the bumpy
faces of chloroplast lamellae.

In conclusion let us discuss our admittedly incomplete knowledge of membrane
structure in terms of the variety of functions in which membranes participate.

What is the meaning of having enzymes in membranes? We have no definite

Fig. 17-29. Comparison of membrane faces of myelin (A) and of chloroplast lamellae
(B) prepared by the freeze-etching technique. Note the smooth faces of the myelin
membranes and the numerous globular subunits on the chloroplast membranes. (From
Branton and Park, 1966 and 1967.)

A   B

 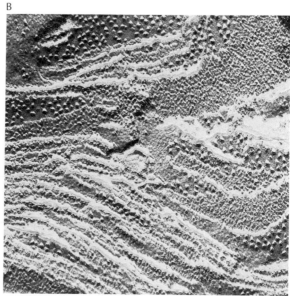

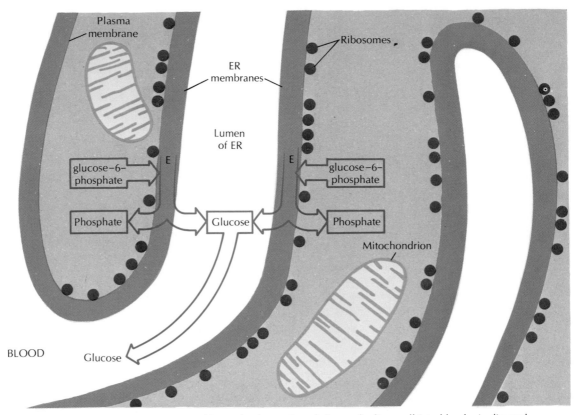

Fig. 17-30. Possible mode of secretion of glucose by liver cell into blood, via directed action of glucose-6-phosphatase (E) localized on endoplasmic reticulum (ER) membranes.

answers, but we can make some good guesses. In Chapter 13 we detailed a possible answer to this question regarding the mitochondrial and chloroplast membranes. As a further example, we can mention the glucose-6-phosphatase found primarily in the liver cell where we think the membrane acts as a vector, as in ion-transport. This enzyme is thought to be responsible for one of the chief roles of the liver: in splitting phosphate off, it allows free glucose to leave the liver cell and enter the circulating blood, keeping the blood glucose at controlled levels. One reason the enzyme might be localized on the endoplasmic reticulum membranes is that the enzyme there acts as a vectorial component, attacking glucose-6-phosphate coming from glycogen breakdown in the soluble cytoplasm, splitting off phosphate, and at the same time, acting as a vector so that phosphate remains in the cytoplasm on one side of the membrane, with the glucose going through the membrane into the lumen of the endoplasmic reticulum on the other side (Fig. 17-30). It is thought that the membranes of the reticulum are in intermittent continuity with the plasma membrane, so that the lumen within the reticulum is in intermittent continuity with the fluid space surrounding the liver cell as depicted in Fig. 17-30. This idea of directedness is not strange if we consider that the two sides of membranes are probably not the same: ribosomes are found on only one side of the endoplasmic reticulum membranes (Fig. 15-2), while the particles of the inner mitochondrial membrane are also only on one side of the membrane (Fig. 13-1C).

Furthermore, we do have some inkling that the rates of enzyme activities can be influenced by the environment in which the enzyme finds itself. It is thought that most tightly bound (see below) membrane enzymes require lipid attached to them for full enzyme activity to become apparent. It could well be that enzymes on membranes are thus more easily subjected to metabolic regulatory mechanisms. As was noted in Chapter 10, enzymes are in an active state only in certain tertiary configurations, and it might well be that this state is attained for certain enzymes only when they are a part of membrane structures; any alteration in these structures would change enzyme activity. Finally, the interiors of membranes are probably excellent hydrophobic environments (Figs. 17-26 and 17-28), and it is possible that enzymic activities of some proteins are attained only in this kind of environment and can be subjected to control mechanisms only in this environment.

Thus, it would seem that we know quite a bit about the specific constituents of membranes, about their chemistry, about their properties, and about their appearance, but there is one aspect where our knowledge is woefully meager, and this is how they are constructed, how they are put together from their constituents, and how they are synthesized, as structures, in the cell. For example, we can isolate membranes and determine their enzymic composition; some of these enzymes can be easily extracted from membranes and, hence, are loosely bound; others are extracted with more difficulty and, hence, are more tightly bound; while still others cannot be extracted at all and, hence, are probably required components of the membrane structure. On the other hand, in the case of some membranes, as the mitochondrial ones, lipid can be almost completely extracted, and the membranes still retain their appearance in the electron microscope. Indeed, the easy extractability of some membrane-bound enzymes might be explained by the hypothesis that these are not really part of the morphologically realized membrane structure, as seen in Figs. 17-26 and 17-28, but instead are loosely, but specifically, attached to certain sites on the outsides of the structures depicted in the above figures.

Recently, a great deal of thought has been given to the idea that in addition to the enzymes found therein, membranes have proteins which are nonenzymatic but are concerned solely with conferring organization to a membrane. These kinds of proteins have been called "structural"; their postulated properties would be to bind certain enzymes on the one hand and certain lipids on the other; they would be "organizers" of the membranes. A good deal of work has been going on recently in attempts to isolate these "structural" proteins; however the work has not been too successful because it is difficult to isolate a protein that has no enzymatic properties and has only certain binding properties that are not yet well defined.

This lack of knowledge of biochemical structure is reflected in our overabundance of ideas concerning the chemical structure of membranes (Fig. 17-28). We need ways of looking at these problems, and two approaches are being attempted: one, as we have already indicated, is the attempt to tear membranes apart into some kind of observable morphological, chemical, and biochemical constituents and then to try to reconstitute them to a recognizable membrane fragment, as has been attempted with the inner mitochondrial membranes (Chapter 13); the other way is to try to observe how known membranes, with known chemical and biochemical properties, are formed in the living cell, in normal, or better yet under abnormal circumstances, and then to try to formulate a picture of membrane structure from these observable processes of membrane formation. Each of these approaches has its difficulties, for we are now beginning to look at the biogenesis of cell structure, which is a harder task than to look at the synthesis of biological molecules, even macromolecules. All the questions raised in this chapter on membrane function will be solved only when we know much more concerning membrane structure than we know at present. This is not the only reason why much work is going on in this field, for once we learn the mechanisms concerning membrane biogenesis, then we can go on to problems relating to the biogenesis of the biochemical morphology of the cell as a whole.

And so we must conclude that biological membranes are variable and complex structures capable of performing a wide variety of biological functions. While the sandwich model might be of relevance to those portions of the membrane that have the function of insulating the cell from its surroundings, other parts of the membrane, no doubt, carry a large variety of different proteins and lipoproteins responsible for the numerous biological functions in which the membrane is engaged. In view of the importance of biological membranes and the interest they have generated we can expect to witness exciting new developments during the next decade.

SUGGESTED READING LIST

Offprints

Baker, P. F., 1966. "The Nerve Axon." *Scientific American* offprints. San Francisco: W. H. Freeman & Co.

Cohen, G. N., and Monod, J., 1957. "Bacterial Permeases." *Bacteriological Reviews*, pp. 169–194. Bobbs-Merrill Reprint Series. Indianapolis: Howard W. Sams & Co.

Hokin, L. E., and Hokin, M. R., 1965. "The Chemistry of Cell Membranes." *Scientific American* offprints. San Francisco: W. H. Freeman & Co.

Holter, H., 1961. "How Things Get into Cells." *Scientific American* offprints. San Francisco: W. H. Freeman & Co.

Katz, B., 1952. "The Nerve Impulse." *Scientific American* offprints. San Francisco: W. H. Freeman & Co.

Katz, B., 1961. "How Cells Communicate." *Scientific American* offprints. San Francisco: W. H. Freeman & Co.

Keynes, R. D., 1958. "The Nerve Impulse and the Squid." *Scientific American* offprints. San Francisco: W. H. Freeman & Co.

Porter, K. R., and Franzini-Armstrong, Clara, 1965. "The Sarcoplasmic Reticulum." *Scientific American* offprints. San Francisco: W. H. Freeman & Co.

Robertson, J. D., 1962. ''The Membrane of the Living Cell.'' *Scientific American* offprints. San Francisco: W. H. Freeman & Co.

Schmidt-Nielsen, K., 1960. ''The Salt Secreting Gland of Marine Birds.'' *Circulation*, pp. 955–967. Bobbs-Merrill Reprint Series. Indianapolis: Howard W. Sams & Co.

Solomon, A. K., 1960. Pores in the Cell Membrane. *Scientific American* offprints. San Francisco: W. H. Freeman & Co.

Solomon, A. K., 1962. ''Pumps in the Living Cell.'' *Scientific American* offprints. San Francisco: W. H. Freeman & Co.

Articles, Chapters, and Reviews

Crane, R. K., 1960. ''Intestinal Absorption of Sugars.'' *Physiological Reviews, 40,* p. 789.

Holter, H., 1959. ''Pinocytosis.'' *Annals of New York Academy of Science, 78,* p. 524.
Lefevre, P. G., 1955. ''Active Transport through Cell Membranes.'' *Protoplasmatologia, 8,* No. 72. Vienna: Springer-Verlag.

Lefevre, P. G., 1959. ''Molecular Structure Factors in Competitive Inhibition of Sugar Transport.'' *Science, 130,* p. 104.

Korn, E. D., 1966. ''Structure of Biological Membranes.'' *Science, 153,* pp. 1491–1498.

Palade, G. E., 1956. ''Endoplasmic Reticulum,'' *Journal of Biophysical and Biochemical Cytology,* Supplement 2, p. 85.

Ponder, E., 1961. ''Cell Membrane and Its Properties.'' In: *The cell,* Vol. 2, p. 1. J. Brachet, and A. E. Mirsky, eds. New York: Academic Press, Inc.

Porter, K. R., 1955–1956. ''Submicroscopic Morphology of Protoplasm,'' *Harvey Lecture Series,* p. 175. New York: Academic Press, Inc.

Porter, K. R., 1961. ''The Ground Substance: Observations from Electron Microscopy.'' In: *The cell,* Vol. 2, p. 621. J. Brachet, and A. E. Mirsky, eds. New York: Academic Press, Inc.

Robertson, R. N., 1960. ''Ion Transport and Respiration.'' *Biological Reviews,* Vol. 35, p. 231.

Wilmer, E. N., 1961. ''Steroids and Cell Surfaces.'' *Biological Reviews,* Vol. 36, p. 368.

Books

Davson, H., and Danielli, J. F., 1952. *The Permeability of Natural Membranes,* 2d ed. London: Cambridge University Press.

Finean, J. B., 1961. *Chemical Ultrastructure in Living Tissues.* Springfield, Ill.: Charles C Thomas, Publisher.

Harris, E. J., 1960. *Transport and Accumulation in Biological Systems.* New York: Academic Press, Inc.

Katz, B., 1966. *Nerve, Muscle and Synapse.* New York: McGraw-Hill Book Co., Inc.

Locke, M. (ed.), 1964. *Cellular Membranes in Development.* New York: Academic Press, Inc.

Stein, W. D., 1967. *The Movement of Molecules Across Cell Membranes.* New York: Academic Press.

Chapter 18

Regulation of Cell Function and Cell Structure

In the last few chapters we have devoted our attention to the functioning of the individual units that make up the living machine. The marvel of the cell is that all its processes are intimately correlated into an overall rhythmical process of cell growth and cell division. How are the myriad processes of the cell tied together in time and space so as to produce the unified phenomenon of life?

We know as yet very few answers to this very important question, so we shall content ourselves in this chapter with drawing attention to some possible avenues of approach and suggest the direction in which future work is likely to proceed. We shall discuss the problem of regulation at two levels: the enzymatic and the architectural.

THE ENZYMATIC LEVEL: GENERAL

By regulation of enzymatic activity we mean those control mechanisms involved in determining the amount of enzyme formed and those involved in controlling the rates of enzyme activities once the enzymes are formed. The former operate at the genetic and translational levels (Chapters 14 and 15) whereas the latter operate at the sites of enzymatic action. As far as we can tell, both procaryotes and eucaryotes have many of the same types of regulatory devices. The above two types of enzymatic control are probably not independent of each other as will be illustrated below; indeed, they seem to be a part of an organization that can arbitrarily be divided into hierarchies that interact with one another. We have chosen to divide this organization of regulatory devices into three general groups according to the speed with which the regulatory devices can be turned on and off and the shortness of the time interval during which the effect can be appreciated by the metabolism of the cell. Thus, the modulations of enzyme activity (devices operating on the enzyme itself) can accomplish a more rapid and sensitive adjustment of metabolic processes than can the modulations of enzyme amount (devices operating at the genetic and ribosomal levels). Finally, enzymes are linked together in metabolic teams, as the electron-transport chains for example, and in eucaryotes further localized in such sub-organelles as chloroplast and mitochondrial membranes. This more complex organization provides another probable site of metabolic control (see Chapter 13), and even though we know very little about the modes of modulation at this level, we can guess that the rate of modulation is much slower than at the other two levels cited above. As will be seen below, these three levels of regulatory control are not self-contained but are constantly interacting so that in many cases an enzymatic reaction is subject to control at several levels, and the degree of control at each level is modified by interactions with another level. Indeed, the more we delve into the nature of these regulatory devices (and we are just at a beginning here), the more we marvel at the beautiful intricacies that are operating within the cell in order to keep the cell operating at a maximal efficiency.

The fact of enzymatic coordination of cellular activity is indicated by the observation that the cell seldom either synthesizes or degrades more than is necessary for normal metabolism and growth. It is indicated further by the finding that the cell is a homeostatic organism: if some reparable damage is done, the cell can right itself. It is even possible that when particular mutations occur in the DNA and hence certain enzymes are not formed, the essential metabolites can still be produced but in some other than "normal" way; this might be the meaning of the alternate metabolic pathways discussed earlier.

The control of cellular metabolism centers around the regulation of enzyme type, enzyme amount, and enzyme activity; the last is determined in large part by the nature of the enzymatic reaction itself. These areas of regulation can be outlined as shown in Fig. 18-1 and numbered here accordingly as follows: (1) control of the type and amount of enzyme by mechanisms acting via the gene or DNA and mediated by the enzyme-forming system (2); (3) control of enzyme activity by factors that are specific for a particular enzyme such as pH, and so forth; and (4) control of enzyme activity by the nature and amount of product that is made. We will discuss each of these below. The rate of a reaction is circumscribed by the amount of active enzyme molecules and by the kinetics of the reaction. The former, in turn, is influenced directly by the enzymatic synthesizing mechanism of the cell, by the DNA, RNA, and ribosome triad mentioned earlier, and also indirectly by the concentrations of the substrates and products of the particular enzymatic reaction. The kinetics are influenced by the conditions prevalent in the reaction milieu, by substrate concentration, the presence or absence of cofactors, activators, inhibitors, the pH, and so forth. We will take up the latter case first, denoted by (3) and (4) in Fig. 18-1.

Modulations of Enzyme Activity: Enzyme Complexes

First, the structure of the enzyme itself may be very important for the control of its activity. While it should be obvious by now that many enzymes exist and act within a

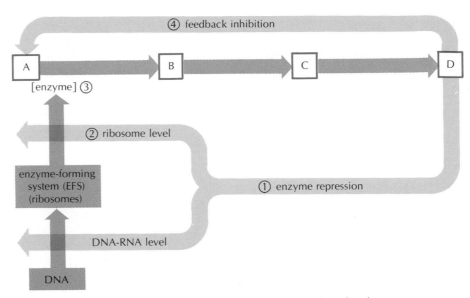

Fig. 18-1. Sites of control of cellular metabolism. Terms are explained in the text.

(See Chapter 10.)

spatial framework of the cell, (and more will be said about this aspect later), it is also becoming increasingly clear that even enzymes that were thought to be single, soluble species can and do interact with other proteins in the cell to form what might be called microstructures. A good example of this is the pyruvate dehydrogenase complex, worked out by Reed; this complex, molecular weight of 4.8×10^6, which catalyzes the oxidative decarboxylation of pyruvate, is made up of 16, 64, and 8 molecules, respectively, of each of three separate enzymes. Because transacetylation and electron transfer are involved, the three molecules perform their function better when tightly coupled to one another; consequently, they probably exist this way in the cell. Indeed, the complex has been isolated and purified, and electron micrographs have been taken of it (Fig. 18-2) since it is a fairly large complex, measuring 300–400 A in diameter.

One of the enzymes in the complex is even made up of smaller subunits, and this condition is another feature of many enzyme molecules. We now know that many enzymes are not single-chain protein molecules but are specific aggregates of specific polypeptide chains that separately have no activity but that when brought together form an active enzyme. The bonding between the separate chains is various; disulfide, hydrogen, and hydrophobic bonding have all been implicated. These findings open up whole new fields of endeavor: the temporal and spatial modes of syntheses of such polypeptide complexes (the cell must regulate the rates of synthesis of the separate chains), how these polypeptides react with one another in a specific manner (see below), and the possible meaning of these complex polypeptide enzymes for enzymatic regulation.

A good example of an enzyme composed of separate chains is the lactic dehydrogenase of striated muscle and of heart muscle. This, like other similar enzymes, has been called an isozyme, that is, it can exist in multiple forms, all of which can be separated from one another by physical means such as electrophoresis, but all of which have the same specific enzymic activity. From the work of Markert and Kaplan, we know that the reason for the existence of the five multiple forms is that the enzyme itself is made up of four polypeptide-chain subunits. Thus, in one form we have HHHH chains (H being the chain found predominantly in heart muscle); in another form we have MMMM chains (M being found predominantly in striated muscle); and the other three forms are the other three possible combinations of the two different chains (HMMM,HHMM,HHHM). Again, each of the chains has no activity by itself, but the four chains, in any combination, exhibit lactic dehydrogenase activity. In this particular case we do have some idea of the physiological meaning of these multiple forms, for the MMMM complex found predominantly in striated muscle and the HHHH form found predominantly in heart muscle do have somewhat different kinetic properties, properties that are compatible with the physiological function of the tissue in question. It should be noted, however, that all forms of the enzyme are found in all tissues of the animal. In summary it appears that at one of the lower levels of enzymatic modulation

Fig. 18-2. Pyruvate dehydrogenase complex of *E. coli*, negatively stained with phosphotungstic acid (× 520,000). (Micrograph courtesy of H. Fernandez-Moran, University of Chicago.) See Chapter 11 for an explanation of the figure.

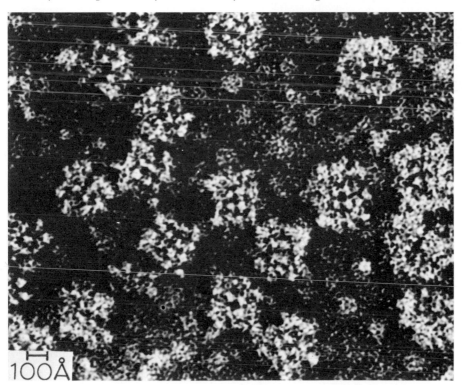

lies an inherent property of many polypeptide chains to form complex proteins, either with similar chains to form a functioning enzyme, or with dissimilar chains to form an multienzyme complex. Thus, one area of the regulation of cellular activity lies in the syntheses and bringing together of these polypeptide chains.

Another example at this lower level is the finding that in some cases more than one enzyme catalyzes the same reaction. Malic dehydrogenase is found both in the mito-chondria, as part of the Krebs cycle, and as a soluble enzyme in the cytoplasm; iso-citric dehydrogenase activity has similar localizations. The malic enzymes of the mitochondria and of the soluble portion of the cell are two entirely different proteins even though they catalyze the same reaction. It is thought that the reason for this difference in localization and in protein structure lies in the kinetics of the enzyme reaction; the rates at which equilibrium is reached may be quite different in the two cases, and this difference would be of great significance to the overall metabolism of the cell (see Fig. 13-7).

Modulations of Enzyme Activity: Enzyme-Metabolite Interactions

The substrates and products of an enzymatic reaction have great influence on the rate of that reaction. As substrate concentration is increased, enzymatic activity is aug-mented. If the reaction is reversible in practice as well as in theory, the mass action effect comes into play, for the increase in product concentration will tend to slow down the reaction, causing it to come to equilibrium. In many cases the product of the reaction inhibits the reaction by competing with the substrate for the active surface of the enzyme; the degree of this inhibition will depend on the relative concentration of substrate and product and on the relative affinities of each of these for the enzyme. Thus, as the product accumulates, the rate of its formation decreases.

Most enzymes seem to be a part of a linear or sequential array of reactions, that is, the product of one reaction is the substrate for the next enzyme in the series. In quite a few cases, it has been found that in a sequence of reactions, A→B→C→D, the product of reaction C→D inhibits the enzymatic reaction A→B. This has been called a "feedback" mechanism: when D increases to a certain concentration, it shuts off the whole series of reactions that lead to its further formation. The intriguing problem here is that in most cases the product of C→D has no steric resemblance to the sub-strate or product involved in the reaction A→B. How can it compete with a substrate it does not resemble for a common site on an enzyme surface? A possible answer is that there may be two sites on the enzyme affecting its activity—one for combination with its substrate and the other for combination with its feedback inhibitor—and the enzyme in the latter state is an inhibited one. That this is the case has now been demonstrated with several proteins. The theory of this mode of changing enzyme activity was first stated in a formal fashion by Monrod, Changeux, and Jacob; we have already discussed it in Chapter 10 in its relationship to the function of enzymes in general and to the enzyme, aspartate transcarbamylase, in particular. We shall summarize their theory briefly and discuss it now in relation to the larger problem of metabolic regulation.

You will remember that some enzymes can be activated or inhibited by specific chemical signalers whose structures may be entirely unrelated to the enzymes' own substrates, coenzymes, or products, and which almost certainly react with the enzyme protein at a site different from the substrate-product site. Proteins whose actions can be modified in this way have been called *allosteric* proteins and the interactions between these proteins and their effectors have been called *allosteric* interactions. We now know quite a few examples of such interactions. In Chapter 10, we discussed the allosteric protein, aspartate transcarbamylase, and the effect of CTP on it. If we once again examine the catalytic pathway of which aspartate transcarbamylase is a member (Fig. 10-22), we notice that this enzyme is the first in a pathway involved in the synthesis of CTP. The pathway begins with a condensation of aspartate with carbamyl phosphate, catalyzed by aspartate transcarbamylase, and then goes through five more steps, catalyzed by different enzymes, before CTP is finally synthesized. This final product, CTP, regulates its own synthesis by reacting allosterically with the first enzyme in the pathway, and in Chapter 10 we have given the details of how this is accomplished. The net result is that no more CTP is synthesized than that amount needed by the cell; when the amount of CTP is increased above this necessary level, this amount is just that necessary to interact with aspartate transcarbamylase, inhibiting this enzyme, and thus shutting off the further synthesis of CTP.

A famous and long-since established case of a change in enzymatic activity by a seemingly arbitrary substance, a change which is now thought to be another instance of allosteric control, is the activation of muscle phosphorylase by AMP; in the absence of AMP the enzyme is totally inactive. Furthermore, AMP does not participate in the chemical reaction, for muscle phosphorylase, you may recall, is the enzyme involved in the equilibrium between glucose-1-phosphate and muscle glycogen, resulting in a phosphorylitic split of glycogen to glucose-1-phosphate molecules. With the findings that phosphorylase b is made up of two subunits, which when separated are each inactive and which when brought together as a dimer have enzymatic activity, and that AMP promotes this dimerization, we now realize that most likely AMP is acting allosterically on the protein itself. However, we still do not know why AMP specifically is acting this way to control the activity of this enzyme. There are by now many examples in which the only explanation for somewhat paradoxical results seems to lie in invoking allosteric effects, and in some instances this has been conditionally verified by noting that physical changes in the enzyme have taken place during the interaction between protein and effector molecule. How widespread the phenomenon might be in controlling rates of reactions, in charting the course of metabolism, we do not know, but it has already been invoked, for example, as an explanation of the effects of the smaller molecular-weight hormones.

Another facet of feedback inhibition is illustrated in the case elucidated by Stadtman, the synthesis by bacteria of the amino acids, lysine, threonine, and methio-

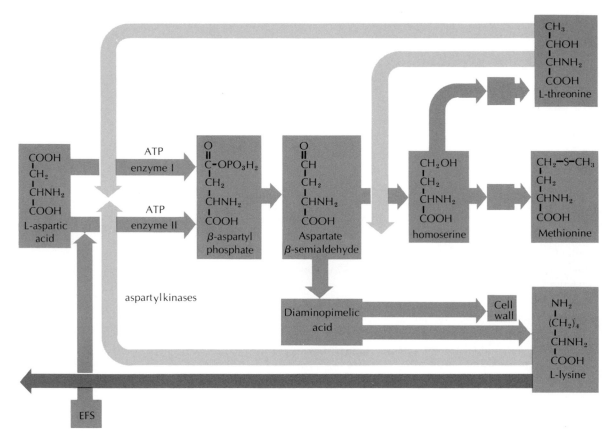

Fig. 18-3. Feedback inhibitions and repression in reactions initiated by aspartylkinases. (After E. R. Stadtman.) Light gray lines indicate sites of product inhibition, while dark gray shows site of repression of enzyme-forming system (EFS). Further details are given in the text.

nine, via a common precursor, aspartic acid (Fig. 18-3). It was found that either threonine or lysine would inhibit the activity of the first enzyme in the series, the aspartokinase. However, as the diagram makes clear, an excessive amount of either one of the products could lead to a decrease in the syntheses of both of them. This dilemma of having nonspecific inhibition was apparently solved by the cell through the scheme of having two different aspartokinases, one responsive to a probably allosteric inhibition by lysine and the other responsive to threonine as shown in Fig. 18-3. In addition, threonine also inhibits the activity of an enzyme situated past the branch point in the series whereas lysine also acts as a repressor (see below) of the enzyme-forming system for the synthesis of one of the aspartokinases. In these ways an excessive amount of only one of the amino acids would dampen the synthesis of only this one amino acid. However, there must be other means of control in the system because methionine can apparently be synthesized, although at low levels, even in the presence of both lysine and threonine, and because diaminopimelic acid, a precursor of cell wall material, has to be readily available at all times.

Figure 18-4 shows a final example, based mainly on the work of Magasanik, of a synthetic cycle in nucleotide metabolism in which are found several cases of enzyme inhibition—some by feedback on enzyme activity, probably by an allosteric effect on the enzyme, and some by repression on enzyme formation. In this cycle, guanosine-5'-phosphate can give rise to adenosine-5'-phosphate through the intermediates inosine-5'-phosphate and adenyl succinate. The adenosine-5'-phosphate can of course be phosphorylated to ATP. The latter compound acts then to repress its own formation by a feedback inhibition on the enzyme forming inosine-5'-phosphate from guanosine-5'-phosphate. Conversely, guanosine-5'-phosphate can arise from inosine-5'-phosphate through xanthosine-5'-phosphate. Here, GMP inhibits the activity of the initial enzymatic reaction, the oxidation of inosine-5'-phosphate to xanthosine-5'-phosphate. Also, as illustrated in the figure, histidine is synthesized via a pathway involving adenosine-5'-phosphate; here again, histidine inhibits the activity of the second enzyme in the series, thus effectively slowing down its own synthesis. In all these cases, it seems that when the final product of a series of enzymatic reactions rises to a certain concentration, it shuts off its own further formation by inhibiting the activity of one of the enzymes, usually an early one, on its synthetic pathway.

Another entirely different point of control derives from the fact that the same compound can serve as a substrate for more than one enzyme. For example, two linear arrays of reactions might intersect at one point; in this case the direction of travel could then be regulated by the relative rates of the slowest reactions, the "pacemaker" reactions, in each series. Any method that can speed up this pacemaker reaction will speed up the entire pathway. If a substrate has greater affinity for binding to one enzyme than to another, this too will influence its direction of metabolism, for if the concentration of a substrate is very low, it will react only with that enzyme for which it has a large affinity (that is, the K_m of the reaction is low). As the concentration is increased, the substrate will begin to react with the second enzyme where the K_m is higher. Another mode of regulation in such a situation was discussed earlier: if the substrate in question is enzymatically coupled to a second compound and if this second compound is available in high concentrations, the substrate will tend to go onto this synthetic pathway. Thus, the concentration of this second coupling molecule can act as a regulating device.

There are many other ways, too numerous to mention here, in which the soluble compounds of a cell, be they substrates, coenzymes, or ions, can influence an enzymatic reaction; most of these theoretical means of regulation have been observed to occur in *in vitro* experiments. For example, in Chapters 13 and 16 we observed that the level of a cofactor, in that case ADP, could regulate enzymatic activity. There are enough instances of the same nature to allow us to generalize that respiration and coupled phosphorylation respond to the need of the cell for ATP for the synthetic machinery. It has been observed that coenzymes like NAD and NADP exist in small

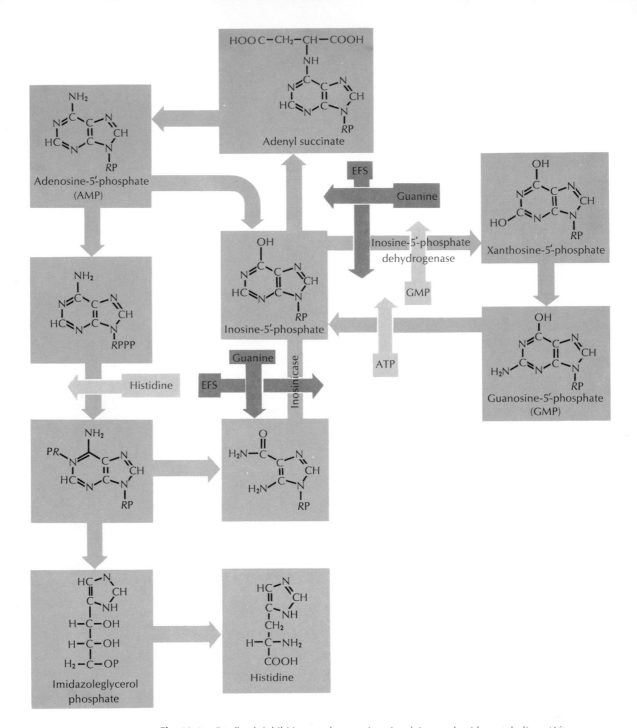

Fig. 18-4. Feedback inhibitions and repressions involving nucleotide metabolism. (After B. Magasanik, *J. Biol. Chem.* 235, 1960.) The dark gray arrows mark sites of repression of enzyme-forming systems (EFS), while light gray arrows mark sites of product inhibitions of various enzymes. Further details are given in the text.

amounts in the cell relative to the number of enzymatic reactions in which they participate; this finding provokes the speculation that the concentrations and conditions of these compounds, be they oxidized or reduced, could determine the direction of a metabolic pathway in the case of these alternate pathways. For example, in fatty acid and steroid syntheses, there is a need for reduced NAD and NADP; the state of reduction of these coenzymes could very well influence the fate of fat in the cell — whether it is to be oxidized for energy or whether it will go to form the higher fatty acids. Of course, the state of reduction of these coenzymes will depend on the relative rates of reduction and oxidation by the dehydrogenases of the cell.

Thus, we believe that the condition of the milieu in which the enzyme acts determines in large part the rate of its activity, and no doubt the cell does use many of the ways outlined above to regulate enzymatic activity. It should also be kept in mind, however, that many enzymes are not soluble in cells; they are bound to structures, and their activities are influenced by this binding. Any loosening, any change in the way an enzyme is tied to a membrane, for example, will influence its activity. Conversely, the fact that some enzymes are within structures within cells means that their substrates may not have ready access to them. A substrate might be attacked by one enzyme in preference to another, not because of any intrinsic properties of the respective substrate-enzyme complexes, but simply because it has easier access to the first enzyme. Indeed, this possibility has been postulated by some authors to account for the Pasteur effect — the effect that, under aerobic conditions, the glycolysis of glucose to lactic acid is inhibited.

At present, the key to what is undoubtedly an important cellular regulatory device, this Pasteur effect, is sought generally — but thus far, unsuccessfully — among the oxidative and glycolytic phosphorylative processes. It has been thought for a long time that since glucose has to be phosphorylated by ATP and hexokinase in order for it to be put onto the glycolytic pathway, the regulation point of the Pasteur effect is precisely at this hexokinase reaction. It may be that during aerobic oxidations, the mitochondria utilize all the available inorganic phosphate of the cell, there being too little left from the glycolytic manufacture of ATP for use by hexokinase. If at the same time, the ATP made in the mitochondria has difficulty in being released from them, the glycolysis will be reduced because of the scarcity of these cofactors. However, glucose phosphorylation is only the initial step in glycolysis, and it may not be the limiting one. Thus, while lack of inorganic phosphate or ADP has been found not to inhibit glucose phosphorylation or the uptake of glucose by the cell, it still does block the formation of lactic acid from glucose. Indeed, it seems also unlikely that the mitochondria, by utilizing inorganic phosphate or ADP in oxidative phosphorylation, can keep the intracellular level so low as to impair glucose utilization or phosphorylation and lactic acid formation. It seems doubtful that the levels of inorganic phosphate or of ADP, even if these be compartmentalized between the mitochondria and the rest of the cell, can solely explain the effect. Similar doubt can be expressed about competition between the glycolytic and oxidative enzymes for those coenzymes common to both, such as NAD or NADP.

Fig. 18-5. Electron micrographs of purified acetyl CoA carboxylase. (Courtesy of A. K. Kleinschmidt, New York University Medical School.) A. In the absence of citrate (× 160,000). B. In the presence of citrate (× 240,000).

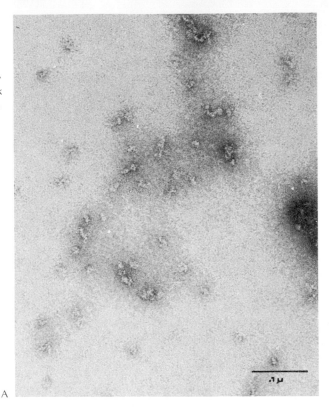

A

B

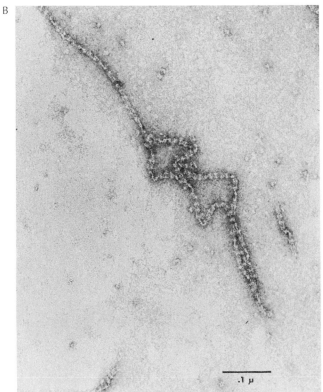

At the present time, the most acceptable explanation for the Pasteur effect is one involving an allosteric effect on the soluble protein, phosphofructokinase, which catalyzes the phosphorylation by ATP of fructose-6-phosphate to fructose-1,6-diphosphate (see Fig. 12-6), and is thus one of the key enzymes in the glycolytic pathway. It has been found that concentrations of ATP higher than that necessary for optimal enzyme activity will actually inhibit the enzyme. Thus, it has been postulated that there are actually two binding sites for ATP on the enzyme, one a substrate site and the other an allosteric site. One can visualize that were the mitochondria to make large amounts of ATP as a result of oxidative phosphorylation, the resultant high ATP level in the cytoplasm would inhibit phosphofructokinase, thus shutting down glycolysis. What makes the theory more attractive as an explanation for the Pasteur effect is that the ATP inhibition can be counteracted by fructose diphosphate itself, and particularly, by high levels of inorganic phosphate and of ADP. Thus, as long as the mitochondria can keep the ATP level in the cytoplasm high, glycolysis will be dampened, but should this level drop and the phosphate and ADP levels increase, the inhibition by ATP will be overcome and glycolysis could proceed once more. The interesting point is that the factor common to mitochondrial and soluble activity, ATP, is a product of mitochondrial activity and is a cofactor in soluble activity; but its effect in control is not through its role as cofactor or substrate, for it would require relatively large changes in ATP levels to influence enzyme activity at the substrate site, but only small changes to influence activity when ATP acts at an allosteric site.

Finally, another striking example of a probably allosteric effect has been recently discovered in the case of acetyl CoA carboxylase, the enzyme which catalyzes the carboxylation of acetyl CoA by CO_2 to form malonyl CoA, the first and key step in fatty acid synthesis. It has been known for some time that citrate increases fatty acid synthesis, and recently it has been found that acetyl CoA carboxylase is the site of this activation by citrate. Moreover, electron-microscopic investigation has revealed an exciting visualization of this chemical event. Figure 18-5A shows the purified enzyme as it appears in the absence of citrate; it has the form of small particles, dimensions from 100 to 300 A, having a sedimentation constant of 20S. In the presence of the amount of citrate that activates the enzyme, a remarkable change takes place (Fig. 18-5B), for the individual enzyme particles apparently line up to form filaments from 70 to 100 A in width and up to 4000 A long, having a sedimentation constant of 46S. Here then is a regulatory step involving a junction point of carbohydrate and fat metabolism (see Krebs cycle in Chapter 12) which occurs by means of organizational changes of macromolecules.

Modulations of Enzyme Amount: Genetic Level

Besides the mechanisms associated with regulation of enzymatic activity, there is another mode of coordination of metabolism in the cell, which concerns the regulation of enzyme amount (denoted by (1) and (2) in Fig. 18-1). First, we will deal with regulation at the DNA-RNA level, called the "transcription" level ((1) in Fig. 18-1); later we will speak of possible regulation at what has been called the "translation" level, the ribosome stage of protein synthesis (see (2) in Fig. 18-1).

In microorganisms, with but very few cases reported for the cells of multicellular organisms, a phenomenon called enzyme induction and repression has been discovered; both have to do with the control of enzyme synthesis. For example, in bacteria the concentrations of quite a few enzymes are strikingly dependent on the conditions of growth, on the compounds in the medium. It has been found in many cases that the addition of a substrate can induce the protein-synthesizing machinery in the bacteria to synthesize many molecules of the enzyme attacking this particular substrate. When the substrate is metabolized and disappears, the synthesis of this enzyme ceases. It has been postulated on this basis that in a linear series of enzymatic reactions, the product of $enzyme_1$, which is the substrate of $enzyme_2$, could induce the synthesis of $enzyme_2$, and so on. Such a mechanism has even been hypothesized to account for new enzyme formation in early embryological development.

In the reverse situation, some compounds can cause a repression of enzyme formation. This repressive effect can be brought about by the product of the enzyme in question. In some cases of a linear series of enzymatic reactions, the product of $enzyme_4$ can cause the repression of formation of $enzyme_1$. Figure 18-4 shows two examples of such a repressor effect. Guanine can act at two sites to suppress the formation of guanosine-5'-phosphate: one, to suppress the formation of enzyme inosinicase, this enzyme being involved in the synthesis of guanosine-5'-phosphate from adenosine-5'-phosphate; and two, to inhibit the synthesis of the enzyme inosine-5'-phosphate dehydrogenase, which is on the pathway of guanosine-5'-phosphate formation.

In the cases of both enzyme induction and repression, the small molecular weight molecules do not act to augment or inhibit enzyme activity, but act at the genetic site that has to do with the synthesis of that particular protein. There is some evidence from biochemical genetic studies that in those bacterial cells that can produce an inducible enzyme, the gene controlling the formation of this enzyme is composed of both genetic material coding for the synthesis of that protein and also a repressor gene, the latter inhibiting the expression of the former. This concept, called the "operon" hypothesis by its formulators, Jacob and Monod, has been very fruitful in an experimental sense. Figure 18-6 shows the main features of the idea. In brief, there is more than one type of genetic material in an operational sense. The structural genes, because they act as templates in the synthesis of messenger RNA, are responsible for the initial steps in the synthesis of the cellular proteins. However, in some cases it has been found that certain genes are linked together operationally, so that if one is active, all are active; hence, the existence of another gene, the operator gene, was postulated to act as an off/on device for the activity of a group of related genes. Also, the many findings of inducible enzyme activity have, in bacteria, led to the concept of a regulator gene, a product of its activity being a specific repressor substance which inhibits the activity of the operator gene. In other words, if the regulator gene (being responsible for the initial steps in the making of repressor substance) is active, the operator

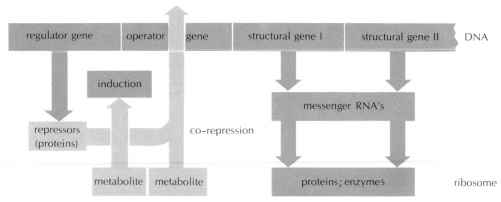

Fig. 18-6. "Operon" hypothesis of Jacob and Monod. Explanation is in the text.

gene, and hence the structural genes, are turned off. The postulated role of the small molecular-weight inducers mentioned above is that these combine tightly with the repressor, inhibiting its activity at the operator gene level, thus allowing the operator gene to function. In some instances, some small molecular-weight substances increase the inhibitory effect; hence these would be corepressors. At first, it was thought that the repressor substance might be some particular RNA molecule, but the evidence now points to its being a protein. Indeed, it appears that a specific repressor for a specific operator has been isolated, and it is a protein. Therefore, the role of the metabolite in repression is to combine allosterically with this protein, changing it so that it can no longer interact with the operator gene. The scheme as presented is a device for dampening genetic activity, hence protein-synthesizing activity, under those conditions where the cell is not called upon to do other than exist. In other words, if the scheme is correct, and in the case of bacterial metabolism it most probably is in its generalities, it says that the metabolic machinery of the cell, with its protein-forming capacity, does not operate at maximal levels under those environmental conditions where the dynamic equilibrium of the cell is sufficient for cell maintenance. Were these conditions to change, the cell is readily responsive by increasing its activity in certain specified ways, primarily at the genetic level. The scheme is thus essentially an energy-conserving mechanism for it says, in effect, that the maximal levels of metabolic operations are not necessarily the optimal ones, that for overall cell metabolism to be optimally efficient, it is not necessary for all the facets of cellular metabolism to be operating at maximal rates.

The work showing that the Jacob-Monod scheme was probably operative in the cell was mostly carried out with *E. coli* bacteria having a region of the DNA called the "lac" operon. As visualized (Fig. 18-7 giving a more specific example of Fig. 18-6), this operon has five genetic sites: the regulator gene that governs the synthesis of the postulated repressor molecule, the operator gene for the postulated operator, and the three structural genes bearing the genetic codes for the three enzymes in this operon, z, y, a. These three, respectively, are known to be β-galactosidase, a permease involved

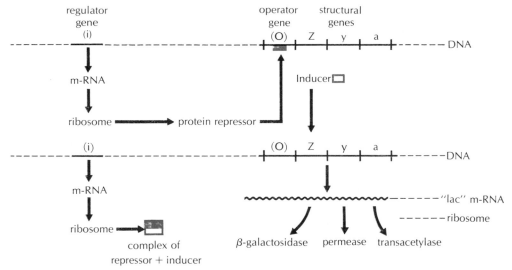

Fig. 18-7. Schematic diagram of the Jacob-Monod operon model particularized to the "lac" operon of *E. coli* DNA. Details are explained in the text.

in the transport of galactosides into the cell, and a galactoside transacetylase. As mentioned, nearly all the evidence for this model has come from genetic analyses using mutants of the wild-type *E. coli,* mutants with defects in some parts of this operon. Also, genetic evidence has been all that we have had until recently to indicate that the repressor is a protein molecule. However, we now have not only biochemical evidence that the repressor is indeed a protein, but also biochemical evidence which is very suggestive that the concepts of Jacob and Monod are correct, that control of genetic activity operates very much like the model they have presented.

Gilbert and his co-workers have collected this evidence by using radioactive inducer molecules, in this case isopropylthiogalactoside (IPTG), which is not metabolized by the cell, and by using a mutant strain of *E. coli* which exhibits tight binding of the inducer to its cells. The repressor of the lac operon was detected by incubating an extract of this mutant inside a dialysis bag with a solution of the radioactive IPTG. After dialysis, more radioactive IPTG was found inside the dialysis bag (evidently bound to something) than was outside in the solution. Using this bound radioactivity as an assay for the repressor, they began to purify it. This binder of IPTG was shown to be a high-molecular weight protein, about 200,000 MW. They proved this protein to be probably the lac operon repressor by showing that when they used an extract from a *E. coli* mutant strain, which genetically had been found to contain a repressor that had lost all its affinity for the inducer, then radioactive IPTG did not accumulate inside the dialysis bag.

Using these results, the same group tested the operon hypothesis more directly. Radioactive-repressor molecules were made by growing *E. coli* cells in a medium containing radioactive sulfate to label all the proteins of the cell. They then purified the radioactive-repressor protein from the mixture using the assay described above (using S^{35}-labeled proteins and tritium-labeled ITPG, they could detect a protein in which both labels were concentrated). This protein was purified and tested for its binding affinity for the lac operon of the DNA. The DNA that was used was what might be called a concentrated-lac operon DNA: it was DNA obtained from a bacteriophage that infected E. coli cells, multiplied, and lysed the cells. In the process, the phage incorporated a region of the E. coli DNA genome into its own DNA, and this region contained the lac-operon segment. Since phage DNA has a much lower molecular weight than E. coli DNA, isolation of it gave a material in which the concentration of E. coli lac-operon per unit weight of DNA was increased. Upon mixing of the radio-active-repressor protein with this phage DNA, it was found that binding of the repressor to the DNA had occurred. That the repressor protein was bound to the operator region of the DNA was inferred from studies showing that another bacteriophage DNA, similar to the first but which by genetic methods had been shown to contain a defective operator region, did not bind the labeled repressor nearly as well as did normal phage DNA containing an unmutated lac-operon region. Finally, if the binding of the labeled repressor protein to the phage DNA containing the lac operon was carried out in the presence of the IPTG inducer, no labeled repressor was found bound to the DNA. Similar results, showing that the repressor is a protein and showing binding to the DNA, have been obtained by Ptashne, using a different experimental system, the transcription of genes of a bacteriophage inside a bacterium.

Thus, it appears that in its essentials the Jacob-Monod scheme is correct: the repressor, a protein molecule, can combine with an operator gene rendering the latter ineffectual in "turning on" the transcription of the structural genes of the operon, and the inducer can bind to the repressor, preventing the binding of the latter to the operator gene, thus allowing expression of the structural genes of the operon (Figs. 18-6 and 18-7). So far not enough evidence has been obtained from mammalian cell studies to indicate whether the same ideas also hold there. However, from what we know concerning the ubiquity of metabolic mechanisms throughout the living world, we would suspect that some scheme similar to the one above is operating in the mammalian cell.

Modulations of Enzyme Amount at the Translation Level

There has been much less work done on the possibilities of regulation of enzyme synthesis at the translational or protein synthesis level; thus, we can only note what these may be. For any regulative process whatsoever to be operative, one dictum must be obeyed, and that is that some component of the system must be limiting in amount or activity, so that any small changes in the amount or activity of this component would cause large changes in the activity of the whole system of which this component is a part. This could be called a process of amplification and is probably the general way hormones act. It has been pointed out that the one component in the ribosomal system of protein synthesis that seems to fulfill this criterion is the amount of specific transfer RNA (tRNA).

One can calculate from data that for every ribosome in the cell there are 20–50 molecules of tRNA; when one takes into account that there must be at least 20 different tRNA's, the above value is reduced to somewhere from 1–3; further, when one considers that many amino acids have more than one tRNA molecule specific to it, this figure is further reduced. All in all, we reach a conclusion that at any one time in the cell, were all ribosomes to participate in protein synthesis at the same time, there would certainly be scrambling for a very limited supply of specific amino acid tRNA's.

Although this theory of ribosome-activity control via tRNA sounds attractive, there is very little evidence as yet to support it. However, it has been known for quite a while that there is some coordinated control of RNA and protein syntheses. For example, in some bacterial mutants that require the addition to the medium of a single amino acid for growth, the omission of this amino acid will not only cut out protein synthesis, but will also greatly reduce RNA synthesis, both ribosomal and tRNA. A possible explanation is that free tRNA, and not aminoacyl tRNA, is an inhibitor of RNA synthesis. In the normal case, where there is an adequate supply of amino acids, most of the tRNA is aminoacylated, and hence RNA synthesis can keep pace with protein synthesis. Should the supply of amino acids drop, tRNA is not aminoacylated, and the resultant increased level of tRNA acts to inhibit RNA synthesis. However, no good evidence has been forthcoming for this hypothesis and it does not appear to be true in this simple form. However, recently another facet of this problem has been uncovered. As mentioned in Chapter 15, there is more than one tRNA molecule specific for a single amino acid (there are species of amino acid-specific tRNA's), and it is probable that there is an aminoacyl synthetase enzyme specific for each of these tRNA's. It has been found that there exist different aminoacyl synthetases for a given amino acid in different animal tissues, that each of these may be specific for a different species of tRNA, and that the level of these synthetases may vary among the different tissues. There seem also to be differences in the tRNA's specific for the same amino acid among different animal cells; also, as mentioned in Chapter 13, the mitochondria have tRNA's and aminoacyl synthetases different from those in the extramitochondrial cytoplasm. One cannot help but conclude that both the tRNA's and the synthetases are involved in the regulation of the syntheses of specific proteins, that for the synthesis of perhaps one type of protein certain of the species of tRNA's and of the synthetases are used, while for the synthesis of another type of protein others of the species are necessary. This view is further strengthened by the findings that during differentiation and development of wheat seedlings and of sea urchin eggs, different species of tRNA's and of aminoacyl synthetases appear, as though during differentiation when new proteins have to be synthesized, new tRNA's and aminoacyl synthetases have to be formed to help in this process. It is obvious that if this were the case, the control of specific enzyme formation could be regulated by the formation of specific tRNA's at the genetic level.

Another facet of control at the ribosome level has to do with modulations of ribosome amount and of ribosome activity. There does seem to be some evidence that the amount of ribosomes per cell can change during various phases of cellular activity and that the activity of a given population of ribosomes in protein synthesis can vary during different phases of cellular activity. For example, it appears that protein synthesis operates at very low levels in unfertilized eggs even though many ribosomes are present, but that after fertilization, there occurs some change in the ribosomes which allows synthesis to proceed. While we can visualize many points at which control can operate such as stability of ribosomes, the fit of messenger RNA's to ribosomes and the movement of mRNA's and ribosomes relevant to each other, we do not as yet have any experimental verification of the extent of control at any of these areas.

Another possibility for a control mechanism linking protein and RNA syntheses is the finding that ribosomes per se have an influence on RNA, probably messenger RNA, synthesis. In *in vitro* experiments it has been found that a deoxyribonucleoprotein complex which is capable of synthesizing RNA (that is, it contains the DNA template in the form of bound DNA and it contains the RNA polymerase) can be further activated by the addition of ribosomes. More RNA is synthesized in the presence of ribosomes than in their absence, and furthermore, contrary to earlier findings (without ribosomes) that the synthesized RNA remains stuck on the DNA template complex, the RNA synthesized in the presence of ribosomes is stripped off the template complex to form a ribosomal-RNA complex. This can take place *in vitro* with bacterial extracts, but whether it does occur *in vivo* in bacteria is not known. Furthermore, we are less sure in the case of the eucaryotic organism where the nucleus is enclosed in a membrane and where the functioning ribosomes are probably all in the cytoplasm. However, it is possible that in the latter case, it is not ribosomes per se, but ribosomal-precursor particles that are involved in the nucleus in this regulation of RNA synthesis; later, in the cytoplasm they may be formed into mature ribosome-RNA complexes. The general idea is that messenger RNA formation is governed by the availability of ribosomes to act in conjunction with the newly formed messenger RNA in protein synthesis, and the simplest way to visualize this is a model in which RNA synthesis is depressed normally because the RNA molecules are bound to their formation site on the polymerase-DNA template, and only when this RNA can come off, via competitive binding with naked ribosomes, will further RNA formation proceed. Ribosomes already containing bound messenger will not work, and hence no more RNA will be synthesized than can be accommodated by the protein-synthesizing machinery. Whether this hypothesis is a valid one, only effort and time can tell.

All in all, we actually know too little as yet concerning the detailed mechanisms of protein synthesis to begin to look at its control, but we are almost certain that control must exist. The synthesis of specific proteins is probably the single most important function of a cell, and therefore there must be more than one level of control of this function, at the genetic (transcription) and at the ribosomal (translation) levels.

Enzymes and Cell Structure

In summary then, we are beginning to grasp how the concentration of small molecules in the cell can control its enzymatic activities. The formation of intermediates takes place in a series of metabolic reactions. In such a series, the rate of the overall reaction is dependent on the rate of the slowest one, this rate in turn dependent on the kinetics of the substrate-enzyme interaction. If this substrate is also common to another metabolic pathway, there must be some means of controlling the direction of metabolic travel. This control could come about if the product of either metabolic pathway acts as an inhibitor of the alternate pathway, either by feedback inhibition or by repressing the formation of that enzyme which first attacks the substrate in question along that pathway. Thus, it is clear that the products of enzymatic reactions have a great deal to do with the rate of their own formation, with the rate of the reactions. Of course, the way these products are partitioned within the cell will have a moderating effect on these interactions. If a product moves into a different part of the cell, it has moved out of the sphere of the enzyme it supposedly influences. In the case of the bacterial cell, where there is no evidence of internal membranes, this does not pose much of a complication. In other cells, however, it is almost certain that what has been observed in bacteria in terms of inhibitions of enzyme formation and activity is modulated by the fact that there are differences in concentration of the same compound within parts of the cell. This concentration of free compounds is governed not only by the intracellular membrane barriers, but also by the probability that many of these compounds are bound noncovalently to other molecules, even to membranes. What we think are compounds freely accessible to enzymes may not really be so within the architectural framework of the cell. We know, for example, that NADH is formed during glycolysis. We also know that this NADH can be used for energy production by the cell, in the formation of ATP. Precisely how the NADH comes to be utilized is not known, but our current visualization is that it somehow gets into the mitochondria, is there oxidized via the electron-transport chain, and thus produces energy equivalents in the form of high-energy phosphates. As Chapter 13 has shown, we are just now beginning to have some inkling how this NADH gets into the mitochondria. It would appear that somehow the mitochondrial membrane has a regulating influence on coenzyme metabolism and, concomitantly, on energy metabolism.

The very fact that individual enzymes, and even whole metabolic cycles, are compartmentalized within the cell strongly indicates that the internal structure of the cell is a regulatory device simply by virtue of being there. In a solution we can measure differences in affinity between the same substrate and various enzymes acting on it, and between the same enzyme acting on slightly varying substrates. These affinities, however, may have no meaning within the partitioned cell. Biochemists have mapped out the sites within a cell where a great many of the individual enzymatic steps occur, but it is not known how these steps are brought together in a smoothly functioning cell. Even the most fundamental of these steps has so far eluded our explanatory experiments—namely, how the pyruvate from glycolysis in the soluble matrix of the cell gets inside the mitochondria to be there oxidized. Thus, the discovery by the cytologists of the wonderful architecture within the cell has opened up for the cell physiologist and the biochemist untold new fields for future research.

REGULATION OF CELL STRUCTURE

We have discussed on a number of occasions how the simple 4-letter language of DNA becomes translated into the much more complex 20-letter language of the proteins and how this linear message is capable of performing some spatial transformations to produce three-dimensional protein molecules capable of highly specific interactions. Some of these protein molecules, which we call enzymes, interact with specific substrates, bringing about specific changes in their covalent structure. This property has given us a technical advantage for studying this class of proteins and the preceeding sections illustrate how rapid the progress has been in studying the regulation of enzyme activity.

There is another vast class of proteins, however, which also have highly specific structures and also interact in a specific manner with other proteins, nucleic acids, lipids, and carbohydrates. These proteins, which we call structural proteins, are as yet poorly understood. In Chapter 11 we have discussed a few cases of structure assembly which are becoming susceptible to experimental analysis. We have seen that our understanding of assembly processes is still at an early stage of development. It should hardly come as a surprise that our understanding of the regulation of assembly processes is virtually nonexistent. And yet we know from observing cells going through cycles of growth and division that structural phenomena are regulated with the greatest of precision. As cells grow and divide, membranes are synthesized, organelles are reproduced, chromosomes are replicated, and so on. All these phenomena are timed in perfect synchrony so that the cell physiologist often has great difficulty in upsetting this synchrony in a specified manner.

In recent years a number of attempts to study the regulation of structure have yielded some results. Jacob, for instance, has been able to find mutants in *E. coli* that appear to act at the level of structure assembly although the mechanism of action of these mutants is as yet not understood. We shall therefore resist the temptation to discuss this fledgeling field any further, but instead we shall express our confidence that 5 years from now there will be much more to say about this important subject.

In summary, we have seen how larger molecules are built from constituent smaller molecules of the cell. The larger molecules, be they proteins, nucleic acids, or lipids, in turn come together with each other or with other similar molecules to form in one case a complex protein system, acting as a single enzyme or as a multienzyme complex, in another case to form DNA complexes, and so forth. In turn, these larger aggregates interact with one another to form structures such as membranes (mostly lipid and protein), or chromosomes (mostly DNA and protein), or ribosomes (mostly RNA and protein). These larger structures can either exist as such separate entities, or they can interact to form still larger specific aggregates where still other molecules are involved such as the membranes of the mitochondria, or of the chloroplast, or as polysomes on membranes, or as the inner-membrane particles of mitochondria. The cell thus exhibits a whole hierarchy of organizational states, each state being relevant for the function of the chemical aggregates involved, each state being circumscribed by certain chemical and physical parameters, but each state also being extended in function to a degree greater than one would expect from the nature of its parts. The machinery of a mitochondrion, for example, is not only the sum of its constituent machines, but is a transcendent expression of an ordered aggregate whose whole is greater than the sum of its parts.

How this machinery functions in cellular regulation and control will be answered with more and more precision in the near future, possibly by some of the readers of this book. But it is even now not too early to begin to ask questions that may lead to satisfactory answers: for example, while we have mentioned extensively intracellular control devices, by means of compounds which can be called intracellular hormones, we have not even mentioned the regulation of one cell's activity by the activity of another cell, via extracellular hormones or signalers.

The reason we have not is that too little is known of the specifics of this interaction. How does cell-to-cell interaction take place? What makes like cells react to each other to form tissues, and what makes specific tissues composed of various types of cells react to other tissues composed of different cells to form organs? What intra-cellular mechanisms respond to extracellular influences and modify intracellular behavior in order that cells be viable in an environment, whether that environment is the media of the bacteria or is one of the environment of the whole multicellular organism? In short, all cells are also parts of populations, and these populations form other levels of organization. On all levels, the overall properties of the organization must modify the activities of its individual parts, for these parts contribute to the dynamic balance of the whole. We have mentioned cells, which form tissues. But tissues in turn form organs, and organs are mobilized into organ systems, which compose an organism. Yet even here we have not reached the top of the hierarchy, for organisms are part of a species the members of which interact with each other and with their environment to form a society. At this point biology begins to merge with the social sciences, and it is thus relevant for biologists to ask questions such as, "To what extent are the workings of a human society dictated by the inescapable biology, and therefore chemistry and physics, of the organisms which constitute it?"

It is appropriate that a book on biology ends with a question.

SUGGESTED READING LIST

Offprints

Cohen, G. N., 1961. "Feedback Inhibition and Repression of Aspartokinase Activity in *Escherichia Coli* and *Saccharomyces Cerevisiae*." *Jour. Biol. Chem.*, pp. 2033–2038. Bobbs-Merrill Reprint Series. Indianapolis: Howard W. Sams & Co.

Changeux, J. P., 1965. "The Control of Biochemical Reactions." *Scientific American* offprints. San Francisco: W.H. Freeman & Co.

Dulbecco, R., 1967. "The Induction of Cancer by Viruses." *Scientific American* offprints. San Francisco: W. H. Freeman & Co.

Gorini, L., and Maas, W. K., 1958. "Feed-back Control of the Formation of Biosynthetic Enzymes." In: *The Chemical Basis of Development*, W. D. McElroy and Bentley Glass, eds., The Johns Hopkins Press, pp. 469–478. Bobbs-Merrill Reprint Series. Indianapolis: Howard W. Sams & Co.

Hogness, D. S., Cohn, M., and Monod, J., 1955. "Studies on the Induced Synthesis of β-galactosidase in *Escherichia Coli*: the kinetics and mechanism of sulfur incorporation." *Biochimica et Biophysica Acta*, pp. 99–116. Bobbs-Merrill Reprint Series. Indianapolis: Howard W. Sams & Co.

Markert, C. L., 1963. "The Origin of Specific Proteins." *The Nature of Biological Diversity*, J. M. Allen, ed. McGraw-Hill Book Co., Inc., pp. 95–119. Bobbs-Merrill Reprint Series. Indianapolis: Howard W. Sams & Co.

Novick, A., and Weiner, M., 1957. "Enzyme Induction as an All-or-none Phenomenon." *PNAS*, pp. 553–566. Bobbs-Merrill Reprint Series. Indianapolis: Howard W. Sams & Co.

Stadtman, E. R., Cohen, G. N., LeBras, G., and de Robichon-Szulmajster, H., 1961. "Feed-back Inhibition and Repression of Aspartokinase activity in *Escherichia Coli* and *Saccharomyces Cerevisiae*." *Jour. Biol. Chem.*, pp. 2033–2038. Bobbs-Merrill Reprint Series. Indianapolis: Howard W. Sams & Co.

Umbarger, H. E., 1961. "Feedback Control by End-product Inhibition." *CSHSQB*, pp. 301–312. Bobbs-Merrill Reprint Series. Indianapolis: Howard W. Sams & Co.

Articles, Chapters, and Reviews

Atkinson, D. E., 1965. "Biological Feedback Control at the Molecular Level." *Science, 150*, pp. 851–857.

Gerhart, J. C., and Schachman, H. K., 1965. "Distinct Subunits for the Regulation and Catalytic Activity of Aspartate Transcarbamylase." *Biochemistry, 4*, pp. 1054–1062.

Monod, J., Changeux, J., and Jacob, F., 1963. "Allosteric Proteins and Cellular Control Systems." *Jour. Mol. Biol., 6*, pp. 306–329.

Umbarger, H. E., 1964. "Intracellular Regulatory Mechanisms." *Science, 145*, pp. 674–679.

Books

Cohen, G. N., 1969. *Cell Regulation*. New York: Holt, Rinehart and Winston, Inc.

Cold Spring Harbor Symposia on Quantitative Biology, 26, 1961. *Cellular Regulatory Mechanisms*. Cold Spring Harbor, L.I., New York.

Ebert, J., 1965. *Interacting Systems in Development.* New York: Holt, Rinehart and Winston, Inc.

Kalmus, H., 1966. *Regulation and Control in Living Systems.* New York: John Wiley & Sons, Inc.

Koningsberger, V. V., and Bosch, L., 1967. *Regulation of Nucleic Acid and Protein Biosynthesis. Biochim. et Biophys. Acta Library,* Vol. 10. Amsterdam: Elsevier Publishing Co.

EPILOGUE

We have attempted in this volume to introduce the beginner to an experimental approach to the biology of the cell. In doing so we have omitted many details of cellular structure, biochemical activity, physiological behavior, and genetic transmission. We have also refrained from speculating a great deal about the periphery of our knowledge where fact and fancy are in rapid and uncertain flux. Instead, we have attempted to provide a coherent picture of the life of the cell, hoping to provide a structural framework around which the student can integrate future studies, both richer in detail and greater in depth. There is much in the biology of the cell which has hitherto escaped understanding in molecular terms. The molecular details of oxidative phosphorylation and of photosynthetic phosphorylation, the precise structure of membranes and the molecular mechanism of active transport, the molecular basis of mechanochemical transductions, the molecular details of the assembly of cellular structures, the regulation of cell growth and cell replication: all these and many other phenomena have just barely begun to be analyzed in molecular terms. In the case of phenomena such as cell differentiation, we have not even begun to conceive of a productive experimental approach. Yet the cell replicates with remarkable precision and predictability; and cell biologists, because of the powerful achievements of the last 20 years, have grown accustomed to viewing the future of molecular biology with confidence and optimism, for if a cell can do it so well we should at least be able to find out how it does it! We sense that we are at the threshold of the greatest revolution in man's history, the comprehension of the nature of life itself. There is no greater object of wonder, no greater thing of beauty than the dynamic order and organized complexity of life. And what we are now witnessing is perhaps the most dramatic event in the slow evolution of life: the human brain scrutinizing itself and its origins, life turning on itself. We who are of nature are evolving to know nature.

Everything that lives is holy
Life delights in life.

BLAKE

Credits

Table 5-1:   From *Biophysical Chemistry,* by Edsall and Wyman: Academic Press, Inc., 1958.

Fig. 8-11A: From Fig. 3, P. Doty, *J. Cell Comp. Physiol.,* **49** (Suppl. 1): 46.

Fig. 8-12:   J. Marmur and P. Doty. *Nature,* **183:** 1427, 1959.

Fig. 8-21:   From *The Organization of Chromosomal Nucleohistone Fibrils,* by Hans Ris, in *Sixth International Congress for Electron Microscopy,* Kyoto (1966): Academic Press, Inc.

Fig. 8-22:   P. Doty, *et al. Proc. Nat. Acad. Sci.,* **45:** 482, 1959.

Fig. 8-24:   G. Brownlee, F. Sanger, and B. Berrell. *Nature,* **215:** 735, 1967.

Figs. 9-6 and 9-7:   From *Ultracentrifugation, Diffusion, Viscometry,* by H. K. Schachman, in *Methods in Enzymology,* **4:** 32, 1957. Academic Press, Inc.

Fig. 9-22:   C. Baglioni, *Biochem. Biophys., Acta,* **48:** 392-356, 1961.

Fig. 9-24:   From the *Atlas of Protein Sequence and Structure 1967–68,* Margaret O. Dayhoff and Richard V. Eck, National Biomedical Research Foundation, Silver Spring, Maryland, 1968.

Fig. 9-27:   From *Currents in Biochemical Research,* by B. Low and J. T. Edsall. Interscience Publishers: John Wiley & Sons, Inc.

Fig. 9-29:   L. Pauling and R. B. Corey. *Proc. Nat. Acad. Sci.,* **37:** 729, 1951.

Fig. 9-32:   From *X-ray Analysis and Protein Structure,* by R. E. Dickerson, in *The Proteins:* Academic Press, Inc., 1954.

Fig. 9-36A: Drawn by Irving Geis originally for *Scientific American,* Vol. 215, No. 5 (copyright by *Scientific American,* November 1966). Adapted by Mr. Geis for the *Atlas of Protein Sequence and Structure 1967–68,* Margaret O. Dayhoff and Richard V. Eck, National Biomedical Research Foundation, Silver Spring, Maryland, 1968.

Fig. 9-39:   Perutz, *et al., Nature,* **185:** 416, 1960.

Fig. 10-7:   From *General Biochemistry,* by Fruton and Simmonds. John Wiley & Sons, Inc., 1958.

Fig. 10-17B: B. W. Mathews, *et al. Nature,* **214:** 652, 1967.

Fig. 10-18:   J. Jolles, J. Jauregui-Adell, and P. Jolles. *Biochem. Biophys., Acta,* **71:** 488, 1963.

Fig. 10-19:   Drawn by Irving Geis originally for *Scientific American,* Vol. 215, No. 5 (copyright by *Scientific American,* November 1966). Adapted by Mr. Geis for the *Atlas of Protein Sequence and Structure 1967–68,* Margaret O. Dayhoff and Richard V. Eck, National Biomedical Research Foundation, Silver Spring, Maryland, 1968.

Fig. 10-20:   From *The Three-Dimensional Structure of an Enzyme Molecule,* by David C. Phillips, *Scientific American,* November 1966. Copyright © 1966 by Scientific American, Inc. and by W. H. Freeman Co. All rights reserved.

Figs. 10-23 and 10-24:   J. C. Gerhart and H. K. Schachman. Reprinted from *Biochemistry 4,* June 1965. Copyright 1965 by the American Chemical Society. Reprinted by permission of the copyright owner.

Fig. 10-25:   J. C. Gerhart and H. K. Schachman. Reprinted from *Biochemistry 7,* February 1968. Copyright 1968 by the American Chemical Society. Reprinted by permission of the copyright owner.

Fig. 10-27:   J. E. Haber and D. E. Koshland. *Proc. Nat. Acad. Sci.,* **58:** 2087, 1967.

Fig. 16-3:   From Lowy and Vibert, *Nature,* **215:** 1254, 1967, and from Huxley and Brown, *J. Molec. Biol.,* **30:** 383, 1967.

Figs. 17-20, 17-21, and 17-22:   From *Nerve, Muscle and Synapse,* by Bernard Katz. Copyright 1966 by McGraw-Hill, Inc. Used with permission of McGraw-Hill Book Company.

Fig. 17-29:   From *Papers on Biological Membrane Structure,* Branton and Park, eds., Little, Brown & Co.

INDEX

INDEX